Fundamentals of the Shoulder

Gazi Huri • Mustafa Özkan
Kerem Bilsel
Editors

Fundamentals of the Shoulder

Springer

Editors
Gazi Huri
Department of Orthopaedics and Traumatology
Hacettepe University School of Medicine
Ankara, Turkey

Mustafa Özkan
Department of Orthopaedics and Traumatology
Dokuz Eylul University School of Medicine
Izmir, Turkey

Kerem Bilsel
Department of Orthopaedics and Traumatology
School of Medicine, Bezmialem Vakif University
Istanbul, Turkey

ISBN 978-3-030-74820-3 ISBN 978-3-030-94702-6 (eBook)
https://doi.org/10.1007/978-3-030-94702-6

This Springer imprint is published by the registered company Springer Nature Switzerland AG
The registered company address is: Gewerbestrasse 11, 6330 Cham, Switzerland

Contents

Part I

Shoulder Basic Science and Principles

Fundamentals of the Shoulder

1

Mete Edizer

1.1 Functional Anatomy of the Shoulder Joint

Shoulder is the area in the proximal part of the arm connecting it to the thoracic skeleton [1–5]. There are connections to the sternum through the scapula and clavicula. Clavicula extremitas lateralis is connected to the acromion region of the scapula with strong ligaments [4–11].

Clavicula is attached to the first costa by the costoclavicular ligament. Also, it is connected to the sternum via sternoclavicular ligament. It gains mobility through the joints at both ends. The acromioclavicular ligament and acromioclavicular joint at its lateral end provide strong connections. Clavicula is not only attached at its medial and lateral ends but also attached to the coracoid process of the scapula via the coracoclavicular ligament from above [12–14].

Acromioclavicular joint and sternoclavicular joint, which are the auxiliary joints of the shoulder girdle, work biomechanically together with the humeroscapular joint, which is a main joint. In a healthy shoulder joint, they take part in a variety of movements at differing rates so as to provide the necessary joint clearance [15–18]. For example, these auxiliary joints are not functional in abduction of the shoulder joint up to 90°, but they participate in the movement when the joint movement exceeds 90°. However, when pathology occurs in the shoulder joint, these joints participate in the movement to compensate for the stiffness and to ensure normal movement. Under normal circumstances, the sliding of the scapula on the thoracic wall allows the range of motion to increase. The scapula can move up and down, or it can slide forward and backward without turning. The scapula can abduct up to 50° in the transverse axis, taking the shape of a wing in the acromioclavicular joint [9, 11, 18].

The shoulder girdle is versatile and highly mobile. It is particularly important for the movements of the arm in the humeroscapular joint. It positions itself to facilitate the movements of this spheroid type of joint. Although the acromioclavicular joint is not very wide, the humeroscapular joint becomes important in terms of range of motion. In the acromioclavicular joint, it can make period movements of 50° towards the clavicula superior, 30–35° from the posterior to the anterior and 35° along its long axis. These angulations are compatible with the movements of the scapula [18].

Rotational movements in the shoulder girdle are also closely related to the position of the humerus. For example, in the neutral position, internal rotation can be performed more frequently than external rotation.

Pulling Up the Scapula with Abduction The patient sits, flexes the arm at 90° in the shoulder

M. Edizer (✉)
Department of Anatomy, Dokuz Eylül University, Izmir, Turkey

G. Huri et al. (eds.), *Fundamentals of the Shoulder*, https://doi.org/10.1007/978-3-030-94702-6_1

joint, and the elbow joint is flexed. When he tries to push his arm forward against a resistance, serratus anterior muscle starts abduction by pulling the anterior scapula forward. This function is important for the abduction of the arm above the 90°.

Adduction of the Scapula The patient lies with face down; the arm is fixed away from the body. The result is positive if the patient can bring his arm closer to the midline. If he cannot, the patient is brought to the sitting position, his arm rests on the table, and he is asked to pull his arm backinward (adduction). If successful, the force is considered medium. If he cannot do that, it is checked whether trapezius muscle, rhomboideus major muscle, and rhomboideus minor muscle are contracted, and it is considered weak.

Elevation of the Scapula The patient is asked to raise one or both shoulders upwards trapezius muscle and levator scapulae muscle, which allow this movement possible, and rhomboideus major muscle and rhomboideus minor muscle, which helps the movement are examined. The existence of contraction in these muscles is determined based on the success of the person's performance against resistance and the absence of resistance. If he fails to perform these tasks, he is asked to lie in face-down position, or palpations are detected, respectively [10, 12].

1.2 Upper Extremity Bones (Ossa Membri Superioris)

Upper extremity bones are placed symmetrically on both sides of the rib cage. They can perform extremely fine movements thanks to the ability of humans using their hands. Bones of the upper limbs are divided into two: the upper extremity free bones and the upper extremity junction bones connecting them to the trunk. Upper extremity junction bones are scapula and clavicula, and free upper extremity bones are humerus, ulna, radius, and hand bones.

The Clavicula This bone extends laterally and horizontally across the root of the neck. Clavicula is S-shaped, and it is 15–17 cm long. The tip of the scapula joining with the sternum is more flattened and is called extremitas sternalis, while the one joining with its acromion is thicker and is called extremitas acromialis [5]. About 2/3 of the inner part of the corpus claviculae between the two ends shows anterior convexity, while the outer 1/3 part shows posterior convexity. It is the bone that first begins to ossify in the fetus and the last to complete ossification. Despite being a long bone, it shows intramembranous ossification between clavicula and the first costa artery passes the branches of subclavia and plexus brachialis. It is the most superficial bone in the body and is most frequently broken one. In adults, it often gets broken as a result of falling on the shoulder. The weakest part of the bone is the junction of the 1/3 middle and 1/3 outer parts [6, 14, 18].

The Scapula It is a flat bone with three corners, three edges, and two sides and is attached to the back of the rib cage. It is between the Th2–7 vertebrae.

Facies Posterior (Back Face) The protrusion called scapular spine extending horizontally on the back of the convex shape ends with acromion, a flat and thick extension from the back to the front. At the tip of the acromion is facies articularis acromii that articulates with the clavicula. Spina scapulae divides the facies posterior into two separate regions, supraspinous fossa and infraspinous fossa.

Anterior Face (Costal) It is the concave front face of the scapula. On this obverse is the depression where the muscle of the same name, called subscapular fossa, is located.

Superior Border It is the shortest edge of the scapula. The notch named suprascapular notch and located close to coracoid process is made into a hole with superior transvers scapular ligament in the organism. From this hole, supracapulary artery and vein pass above the ligament. It is an important point in peripheral nerve compression.

Medial Border On this edge is the triangular space, which is the starting point of the spina scapulae.

Lateral Border It is the thickest edge.

Superior Angle It is the corner where the superior angle and medial border meet.

Inferior Angle It is the corner where the medial border and lateral border meet.

Lateral Angle It is the corner where the lateral border and superior border meet. Lateral angle has an oval articular surface called glenoid cavity where the humeral head is located. There is supraglenoidal tubercle on the upper edge of glenoid cavity and infraglenoid tubercle on the lower edge.

The protrusion in the form of a bird's beak that starts with a broad root between the supraglenoid tubercle and the suprascapular notch is called coracoid process [14, 18].

The Humerus It is a single bone of the arm region in the upper extremity. Humerus has upper and lower ends and a body.

Extremitas Proximalis The round joint face of the scapula, called the humeral head, articulating with the glenoid cavity, is separated from the two lower reliefs by a shallow groove, the anatomical neck. The larger one on the outer side is called greater tubercle, and the smaller one on the anterior and inner side is called lesser tubercle. The bony protrusions extending down from these ridges are called crest of greater tubercle and crest of lesser tubercle, respectively. Through the groove called intertubercular groove among the crest, the tendon of the long head of the biceps brachii passes. The collum chirurgicum, the weakest part of the bone is located where the greater tubercle and lesser tubercle end. It is the most common place where humerus fractures are observed and axillar nerve is injured due to close proximity [14, 18].

Body of Humeri The body of the humerus is round at the top and nearly triangular at the bottom. Its edges are medial border, lateral border, and anterior border. Its faces are anteromedial face, anterolateral face, and posterior face. On the upper part of the anterolateral facies, which is a rough field called deltoid tuberosity, deltoid muscle ends. From the groove called radialis nerve groove on the back of the body of humeri, both radial nerve and deep brachial artery pass. The most common complication of body of humeri fractures is radial nerve injury [6, 14, 18].

Extremitas Distalis The face of the roller-shaped joint that connects with the ulna on the inner side of the lower end of the humerus is called the trochlea humeri, and the small convex joint surface that connects with the radius on the outer side is called the capitulum humeri. On the front face is located coronoid fossa where coronoid process of ulna is situated in forearm flexion. The olecranon fossa, where the olecranon part of the ulna sits, and the radial fossa, where the radial head part of the radius sits, are located at the forearm extension on the back. At the lower end, there are two large lump-shaped projections called medial condylus and lateral condylus. The smaller protrusion on the inner one is called medial epicondylus, and the outer one is called lateral epicondylus, and both can be palpated manually. Below medial epicondylus, there is a notch called sulcus nervi ulnaris through which ulnar nerve passes. In the distal end fractures of the humerus, ulnar nerve, located in front median nerve and brachial artery may be damaged [6, 10, 12, 18].

1.3 Shoulder Joints

The acromioclavicular joint and sternoclavicular joint upper extremity junction joints and joints in the upper extremity starting from the shoulder joint are known as free upper extremity joints.

Acromioclavicular Joint It is between clavicula (articular face of acromial) and scapula (articular face of acromion). The joint faces are covered with fibrous cartilage and are of joint plana type. It allows the scapula to slide and rotate on the clavicula. There is an articular disc located in the joint. It strengthens articular capsule by acromioclavicular ligament. It consists of coracoclavicular ligament, trapezoid ligament, and conoid ligament and plays the most effective role in joint stabilization [5].

Sternoclavicular Joint It is the sellar type of joint between the sternum (clavicular notch) and the clavicula (articularis face of sternum). It connects the upper limb to the trunk. Since the joint faces do not fit together, there is articular disc between them, which completely divides the joint space into two. The joint is strengthened by anterior sternoclavicular ligament and posterior sternoclavicular ligament [14, 18].

Shoulder Joint (Glenohumeral Joint) The glenohumeral joint is of the spheroid type, and the humeral head sits on the shallow joint surface on the scapula. It performs flexion-extension in the transverse axis, abduction-adduction in the sagittal axis and internal and external rotation movements in the vertical axis. Since it can move in three axes, it can make a circumduction movement. The shoulder joint, which is the most mobile in the body, can move in three axes and is the joint with the most common dislocation.

Since the ligaments that strengthen and stabilize the joint are not sufficient, the rotator cuff muscles also support this process. The surface of the joint pit, called glenoid cavity, is approximately 6 cm^2. The glenoid cavity is deepened by the limbus in the fibrous cartilage structure called articular labrum and glenoidal labrum. The synovial layer of the joint capsule adheres to the articular labrum. The joint capsule is quite loose, forming a pocket called the axillar recess medially on the arm standing in neutral position. Joint capsule is strengthened by the coracohumeral ligament proximally and the remaining part by the glenohumeral ligament. In the arm standing in a neutral position, the upper half of the humeral head touches the joint capsule, and the lower half touches the glenoidal cavity. The effect that stabilizes the humeral head to the joint pit consists of the combination of muscle strength and arm weight. For example, the supraspinatus muscle and the middle part of the deltoid muscle, which are activated at the initiation of the abduction movement of the arm, affect the joint [14, 18].

The part of the spherical humeral head covered with cartilage is 2.5 cm in diameter. The rotation center, which changes according to movements, is very important in functional anatomy. For example, in abduction up to 50°, the center shifts medially when it reaches 50–90°. There is only one turning point in front and back elevation. The anterior-underside of the joint capsule supported by the glenohumeral ligament and coracohumeral ligament is weak, and the dislocations are often forward-downward. The stability of the joint (especially during abduction) provides "rotator cuff muscles," involving supraspinatus muscle, infraspinatus muscle, teres minor muscle and subscapularis muscle. The tendon of the long head of biceps brachii muscle passes through the shoulder joint and attaches to the scapula. Two burs, called bursa subacromialis and bursa subtendinea subscapularis muscle, are connected to the joint space [7, 14, 18].

1.4 Muscles that Affect the Movement of the Shoulder Girdle

Pectoralis Major Muscle

- Origin: Clavicular head: Anterior surface of the medial half of clavicle.
- Sternocostal head: Anterior surface of sternum, superior six costal cartilage, aponeurosis of external oblique muscle.
- Insertion: Lateral lip of intertubercular groove of humerus.
- Function: Adducts and medially rotates humerus.
- Nerve: Lateral and medial pectoral nerves (clavicular head C5–6, sternocostal head C7, C8, and T1).

Pectoralis Minor Muscle
- Origin: Ribs 3–5 near the costal cartilages.
- Insertion: Medial border and superior surface of coracoid process of scapula.
- Function: Stabilizes scapula by drawing inferiorly and anteriorly against thoracic wall.
- Nerve: Medial pectoral nerve (C8 and T1).

Subclavius Muscle
- Origin: Junction of rib 1 and its costal cartilage.
- Insertion: Inferior surface of middle third of clavicle.
- Function: Draws clavicle medially.
- Nerve: Nerve to subclavius (C5 and C6).

Serratus Anterior Muscle
- Origin: External surface of lateral parts of ribs 1–8.
- Insertion: Anterior surface of medial border of scapula.
- Function: Protracts scapula and holds it against thoracic wall; rotates scapula.
- Nerve: Long thoracic nerve (C5, C6, and C7).

1.5 Muscles Connecting the Upper Limb to the Vertebral Column

Trapezius Muscle
- Origin: Medial third of superior nuchal line, external occipital protuberance, ligamentum nuchae, spinosus processes of C7 to T12 vertebrae, and lumbar and sacral spinosus processes.
- Insertion: lateral third of clavicle, acromion, and spine of scapula.
- Function: Elevates, retracts, and rotates scapula, superior fibers elevate, middle fibers retract, and inferior fibers depress scapula; superior and inferior fibers act together in superior rotation of scapula.
- Nerve: Spinal root of accessory nerve and cervical nerves (C3 and C4).

Latissimus Dorsi Muscle
- Origin: Spinosus processes of the inferior six thoracic vertebrae, thoracolumbar fascia, iliac crest and inferior 4 ribs.
- Insertion: Floor of intertubercular groove of humerus.
- Function: Extends, adducts, medially rotates humerus, raises body toward arms during climbing.
- Nerve: Thoracodorsal nerve (C6, C7, and C8).

Levator Scapulae Muscle
- Origin: Posterior tubercles of transverse process of C1–C4 vertebrae.
- Insertion: Superior part of medial border of scapula.
- Function: Elevates scapula and tilts its glenoid cavity inferiorly by rotating scapula.
- Nerve: Dorsal scapular nerve (C5) and cervical nerves (C3 and C4).

Rhomboid Major Muscle
- Origin: Spinosus process of T2–T5 vertebrae.
- Insertion: Medial border of the scapula from level of spine to inferior angle.
- Function: Retracts scapula and rotates it to depress glenoid cavity, fixes scapula to thoracic wall.
- Nerve: Dorsal scapular nerve (C4 and C5).

Rhomboid Minor Muscle
- Origin: Ligamentum nuchae and spinosus process of C7 to T1 vertebrae.
- Insertion: Medial border of the scapula from level of spine to superior angle.
- Function: Retracts scapula and rotates it to depress glenoid cavity, fixes scapula to thoracic wall.
- Nerve: Dorsal scapular nerve (C5) and cervical nerves (C3 and C4).

1.6 The Scapular Muscles

Deltoid Muscle

- Origin: Lateral third of clavicle, acromion, and spine of scapula.
- Insertion: Deltoid tuberosity of humerus.
- Function: Anterior part: flexes and medially rotates arm.
  - Middle part: abducts arm.
  - Posterior part: Extends and laterally rotates arm.
- Nerve: Axillary nerve (C5 and C6).

Supraspinatus Muscle

- Origin: Supraspinous fossa of scapula.
- Insertion: Superior facet on greater tubercle of humerus.
- Function: Helps deltoid to abduct arm and acts with rotator cuff muscles.
- Nerve: Suprascapular nerve (C4, C5, and C6).

Infraspinatus Muscle

- Origin: Infraspinous fossa of scapula.
- Insertion: Middle facet on greater tubercle of humerus.
- Function: Laterally rotates arm.
- Nerve: Suprascapular nerve (C5, C6).

Teres Minor Muscle

- Origin: Superior part of lateral border of scapula.
- Insertion: Inferior facet on greater tubercle of humerus.
- Function: Laterally rotates arm.
- Nerve: Axillary nerve (C5, C6).

Teres Major Muscle

- Origin: Dorsal surface of inferior angle of scapula.
- Insertion: Medial lip of intertubercular Groove of humerus.
- Function: Adducts and medially rotates arm.
- Nerve: Lower subscapular nerve (C6 and C7).

Subscapularis Muscle

- Origin: Subscapular fossa.
- Insertion: Lesser tubercle of humerus.
- Function: Medially rotates arm and adducts it; helps to hold humeral head in glenoid cavity.
- Nerve: Upper and lower subscapular nerve (C5, C6, and C7).

In the flexion of the arm, the coracobrachialis muscle and the anterior part of the deltoid muscle work. Deltoid muscle middle part, pectoralis major muscle, and biceps brachii muscle also help them. Flexion range of motion of the arm is approximately 150–170°.

In the extension movement of the arm, the Latissimus dorsi muscle, teres major muscle, deltoid muscle posterior part work as well as teres minor muscle and triceps brachii muscle as auxiliary muscles. The extension opening of the arm is approximately 40°.

During the first 15° part of the abduction movement of the arm, the supraspinatus muscle takes part, then the middle part of the deltoid muscle takes over. The supraspinatus muscle also lends support under the deltoid muscle and presses the humeral head toward the joint surface to keep it within the glenoid cavity, preventing the humeral head from slipping downwards.

In supraspinatus muscle paralysis or weakness, arm abduction cannot be initiated. Also, formation of scapula alatae and spontaneous dislocations occur. The long head of the biceps brachii muscle, the serratus anterior muscle, and the trapezius muscle also help the abduction of the arm. For abduction to reach over 90°, external rotation of the humerus and rotation of the scapula are required. The full width of the abduction is about 180°.

The adduction movement of the arm is made by the pectoralis major muscle, the caput longum of the triceps brachii muscle, the teres major muscle, the caput breve of the biceps brachii muscle, and the anterior part and posterior parts of the deltoid muscle. The full width of adduction is 20°–40°.

While the outer rotation of the arm is mainly performed by the infraspinatus muscle and teres minor muscle, the posterior part of the deltoid muscle also helps. While performing external rotation, the trapezius muscle, rhomboideus major muscle, and rhomboideus minor muscle pull the clavicula and scapula posteriorly. Thus, movements occur in the sternoclavicular joint and acromioclavicular joint.

The internal rotation of the arm is performed by subscapularis muscle, pectoralis major muscle, caput longum of biceps brachii muscle, anterior part of deltoid muscle, teres major muscle, and latissimus dorsi muscle. In the neutral position, the outer rotation is 40°–60°, and the inner rotation is 95°. In the 90° abduction of the shoulder joint, internal and external rotation values are up to 90°.

References

1. Apoil A, Augereau B. Réparation par lambeau deltoidien des grandes pertes de substance de la coiffe des rotateurs de l'épaule. Rev Chir Orthop. 1988;74:298–301.
2. Ball CM, Steger T, Galatz LM, Yamaguchi K. The posterior branch of the axillary nerve: an anatomic study. J Bone Joint Surg Am. 2003;85:1497.
3. Bakhsh W, Nicardri G. Anatomy and physical examination of the shoulder. Sports Med Arthrosc Rev. 2018;26(3):e10–22. (Pubmed:30059442)
4. Bhatia DN, de Beer JF, van Rooyen KS, du Toit DF. Arthroscopic suprascapular nerve decompression at the suprascapularnotch. Arthroscopy. 2006;22:1009–13.
5. Connor PM, Yamaguchi K, Pollock RG, Flatow EL, Bigliani LU. Comparison of arthroscopic and open revision decompression for failed anterior acromioplasty. Orthopedics. 2000;23(6):549–54.
6. Cormack CC, Lamberty BGH. The arterial anatomy of skin flaps. 2nd ed. Edinburgh: Churchill Livingstone; 1994. p. 182–4, 404.
7. Culham E, Peat M. Functinal anatomy of the shoulder complex. J Orthop Sports Phys Ther. 1993;18(1): 342–50. https://doi.org/10.2519/jospt.1993.18.1.342. PMID: 8348135 Review
8. Das S, Suri R, Kapur V. Ossification of superior transverse scapular ligament and its clinical implications. Sultan Qaboos Univ Med J. 2007;7(2):157–60.
9. Greiner SH, Back DA, Herrmann S, et al. Degenerative changes of the deltoid muscle have impact on clinical outcome after reversed total shoulder arthroplasty. Arch Orthop Trauma Surg. 2010;130(2):177–83.
10. Henninger HB, Barg A, Andrew AE, et al. Effect of lateral offset center of rotation in reverse total shoulder arthroplasty: a biomechanical study. J Shoulder Elb Surg. 2012;21:1128–35.
11. Wang H-J, Chen C, Li-Ping W, Pan C-Q, Zhang W-J, Li Y-K. Variable morphology of the suprascapular notch: an investigation and quantitative measurements in Chinese population. Clin Anat. 2011;24:47–55.
12. Jobin CM, Brown GD, Bahu MJ, et al. Reverse total shoulder arthroplasty for cuff tear arthropathy: the clinical effect of deltoid lengthening and center of rotation medialization. J Shoulder Elb Surg. 2012;21:1269–77.
13. Mathes SJ, Nahai F. Reconstructive surgery: principles, anatomy, and technique. New York: Churchill Livingstone; 1997. p. 703.
14. Moore KL. Clinically oriented anatomy. 3rd ed. Baltimore: Williams and Wilkins; 1992. p. 525–7.
15. Serajin D. Atlas of microsurgical composite tissue transplantation. Philadelphia: WBSaunders; 1996. p. 153.
16. Ticker JB, Djurasovic M, Strauch RJ, April EW, Pollock RG, Flatow EL, Bigliani LU. The incidence of ganglion cysts and other variations in anatomy along the course of the suprascapular nerve. J Should Elb Surg. 1998;7:472–8.
17. Polgu M, Jedrzejewski K, Podgorski M, Topol M. Morphometric study of the suprascapular notch: proposal of classification. Surg Radiol Anat. 2011;33:781–7.
18. Williams LP, Bannister LH, Berry MM, Collins P, Dyson M, Dussek JE, Ferguson MWJ. Gray's anatomy. 38th ed. London: Churchill Livingstone; 1995. p. 856.

Molecular Biology and Genetics in Shoulder Pathologies

2

Ahmet Emre Paksoy and Baris Kocaoglu

The studies conducted in the field of molecular biology and genetics have focused mostly on rotator cuff tears and partly shoulder dislocations among shoulder diseases. The rate of re-tear is high in rotator cuff tears, and despite the improved surgical repair techniques and successful physiotherapy models for patients with rotator cuff tear (RCT), high rates of re-tear have led to the view of considering individual genetic differences in the treatment of this disease [1]. The most common cause of anterior shoulder dislocations is trauma. However, some patients who receive the same treatment had recurrent dislocations due to capsular deformation, while others have a good recovery and do not have recurrent dislocations. This difference in prognosis among patient groups with the same diagnosis and similar demographic features increases the interest in molecular biology and genetics.

The etiopathogenesis of atraumatic rotator cuff tear is multifactorial and has still not been fully understood. Although factors such as impingement, overusing, aging, and smoking are effective in its etiology, there are studies showing the presence of familial predisposition. Defining the genetic dimension of RCT can help us better understand the pathogenesis of the disease. The expression of events from the formation of a tenocyte in the cuff to the degeneration and death is in the nucleus and is far from the arthroscope. Processes occurring in the nucleus, protein synthesis, and degradation are responsible for events ranging from muscle atrophy to tendon calcification and even re-rupture. After all, atraumatic RCT is an imbalance between protein synthesis and protein degradation.

Post-RCT repair recovery occurs in the tendon to bone interface. RCT treatment has high re-tear rates. For this reason, molecules that enhance recovery in the interface may affect the success of treatment. Matrix metalloproteinase (MMP) regulates the reparative or degenerative balance in the extracellular matrix. Supporting this balance in favor of reparation will improve the collagen organization required for recovery. Affecting it in favor of degeneration will result in re-tear. In their study, Robertson et al. sampled from the rupture region during the surgery of RCT patients and found MMP1 and MMP9 expression significantly higher in patients who had re-tear in the follow-ups [2]. Tetracyclines inhibit MMP. In their experimental study, Bedi et al. obtained larger fibrocartilage tissue and larger collagen tissue in the repair zone by reducing the MMP13 activity with doxycycline. As a

A. E. Paksoy
Ataturk University School of Medicine, Orthopaedics and Traumatology Department, Erzurum, Turkey

B. Kocaoglu (✉)
Acibadem University School of Medicine, Orthopaedics and Traumatology Department, Istanbul, Turkey

G. Huri et al. (eds.), *Fundamentals of the Shoulder*, https://doi.org/10.1007/978-3-030-94702-6_2

matter of fact, their biomechanical analysis revealed that load to failure was also significantly higher in the doxycycline-treated group [3]. Local use of doxycycline exhibits similar effects. Namely, it has been shown in Achilles repair that doxycycline-coated sutures have higher pull out values than uncoated ones [4]. Gotoh et al. demonstrated a re-tear rate of 25% in patients who underwent RCT repair and increased expression of MMP3 and tissue inhibitor of MMP (TIMP1) in this re-tear group [5].

The frequency of having rotator cuff surgery is higher in the family of patients who had undergone rotator cuff surgery compared to the normal population. A high level of genetic predisposition associated with this disease has been reported, especially in close and distant relatives of young adult rotator cuff tear patients [6]. Gvylim et al. showed that tears were more symptomatic and more easily progressive in the presence of genetic predisposition [7].

The increasing interest in the genetic origin of shoulder diseases in the last 20 years could demonstrate the relationship of shoulder diseases with some genes. The regulation of genes that regulate metabolic pathways such as apoptosis, inflammation, blood supply, which may be effective in the formation and progression of a rotator cuff tear, has been attempted to be studied.

Apoptosis (programmed cell death) has been shown to play an important role in the pathogenesis of RCT [8, 9]. Lundgreen et al. showed an increase in the apoptotic index, an increase in P53 gene activity-inducing apoptosis, and a decrease in P53 inhibitory proteins [10]. In their genome-wide association study, Tashjian et al. showed that single-nucleotide polymorphism (SNP) in the genes called *SAP30B* and *SASH1*, which are effective in apoptosis, are associated with rotator cuff tear [11]. Genetic changes in the apoptotic pathway also appear to affect the outcomes of RCT surgery. An SNP within the estrogen-related receptor beta (*ESRRB*) gene is associated with postoperative lateral failure type re-tear [12]. ESRRB regulates the function of the hypoxia-inducible factor (HIF). Irregularity in the ESRRB/HIF system caused by this SNP causes tenocyte apoptosis [13].

Muscle atrophy results in fatty degeneration of the muscle and loss of function in the shoulder. Atrophic changes in the muscle are critical in the selection of the treatment for rotator cuff tear and prediction of outcomes. In skeletal muscle atrophy, a decrease in protein synthesis and an increase in protein degradation concurrently occur. A decrease in signal proteins that send survival signals to the cell nucleus activates the cell apoptosis. Calpains and cathepsins, which regulate proteolytic systems in cell death, lead to atrophy of the muscle cell. Indeed, Schmutz et al. showed increased expression of *CAPN1*, *UBE2B*, and *UBE3A* genes in massive RCT compared to smaller tears or individuals without RCT [14].

In the etiology of RCT, hypovascularity is an etiological factor that should be emphasized. Indeed, a hypovascular zone has been shown to exist in the attachment site of the rotator cuff [15, 16]. Rotator cuff tear is mostly middle and advanced age disease and a decrease in blood supply to the rotator cuff has been found with increasing age [17]. Hypoxia induces apoptotic pathways. Apoptosis is induced in the cell with the increase of hypoxia-inducible factor 1 α (HIF1α) [18].

Stiffness is an important issue in postoperative rehabilitation of RCT treatment. The presence of some genetic variations has been associated with increased postoperative stiffness. Ling et al. performed a mini-open repair on patients with RCT and found that stiffness was significantly higher in the group with SNPs in the IL6 and *MMP3* genes [19].

Anterior shoulder dislocations occur as a result of major trauma or more minor traumas due to the accumulation of changes in capsular and ligamentous structures caused by repetitive microtraumas. The most affected area in recurrent dislocations of the shoulder is the anteroinferior capsule, and plastic deformation occurs in this area in the case of shoulder instability. Collagen is one of the important building blocks of the shoulder capsule. Indeed, shoulder dislocations are common in Ehler Danlos syndrome, a disease in which collagen production is impaired.

In their studies, Belangero et al. showed decreased expression of the *COL5A1* gene in the anteroinferior capsule in patients with anterior shoulder instability [20]. The expression of this gene in the anteroinferior capsule was significantly different from the anterosuperior capsule that was not affected by instability. Although the *COL1A1* gene has been attempted to be associated with shoulder instability [21], this relationship could not be confirmed in repeated studies [20, 22]. In their genome-wide association study, Kim et al. showed that shoulder dislocations were associated with the presence of no rs12913965 SNP of the *TICRR* gene, which is effective in the replication phase of the cell cycle [22]. Although the current information on the genetic aspect of shoulder instability is limited, we can speculate that the interest in this subject is increasing.

In summary, it is clear that some genes are directly associated with rotator cuff disease and shoulder instability. It can be thought that decoding the genetic codes of shoulder diseases will be effective in the long term, from treatment to rehabilitation, and even prevention. It can be anticipated that genetic studies will be used to screen for shoulder diseases in the future. In the near future, it seems that we will meet with preventive rehabilitation programs and regenerative medicine applications for individuals with genetic characteristics that are the precursor of shoulder diseases.

References

1. Dabija DI, Gao C, Edwards TL, Kuhn JE, Jain NB. Genetic and familial predisposition to rotator cuff disease: a systematic review. J Shoulder Elb Surg. 2017;26(6):1103–12.
2. Robertson CM, Chen CT, Shindle MK, Cordasco FA, Rodeo SA, Warren RF. Failed healing of rotator cuff repair correlates with altered collagenase and gelatinase in supraspinatus and subscapularis tendons. Am J Sports Med. 2012;40(9):1993–2001.
3. Bedi A, Fox AJ, Kovacevic D, Deng X-h, Warren RF, Rodeo SA. Doxycycline-mediated inhibition of matrix metalloproteinases improves healing after rotator cuff repair. Am J Sports Med. 2010;38(2):308–17.
4. Pasternak B, Missios A, Askendal A, Tengvall P, Aspenberg P. Doxycycline-coated sutures improve the suture-holding capacity of the rat Achilles tendon. Acta Orthop. 2007;78(5):680–6.
5. Gotoh M, Mitsui Y, Shibata H, Yamada T, Shirachi I, Nakama K, et al. Increased matrix metalloprotease-3 gene expression in ruptured rotator cuff tendons is associated with postoperative tendon retear. Knee Surg Sports Traumatol Arthrosc. 2013;21(8):1807–12.
6. Tashjian RZ, Farnham JM, Albright FS, Teerlink CC, Cannon-Albright LA. Evidence for an inherited predisposition contributing to the risk for rotator cuff disease. J Bone Joint Surg Am. 2009;91(5):1136.
7. Gwilym S, Watkins B, Cooper C, Harvie P, Auplish S, Pollard T, et al. Genetic influences in the progression of tears of the rotator cuff. J Bone Joint Surg. 2009;91(7):915–7.
8. Yuan J, Murrell GA, Wei AQ, Wang MX. Apoptosis in rotator cuff tendonopathy. J Orthop Res. 2002;20(6):1372–9.
9. Maffulli N, Longo UG, Berton A, Loppini M, Denaro V. Biological factors in the pathogenesis of rotator cuff tears. Sports Med Arthrosc Rev. 2011;19(3):194–201.
10. Lundgreen K, Lian ØB, Engebretsen L, Scott A. Tenocyte apoptosis in the torn rotator cuff: a primary or secondary pathological event? Br J Sports Med. 2011;45(13):1035–9.
11. Tashjian RZ, Granger EK, Farnham JM, Cannon-Albright LA, Teerlink CC. Genome-wide association study for rotator cuff tears identifies two significant single-nucleotide polymorphisms. J Shoulder Elb Surg. 2016;25(2):174–9.
12. Tashjian RZ, Granger EK, Zhang Y, Teerlink CC, Cannon-Albright LA. Identification of a genetic variant associated with rotator cuff repair healing. J Shoulder Elb Surg. 2016;25(6):865–72.
13. Teerlink CC, Cannon-Albright LA, Tashjian RZ. Significant association of full-thickness rotator cuff tears and estrogen-related receptor-β (ESRRB). J Shoulder Elb Surg. 2015;24(2):e31–e5.
14. Schmutz S, Fuchs T, Regenfelder F, Steinmann P, Zumstein M, Fuchs B. Expression of atrophy mRNA relates to tendon tear size in supraspinatus muscle. Clin Orthop Relat Res. 2009;467(2):457–64.
15. Brooks C, Revell W, Heatley F. A quantitative histological study of the vascularity of the rotator cuff tendon. J Bone Joint Surg. 1992;74(1):151–3.
16. Lohr J, Uhthoff H. The microvascular pattern of the supraspinatus tendon. Clin Orthop Relat Res. 1990;254:35–8.
17. Rudzki JR, Adler RS, Warren RF, Kadrmas WR, Verma N, Pearle AD, et al. Contrast-enhanced ultrasound characterization of the vascularity of the rotator cuff tendon: age-and activity-related changes in the intact asymptomatic rotator cuff. J Shoulder Elb Surg. 2008;17(1):S96–S100.
18. Vasagiri N, Kutala VK. Structure, function, and epigenetic regulation of BNIP3: a pathophysiological relevance. Mol Biol Rep. 2014;41(11):7705–14.
19. Ling Y, Peng C, Liu C, Zhang N, Yue S. Gene polymorphism of IL-6 and MMP-3 decreases passive range of motion after rotator cuff repair. Int J Clin Exp Pathol. 2015;8(5):5709.
20. Belangero PS, Leal MF, Pochini AC, Andreoli CV, Ejnisman B, Cohen M. Profile of collagen gene expression in the glenohumeral capsule of patients

with traumatic anterior instability of the shoulder. Rev Bras Ortop. 2014;49(6):642–6.
21. Collins M. Genetic risk factors for soft-tissue injuries 101: a practical summary to help clinicians understand the role of genetics and 'personalised medicine'. Br Assoc Sport Excer Med. 2010;44(13):915–7.
22. Kim S, Kleimeyer JP, Ahmed MA, Avins AL, Fredericson M, Dragoo JL, et al. A genetic marker associated with shoulder dislocation. Int J Sports Med. 2017;38(07):508–14.

Animal Models for Research on Shoulder Pathologies

3

Arel Gereli

The use of animals in shoulder research is an important bridge between in vitro studies and clinical trials. A validated animal model is useful to understand the etiology, molecular mechanisms, and potential treatments of shoulder pathologies. Although the animal shoulder does not exactly replicate the human counterpart, anatomical, histological, and biomechanical similarities make them useful in the shoulder research. They can provide the variation control in a homogeneous population to isolate the effect of a single factor. Interventions in animal models can be monitored in quantifiable manner, and it is possible to obtain living tissue from all desirable locations at various stages of pathology. Despite the pace of research on rotator cuff injury and repair, shoulder pathologies remain one of the least studied conditions in the medical science. The lack of suitable animal models for the study of shoulder pathologies limits the research in this field. Different animals have been studied to develop an appropriate shoulder model such as rat, mouse, rabbit, dog, sheep, goat, calf, and primates. Each model has inherent strengths and limitations that should be considered to answer specific research questions. In this section, the most commonly utilized large and small animal models are reviewed.

A. Gereli (✉)
Department of Orthopedic Surgery and Traumatology, Acibadem University, Istanbul, Turkey

3.1 Evaluation of Animal Models

An ideal animal shoulder model should have morphologic, histologic, and biomechanical features that closely resemble of human shoulder. Beside the appropriateness of the model to the human condition, cost, housing requirements, and ease of manipulation of the animal should also be considered.

The morphologic characteristics of a model should include shoulder musculature containing rotator cuff, deltoid, and biceps with the neurovascular supply similar in human; bony anatomy of the acromion, coracoid, clavicle and humerus; articulations of the glenohumeral, subacromial, and acromioclavicular regions [1–3]. The presence of the coracoacromial arch and a prominent tendon passing under this arch is an important structure for the rotator cuff research [3, 4]. In addition, tendon size should be large enough to allow research on repair techniques. Other morphological features include laterally directed and relatively small glenoid fossa articulating with a larger head of humerus, oblique orientation of the scapula, longer clavicle, and locations of greater and lesser tuberosities [5].

Histologically, ideal animal model has to show no spontaneous healing or scar formation after injury. Tissue healing is generally faster in animals than in humans. This should be considered when the sacrification time planned. Human rotator cuff is organized as blending of individual

G. Huri et al. (eds.), *Fundamentals of the Shoulder*, https://doi.org/10.1007/978-3-030-94702-6_3

tendons to form a common insertion. An ideal model should mimic this organization and also provides intrasynovial healing environment. Rotator cuff or suprascapular nerve injury should cause muscle atrophy, stiffness, and fat accumulation [6].

An ideal animal model should also exhibit biomechanic properties similar to that of the human shoulder such as being able to perform overhead activities while standing on their legs, arm elevation and rotation in various planes, synergistic muscle contraction of the deltoid with the rotator cuff [1–3].

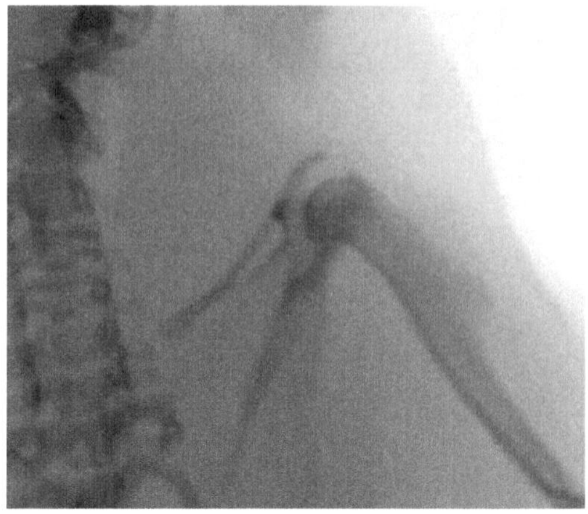

Fig. 3.1 X-ray view of a rat shoulder shows similarity to humans. Small glenoid fossa, larger humeral head, longer clavicle and greater tuberosity are seen (The photo was taken in the Acibadem University Animal Laboratory)

3.2 Animal Models

While small animal models are more suitable for studying molecular pathways and biologic bases of diseases, large animals allow evaluation of surgical techniques and related issues. Selection of an animal model must be appropriate for the condition being studied. No single model can be superior in all situations.

3.2.1 Rat

Rats have the apparent anatomical similarity to humans due to the presence of an acromial arch (Fig. 3.1). The supraspinatus tendon and muscle are located immediately below this arch and when the rat reaches overhead, tendon moves closely to the acromion. This makes them useful to study the pathogenesis of the rotator cuff disease especially extrinsic damage [2, 3, 7]. Rats have also been used to study intrinsic factors and overuse injury on rotator cuff tendinopathy [8]. Genomic similarity between the human and rodents provide useful experimental data on the gene expression, mechanism, augmentation, and regenerative strategies of rotator cuff healing after acute tendon to bone repair [9–12]. The analogy of the suprascapular nerve and scapula between the rats and human allows to research of the effect of nerve injury and scapular dyskinesis on the rotator cuff [13–15]. Rats produce atrophy and fatty degeneration of the rotator cuff muscle similar to human muscles after chronic cuff tears [16, 17]. Rotator cuff tears also decrease the range of motion and result in cartilage degeneration on the humeral head and glenoid in rats [18, 19]. The majority of the studies that have used rat as a model have focused on the biologic aspect of rotator cuff healing. Rat models have less frequently been used to study shoulder instability labral lesions and adhesive capsulitis [20–22]. Availability, low maintenance requirements, and ease of handling are other peculiar advantages of rats. However, rat has a few major limitations for its use. The ability of spontaneous healing, lack of irreversible fatty atrophy, and no re-tear make it less suitable model to evaluate repair techniques [1].

3.2.2 Mice

The advantages of rodent models (rats and mice) are high level of anatomical resemblance and genetic similarity to human [23]. Rapid growth rates and short life spans make them preferable. Histological analyses confirmed that rodent rotator cuff insertion site exhibits fibrocartilaginous transition zone similar to that of humans [24]. Hence, this makes the rodents suitable model for evaluating molecular mechanism, gene expression, and regenerative strategies of healing after tendon to bone repair [25]. The main superiority of mice compared with rats is the availability of genetically engineered strains for investigation of biological mechanisms that underlie rotator cuff insertion site development, degeneration, and healing [23, 26]. Another advantage of the mouse model is the ability to perform molecular imaging and microCT in the live animal [4]. Recently, researchers demonstrated a mouse model for shoulder implant infection [27]. This model is capable of producing real-time, quantifiable data on bacterial load and possibility to analyze the host immune response to shoulder prosthetic infection. Mice are also useful to demonstrate reproducible muscle atrophy, fibrosis, and fatty degeneration [28, 29]. Genetically altered strains are particularly advantageous to understand specific signaling pathways and molecules on fatty degeneration of rotator cuff. However, rapid healing capacity, small size, and difficult functional analysis are the major limitations.

3.2.3 Rabbit

Rabbit is a frequently used animal in shoulder research (Fig. 3.2). Previous studies on rabbit shoulder model focused on the structural muscular changes associated with rotator cuff tears and neural injury including decreased mechanosensitive units, sarcomere length, and fiber length [30–33]. Rabbit models have provided valuable information about healing response, enthesis formation, growth factor expression, fatty degeneration, and atrophy after rotator cuff injury [34–37]. Rabbits have the tendon large enough to allow easy manipulation during the surgical technique experiments, and they have been used to determine the healing process after glenoid labral lesion [38]. Although rabbits have been shown to a reliable and reproducible model for rotator cuff pathology, they have disadvantage to be less able to tolerate surgical stress and susceptibility to spinal fracture dislocation when frightened [39]. Special precautions should be taken for rabbits that are used in the experimental setting. Most of the rabbit rotator cuff studies have been conducted on the supraspinatus tendon, but recent studies have shown that the subscapularis tendon may be more comparable to the human supraspinatus. The rabbit subscapularis tendon passes under the tuberculum supraglenoidale and inserts on the lesser tubercule of the humerus in an analogous manner to human [40].

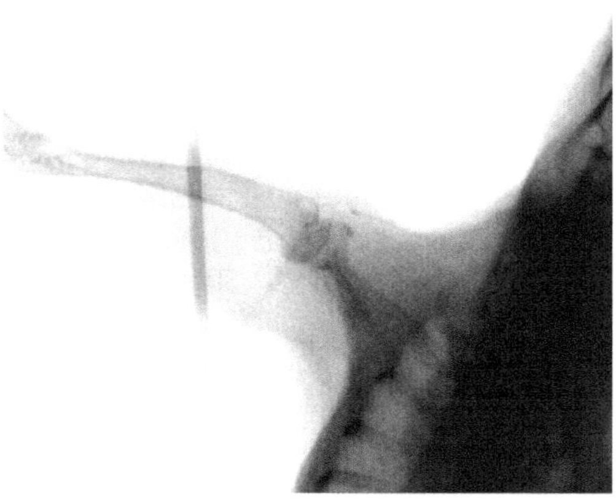

Fig. 3.2 X-ray of a rabbit shoulder shows analogy of glenohumeral joint to human. (The photo was taken in the Acibadem University Animal Laboratory)

3.2.4 Calf

Calf rotator cuff tendon is similar in size to human, and it has been used in mainly biomechanical experiments. Differences in response to cyclic loading after acute repair and common failure modes of sutures and anchors have been studied [41–43]. Larger dimensions of rotator cuff make them useful to evaluate repair techniques. Moving from smaller to larger animals, maintenance can become inhibiting factors for extensive studies.

3.2.5 Sheep and Goat

Sheep and goat shoulder models are frequently used in rotator cuff studies. In lesser extent, they have also been used for research on shoulder instability [44]. However, it is not recommended to use them for experimental shoulder instability studies because of the different orientation and size of the scapula and tuberosities when compared with the human. Although it does not mimic the human condition, transecting the infraspinatus tendon of goats and sheep then performing an immediate repair to the bone is useful to address the biomechanical, histologic, and biochemical process of the rotator cuff repair. The sheep infraspinatus is similar to the human supraspinatus in size [45]. Larger dimensions of the sheep infraspinatus tendon make them viable option to test different suture anchors, suture patterns, scaffolds, growth factors, and other biologic agents to enhance healing response [46–52]. Human supraspinatus tendon is intraarticular, whereas in the sheep, it is extraarticular. Although the healing environment of the goat and sheep infraspinatus tendon is not fully comparable to human, sheep infraspinatus tendon lies on a bursa, and therefore, the repaired tendon has some contact with synovial fluid [53]. It has been demonstrated that rotator cuff muscles undergo degradative changes including fibrosis and fatty infiltration, and sheep has also been used to simulate a chronic retracted tear by release of the tendon, followed by delayed repair [54]. However, robust scar formation between the retracted tendon and the bone can prevent distinguishing the scar tissue from normal tendon and direct reattachment. Wrapping the end of the tendon by the silicone tube or specific separator at the initial detachment may allow to clearly identify the cut tendon during the reattachment procedure [55, 56]. In addition, it is recommended that the reattachment surgery can be scheduled as soon as 4 weeks if tendon to bone healing is to be evaluated and 8 weeks after detachment and covering if the bioimplant or scaffold is to be evaluated [45]. One of the major shortcomings of the use of the sheep infraspinatus tendon is the difficulties in controlling the postoperative loads on the repaired tendon [6]. Because slinging is poorly tolerated in sheep, a rubber ball can be positioned under the hoof of the operated limb for 5 weeks, but this is not a method to immobilize the repaired shoulder.

3.2.6 Dog

The anatomy of the dog shoulder is similar to that of the sheep, and the infraspinatus tendon is most commonly used [57]. They have been utilized for the studies of acute full thickness rotator cuff tendon injury and repair [58, 59]. Dog also mimics the human condition in that muscle stiffness increases, and atrophy and fat accumulation occur and persist in chronically detached tendons [60]. Since dogs can tolerate the casting, slinging, and treadmill running, they have been preferred in the evaluation of various postoperative rehabilitation protocols [61, 62]. They proved to be a good model for arthroplasty experiments because bone remodeling kinetics in the dog have been well characterized and are similar to that of human [63–65]. This is the reason the dog is the primary animal model for studying total hip and total knee prosthesis. Anatomically dog shoulder is a moderately congruent joint with a shallow glenoid and large humeral articulation. Stability is primarily dependent on the joint capsule as in the human [1, 64]. However, shoulder arthroplasty studies are less common due to criticism that dog forelimb is a major load bearing joint.

3.2.7 Primates

A true rotator cuff is organized as blending of individual tendons to form a common insertion with intertendinous connection. Microstructure of the rotator cuff consists of layers arranged orthogonally to dissipate the multidirectional stresses created by the participating tendons [5]. This form is only found in advanced primates. The Baboon is the best primate model to study of the rotator cuff repair [66]. In this study, researchers described a model for supraspinatus repair in Baboons. Healing of the Baboon supraspinatus tendon is similar to that of human. Although macroscopically the repaired tendon seems to have healed at 8 weeks, the Sharpey fibers that provide strength of the enthesis are not sufficient until 12 weeks. These results support the use of a postoperative rehabilitation program in human, which preserves the enthesis for at least 12 weeks. Another primate, the African Green Monkey shows more than 98% genomic homology with humans and can be used to study of immunologic reaction after allograft augmentation of rotator cuff repair [67]. Despite the anatomic similarity, the primate has only been used in a few studies of rotator cuff healing, likely owing to economic and ethical concerns.

3.3 Conclusion

Results of animal model studies on shoulder research have not indicated a single model that solves all problems. Therefore, researchers should determine the most suitable animal model for the purpose, when designing the study. Commonly, larger animals are more appropriate for technical studies, small animals allow biologic and molecular research. Although animals can be used to address many clinically relevant questions, further studies are still required to find a validated model for shoulder research.

References

1. Longo UG, Forriol F, Campi S, Maffulli N, Denaro V. Animal models for translational research on shoulder pathologies: from bench to bedside. Sports Med Arthrosc Rev. 2011;19(3):184–93.
2. Edelstein L, Thomas SJ, Soslowsky LJ. Rotator cuff tears: what have we learned from animal models? J Musculoskelet Neuronal Interact. 2011;11(2):150–62.
3. Soslowsky LJ, Carpenter JE, DeBano CM, et al. Development and use of an animal model for investigations on rotator cuff disease. J Shoulder Elb Surg. 1996;5(5):383–92.
4. Lebaschi A, Deng XH, Zong J, et al. Animal models for rotator cuff repair. Ann N Y Acad Sci. 2016;1383(1):43–57.
5. Sonnabend DH, Young AA. Comparative anatomy of the rotator cuff. J Bone Joint Surg Br. 2009;91(12):1632–7.
6. Deprés-Tremblay G, Chevrier A, Snow M, Hurtig MB, Rodeo S, Buschmann MD. Rotator cuff repair: a review of surgical techniques, animal models, and new technologies under development. J Shoulder Elb Surg. 2016;25(12):2078–85.
7. Carpenter JE, Flanagan CL, Thomopoulos S, et al. The effects of overuse combined with intrinsic or extrinsic alterations in an animal model of rotator cuff tendinosis. Am J Sports Med. 1998;26:801–7.
8. Soslowsky LJ, Thomopoulos S, Tun S, et al. Overuse activity injures the supraspinatus tendon in an animal model: a histologic and biomechanical study. J Shoulder Elb Surg. 2000;9:79–84.
9. Jelinsky SA, Lake SP, Archambault JM, et al. Gene expression in rat supraspinatus tendon recovers from overuse with rest. Clin Orthop Relat Res. 2008;466:1612–7.
10. Galatz LM, Sandell LJ, Rothermich SY, et al. Characteristics of the rat supraspinatus tendon during tendon-to-bone healing after acute injury. J Orthop Res. 2006;24:541–50.
11. Gereli A, Kocaoglu B, Ulku TK, et al. Completion repair exhibits increased healing characteristics compared with in situ repair of partial thickness bursal rotator cuff tears. Knee Surg Sports Traumatol Arthrosc. 2018;26(8):2498–504.
12. Murray DH, Kubiak EN, Jazrawi LM, et al. The effect of cartilage-derived morphogenetic protein 2 on initial healing of a rotator cuff defect in a rat model. J Shoulder Elb Surg. 2007;16:251–4.
13. Sasaki Y, Ochiai N, Hashimoto E, et al. Relationship between neuropathy proximal to the suprascapular nerve and rotator cuff tear in a rodent model. J Orthop Sci. 2018;23(2):414–9.
14. Gereli A, Uslu S, Okur B, Ulku TK, Kocaoğlu B, Yoo YS. Effect of suprascapular nerve injury on rotator cuff enthesis. J Shoulder Elb Surg. 2020;29(8):1584–9.
15. Reuther KE, Thomas SJ, Tucker JJ, et al. Scapular dyskinesis is detrimental to shoulder tendon properties and joint mechanics in a rat model. J Orthop Res. 2014;32(11):1436–43.
16. Kim HM, Galatz LM, Lim C, Havlioglu N, Thomopoulos S. The effect of tear size and nerve injury on rotator cuff muscle fatty degeneration in a rodent animal model. J Shoulder Elb Surg. 2012;21(7):847–58.
17. Liu X, Joshi SK, Samagh SP, et al. Evaluation of Akt/mTOR activity in muscle atrophy after rotator cuff tears in a rat model. J Orthop Res. 2012;30(9):1440–6.
18. Kramer EJ, Bodendorfer BM, Laron D, et al. Evaluation of cartilage degeneration in a rat model of rotator cuff tear arthropathy. J Shoulder Elb Surg. 2013;22(12):1702–9.
19. Perry SM, Getz CL, Soslowsky LJ. Alterations in function after rotator cuff tears in an animal model. J Shoulder Elb Surg. 2009;18(2):296–304.
20. Okajima SM, Cubria MB, Mortensen SJ, et al. Rat model of adhesive capsulitis of the shoulder. J Vis Exp. 2018;139:58335.
21. Packer JD, Varthi AG, Zhu DS, et al. Ibuprofen impairs capsulolabral healing in a rat model of anterior glenohumeral instability. J Shoulder Elb Surg. 2018;27(2):315–24.
22. Mulcahey MK, Marshall M, Gallacher SE, Kaback LA, Blaine TA. Factors expressed in an animal model of Anteroinferior Glenohumeral instability. Orthop J Sports Med. 2015;3(8):2325967115599733.
23. Warden SJ. Animal models for the study of tendinopathy. Br J Sports Med. 2007;41:232–40.
24. Bell R, Taub P, Cagle P, Flatow EL, Andarawis-Puri N. Development of a mouse model of supraspinatus tendon insertion site healing. J Orthop Res. 2015;33:25–32.

25. Wang D, Tan H, Lebaschi AH, et al. Kartogenin enhances collagen organization and mechanical strength of the repaired Enthesis in a murine model of rotator cuff repair. Arthroscopy. 2018;34(9):2579–87.
26. Abbott A. Laboratory animals: the renaissance rat. Nature. 2004;428:464–6.
27. Sheppard WL, Mosich GM, Smith RA, et al. Novel in vivo mouse model of shoulder implant infection. J Shoulder Elb Surg. 2020;29(7):1412–24.
28. Liu X, Laron D, Natsuhara K, Manzano G, Kim HT, Feeley BT. A mouse model of massive rotator cuff tears. J Bone Joint Surg Am. 2012;94(7):e41.
29. Wang Z, Liu X, Davies MR, Horne D, Kim H, Feeley BT. A mouse model of delayed rotator cuff repair results in persistent muscle atrophy and fatty infiltration. Am J Sports Med. 2018;46(12):2981–9.
30. Fabis J, Danilewicz M, Omulecka A. Rabbit supraspinatus tendon detachment: effects of size and time after tenotomy on morphometric changes in the muscle. Acta Orthop Scand. 2001;72:282–6.
31. Fabis J, Kordek P, Bogucki A, et al. Function of the rabbit supraspinatus muscle after large detachment of its tendon: 6- week, 3-month, and 6-month observation. J Shoulder Elb Surg. 2000;9:211–6.
32. Fabis J, Kordek P, Bogucki A, et al. Function of the rabbit supraspinatus muscle after detachment of its tendon from the greater tubercle. Observations up to 6 months. Acta Orthop Scand. 1998;69:570–4.
33. Bjorkenheim JM. Structure and function of the rabbit's supraspinatus muscle after resection of its tendon. Acta Orthop Scand. 1989;60:461–3.
34. Bjorkenheim JM, Paavolainen P, Ahovuo J, Slatis P. Resistance of a defect of the supraspinatus tendon to intraarticular hydrodynamic pressure: an experimental study on rabbits. J Orthop Res. 1990;8: 175–9.
35. Choi HR, Kondo S, Hirose K, Ishiguro N, Hasegawa Y, Iwata H. Expression and enzymatic activity of MMP-2 during healing process of the acute supraspinatus tendon tear in rabbits. J Orthop Res. 2002;20:927–33.
36. Pulatkan A, Anwar W, Ayık O, Bozdag E, Yildirim AN, Kapicioglu M, Tuncay I, Bilsel K. Tear completion versus in situ repair for 50% partial-thickness bursal-side rotator cuff tears: a biomechanical and histological study in an animal model. Am J Sports Med. 2020;48(8):1818–25.
37. Rowshan K, Hadley S, Pham K, Caiozzo V, Lee TQ, Gupta R. Development of fatty atrophy after neurologic and rotator cuff injuries in an animal model of rotator cuff pathology. J Bone Joint Surg Am. 2010;92(13):2270–8.
38. Abe H, Itoi E, Yamamoto N, Minagawa H, Tomioka T, Kijima H, Shimada Y. Healing processes of the glenoid labral lesion in a rabbit model of shoulder dislocation. Tohoku J Exp Med. 2012;228(2):103–8.
39. Lui PP, Maffulli N, Rolf C, Smith RK. What are the validated animal models for tendinopathy? Scand J Med Sci Sports. 2011;21(1):3–17.
40. Gupta R, Lee TQ. Contributions of the different rabbit models to our understanding of rotator cuff pathology. J Shoulder Elb Surg. 2007;16(5 Suppl):S149–57.
41. Mahar A, Tamborlane J, Oka R, et al. Single-row suture anchor repair of the rotator cuff is biomechanically equivalent to double-row repair in a bovine model. Arthroscopy. 2007;23:1265–70.
42. Koganti AK, Adamson GJ, Gregersen CS, et al. Biomechanical comparison of traditional and locked suture configurations for arthroscopic repairs of the rotator cuff. Am J Sports Med. 2006;34:1832–8.
43. Bisson LJ, Manohar LM, Wilkins RD, et al. Influence of suture material on the biomechanical behavior of suturetendon specimens: a controlled study in bovine rotator cuff. Am J Sports Med. 2008;36:907–12.
44. Kelly BT, Turner AS, Bansal M, Terry M, Wolf BR, Warren RF, Altchek DW, Allen AA. In vivo healing after capsular plication in an ovine shoulder model. Iowa Orthop J. 2005;25:95–101.
45. Turner AS. Experiences with sheep as an animal model for shoulder surgery: strengths and shortcomings. J Shoulder Elb Surg. 2007;16(5 Suppl):S158–63.
46. Peterson DR, Ohashi KL, Aberman HM, Piza PA, Crockett HC, Fernandez JI, Lund PJ, Funk KA, Hawes ML, Parks BG, Mattern RH. Evaluation of a collagen-coated, resorbable fiber scaffold loaded with a peptide basic fibroblast growth factor mimetic in a sheep model of rotator cuff repair. J Shoulder Elb Surg. 2015;24(11):1764–73.
47. Schlegel TF, Hawkins RJ, Lewis CW, et al. The effects of augmentation with swine small intestine submucosa on tendon healing under tension: histologic and mechanical evaluations in sheep. Am J Sports Med. 2006;34:275–80.
48. Cummins CA, Appleyard RC, Strickland S, et al. Rotator cuff repair: an ex vivo analysis of suture anchor repair techniques on initial load to failure. Arthroscopy. 2005;21:1236–41.
49. Nicholson GP, Breur GJ, Van Sickle D, et al. Evaluation of a cross-linked acellular porcine dermal patch for rotator cuff repair augmentation in an ovine model. J Shoulder Elb Surg. 2007;16:S184–90.
50. Brassart N, Sanghavi S, Hansen UN, et al. Loss of rotator cuff tendon-to-bone interface pressure after reattachment using a suture anchor. J Shoulder Elb Surg. 2008;17:784–9.
51. Baums MH, et al. Biomechanical and magnetic resonance imaging evaluation of a single- and double-row rotator cuff repair in an in vivo sheep model. Arthroscopy. 2012;28:769–77.
52. Seeherman HJ, et al. rhBMP-12 accelerates healing of rotator cuff repairs in a sheep model. J Bone Joint Surg Am. 2008;90:2206–19.
53. Gerber C, Schneeberger AG, Beck M, et al. Mechanical strength of repairs of the rotator cuff. J Bone Joint Surg Br. 1994;76:371–80.
54. Luan T, et al. Muscle atrophy and fatty infiltration after an acute rotator cuff repair in a sheep model. Muscles Ligaments Tendons J. 2015;5:106–12.

55. Gerber C, Meyer DC, Schneeberger AG, et al. Effect of tendon release and delayed repair on the structure of the muscles of the rotator cuff: an experimental study in sheep. J Bone Joint Surg Am. 2004;86-A:1973–82.
56. Coleman SH, Fealy S, Ehteshami JR, et al. Chronic rotator cuff injury and repair model in sheep. J Bone Joint Surg Am. 2003;85-A:2391–402.
57. Derwin KA, Baker AR, Codsi MJ, et al. Assessment of the canine model of rotator cuff injury and repair. J Shoulder Elb Surg. 2007;16:S140–8.
58. Ji X, Bao N, An KN, Amadio PC, Steinmann SP, Zhao C. A canine non-weight-bearing model with radial Neurectomy for rotator cuff repair. PLoS One. 2015;10(6):e0130576.
59. Derwin KA, Codsi MJ, Milks RA, Baker AR, McCarron JA, Iannotti JP. Rotator cuff repair augmentation in a canine model with use of a woven poly-L-lactide device. J Bone Joint Surg Am. 2009;91(5):1159–71.
60. Safran O, Derwin KA, Powell K, et al. Changes in rotator cuff muscle volume, fat content, and passive mechanics after chronic detachment in a canine model. J Bone Joint Surg Am. 2005;87:2662–70.
61. Setton LA, Mow VC, Müller FJ, Pita JC, Howell DS. Mechanical behavior and biochemical composition of canine knee cartilage following periods of joint disuse and disuse with remobilization. Osteoarthr Cartil. 1997;5(1):1–16.
62. Anderst WJ, Les C, Tashman S. In vivo serial joint space measurements during dynamic loading in a canine model of osteoarthritis. Osteoarthr Cartil. 2005;13(9):808–16.
63. Harris WH, Jackson RH, Jowsey J. The in vivo distribution of tetracycline in canine bone. J Bone Joint Surg. 1962;44:1308–20.
64. Wirth MA, Korvick DL, Basamania CJ, Toro F, Aufdemorte TB, Rockwood CA Jr. Radiologic, mechanical, and histologic evaluation of 2 glenoid prosthesis designs in a canine model. J Shoulder Elb Surg. 2001;10(2):140–8.
65. Matsen FA 3rd, Clark JM, Titelman RM, et al. Healing of reamed glenoid bone articulating with a metal humeral hemiarthroplasty: a canine model. J Orthop Res. 2005;23(1):18–26.
66. Sonnabend DH, Howlett CR, Young AA. Histological evaluation of repair of the rotator cuff in a primate model. J Bone Joint Surg Br. 2010;92:586–94.
67. Xu H, Sandor M, Qi S, Lombardi J, Connor J, McQuillan DJ, Iannotti JP. Implantation of a porcine acellular dermal graft in a primate model of rotator cuff repair. J Shoulder Elb Surg. 2012 May;21(5):580–8.

4 Shoulder Kinematics and Biomechanics

Alper Yataganbaba, Erman Ceyhan, and Gazi Huri

After the human race evolved to be bipedal, the scapulohumeral complex also adapted. The bone continuity required in the weight-bearing joints is compromised to perform more complex movements with the upper limb and increase the range of motion. This adaptation in bony structures of shoulder complex increased the importance of soft tissue in joint stability [1, 2]. Thus, more unstable but the most flexible joint in our body has been formed. This is called "mobility-stability trade-off" [3].

The shoulder complex consists of four joints: glenohumeral joint (GH), acromioclavicular joint (AC), sternoclavicular joint (SC) joint, and scapulothoracic (ST) joint.

The GH joint is the main component of the shoulder complex. It connects the humerus and the scapula and is the joint with the widest range of motion in the human body. The mismatch between the humeral head and the relatively smaller glenoid creates instability, which provides a wide range of motion [4]. The GH joint can perform 180° of vertical abduction and 40° of vertical adduction (a), 180° of flexion and 55° of extension in the sagittal plane (b), 130° of horizontal abduction and 40° of horizontal adduction (c), 70° of internal rotation and 90° of external rotation movements around the long axis of the humerus (d). The glenohumeral joint also allows translation in all directions, which also increases the shoulder range of motion [5] (Fig. 4.1).

Although the shoulder complex constitutes most of the upper limb, they are connected to the axial skeleton by a single joint, the sternoclavicular (SC) joint [6]. Keeping the shoulder complex steady in the trunk is done mainly with muscle strength than this single joint. The sternoclavicular joint is a plane synovial joint that allows elevation/depression, protraction/retraction, and axial rotation movements. The position of the lateral end of the clavicle defines elevation/depression and protraction/retraction movements; the rotation is around the long axis of the clavicle. Besides, the medial end of the clavicle can translate in the anterior/posterior, superior/inferior, and medial/lateral directions on the sternum. Stability is provided by a synovial capsule, joint disc, and three major ligaments [6, 7]. Since the clavicle is connected laterally to the scapula with the acromioclavicular (AC) joint, the SC joint is also involved in the movement of the scapula [8, 9].

The acromioclavicular joint is the synovial plane joint between the lateral end of the clavi-

A. Yataganbaba
Hacettepe University School of Medicine, Department of Orthopaedics and Traumatology, Ankara, Turkey

E. Ceyhan
Department of Orthopaedics and Traumatology, Ankara City Hospital, Ankara, Turkey

G. Huri (✉)
Department of Orthopaedics and Traumatology, Hacettepe University School of Medicine, Ankara, Turkey

G. Huri et al. (eds.), *Fundamentals of the Shoulder*, https://doi.org/10.1007/978-3-030-94702-6_4

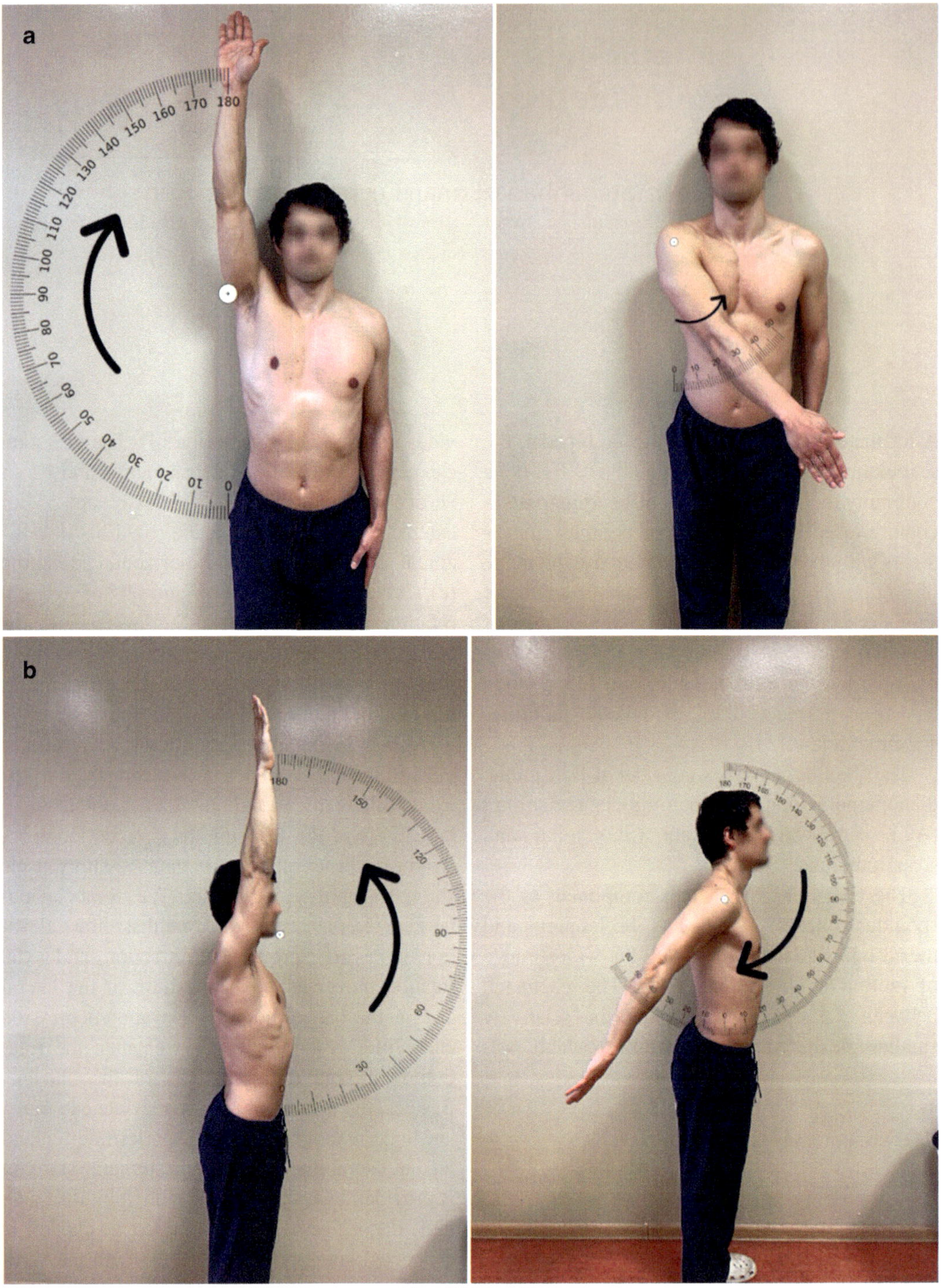

Fig. 4.1 Glenohumeral Joint movements

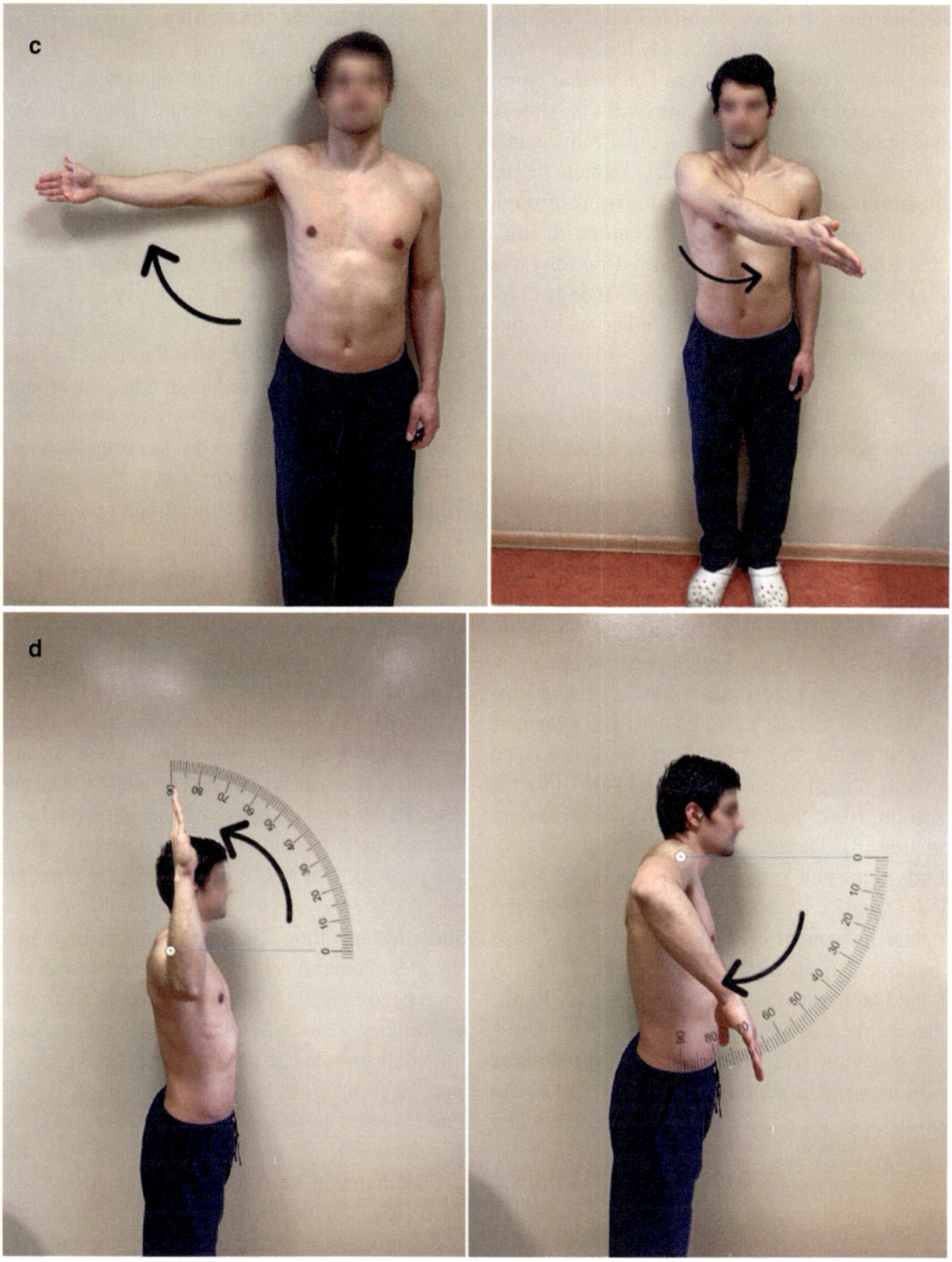

Fig. 4.1 (continued)

cle and the acromion. Similar to the SC joint, stability is ensured by capsule, ligaments, and joint disc [10]. The acromioclavicular joint helps scapula move in harmony with the thorax that changes shape during shoulder movements [11]. It also allows the forces applied to the upper limb to be transferred to the trunk through the clavicle and are more susceptible to injuries. Joint movements are limited because the joint surfaces between the scapula and the clavicle are incongruent. The number of studies describing the movements of this joint is limited.

The scapulothoracic joint forms the connection between the scapula and thorax. Still, the ST joint is not a real joint where the bone segments are connected by fibrous, synovial, and cartilage tissue. Thus, the ST joint is often referred to as "functional joint" in the literature [12]. Accordingly, the shoulder is mainly kept stable on the thorax by muscle contractions. The scapula is attached to the clavicle with an AC joint. Therefore, every movement of the scapula affects the AC joint and SC joint [12–14]. The scapula is located on the thorax between the second and seventh ribs. It is positioned in 35–45° internal rotation, 10–15° anteriorly tilted, and 10° upward rotation [8]. The glenohumeral joint forms two-thirds of the total range of motion of the shoulder, and scapula movements create one third. The regular movement of the scapula includes three components: upward and downward rotation around a horizontal axis perpendicular to the plane of the scapula (a), abduction and adduction (b), elevation and depression(c). During these movements, protraction and retraction occur with the help of the clavicle and acromioclavicular joint [15, 16] (Fig. 4.2).

The subacromial space, which is part of the glenohumeral joint, can also be considered another "functional joint." The movements of this joint are essential in shoulder functions [12].

The shoulder complex allows more complicated movements than other parts of the body as different types of joints work together in harmony. This large range of motion is allowed by a balanced interaction between static and dynamic stabilizers.

4.1 Shoulder Stability

Stability is the state that remains unchanged in the presence of forces that would change the current situation [17]. Shoulder stability can be analyzed in two parts: glenohumeral stability and scapulothoracic stability.

4.1.1 Glenohumeral Stability

Glenohumeral stability is that the humeral head remains in the glenoid and maintains its anatomic alignment during and after shoulder movements. Glenohumeral joint instability has been the most studied shoulder problem since the time of Hippocrates [18]. The stabilization of the joint is analyzed in two parts: static stabilization and dynamic stabilization [19].

4.1.1.1 Static Stabilization

Bony Static Stabilizers

Although the continuity between the humeral head and glenoid is low, bony structures are essential in ensuring shoulder stability. During rest, the inferior surface of the humeral head touches only a small area in the inferior part of the glenoid. Only 30% of the humeral articular surface is in contact with the glenoid articular surface at any time [20, 21]. Abduction increases the glenohumeral contact, and the pressure in the joint decreases [22]. When the pressure increases at the glenohumeral contact point, the humeral chondral surface can penetrate the glenoid chondral surface up to 1.2 mm. However, it is still controversial in which movements the pressure increases [23].

The humeral head forms the distal joint surface. The humeral head faces medially, superiorly, and posteriorly with regard to the humeral shaft and the condyles. The humeral head is retroverted on average 19° (range 9–31°) and inclined on average 41° (range 34–47°); head radius measures 23 mm (range 17–28 mm), and medial and posterior head center offsets are on average 7 mm (range 4–12 mm) and 2 mm (range 1–8 mm), respectively [24, 25].

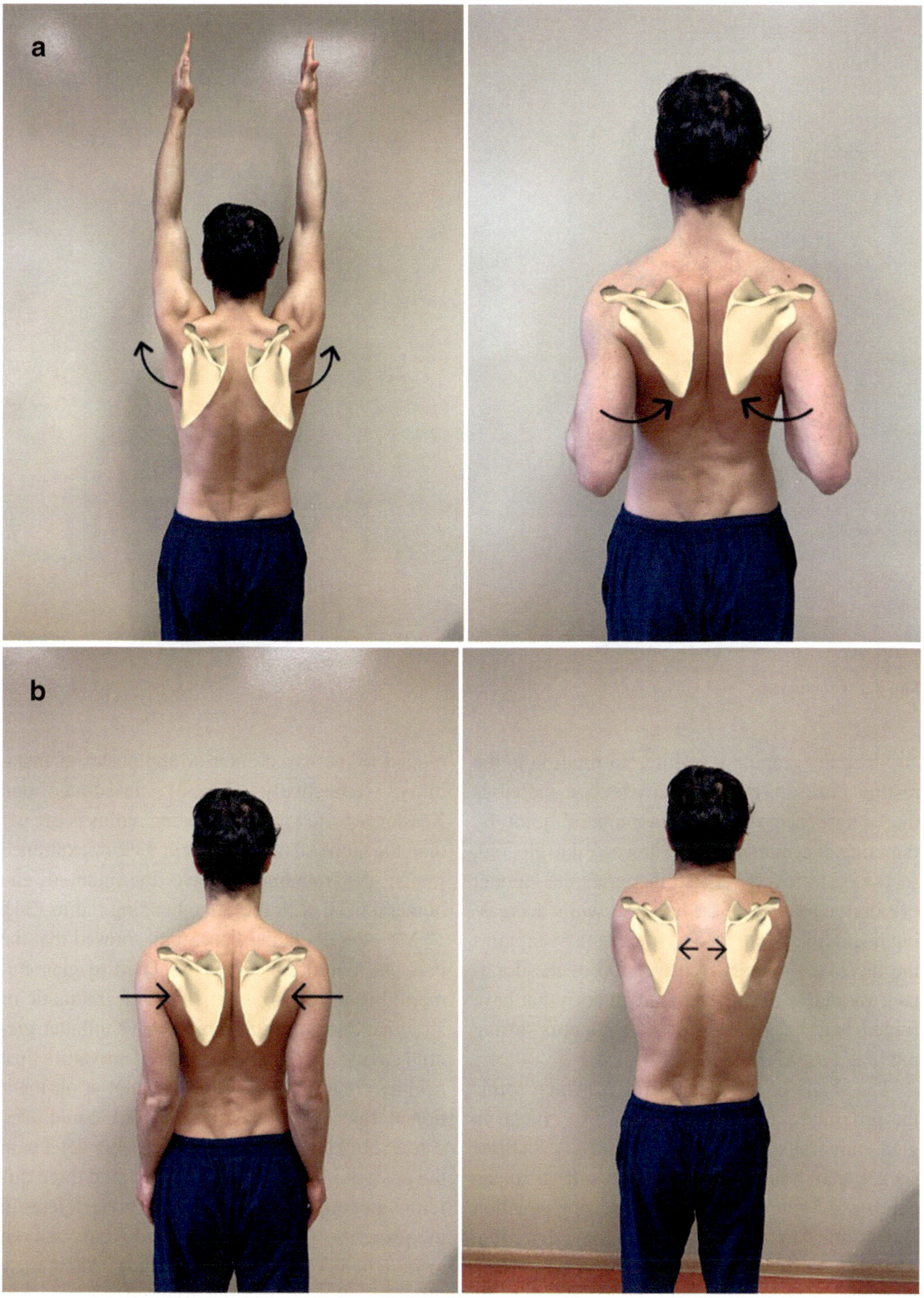

Fig. 4.2 Scapular movements

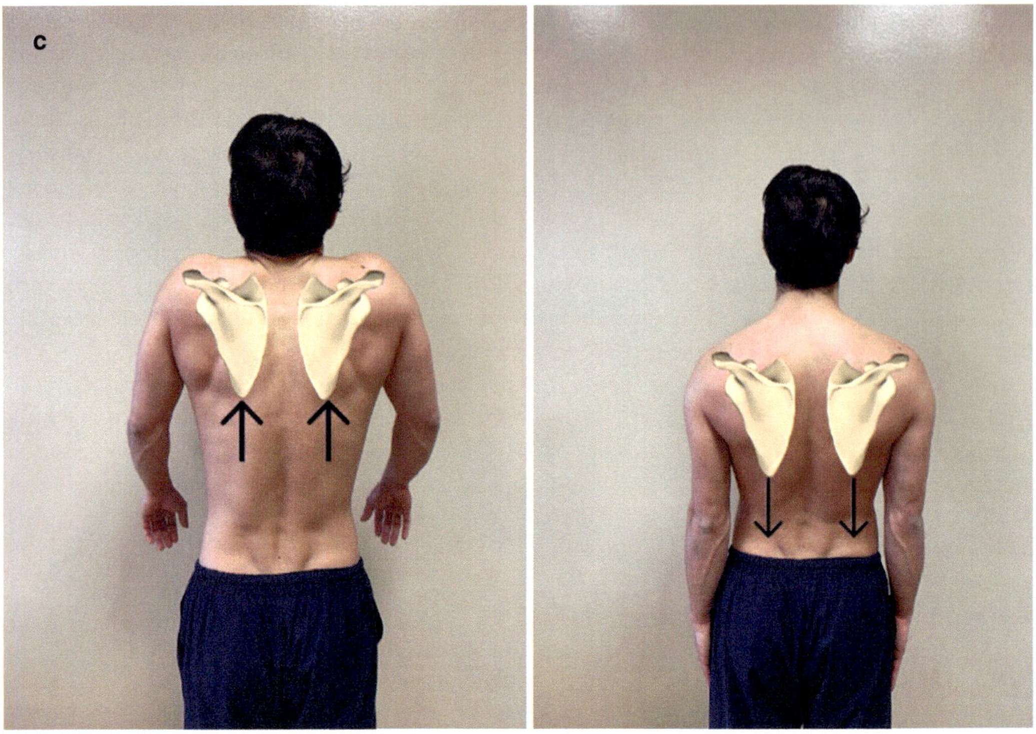

Fig. 4.2 (continued)

Since the scapula is in internal rotation in the resting position, humeral retroversion increases the congruence of the glenohumeral joint by directing the humeral head toward the glenoid. Increased retroversion also increases the amount of external rotation of the humerus while decreasing its internal rotation. This mechanism explains the increased humeral retroversion of the dominant shoulders of the overhead athletes that have forced external rotation of the humerus during pitching [26, 27].

The shape of the glenoid fossa, which forms the proximal part of the joint, is also crucial in glenohumeral stability. The glenoid is a shallow socket that holds the humeral head; its mean depth is 2.5 mm on the anteroposterior direction and 9 mm in the superior-inferior direction. Therefore, different amounts of displacing forces must be applied to dislocate the shoulder in different directions [28, 29]. It is retroverted on average, 1.23° (range 9.5° of anteversion to 10.5° of retroversion), and inclined superiorly, on average 4.2° (range, 7° of inferior inclination to 15.8° of superior inclination) [30]. More than 10° of anteversion and more than 15° of retroversion is related to increased anterior and posterior instability, respectively [31–33]. Friedman and Kessler reported that its bending radius is greater than the humeral head radius in 93% of examined joints; the remainders have the glenoid and humeral head with the same bending radius [34].

Moroder et al. and Peltz et al. showed that the loss in glenoid concavity is related to glenohumeral instability. And patients with traumatic or atraumatic shoulder instability have a flatter glenoid cavity with a higher radius of curvature than healthy controls [35–37]. Weishaupt et al. mentioned that the dysplastic glenoid could also cause shoulder instability due to bone defects in the posterior glenoid rim. They defined three different glenoid forms according to bone defects in the posterior glenoid rim: pointed form (without any deficiency), rounded glenoid deficiency ("lazy J" form), and the triangular bony deficiency ("delta" form) [38, 39].

Bone loss is also an important factor in shoulder instability. It usually occurs traumatically. In most cases, forced abduction and external rotation force cause the humeral head to dislocate anterior-inferiorly [31]. Most important bony lesions that result in instability occur after traumatic events and involve the anterior-inferior glenoid rim (Bony Bankart lesion) and the posterolateral aspect of the humeral head (Hill-Sachs lesion).

Bony Bankart lesions are significant if they involve more than 20% of the length of the glenoid. In this case, if the correct soft tissue repair is not performed, there is a high probability of recurrence. If Bony Bankart involves more than 50% of the glenoid, there will be more than a 30% reduction in shoulder stability [40]. Bony Bankart lesions are classified according to Bigliani et al.: type I, a displaced avulsion fracture with attached capsule; type II, a medially displaced fragment mal-united to the glenoid rim; type III, an erosion of the glenoid rim lower than 25% (III A) and more than 25% (III B) [41]. The PICO method suggested by Baudi et al. could be used to calculate glenoid bone defect [42].

Hill-Sachs lesion occurs after anterior shoulder dislocation due to a compression fracture involving the posterior-lateral part of the humeral head. The effect of the lesion on shoulder stability depends on the size and location. There are different classification methods. Calandra classification, which uses arthroscopy to measure the depth of the lesion, is the most frequently used method [43]. Apart from this, classification can be made according to radiography or magnetic resonance imaging [44, 45].

It is necessary to evaluate bone defects that cause instability in glenoid and humerus together and not to ignore injuries in soft tissue other than bone defects [46]. Glenoid track concept and its association with the concept of "engaging" and "non-engaging" lesions showed that the relationship between the humerus and glenoid lesions determines stability [47, 48].

Posterior shoulder dislocations are much rarer. It usually occurs after direct trauma or seizure. It usually occurs after direct trauma or seizure. In this case, a compression fracture occurs in the anterior superior of the humeral head (Reverse Hill-Sachs Lesion or McLaughlin lesion), and another fracture may occur in the posteroinferior rim of the glenoid (Reverse Bankart Lesion) [49–51].

Soft Tissue Static Stabilizers

Soft tissue static stabilizers include glenoid labrum, glenohumeral capsule, glenohumeral ligaments, rotator interval, negative intracapsular pressure, and adhesion cohesion mechanism.

Glenoid Labrum

The glenoid labrum is a triangular section ring around the glenoid rim, deepening the relatively flat glenoid cavity. The upper part is more mobile than the lower part, which is more tightly attached to the glenoid rim [52]. The superior part joins the structure of the biceps anchor, and the long head of the biceps tendon.

The glenoid labrum increases the depth of the glenoid cavity by 50% and increases its congruity with the humeral head and contributes to the negative pressure required for shoulder stability [28]. It increases the contact surface between the humerus and the glenoid by 2 mm anteroposteriorly and 4.5 mm supero-inferiorly [53].

The negative pressure in the glenohumeral joint is 32 mmHg. This pressure is particularly effective against traction force, while it is less effective against shear forces [54]. The contribution of negative pressure to joint stability is higher in the hanging arm position, while it decreases with shoulder abduction [55]. Loss of intracapsular negative pressure can manifest itself as an anterior translation of the humeral head. The labrum creates an attachment site around the glenoid rim for the glenohumeral ligaments and joint capsule. It also acts as an anti-shear bumper during mid-range movements [21].

When defining lesions in the labrum, it is necessary to analyze anatomical variants such as sublabral foramen, meniscoid labrums, and cord-like middle glenohumeral ligament do not require surgery [56].

The most common glenoid labrum injury is Bankart lesion. It accompanies 90% of traumatic anterior shoulder instability [57]. It is defined as a detachment of the anteroinferior aspect of the labrum and capsule. It occurs due to the detachment of the middle glenohumeral ligament and inferior glenohumeral ligament from the glenoid. Despite its frequency, it cannot be considered as an isolated cause of instability [58].

Green and Christensen classified Bankart lesions in five arthroscopic types: type 1 refers to the entire labrum; type 2 is a simple detachment of labrum with no other significant lesions; type 3 is an intra-parenchymal labrum tear; type 4 and

5 are complex tears with significant or complete degeneration of inferior glenohumeral ligament, respectively [59]. This classification also has a prognostic value: type 4 and 5 lesions have a high chance of recurrent instability after arthroscopic Bankart procedure of 87%.

Another lesion involving the anteroinferior aspect of the labrum is the ALPSA lesion (anterior labroligamentous periosteal sleeve avulsion). The anterior labroligamentous complex rolls up in a sleeve-like fashion and becomes displaced medially and inferiorly on the glenoid neck [60].

The redislocation rate in ALPSA lesions and the probability of engaging the Hill-Sachs lesion are higher than those of Bankart lesions. Besides, the external rotation limitation developed after ALPSA lesion repair is another crucial problem [61].

Specular lesions can be described for the posterior aspect of the labrum. Reverse Bankart lesion involves the posterior labrum and the posterior band of the inferior glenohumeral ligament. POLPSA is the posterior labroligamentous sleeve avulsion. In chronic conditions, Bennett lesions may occur (an extra-articular calcification along the posteroinferior glenoid neck close to the posterior band of the glenohumeral ligament) [62, 63].

Reverse Bankart lesion is frequent in athletes, such as rugby players, with a 20% incidence reported in a study of 142 elite rugby player shoulder arthroscopy [64]. The injury mechanism could be traced to a direct blow to the anterior and lateral aspects of the shoulder, while the arm is adducted; a rare mechanism of injury is a posterior blow to the arm while holding a tackle shield [65].

Concerning superior labrum, a prevalent lesion in throwing overhead athletes is SLAP (superior labrum anterior and posterior) tear. This lesion is described for the first time by Snyder et al. [66]. Snyder classified SLAP tears into four types. Type 2 and type 4 are more likely to create instability as they involve both the labrum and the long head of the biceps.

Moreover, SLAP lesions are common in contact sports. Funk and Snow have reported a 35% incidence of SLAP tears, arthroscopically diagnosed, in 51 rugby players' shoulders [67].

Capsuloligamentous Structures

Capsuloligamentous structures include joint capsule and glenohumeral ligaments (superior, middle, and inferior). There are many cadaveric and clinical studies investigating the biomechanical properties of these structures.

The constitutional trait of laxity facilitates extensive motion in multiple planes and may be essential to athletic performance. On the other hand, capsular stretching is noted along with a Bankart lesion is up to 28% of patients with recurrent anterior instability [68].

Superior and middle glenohumeral ligaments, together with coracohumeral ligament, long head of the biceps, and a thin layer of capsule, help to form rotator interval, and they will be treated in detail later.

The inferior glenohumeral ligament is also called the inferior glenohumeral ligament complex (IGHLC). It comprises three parts: two thicker bands on anterior and posterior and a thinner recess. During the abduction and external rotation, extension IGHLC moves anteriorly, forming a restraint to anterior translation of the humeral head.

During adduction, flexion, and internal rotation, IGHLC moves posteriorly, forming a restraint to posterior translation. IGHLC suffers from initial plastic deformation during the initial dislocation, but the damage becomes more critical after several episodes [69]. The lesion could more frequently occur at the glenoid insertion (anteroinferior glenoid rim) and in the middle part or at the humeral insertion [70].

Capsular stretching is often noted along with a Bankart lesion in up to 28% of patients with recurrent anterior instability [68]. The posterior capsule can also be injured; repetitive subluxations may lead to posterior instability by causing posterior capsular redundancy and increased joint volume.

The Rotator Interval

The rotator interval is a triangular space in the anterosuperior of the shoulder. It was first described by Neer in 1970 [71]. It creates resistance against extreme flexion, extension, adduction, and external rotation movements, limits inferior translation of the humeral head during adduction, and limits posterior translation of the humeral head during flexion or external rotation with abduction [72].

Furthermore, the synovial fluid provides to generate adhesion cohesion mechanism. The force formed between the wet surfaces of the humeral head and the glenoid contributes to stability [4].

4.1.1.2 Dynamic Stabilization

Dynamic stabilization provides a wide range of motion while securing stability during the normal function of the joint. There is a delicate balance between stability in the shoulder and range of motion. The muscles surrounding the shoulder and the neuromuscular balance between them ensure the dynamic stability of the joint. The muscles surrounding the shoulder and the neuromuscular balance between them provide the dynamic stability of the joint.

Proprioception

We know that capsuloligamentous structures also contribute to shoulder stability with their sensorimotor properties in addition to their mechanical functions. There are mechanoreceptors, especially in the anterior-inferior of the glenohumeral joint capsule. Proprioceptive information obtained from these structures contributes to shoulder stability by coordinating motor movements, reflexes, and joint stiffness.

As a result of the injuries in these structures, the decrease in proprioceptive information causes shoulder instability [73, 74]. Besides direct injury, capsular laxity has also been shown to cause a decrease in proprioception, leading to instability [75, 76].

Repairing of the capsuloligamentous structures restores the mechanical functions and tension of these tissues [77]. Retention allows joint capsule and ligamentous structures to sense mechanical stimulation and to facilitate proprioceptive feedback [74, 78].

Rotator Cuff Muscles

The rotator cuff is the common name of the structure consisting of muscles and tendons that contributes to shoulder stability. The rotator cuff consists of four muscles. These are supraspinatus (SSP), infraspinatus (ISP), teres minor (TM), and subscapularis (SSC).

Rotator cuff muscles provide fine control of shoulder movement. They play an essential role in dynamic stability, as well as contribute to proprioception [21].

Rotator cuff muscles compress the humeral head toward the glenoid and make an essential contribution to dynamic stabilization during shoulder movements. While symmetric rotator cuff contraction provides concavity compression, asymmetric contractions during shoulder movements rotate the humeral head. Joint reaction force decreases in rotator cuff tears [29, 79]. This stabilizing effect depends on the force couple formed by coordinated activation of the anterior and posterior fibers of the rotator cuff [80]. They act as an anti-shear force with the help of their mechanoreceptors. During the abduction, the rotator cuff tendon acts as a depressor for the humeral head and balances the pull of the deltoid muscle superiorly. Since this balance is disrupted after rotator cuff tears, the humeral head may be migrated superiorly [81]. A 50% reduction in rotator cuff force increases anterior dislocation by 46% and posterior dislocation by 31% [82].

The SSC is larger than the other three rotator cuff muscles and alone creates as much force as the sum of SSP, ISP, and TM [83]. The attachments of the muscles can be as tendons or muscle bodies [84, 85]. Therefore, the symptoms vary depending on the location and size of the rupture [86].

Long Head of the Biceps

The long head of the biceps (LHB) is a secondary stabilizer with a predominant role in the rotator cuff or capsuloligamentous deficiency. This tendon, originating from the supraglenoid tubercle and passing through the bicipital groove, acts as an anterior stabilizer during external rotation. During the late throwing phase, LHB reduces anterior translation, helping to prevent excessive torsion of the glenohumeral joint with a flexing elbow. These concepts can explain why type II or IV SLAP lesions are widespread in throwing athletes. Also, patients with rotator cuff insufficiency have hypertrophy in the tendon due to increased tension [87].

4.1.2 Scapulothoracic Joint Stability

The contribution of the scapula to upper extremity movements is better understood, especially in the last two decades [88]. The scapula provides a base to support the glenohumeral joint for regular upper limb movements [89, 90]. Since the scapulothoracic joint is not a real joint, its stability is provided only by dynamic stabilizers. The agonist, antagonist, and synergist contraction of the muscles adhering to both the thorax and the scapula ensures scapulothoracic joint stability. Scapular muscles dynamically coordinate the position of the glenoid, helping to create an effective glenohumeral joint movement. This harmonious relationship between the scapula and the humerus is called "scapulohumeral rhythm" [90, 91].

Upper and lower trapezius muscles, the serratus anterior and rhomboids (major and minor) are the structures that contribute most to scapulothoracic stability [15, 92].

Trapezius, together with the serratus anterior, initiates the upward rotation and posterior tilt movement of the scapula. Lower fibers of the trapezius contribute to the stability of the scapulothoracic joint during the descending of the arm from maximum elevation [15].

The serratus anterior muscle pulls the scapula toward the thoracic wall and makes a protraction movement. It provides stability, especially during abduction and pushing or punching type activities [91].

The rhomboids (major and minor) are especially active during adduction and retraction. They control the medial border of the scapula. It is quite active during swimming strokes and pulling [88]. It also takes part in the overhead throwing, both by reducing the stress on the anterior structures by fully retracting the scapula and braking by contracting eccentrically during the follow-through phase of pitching [93, 94].

Most abnormal biomechanics and overuse injuries in the shoulder girdle can be attributed to scapulothoracic joint instability [95, 96]. Alterations in joint movements due to weakness in scapular stabilizing muscles are called scapular dyskinesis [15, 97].

4.2 The Thrower's Shoulder

Throwing consists of six stages: the windup, early cocking, late cocking, acceleration, deceleration, and follow-through. During throwing, large muscle groups work together [98]. The transition between late cocking and acceleration is critical, and most of the injuries occur in this segment. During the late cocking, the shoulder is in abduction, and external rotation, the anterior capsule, and the coracohumeral ligament are under tension. Repetitive stress may cause stains or tensile failure in these structures, causing anterior shoulder instability [99, 100]. When the shoulder is in the 90°–90° position, the posterosuperior rotator cuff can be trapped between the greater tuberosity and the glenoid labrum, causing internal impingement. Shear forces also act on the posterosuperior labrum and biceps anchor in this position [101, 102]. In late cocking, structures in the posterior contract, when leading from late cocking to acceleration, the opposite happens. The anterior structures contract rapidly, allowing energy to be transferred to the ball. In acceleration, mainly the pectoralis major, latissimus dorsi, triceps, and serratus anterior muscles contract. The rotator cuff contracts during deceleration. During follow-through, the posterior capsule and the posterior rotator cuff are under eccentric stress. In repetitive stress, posterior rotator cuff failure, thickening in the capsule, and decrease in compliance may occur [103, 104].

References

1. Rockwood CA, Matsen FA, Wirth MA, Lippitt SB, Fehringer EV, Sperling JW. Rockwood and Matsen's the shoulder. Philadelphia, PA: Elsevier; 2017.
2. Arias-Martorell J. The morphology and evolutionary history of the glenohumeral joint of hominoids: a review. Ecol Evol. 2018;9:703–22. https://doi.org/10.1002/ece3.4392.

3. Veeger HEJ, van der Helm FCT. Shoulder function: the perfect compromise between mobility and stability. J Biomech. 2007;40:2119–29. https://doi.org/10.1016/j.jbiomech.2006.10.016.
4. Terry GC, Chopp TM. Functional anatomy of the shoulder. J Athl Train. 2000;35:248–55.
5. Karduna AR, Williams GR, Williams JL, Iannotti JP. Kinematics of the glenohumeral joint: influences of muscle forces, ligamentous constraints, and articular geometry. J Orthop Res Off Publ Orthop Res Soc. 1996;14:986–93. https://doi.org/10.1002/jor.1100140620.
6. Dempster WT. Mechanisms of shoulder movement. Arch Phys Med Rehabil. 1965;46:49–70.
7. Spencer EE, Kuhn JE, Huston LJ, Carpenter JE, Hughes RE. Ligamentous restraints to anterior and posterior translation of the sternoclavicular joint. J Shoulder Elb Surg. 2002;11:43–7. https://doi.org/10.1067/mse.2002.119394.
8. Ludewig PM, Phadke V, Braman JP, Hassett DR, Cieminski CJ, LaPrade RF. Motion of the shoulder complex during multiplanar humeral elevation. J Bone Joint Surg Am. 2009;91:378–89. https://doi.org/10.2106/JBJS.G.01483.
9. Fung M, Kato S, Barrance PJ, Elias JJ, McFarland EG, Nobuhara K, et al. Scapular and clavicular kinematics during humeral elevation: a study with cadavers. J Shoulder Elb Surg. 2001;10:278–85. https://doi.org/10.1067/mse.2001.114496.
10. Saccomanno MF, Ieso DE, C, Milano G. Acromioclavicular joint instability: anatomy, biomechanics and evaluation. Joints. 2014;2:87–92. https://doi.org/10.11138/jts/2014.2.2.087.
11. Teece RM, Lunden JB, Lloyd AS, Kaiser AP, Cieminski CJ, Ludewig PM. Three-dimensional acromioclavicular joint motions during elevation of the arm. J Orthop Sports Phys Ther. 2008;38:181–90. https://doi.org/10.2519/jospt.2008.2386.
12. Levangie PK, Norkin CC. Joint structure and function: a comprehensive analysis. Philadelphia: F.A. Davis Company; 2011.
13. Peat M. Functional anatomy of the shoulder complex. Phys Ther. 1986;66:1855–65. https://doi.org/10.1093/ptj/66.12.1855.
14. Bigliani LU, Codd TP, Connor PM, Levine WN, Littlefield MA, Hershon SJ. Shoulder motion and laxity in the professional baseball player. Am J Sports Med. 1997;25:609–13. https://doi.org/10.1177/036354659702500504.
15. Roche SJ, Funk L, Sciascia A, Ben KW. Scapular dyskinesis: the surgeon's perspective. Shoulder Elb. 2015;7:289–97. https://doi.org/10.1177/1758573215595949.
16. Ben KW, Sciascia A. Current concepts: scapular dyskinesis. Br J Sports Med. 2010;44:300–5. https://doi.org/10.1136/bjsm.2009.058834.
17. Myers JB, Wassinger CA, Lephart SM. Sensorimotor contribution to shoulder stability: effect of injury and rehabilitation. Man Ther. 2006;11:197–201. https://doi.org/10.1016/j.math.2006.04.002.
18. Wang VM, Flatow EL. Pathomechanics of acquired shoulder instability: a basic science perspective. J Shoulder Elb Surg. 2005;14:2S–11S. https://doi.org/10.1016/j.jse.2004.10.002.
19. Abboud JA, Soslowsky LJ. Interplay of the static and dynamic restraints in glenohumeral instability. Clin Orthop Relat Res. 2002:48–57. https://doi.org/10.1097/00003086-200207000-00007.
20. Soslowsky LJ, Flatow EL, Bigliani LU, Mow VC. Articular geometry of the glenohumeral joint. Clin Orthop Relat Res. 1992;285:181–90.
21. Lugo R, Kung P, Ma CB. Shoulder biomechanics. Eur J Radiol. 2008;68:16–24. https://doi.org/10.1016/j.ejrad.2008.02.051.
22. Warner JJ, Bowen MK, Deng XH, Hannafin JA, Arnoczky SP, Warren RF. Articular contact patterns of the normal glenohumeral joint. J Shoulder Elb Surg. 1998;7:381–8. https://doi.org/10.1016/s1058-2746(98)90027-1.
23. Massimini DF, Warner JJP, Li G. Glenohumeral joint cartilage contact in the healthy adult during scapular plane elevation depression with external humeral rotation. J Biomech. 2014;47:3100–6. https://doi.org/10.1016/j.jbiomech.2014.06.034.
24. Poppen NK, Walker PS. Normal and abnormal motion of the shoulder. J Bone Joint Surg Am. 1976;58:195–201.
25. Robertson DD, Yuan J, Bigliani LU, Flatow EL, Yamaguchi K. Three-dimensional analysis of the proximal part of the humerus: relevance to arthroplasty. J Bone Joint Surg Am. 2000;82:1594–602.https://doi.org/10.2106/00004623-200011000-00013.
26. Kronberg M, Broström LA, Söderlund V. Retroversion of the humeral head in the normal shoulder and its relationship to the normal range of motion. Clin Orthop Relat Res. 1990:113–7.
27. Reagan KM, Meister K, Horodyski MB, Werner DW, Carruthers C, Wilk K. Humeral retroversion and its relationship to glenohumeral rotation in the shoulder of college baseball players. Am J Sports Med. 2002;30:354–60. https://doi.org/10.1177/03635465020300030901.
28. Howell SM, Galinat BJ. The glenoid-labral socket. A constrained articular surface. Clin Orthop Relat Res. 1989;243:122–5.
29. Lippitt SB, Vanderhooft JE, Harris SL, Sidles JA, Harryman DT 2nd, Matsen FA 3rd. Glenohumeral stability from concavity-compression: a quantitative analysis. J Shoulder Elb Surg. 1993;2:27–35. https://doi.org/10.1016/S1058-2746(09)80134-1.
30. Churchill RS, Brems JJ, Kotschi H. Glenoid size, inclination, and version: an anatomic study. J Shoulder Elb Surg. 2001;10:327–32. https://doi.org/10.1067/mse.2001.115269.
31. Di Giacomo G, Piscitelli L, Pugliese M. The role of bone in glenohumeral stability. EFORT Open Rev. 2018;3:632–40. https://doi.org/10.1302/2058-5241.3.180028.

32. Bradley JP, Forsythe B, Mascarenhas R. Arthroscopic management of posterior shoulder instability: diagnosis, indications, and technique. Clin Sports Med. 2008;27:649–70. https://doi.org/10.1016/j.csm.2008.06.001.
33. Brewer BJ, Wubben RC, Carrera GF. Excessive retroversion of the glenoid cavity. A cause of non-traumatic posterior instability of the shoulder. J Bone Joint Surg Am. 1986;68:724–31.
34. Friedman R, An Y, Chokeski R, Kessler L. Anatomic and biomechanical study of glenohumeral contact. J Shoulder Elb Surg. 1994;3:S35.
35. Moroder P, Ernstbrunner L, Pomwenger W, Oberhauser F, Hitzl W, Tauber M, et al. Anterior Shoulder instability is associated with an underlying deficiency of the bony glenoid concavity. Arthrosc J Arthrosc Relat Surg Off Publ Arthrosc Assoc North Am Int Arthrosc Assoc. 2015;31:1223–31. https://doi.org/10.1016/j.arthro.2015.02.009.
36. Moroder P, Haniel F, Quirchmayr M, Schulz E, Eppel M, Matis N, et al. Effect of glenoid concavity loss on shoulder stability- a case report in a professional wrestler. BMC Musculoskelet Disord. 2016;17:357. https://doi.org/10.1186/s12891-016-1210-9.
37. Peltz CD, Zauel R, Ramo N, Mehran N, Moutzouros V, Bey MJ. Differences in glenohumeral joint morphology between patients with anterior shoulder instability and healthy, uninjured volunteers. J Shoulder Elb Surg. 2015;24:1014–20. https://doi.org/10.1016/j.jse.2015.03.024.
38. Weishaupt D, Zanetti M, Nyffeler RW, Gerber C, Hodler J. Posterior glenoid rim deficiency in recurrent (atraumatic) posterior shoulder instability. Skelet Radiol. 2000;29:204–10. https://doi.org/10.1007/s002560050594.
39. Inui H, Sugamoto K, Miyamoto T, Yoshikawa H, Machida A, Hashimoto J, et al. Glenoid shape in atraumatic posterior instability of the shoulder. Clin Orthop Relat Res. 2002:87–92. https://doi.org/10.1097/00003086-200210000-00014.
40. Piasecki DP, Verma NN, Romeo AA, Levine WN, Bach BRJ, Provencher MT. Glenoid bone deficiency in recurrent anterior shoulder instability: diagnosis and management. J Am Acad Orthop Surg. 2009;17:482–93. https://doi.org/10.5435/00124635-200908000-00002.
41. Bigliani LU, Newton PM, Steinmann SP, Connor PM, Mcllveen SJ. Glenoid rim lesions associated with recurrent anterior dislocation of the shoulder. Am J Sports Med. 1998;26:41–5. https://doi.org/10.1177/03635465980260012301.
42. Baudi P, Righi P, Bolognesi D, Rivetta S, Rossi Urtoler E, Guicciardi N, et al. How to identify and calculate glenoid bone deficit. Chir Organi Mov. 2005;90:145–52.
43. Shibayama K, Iwaso H. Hill-Sachs lesion classification under arthroscopic findings. J Shoulder Elb Surg. 2017;26:888–94. https://doi.org/10.1016/j.jse.2016.10.017.
44. Rowe CR, Zarins B, Ciullo JV. Recurrent anterior dislocation of the shoulder after surgical repair. Apparent causes of failure and treatment. J Bone Joint Surg Am. 1984;66:159–68.
45. Richards RD, Sartoris DJ, Pathria MN, Resnick D. Hill-Sachs lesion and normal humeral groove: MR imaging features allowing their differentiation. Radiology. 1994;190:665–8. https://doi.org/10.1148/radiology.190.3.8115607.
46. Di Giacomo G, Itoi E, Burkhart SS. Evolving concept of bipolar bone loss and the hill-Sachs lesion: from "engaging/non-engaging" lesion to "on-track/off-track" lesion. Arthrosc J Arthrosc Relat Surg Off Publ Arthrosc Assoc North Am Int Arthrosc Assoc. 2014;30:90–8. https://doi.org/10.1016/j.arthro.2013.10.004.
47. Burkhart SS, De Beer JF. Traumatic glenohumeral bone defects and their relationship to failure of arthroscopic Bankart repairs: significance of the inverted-pear glenoid and the humeral engaging hill-Sachs lesion. Arthrosc J Arthrosc Relat Surg Off Publ Arthrosc Assoc North Am Int Arthrosc Assoc. 2000;16:677–94. https://doi.org/10.1053/jars.2000.17715.
48. Shaha JS, Cook JB, Rowles DJ, Bottoni CR, Shaha SH, Tokish JM. Clinical validation of the glenoid track concept in anterior Glenohumeral instability. J Bone Joint Surg Am. 2016;98:1918–23. https://doi.org/10.2106/JBJS.15.01099.
49. Fronek J, Warren RF, Bowen M. Posterior subluxation of the glenohumeral joint. J Bone Joint Surg Am. 1989;71:205–16.
50. McLaughlin HL. Posterior dislocation of the shoulder. J Bone Joint Surg Am. 1952;24 A:584–90.
51. Goudie EB, Murray IR, Robinson CM. Instability of the shoulder following seizures. J Bone Joint Surg Br. 2012;94:721–8. https://doi.org/10.1302/0301-620X.94B6.28259.
52. Cooper DE, Arnoczky SP, O'Brien SJ, Warren RF, DiCarlo E, Allen AA. Anatomy, histology, and vascularity of the glenoid labrum. An anatomical study. J Bone Joint Surg Am. 1992;74:46–52.
53. Clavert P. Glenoid labrum pathology. Orthop Traumatol Surg Res. 2015;101:S19–24. https://doi.org/10.1016/j.otsr.2014.06.028.
54. Warner JJ, McMahon PJ. The role of the long head of the biceps brachii in superior stability of the glenohumeral joint. J Bone Joint Surg Am. 1995;77:366–72. https://doi.org/10.2106/00004623-199503000-00006.
55. Halder AM, Kuhl SG, Zobitz ME, Larson D, An KN. Effects of the glenoid labrum and glenohumeral abduction on stability of the shoulder joint through concavity-compression : an in vitro study. J Bone Joint Surg Am. 2001;83:1062–9. https://doi.org/10.2106/00004623-200107000-00013.
56. Rao AG, Kim TK, Chronopoulos E, McFarland EG. Anatomical variants in the anterosuperior aspect of the glenoid labrum: a statistical analysis of seventy-three cases. J Bone

Joint Surg Am. 2003;85:653–9. https://doi.org/10.2106/00004623-200304000-00011.
57. Owens BD, Nelson BJ, Duffey ML, Mountcastle SB, Taylor DC, Cameron KL, et al. Pathoanatomy of first-time, traumatic, anterior glenohumeral subluxation events. J Bone Joint Surg Am. 2010;92:1605–11. https://doi.org/10.2106/JBJS.I.00851.
58. Brand RA. Recurrent dislocation of the shoulder joint. Clin Orthop Relat Res. 2008;466:520–1. https://doi.org/10.1007/s11999-007-0105-3.
59. Green MR, Christensen KP. Arthroscopic Bankart procedure: two- to five-year followup with clinical correlation to severity of glenoid labral lesion. Am J Sports Med. 1995;23:276–81. https://doi.org/10.1177/036354659502300304.
60. Neviaser TJ. The anterior labroligamentous periosteal sleeve avulsion lesion: a cause of anterior instability of the shoulder. Arthrosc J Arthrosc Relat Surg Off Publ Arthrosc Assoc North Am Int Arthrosc Assoc. 1993;9:17–21. https://doi.org/10.1016/s0749-8063(05)80338-x.
61. Lee BG, Cho NS, Rhee YG. Anterior labroligamentous periosteal sleeve avulsion lesion in arthroscopic capsulolabral repair for anterior shoulder instability. Knee Surg Sports Traumatol Arthrosc. 2011;19:1563–9. https://doi.org/10.1007/s00167-011-1531-z.
62. Ciccone WJ 2nd. Arthroscopic posterior labral repair and capsular shift with a lateralized posterior portal. Arthrosc Tech. 2013;2:e323–6. https://doi.org/10.1016/j.eats.2013.05.005.
63. Van Tongel A, Karelse A, Berghs B, Verdonk R, De Wilde L. Posterior shoulder instability: current concepts review. Knee Surg Sports Traumatol Arthrosc. 2011;19:1547–53. https://doi.org/10.1007/s00167-010-1293-z.
64. Badge R, Tambe A, Funk L. Arthroscopic isolated posterior labral repair in rugby players. Int J Shoulder Surg. 2009;3:4–7. https://doi.org/10.4103/0973-6042.50875.
65. McDonough A, Funk L. Critical reflection of the advanced rehabilitation of an elite rugby league player sustaining a posterior Bankart lesion. Phys Ther Sport Off J Assoc Chart Physiother Sport Med. 2013;14:60–7. https://doi.org/10.1016/j.ptsp.2012.01.002.
66. Snyder SJ, Karzel RP, Del Pizzo W, Ferkel RD, Friedman MJ. SLAP lesions of the shoulder. Arthrosc J Arthrosc Relat Surg Off Publ Arthrosc Assoc North Am Int Arthrosc Assoc. 1990;6:274–9. https://doi.org/10.1016/0749-8063(90)90056-j.
67. Funk L, Snow M. SLAP tears of the glenoid labrum in contact athletes. Clin J Sport Med Off J Can Acad Sport Med. 2007;17:1–4. https://doi.org/10.1097/JSM.0b013e31802ede87.
68. Rowe CR, Patel D, Southmayd WW. The Bankart procedure: a long-term end-result study. J Bone Joint Surg Am. 1978;60:1–16.
69. Robinson CM, Dobson RJ. Anterior instability of the shoulder after trauma. J Bone Joint Surg Br. 2004;86:469–79. https://doi.org/10.1302/0301-620x.86b4.
70. Bigliani LU, Pollock RG, Soslowsky LJ, Flatow EL, Pawluk RJ, Mow VC. Tensile properties of the inferior glenohumeral ligament. J Orthop Res Off Publ Orthop Res Soc. 1992;10:187–97. https://doi.org/10.1002/jor.1100100205.
71. Neer CS 2nd. Displaced proximal humeral fractures. I. Classification and evaluation. J Bone Joint Surg Am. 1970;52:1077–89.
72. Frank RM, Taylor D, Verma NN, Romeo AA, Mologne TS, Provencher MT. The rotator interval of the Shoulder: implications in the treatment of Shoulder instability. Orthop J Sport Med. 2015;3:2325967115621494. https://doi.org/10.1177/2325967115621494.
73. Riemann BL, Lephart SM. The sensorimotor system, part I: the physiologic basis of functional joint stability. J Athl Train. 2002;37:71–9.
74. Tibone JE, Fechter J, Kao JT. Evaluation of a proprioception pathway in patients with stable and unstable shoulders with somatosensory cortical evoked potentials. J Shoulder Elb Surg. 1997;6:440–3. https://doi.org/10.1016/s1058-2746(97)70050-8.
75. Allegrucci M, Whitney SL, Lephart SM, Irrgang JJ, Fu FH. Shoulder kinesthesia in healthy unilateral athletes participating in upper extremity sports. J Orthop Sports Phys Ther. 1995;21:220–6. https://doi.org/10.2519/jospt.1995.21.4.220.
76. Blasier RB, Carpenter JE, Huston LJ. Shoulder proprioception. Effect of joint laxity, joint position, and direction of motion. Orthop Rev. 1994;23:45–50.
77. Rokito AS, Birdzell MG, Cuomo F, Di Paola MJ, Zuckerman JD. Recovery of shoulder strength and proprioception after open surgery for recurrent anterior instability: a comparison of two surgical techniques. J Shoulder Elb Surg. 2010;19:564–9. https://doi.org/10.1016/j.jse.2009.09.010.
78. Lephart SM, Warner JJ, Borsa PA, Fu FH. Proprioception of the shoulder joint in healthy, unstable, and surgically repaired shoulders. J Shoulder Elb Surg. 1994;3:371–80. https://doi.org/10.1016/S1058-2746(09)80022-0.
79. Parsons IM, Apreleva M, Fu FH, Woo SLY. The effect of rotator cuff tears on reaction forces at the glenohumeral joint. J Orthop Res Off Publ Orthop Res Soc. 2002;20:439–46. https://doi.org/10.1016/S0736-0266(01)00137-1.
80. Thompson WO, Debski RE, Boardman ND 3rd, Taskiran E, Warner JJ, Fu FH, et al. A biomechanical analysis of rotator cuff deficiency in a cadaveric model. Am J Sports Med. 1996;24:286–92. https://doi.org/10.1177/036354659602400307.
81. Mura N, O'Driscoll SW, Zobitz ME, Heers G, Jenkyn TR, Chou S-M, et al. The effect of infraspinatus disruption on glenohumeral torque and superior migration of the humeral head: a biomechanical study. J Shoulder Elb Surg. 2003;12:179–84. https://doi.org/10.1067/mse.2003.9.

82. Wuelker N, Korell M, Thren K. Dynamic glenohumeral joint stability. J Shoulder Elb Surg. 1998;7:43–52. https://doi.org/10.1016/s1058-2746(98)90182-3.
83. Keating JF, Waterworth P, Shaw-Dunn J, Crossan J. The relative strengths of the rotator cuff muscles. A cadaver study. J Bone Joint Surg Br. 1993;75:137–40.
84. Cleeman E, Brunelli M, Gothelf T, Hayes P, Flatow EL. Releases of subscapularis contracture: an anatomic and clinical study. J Shoulder Elb Surg. 2003;12:231–6. https://doi.org/10.1016/s1058-2746(02)00035-6.
85. Collin P, Lädermann A, Le Bourg M, Walch G. Subscapularis minor--an analogue of the Teres minor? Orthop Traumatol Surg Res. 2013;99:S255–8. https://doi.org/10.1016/j.otsr.2013.03.003.
86. Collin P, Matsumura N, Lädermann A, Denard PJ, Walch G. Relationship between massive chronic rotator cuff tear pattern and loss of active shoulder range of motion. J Shoulder Elb Surg. 2014;23:1195–202. https://doi.org/10.1016/j.jse.2013.11.019.
87. Pagnani MJ, Deng XH, Warren RF, Torzilli PA, O'Brien SJ. Role of the long head of the biceps brachii in glenohumeral stability: a biomechanical study in cadavera. J Shoulder Elb Surg. 1996;5:255–62. https://doi.org/10.1016/s1058-2746(96)80051-6.
88. Paine R, Voight ML. The role of the scapula. Int J Sports Phys Ther. 2013;8:617–29.
89. Voight ML, Thomson BC. The role of the scapula in the rehabilitation of shoulder injuries. J Athl Train. 2000;35:364–72.
90. Ben KW, Ludewig PM, McClure PW, Michener LA, Bak K, Sciascia AD. Clinical implications of scapular dyskinesis in shoulder injury: the 2013 Consensus statement from the "scapular summit". Br J Sports Med. 2013;47:877–85. https://doi.org/10.1136/bjsports-2013-092425.
91. McClure PW, Michener LA, Sennett BJ, Karduna AR. Direct 3-dimensional measurement of scapular kinematics during dynamic movements in vivo. J Shoulder Elb Surg. 2001;10:269–77. https://doi.org/10.1067/mse.2001.112954.
92. Smith J, Dietrich CT, Kotajarvi BR, Kaufman KR. The effect of scapular protraction on isometric shoulder rotation strength in normal subjects. J Shoulder Elb Surg. 2006;15:339–43. https://doi.org/10.1016/j.jse.2005.08.023.
93. Digiovine NM, Jobe FW, Pink M, Perry J. An electromyographic analysis of the upper extremity in pitching. J Shoulder Elb Surg. 1992;1:15–25. https://doi.org/10.1016/S1058-2746(09)80011-6.
94. Provencher CMT, Makani A, McNeil JW, Pomerantz ML, Golijanin P, Gross D. The role of the scapula in throwing disorders. Sports Med Arthrosc. 2014;22:80–7. https://doi.org/10.1097/JSA.0000000000000023.
95. Moseley JBJ, Jobe FW, Pink M, Perry J, Tibone J. EMG analysis of the scapular muscles during a shoulder rehabilitation program. Am J Sports Med. 1992;20:128–34. https://doi.org/10.1177/036354659202000206.
96. Kuhn JE, Plancher KD, Hawkins RJ. Scapular winging. J Am Acad Orthop Surg. 1995;3:319–25. https://doi.org/10.5435/00124635-199511000-00002.
97. Burn MB, McCulloch PC, Lintner DM, Liberman SR, Harris JD. Prevalence of scapular dyskinesis in overhead and nonoverhead athletes: a systematic review. Orthop J Sport Med. 2016;4:2325967115627608. https://doi.org/10.1177/2325967115627608.
98. Ben KW, Wilkes T, Sciascia A. Mechanics and pathomechanics in the overhead athlete. Clin Sports Med. 2013;32:637–51. https://doi.org/10.1016/j.csm.2013.07.003.
99. Kuhn JE, Bey MJ, Huston LJ, Blasier RB, Soslowsky LJ. Ligamentous restraints to external rotation of the humerus in the late-cocking phase of throwing. A cadaveric biomechanical investigation. Am J Sports Med. 2000;28:200–5. https://doi.org/10.1177/03635465000280021001.
100. Mihata T, McGarry MH, Neo M, Ohue M, Lee TQ. Effect of anterior capsular laxity on horizontal abduction and forceful internal impingement in a cadaveric model of the throwing Shoulder. Am J Sports Med. 2015;43:1758–63. https://doi.org/10.1177/0363546515582025.
101. Gelber JD, Soloff L, Schickendantz MS. The Thrower's shoulder. J Am Acad Orthop Surg. 2018; https://doi.org/10.5435/JAAOS-D-15-00585.
102. Burkhart SS, Morgan CD, Ben KW. The disabled throwing shoulder: spectrum of pathology Part I: pathoanatomy and biomechanics. Arthrosc J Arthrosc Relat Surg Off Publ Arthrosc Assoc North Am Int Arthrosc Assoc. 2003;19:404–20. https://doi.org/10.1053/jars.2003.50128.
103. Werner SL, Gill TJ, Murray TA, Cook TD, Hawkins RJ. Relationships between throwing mechanics and shoulder distraction in professional baseball pitchers. Am J Sports Med. 2001;29:354–8. https://doi.org/10.1177/03635465010290031701.
104. Thomas SJ, Swanik CB, Higginson JS, Kaminski TW, Swanik KA, Bartolozzi AR, et al. A bilateral comparison of posterior capsule thickness and its correlation with glenohumeral range of motion and scapular upward rotation in collegiate baseball players. J Shoulder Elb Surg. 2011;20:708–16. https://doi.org/10.1016/j.jse.2010.08.031.

Biomechanics of Anterior Shoulder Instability

5

Nobuyuki Yamamoto and Eiji Itoi

5.1 Dislocation Position

It has been believed that a shoulder dislocation occurs with the arm in 90° of abduction and maximum external rotation (so-called dislocation position). Krøner et al. [1], in a large-scale epidemiological study of 250,000 patients, investigated the mechanism of injury in 216 patients with anterior shoulder dislocation. They found that the most common mechanism of injury was a blow against the shoulder, which accounted for 46% of all the cases. Recently, video analysis of rugby players demonstrated the arm positions at the time of dislocation. Crichton et al. [2] reported that, out of 16 rugby players, injury was sustained with the arm in greater than 90° of elevation in eight players, with the arm at 90° of abduction in three players, and with the arm in internal rotation in either abduction or adduction in three players. However, there have been only a limited number of reports regarding the arm position at the time of anterior shoulder dislocation.

In order to clarify the arm position when a Hill-Sachs lesion was created, we performed a CT study [3]. A Hill-Sachs lesion, a common injury associated with anterior shoulder dislocation, is a compression fracture created by the anterior rim of the glenoid. If a Hill-Sachs lesion is created with the arm in abduction and external rotation at the time of dislocation, their arms should be in abduction and external rotation when they come to the emergency room. In fact, their arms are usually in adduction and neutral rotation. We hypothesized that a Hill-Sachs lesion was created not at the end range of motion (abduction and external rotation) but in the mid-range of motion. Using 3D CT images, we reproduced the arm position where the Hill-Sachs lesion and the anterior rim of the glenoid best fit with each other (Fig. 5.1) because this arm position was considered to be the one in which the Hill-Sachs lesion had been created. Our data showed that the arm position when a Hill-Sachs lesion best fit the anterior rim of the glenoid was 74° of abduction, 27° of external rotation, and 3° of horizontal flexion (Fig. 5.2). This arm position was not so-called dislocation position but a position in much lower angle of abduction and external rotation. This might have been a position of dislocation. However, we need to consider another possibility that a Hill-Sachs lesion was created not at the time of dislocation (the moment the humeral head overrode the anterior rim of the glenoid), but sometime later when the patients lowered and internally rotated the arm after dislocation. There have been only a few reports regarding the dislocation position and mechanisms of anterior shoulder dislocation.

N. Yamamoto · E. Itoi (✉)
Department of Orthopaedic Surgery, Tohoku University School of Medicine, Sendai, Japan
e-mail: koyomoe@med.tohoku.ac.jp; itoi-eiji@med.tohoku.ac.jp

G. Huri et al. (eds.), *Fundamentals of the Shoulder*, https://doi.org/10.1007/978-3-030-94702-6_5

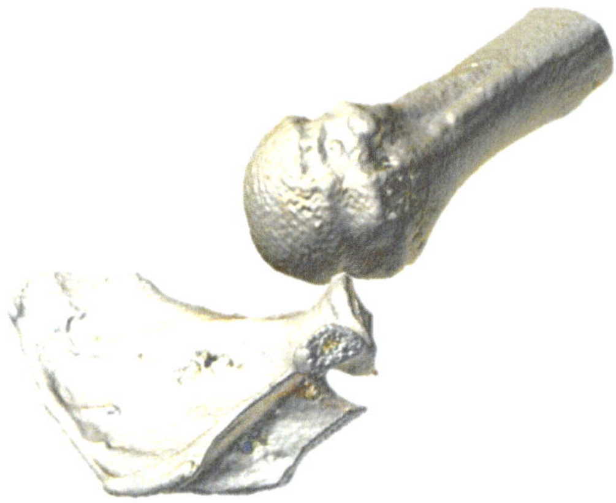

Fig. 5.1 Engagement in the 3D CT image. We reproduced the arm position where the Hill-Sachs lesion and the anterior rim of the glenoid best fit with each other using 3D CT images

5.2 Remplissage Procedure

The remplissage procedure [4] which has tenodesis effect of the infraspinatus tendon is one of surgical procedures for a large Hill-Sachs lesion. Recently, the remplissage procedure has rapidly gained popularity because it is relatively easy to perform arthroscopically.

5.2.1 Effectiveness on Off-Track Hill-Sachs Lesions

A Hill-Sachs lesion which has a risk of engagement with the anterior rim of the glenoid is defined an "off-track" lesion [5], which means the Hill-Sachs lesion is out of the glenoid track [6, 7]. There are several clinical reports investigating the effectiveness of the remplissage procedure on an off-track Hill-Sachs lesion. Also, some biomechanical studies clarified the usefulness of this procedure.

5.2.2 Recurrence and Return to Sports

In order to prove that the engagement can be avoided by performing the remplissage procedure for an "off-track" Hill-Sachs lesion, we need to compare two procedures: arthroscopic Bankart repair alone and arthroscopic Bankart repair with remplissage procedure. Looking at clinical reports comparing two procedures, all reported lower recurrence rates and most showed better shoulder function after a Bankart repair with remplissage compared with an isolated Bankart repair [8–13]. Only three reports [8–10] described the return-to-sport rates and they were mostly similar.

5.2.3 Restriction of Range of External Rotation Motion

Theoretically, the range of motion, especially external rotation and horizontal extension should be restricted after the remplissage procedure because this procedure makes the Hill-Sachs lesion an extra-articular structure, preventing the glenoid to move over the Hill-Sachs lesion. In some biomechanical studies, it was already demonstrated that this procedure caused significant restrictions in the range of abduction and external rotation. Although there are some differences of placing the anchor in the rim or valley of the humeral defect or the size of Hill-Sachs lesions simulated among those reports, an average restriction of 5–15° in the range of external rotation with the humerus in abduction was observed [14–16]. On the other hand, looking at the clinical reports, most of the differences in the clinical studies did not exceed the minimal detectable change in measuring the range of motion, which is 3–5° and none of these differences reached statistical significance [9–12]. Why does this difference occur? There are several reasons which we can think about. First, the scapula is fixed to the experimental device and the humerus is moved to measure the range of motion in the biomechanical studies. However, in the clinical setting, the scapula is not fixed and the range of motion relative to the trunk is measured. The scapular motion might compensate the restricted motion of the glenohumeral joint. Second, the location of the suture anchor placed may be different. The restriction of motion should be different between the shoulders with the anchors placed in the rim and those placed in the valley of the humeral defect (Fig. 5.3). Theoretically, the engagement

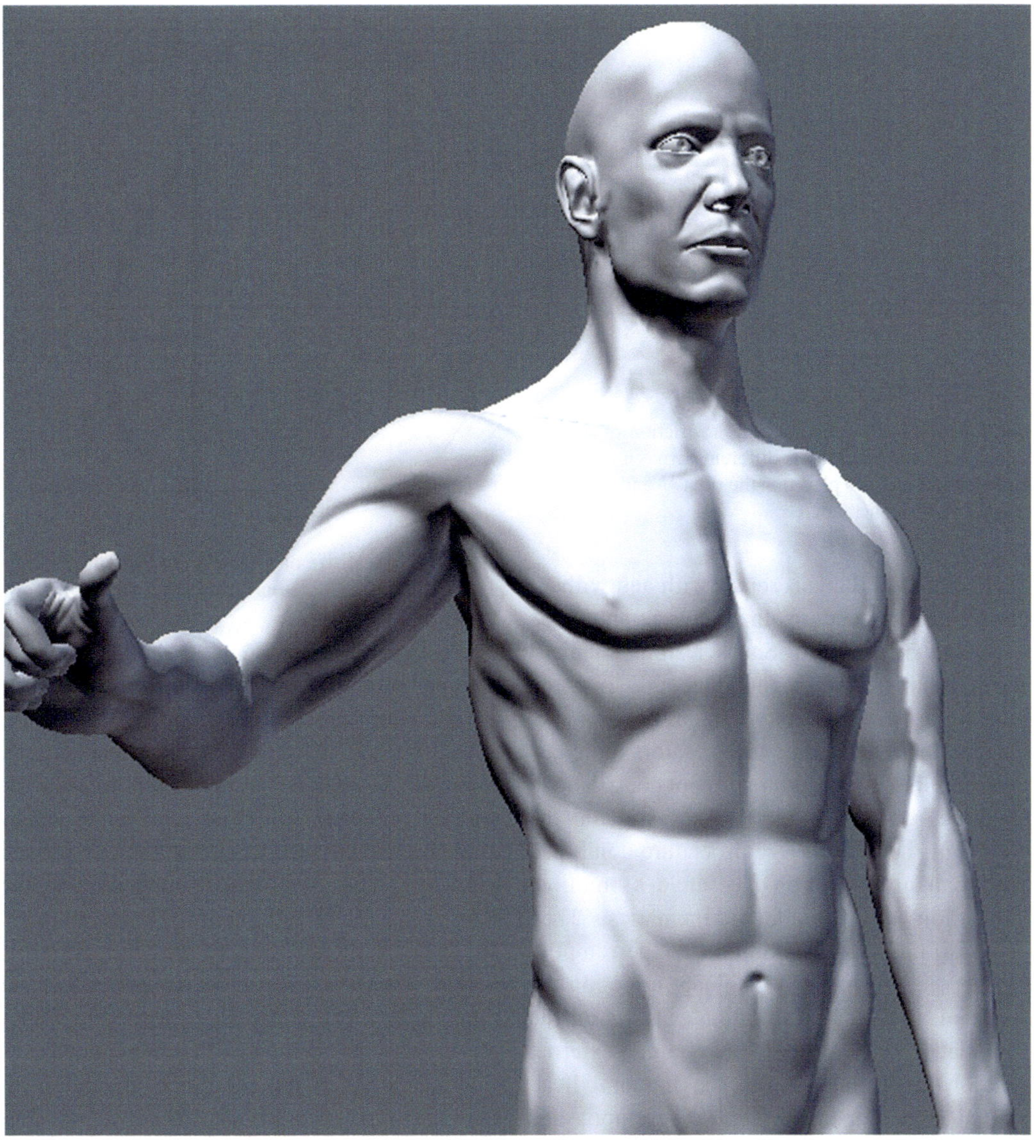

Fig. 5.2 The arm position where a Hill-Sachs lesion is created. Our data showed that the arm position when a Hill-Sachs lesion and the anterior rim of the glenoid best fit was 74° of abduction, 27° of external rotation, and 3° of horizontal flexion

against the glenoid can be best avoided if the anchor is placed at the medial rim of the humeral defect, which in turn may cause the greatest restriction of motion. Most surgeons are likely to insert the anchors not at the rim but closer to the valley of the lesion because it is easier. Third, biomechanical studies are done at time zero, whereas the clinical study is usually performed years after surgery. Soft tissue elongation and adjustment might have occurred.

5.2.4 Contribute to Stability in the Mid-Range Position?

Does the remplissage procedure contribute to stability? The surgical indication of this procedure is becoming wider in the current literature. The recent biomechanical studies investigated the effect of remplissage procedure on stability. Some studies support the effectiveness of this procedure, but some do not. Grimberg et al.

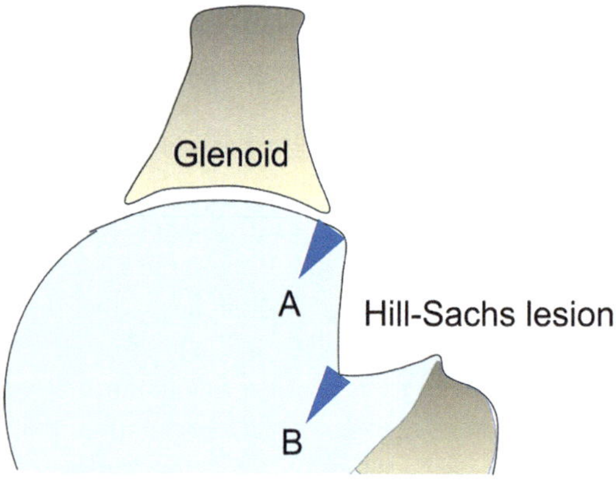

Fig. 5.3 Anchor location and the restriction of motion. The restriction of motion is different between the shoulders with the anchors placed in the rim (A) and in the valley (B) of the humeral defect

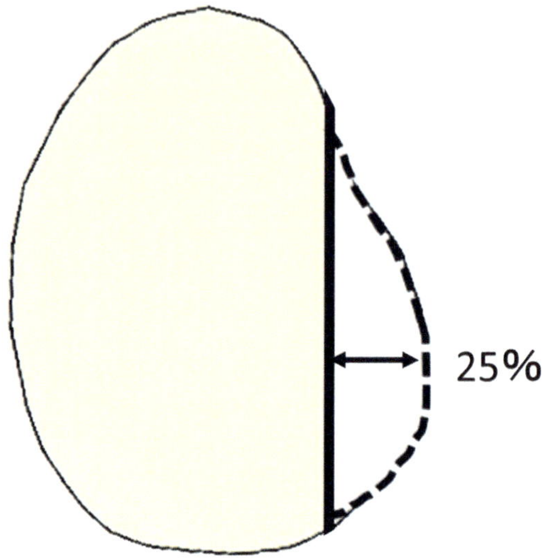

Fig. 5.4 Critical size of glenoid bone loss. Greater than 25% glenoid bone loss is an indication for bone grafting

[17] compared Bankart repair with and without remplissage procedure. In cadaveric shoulders with Bankart and Hill-Sachs lesions, Bankart repair with remplissage procedure restored joint stability compared to Bankart repair alone. On the other hand, some investigators indicated that remplissage procedure did not contribute to stability. Arginter et al. [15] investigated the effects of the remplissage procedure combined with Bankart repair on the range of motion, translation, and shoulder kinematics. They demonstrated that the addition of the remplissage procedure had no significant effect on the range of motion or translation but altered the shoulder kinematics: at maximum external rotation at 60° abduction, the humeral head shifted posterior-inferiorly with remplissage procedure.

From the viewpoint of shoulder biomechanics, shoulder stability can be divided into two: stability in the mid-range of motion (mid-range stability) and the stability at the end-range of motion (end-range stability). At the end-range of motion such as abduction and maximum external rotation, the capsule-labrum complex contributes to stability because it is taught. In the mid-range of motion, the capsule-labrum complex is lax, and the humeral head is kept in the concave glenoid fossa by the compressive force generated by the surrounding muscles (concavity-compression effect). It was already demonstrated that the anteroinferior stability of the shoulder increased after Bankart repair in abduction and external rotation (end-range). However, there have been no reports demonstrating that the mid-range stability increases after the remplissage procedure. As mentioned previously, the remplissage procedure fills the Hill-Sachs defect with the infraspinatus tendon so that the glenoid would not fall into this defect when it comes close to the defect. In other words, the remplissage works by filling the defect and preventing the glenoid to override the defect at the end-range of motion. In the mid-range of motion, the shoulder stability seems not to be affected by the remplissage.

5.3 New Evaluation of the Glenoid Bone Loss

Glenoid bone loss is also a common injury associated with anterior shoulder instability. It has been reported that greater than 25% [18–20] glenoid bone loss has been an indication for bone grafting (Fig. 5.4). However, there is one question. Can this borderline be drawn clearly? Is bone grafting unnecessary for 24% glenoid bone loss? Some surgeons prefer to perform bone grafting when a defect is very close to the critical size, whereas others perform arthroscopic Bankart repair. Thus, the decision is upon the surgeon's discretion.

5.3.1 Subcritical Bone Loss

Shaha et al. [21] indicated how we should judge for such a borderline bone loss. They showed that a bone loss of 13.5–20% led to a clinically significant decrease in their quality of life, even though the patients did not sustain a recurrent instability. They proposed that such a bone loss be treated by bone grafting rather than arthroscopic Bankart repair as a "subcritical bone loss." All the subjects in Shaha's [21] study were military patients with high activity level. Hypothesizing that the subcritical bone loss might vary according to activity levels of patients, we performed a clinical study [22] investigating the size of subcritical bone loss, if any, in our series of general population. The Western Ontario Shoulder Instability Index (WOSI), which was a questionnaire for disease-specific quality-of-life scoring system for shoulder instability, was significantly lower for a bone loss of 17–25% (Fig. 5.5). This value was a little greater than the value reported by Shaha et al. [21] (13.5–20%). In our study, there were two factors which decreased the WOSI score: male and contact athlete. Thus, for the male contact athlete with the bone loss of 17–25%, bone grafting such as Latarjet procedure rather than arthroscopic Bankart repair may as well be chosen.

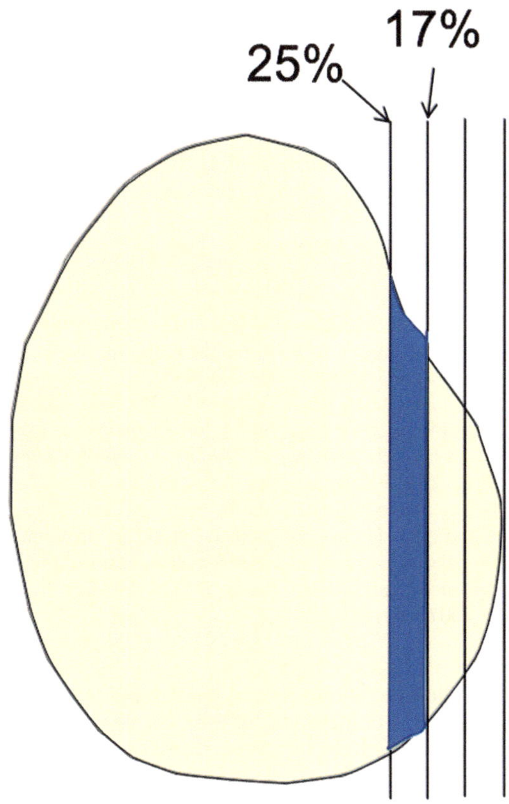

Fig. 5.5 "Subcritical bone loss". Glenoid bone loss between 17% and 25% is considered to be a "subcritical bone loss" in our series

5.3.2 Does the Size Measurement Really Tell It All?

Recently, one interesting article reporting the evaluation of the glenoid bone loss was published. Previous biomechanical or clinical investigations described the size of the glenoid bone loss such as percentage of glenoid width or length. However, the concave shape of the glenoid articular surface (especially, depth or shape of concavity) is different in each patient. We should highlight the 3D concave shape of the glenoid articular surface, and the individual constitutional glenoid concavity shape needs to be accounted for when estimating the effect of glenoid bone loss on shoulder stability.

Morodor et al. [23] performed a patient-specific analysis of the effect of individual glenoid concavity shapes and various degrees of bone loss on glenoid bone–mediated shoulder stability with finite element analysis of CT-based 3D shoulder models of patients with anterior shoulder instability. They concluded that current glenoid bone loss measurements are unable to provide an adequate estimation on the actual biomechanical effect of glenoid defects because the relation between the glenoid defect size and its biomechanical effect was nonlinear, and patients with shoulder instability have constitutional difference in the glenoid concavity shape.

References

1. Krøner K, Lind T, Jensen J. The epidemiology of shoulder dislocations. Arch Orthop Trauma Surg. 1989;108(5):288–90.
2. Crichton J, Jones DR, Funk L. Mechanisms of traumatic shoulder injury in elite rugby players. Br J Sports Med. 2012;46(7):538–42.
3. Kawakami J, Yamamoto N, Hatta T, Shinagawa K, Itoi E. In which arm position is a hill-Sachs lesion created? Am J Sports Med. 2019;47(10):2464–8.
4. Wolf EM, Pollack ME. Hill-Sachs "Remplissage": an arthroscopic solution for the engaging hill-Sachs lesion. Arthroscopy. 2004;20:e14–5.
5. Di Giacomo G, Itoi E, Burkhart SS. Evolving concept of bipolar bone loss and the hill-Sachs lesion: from "engaging/non-engaging" lesion to "on-track/off-track" lesion. Arthroscopy. 2014;30(1):90–8.
6. Yamamoto N, Itoi E, Abe H, et al. Contact between the glenoid and the humeral head in abduction, external rotation, and horizontal extension: a new concept of glenoid track. J Shoulder Elb Surg. 2007;16(5):649–56.
7. Omori Y, Yamamoto N, Koishi H, et al. Measurement of the glenoid track in vivo as investigated by 3-dimensional motion analysis using open MRI. Am J Sports Med. 2014;42(6):1290–5.
8. Garcia GH, Park MJ, Zhang C, Kelly JD, Huffman GR. Large hill-Sachs lesion: a comparative study of patients treated with arthroscopic Bankart repair with or without remplissage. HSS J. 2015;11:98–103.
9. Cho NS, Yoo JH, Juh HS, et al. Anterior shoulder instability with engaging hill-Sachs defects: a comparison of arthroscopic Bankart repair with and without posterior capsulodesis. Knee Surg Sport Traumatol Arthrosc. 2016;24:3801–8.
10. Franceschi F, Papalia R, Rizzello G, et al. Remplissage repaired new frontiers in the prevention of recurrent shoulder instability: a 2-year follow-up comparative study. Am J Sports Med. 2012;40:2462–9.
11. Ko S-H, Cha J-R, Lee C-C, et al. The influence of arthroscopic remplissage for engaging hill-Sachs lesions combined with Bankart repair on redislocation and shoulder function compared with Bankart repair alone. Clin Orthop Surg. 2016;8:428–36.
12. Nourissat G, Kilinc AS, Werther JR, et al. A prospective, comparative, radiological, and clinical study of the influence of the "remplissage" procedure on shoulder range of motion after stabilization by arthroscopic Bankart repair. Am J Sports Med. 2011;39:2147–52.
13. Miyamoto R, Yamamoto A, Shitara H, et al. Clinical outcome of arthroscopic remplissage as augmentation during arthroscopic Bankart repair for recurrent anterior shoulder instability. Open Orthop J. 2017;11:1268–76.
14. Elkinson I, Giles JW, Boons HW, et al. The shoulder remplissage procedure for hill-Sachs defects: does technique matter? J Shoulder Elb Surg. 2013;22:835–41.
15. Argintar E, Heckmann N, Wang L, et al. The biomechanical effect of shoulder remplissage combined with Bankart repair for the treatment of engaging hill-Sachs lesions. Knee Surg Sport Traumatol Arthrosc. 2016;24:585–92.
16. Omi R, Hooke AW, Zhao KD, et al. The effect of the remplissage procedure on shoulder range of motion: a cadaveric study. Arthroscopy. 2014;30(2):178–87.
17. Grimberg J, Diop A, Ghosn RB, et al. Bankart repair versus Bankart repair plus remplissage: An in vitro biomechanical comparative study. Knee Surg Sport Traumatol Arthrosc. 2016;24:374–80.
18. Itoi E, Lee SB, Berglund LJ, Berge LL, An KN. The effect of a glenoid defect on anteroinferior stability of the shoulder after Bankart repair: a cadaveric study. J Bone Joint Surg Am. 2000;82(1):35–46.
19. Yamamoto N, Muraki T, Sperling JW, et al. Stabilizing mechanism in bone-grafting of a large glenoid defect. J Bone Joint Surg Am. 2010;92(11):2059–66.
20. Yamamoto N, Itoi E, Abe H, et al. Effect of an anterior glenoid defect on anterior shoulder stability: a cadaveric study. Am J Sports Med. 2009;37(5):949–54.
21. Shaha JS, Cook JB, Song DJ, et al. Redefining "Critical" bone loss in shoulder instability: functional outcomes worsen with "Subcritical" bone loss. Am J Sports Med. 2015;43(7):1719–25.
22. Yamamoto N, Kawakami J, Hatta T, et al. Effect of subcritical glenoid bone loss on activities of daily living in patients with anterior shoulder instability. Orthop Traumatol Surg Res. 2019;105(8):1467–70.
23. Moroder P, Damm P, Wierer G, et al. Challenging the current concept of critical glenoid bone loss in shoulder instability: does the size measurement really tell it all? Am J Sports Med. 2019;47(3):688–94.

Pathomechanics in CTA and Rationale of RSA

6

Gokhan Karademir, Onur Tunalı, and Ata Can Atalar

6.1 Pathomechanics in Cuff Tear Arthropathy

6.1.1 Introduction

The term cuff tear arthropathy [1] was first invented by Charles Neer in 1977 [2]. CTA was described as a rare pathologic entity mainly characterized by chronic full-thickness rotator cuff tear, degenerative changes in glenohumeral joint, superior migration of the humeral head. Other characteristic findings that may see are humeral head collapse, acetabularization of the acromion, femoralization of the humeral head, erosion of the superior glenoid or acromion, and subdeltoid effusion.

Although numerous pathologic mechanisms for the development of CTA have been postulated, the exact etiology of CTA is still unclear [3]. None of these pathomechanic theories could explain why only some patients with a massive rotator cuff tear progress to CTA. Neer reported that CTA may only develop in approximately 4% of patients with complete rotator cuff tear [2].

G. Karademir · A. C. Atalar (✉)
Department of Orthopedics and Traumatology, School of Medicine, Acıbadem MAA University, Istanbul, Turkey

Department of Orthopedics and Traumatology, Acıbadem Maslak Hospital, Istanbul, Turkey
e-mail: dr@gokhankarademir.com; atacan.atalar@acibadem.edu.tr

O. Tunalı
Department of Orthopedics and Traumatology, Acıbadem Maslak Hospital, Istanbul, Turkey

6.1.2 Historical Review

In the literature, several authors described the same clinical syndrome using different terms. The earliest description of the pathologic features of the CTA was performed by Adams [4] and Smith [5] in the nineteenth century. Adams described two types of rheumatoid arthritis: a generalized form and localized form involving the shoulder which had the morphological characteristics of CTA. Codman reported a case of rotator cuff-mediated hygroma and advanced shoulder arthritis in his 1934 self-published monograph [6]. He described recurrent swelling of the shoulder, absence of rotator cuff, cartilaginous bodies attached to synovial tissue and, severe destructive glenohumeral arthritis.

In 1968 De Seze described a similar clinic entity as "hemorrhagic shoulder of the elderly" in three women [7]. In the following years, single case reports or small case series were published about spontaneous hemarthrosis of the glenohumeral joint in the literature. In 1981, the term Milwaukee shoulder was described by Halverson in four elderly women who had recurrent bilateral shoulder effusions, severe destructive glenohumeral arthritis, and massive rotator cuff tear [8].

G. Huri et al. (eds.), *Fundamentals of the Shoulder*, https://doi.org/10.1007/978-3-030-94702-6_6

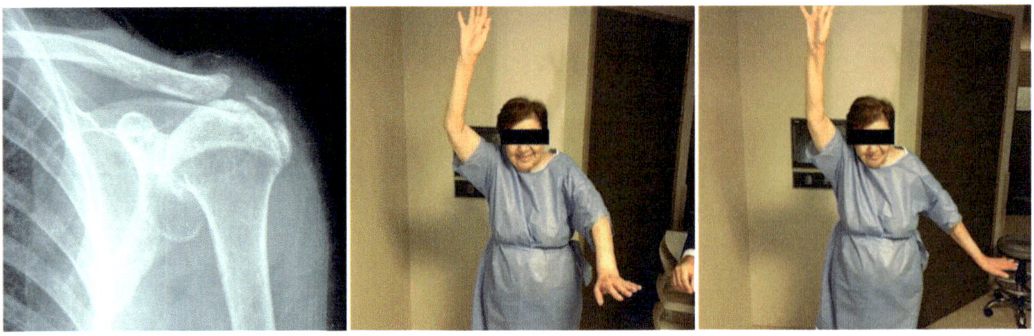

Fig. 6.1 A 79-year-old female patient diagnosed with cuff tear arthropathy who has restricted shoulder motion on the left side

The term cuff tear arthropathy was coined by Neer in 1977 [2]. He described the clinical and pathological findings of CTA in 26 patients. Main pathological findings are chronic full-thickness rotator cuff tear, erosions of the osseous structures, humeral osteopenia, and restricted shoulder motion (Fig. 6.1) [9]. The term CTA has been generally accepted for this clinical entity.

6.1.3 Biomechanics of the Shoulder

Before the pathomechanic of the CTA, we should mention briefly about the role of rotator cuff in the biomechanics of the shoulder. Rotator cuff plays a critical role in dynamic stability of the glenohumeral joint. Osseous geometry of the glenohumeral joint provides little intrinsic stability. Convex humeral head has a small contact area on concave glenoid. Rotator cuff actively compress the humeral head within the glenoid fossa, especially during the middle and end of the range of shoulder motion where the capsuloligamentous structures are lax and become the primary stabilizer of the joint [10]. This mechanism has been coined concavity-compression [11]. Alterations in the compressive force created by the rotator cuff cause instability and translation [12].

In vertical plane, rotator cuff provides inferiorly directed and compressive force, and deltoid provides superiorly directed force resulting in a net force balance. Loss of supraspinatus integrity causes unbalanced vertical force couple, and deltoid pulls the humeral head superiorly [13] In horizontal plane, subscapularis muscle anteriorly and infraspinatus and teres minor posteriorly balanced each other and compress the humeral head. Burkhart performed a fluoroscopic comparison of kinematic patterns in massive rotator cuff tears [14] and describe the importance of the transverse plane force couple. He described a suspension bridge model (Fig. 6.2). Although patients have large or massive supraspinatus tear, if the transverse plane force couple is intact, they have a stable fulcrum, and they can actively elevate the shoulder. Loss of transverse plane force couple balance creates an unstable fulcrum leading to superior migration of the humeral head and instability of the glenohumeral joint.

6.2 Pathogenesis of CTA

6.2.1 Neer Cuff Tear Arthropathy Theory

Neer introduced the term "cuff tear arthropathy" in 1977. He reported the clinical and pathological findings in 26 patients in 1983. Clinical symptoms were described as long-standing and progressively increasing pain that is worse at night and exacerbated by activity, inability to actively elevate and externally rotate the shoulder, swelling, atrophy of the supraspinatus and infraspinatus muscles, tenderness, and crepitus at the glenohumeral joint line posteriorly. The radiologic findings were an area of collapse of the proximal aspect of the humeral articular surface

Fig. 6.2 The suspension bridge model described by Burkhart et al

(this was a requirement for the diagnosis of CTA), anterior or posterior dislocation or subluxation, sclerosis, and small osteophytes at the contact area of humeral head and glenoid, rounding-off of the greater tuberosity, decrease in acromio-humeral distance, and erosion of the undersurface of the anterior third of the acromion and the acromioclavicular joint (Fig. 6.3) [2].

Neer noted three main histological changes. The atrophic articular cartilage of the humeral head usually became covered with a disorderly fibrous membrane containing scattered connective tissue cells. In these areas, the spongiosa of the head was osteoporotic and hypervascular. At points of fixed contact between humeral head and scapula, the cartilage was completely denuded, and the subchondral bone was sclerotic. Third, fragments of articular cartilage were found in the subsynovial layers [2].

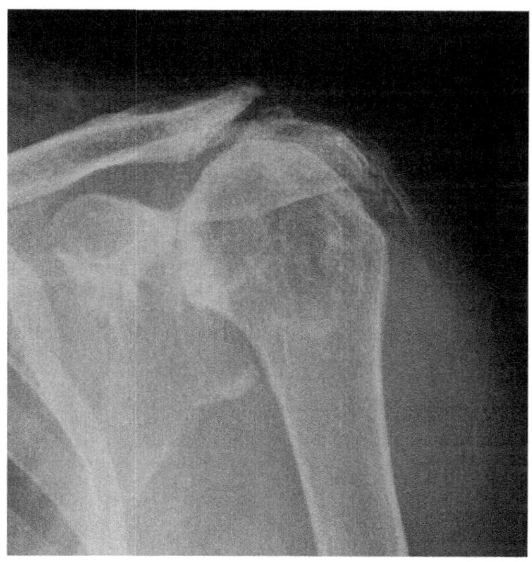

Fig. 6.3 An AP X-raygraphy of shoulder joint in a patient with cuff tear arthropathy

Mechanically, massive rotator cuff tear leads to unbalanced force coupling and loss of the concavity-compression mechanism, resulting in upward migration of the humeral head and instability of the glenohumeral joint. Anterior and posterior subluxations and dislocations produce abnormal trauma and injury to the articular surfaces. Excessive upward migration causes the erosion of the superior glenoid, anterior part of the acromion, acromioclavicular joint, and the outer aspect of the clavicle. Neer mentioned that instability of the head seems to be essential for the development of its collapse.

6.2.2 Crystal-Mediated Theory (Milwaukee Shoulder Syndrome)

In 1981, Halverson coined the term "Milwaukee shoulder syndrome" and identified the association between the presence of calcium phosphate crystals within the shoulder joint and CTA [8]. They hypothesized that a hydroxyapatite-mineral phase develops in the altered capsule, synovial tissue, or degenerative articular cartilage and releases basic calcium phosphate crystals into the synovial fluid. These crystals are composed of carbonate-substituted hydroxyapatite, octacalcium phosphate, or more rarely tricalcium phosphate [1]. These crystals are phagocytosed by synovial cells, forming calcium phosphate crystal microspheroids, which stimulate the release of proteolytic enzymes including collagenase and protease. These enzymes lead to articular, capsular, and rotator cuff destruction. Tissue degeneration resulted in the release of additional crystals, propagating a cycle of increasing joint degeneration and instability [3, 15].

The role of two genes which encode the proteins which regulate the extracellular concentration of inorganic pyrophosphate, fluctuations of which can lead to calcium crystal formation [16]. They found that CTA is associated with variants in ANKH (the human homologue of the mouse progressive ankylosis gene) and TNAP (tissue nonspecific alkaline phosphatase) that alter extracellular inorganic pyrophosphate concentrations causing calcium crystal deposition. This supports a theory that genetic variants predispose patients to primary crystal deposition which when combined with a massive RCT leads to the development of arthritis.

On the other hand, Jensen reported in his review that the exact origin of basic calcium phosphate crystal remains unclear, and whether these crystals are the cause or the result of arthritis remains unanswered [17].

In 1997, Collins and Harryman described the pathogenesis of CTA as a combination of these two hypotheses [18]. They stated that progression of a chronic cuff tear leads to superior migration of the humeral head. Impingement of the remaining cuff against acromion occurs, resulting in articular cartilage wear. Cartilage fragmentation causes synovial thickening and effusion as well as the generation of calcium crystals. The enzymatic response to these crystals furthers the damage to the remaining cuff and articular cartilage [12]. Although numerous pathologic mechanisms for the development of cuff tear arthropathy have been reported, it still remains unclear why only a percentage of patients with a massive rotator cuff tear progress to CTA or why patients with similar radiologic findings may have widely variable clinical presentations [3, 9, 12].

6.2.3 Histopathological Changes in Cuff Tear Arthropathy

Neer noted three main histological changes. The atrophic articular cartilage of the humeral head usually became covered with a disorderly fibrous membrane containing scattered connective tissue cells. In these areas, the spongiosa of the head was osteoporotic and hypervascular. At points of fixed contact between humeral head and scapula, the cartilage was completely denuded, and the subchondral bone was sclerotic. Third, fragments of articular cartilage were found in the subsynovial layers [2].

Takashi et al. compared the histopathological changes in the humeral head of the patients with osteoarthritis, humeral neck fracture, and CTA [19]. Fibrillation, thinning, and tearing of the cartilage were observed in the superior area of the humeral head. Multiple clusters of chondrocytes, an irregular calcified zone, and thickening of the cartilage were observed from the middle to the inferior part of the humeral head. The cartilage in the middle of the humeral head was thicker in CTA group than in osteoarthritic group. The safranin-O-stained zone was larger, and trabecular bone was thinner in CTA group than humeral neck fracture group.

These histopathological changes were also investigated in animal models, and histopathologic scoring system was developed for evaluating shoulder arthritis in a reproducible murine model [20]. Kramer et al. evaluate the cartilage degeneration in a rat model of rotator cuff tear arthropathy. CTA model was created by either infraspinatus and supraspinatus tenotomy or suprascapular nerve transection. They found similar cartilage degeneration on both humeral head and the glenoid in two groups. Because the joint capsule was intact in the denervation group, it is unlikely the nutrient factors in the joint fluid were lost in the denervated joint. Thus, they mentioned that altered biomechanical loading in the glenohumeral joint surface, rather than the loss of joint fluid plays a critical role in CTA [21].

6.3 Rationale of the Reverse Shoulder Arthroplasty

6.3.1 History of Reverse Shoulder Arthroplasty Rationale

A total shoulder arthroplasty was first carried out by Dr. Jules Emile Pean in 1893 [22]. Unlike its antecedents, a total shoulder prosthesis (Neer I) with longer survival was introduced by Neer in the 1950s [23]. This was followed by the introduction of the Neer 2 prosthesis, which forms the basis of the modern total shoulder prosthesis used today [24]. The first design shoulder prostheses had promising treatment outcomes in indications such as proximal humeral fractures, glenohumeral arthrosis [25]. However, the outcomes of treatment with these designs that mimic a healthy shoulder joint were quite poor in cases with an underlying unhealthy rotator cuff [26–28]. In patients with insufficient rotator cuff, it was noticed that the shoulder joint could not move effectively given the rotator cuff did not have a stabilizing effect acting on the shoulder joint. The concept of reverse total shoulder arthroplasty (RTSA) was first introduced by Neer in 1970s. Neer believed that aforementioned issues could be overcome with a constrained reverse prosthetic design. He designed a series of reverse shoulder prostheses with a fixed fulcrum: Mark I, Mark II, and Mark III. The Mark 1 design had a glenoid component with a neck and a large ball. This glenoid component, which was designed to be large in purpose to increase stability, precluded the rotator cuff tendons reattachments to the tubercles. In order to be able to re-attach the rotator cuff tendons, a smaller ball-shaped glenoid component was used in the Mark 2 design. In this design, the distance that the humeral component can move on the spherical glenoid component was decreased. It led to a considerable restriction in range of motion. For this reason, Mark 2 design abandoned in a short time [29]. In the hope of reducing this limitation, though the axial rotation feature was added to the humeral component in the Mark 3 design, the desired effect could not be achieved in increasing the range of motion. However, the prominent issue in almost all of these prosthesis designs was early loosening. It was thought that this early loosening seen especially in the glenoid component was due to the high shear force at the glenoid component–bone interface. Then, this issue was addressed in the following designs: It was aimed to prevent early loosening of the glenoid component utilizing the glenoid component with a divergent threaded peg by Leed et al., utilizing the central glenoid screw by Kessel et al., utilizing a larger thread and hydroxyapatite-coated central screw by Bayley-Walker et al. On the other hand, it was noticed that scapular spine fractures occur more frequently as the glenoid was better fixed and prosthetic designs became increasingly constrained. The idea was to design prostheses that permit easier dislocation in order to prevent the increased stress at the end of the joint range of motion from causing scapular fractures. The design, in which Kolbel and Friedebold added a flange to attach the scapular spine in a wider area, aimed to distribute the stress occurring in a limited area of the scapula, reducing scapular fractures while allowing easier dislocation.

Another remarkable milestone in the development of reverse prosthesis design was the suggestion that more deltoid muscle power could be utilized. In the prosthesis designed by Fenlin et al., the tension of the deltoid was increased by utilizing a large glenosphere. Although it has better functional results in the early period, this design was not widely used due to the loosening of the glenoid during early follow-up and implant failure. The demographic data and clinical outcomes of the patients treated with these arthroplasties have not been comparatively reported, since these arthroplasties were performed on a very limited number of patients and the clinical outcomes of each prosthesis cannot be evaluated effectively, and they did not have sufficient follow-up period.

The prosthesis designs mentioned so far had a small ball and large socket, inspired by the hip prosthesis designs. The revolutionary design in reverse shoulder prosthesis was developed by Paul Grammont [30]. The fact that Grammont's design is the pioneer of the modern reverse shoulder prosthesis designs used today may be based on it was developed regarding following innovative principles with proven effectiveness today: (1) fixed fulcrum, (2) inherently stable prosthesis, (3) the center of the sphere that localized at or within the glenoid neck, (4) medialized and distalized center of rotation, (5) design with convex weight-bearing part and concave supported part, (6) large glenosphere, (7) small humeral cup [31]. Although the initial designs had brought the fixed center of rotation (COR) innovation, they had poor clinical outcomes owing to containing some biomechanical drawbacks [31]. These problems were significantly overcome by Paul Grammont design. Unlike previous reverse prothesis designs, the absence of early glenoid component loosening in the DELTA prosthesis designed by Grammont was owing to the medialization of the CoR and that the center of the glenosphere was within the glenoid neck. In this way, the shear forces generated at the glenoid component bone interface were able to be converted into compression forces in this interface. On the other hand, by increasing the length of the lever arm of the deltoid with the medialization of the center of rotation, the moment produced by the deltoid was increased. Deltoid tension was increased with the distalization of the center of rotation, and it was possible for the deltoid muscle fibers, which were previously in a relatively loose position, to contract more effectively. This prosthetic design was designated as DELTA, inspired by the deltoid muscle. Of course, all these innovations did not emerge with a single design. Grammont's first design consisted of a metallic or ceramic glenosphere making up 2/3 of a sphere placed with cement and press-fitly, and a humeral component stem made entirely of polyethylene which was inserted into medullar canal with cement. In their series consisting of eight patients, the authors stated that pain was resolved in all patients, anterior flexion was above 100 degrees in five patients, and less than 60 degrees in three patients. Grammont's second-generation design featured a metal single-piece humeral component and a glenosphere that made up 1/2 of the sphere. In addition, another important feature of the second generation was that it had a glenoid baseplate fixed with a divergent directed trans fixation screw instead of a peg baseplate placed press-fitly without cement. Delta III (DePuy International Limited, Leeds, England) reverse shoulder prosthesis, a second-generation Grammont design, had been widely used in Europe with successful outcomes except scapular notching [32]. In the third generation, it was aimed to address scapular notching, medial impingement, and instability. Delta III was composed of glenosphere, a glenoid base plate, a polyethylene humeral cup, and a humeral component consisted of a humeral stem and a metaphyseal block with different options attached to each other with a screw. In order to eliminate possible misunderstandings: Delta I is a hemiarthroplasty, Delta II is a total shoulder arthroplasty with polyethylene glenoid component, Delta III is a reverse shoulder arthroplasty, whereas DELTA XTEND™ is an enhanced version of Delta III, which has some improvements and is the commercially available version today. DELTA XTEND™ (Fig. 6.4) includes the following improvements: (1) any glenoid size can be used with any humeral size, (2) a metaglene with a curved back surface in purpose to gain better stability and to preserve bone stock, (3) adjustable and cannulated screw fixation, (4) eccentric humeral epiphysis in purpose to preserve the

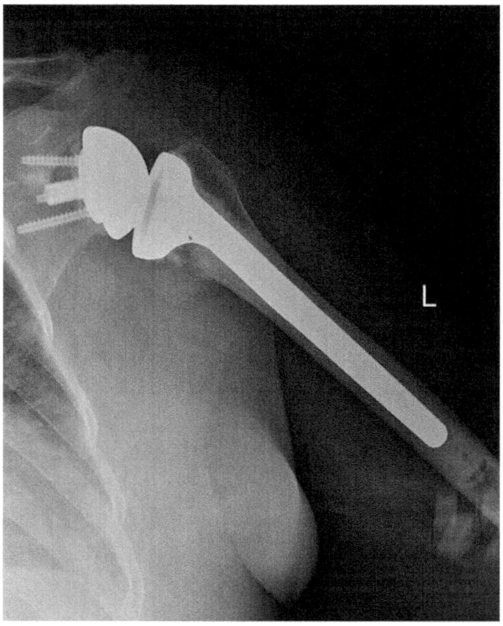

Fig. 6.4 DELTA XTEND™ DePuy Synthes reverse total shoulder arthroplasty

proximal humeral bone stock, (5) long peg metaglenes for impaired bone stock in revision surgeries, (6) additional features for fracture cases.

6.3.2 Biomechanics and Basics of Reverse Shoulder Arthroplasty

In the setting of the normal shoulder biomechanics, the deltoid muscle and rotator cuff (RC) muscles work in concordance, thus the combined force couple vector formed during the glenohumeral (GH) joint movement makes an angle from 0° to 30° with the glenoid surface. And, as long as the force couple vector is in this range, GH joint stability is maintained. In the RC tears in which rotator cable is impaired, deltoid fibers, which provide abduction in normal biomechanics, cause superior migration of the humeral head, which causes a change of the CoR's position constantly during the joint movement (Fig. 6.5). The most important innovation brought by the RTSA designed by Neer that addresses this problem is the fixed fulcrum. However, high torsional and shear forces in reverse ball-and-socket design caused early loosening in the glenoid component–bone interface and scapula fractures. In the Grammont design in which a hemispheric glenosphere without a neck utilizing one-third of a sphere used, high torsional and high shear forces are reduced owing to medialization and distalization of the CoR [33]. Thus, the amount of micro-movements that occur at the glenosphere and bone interface is reduced, and early glenoid component loosening can be avoided [33]. Also, by the reduction of torsional forces, the risk of scapular fracture is significantly reduced. In modern RTSA designs, it is aimed to increase the stability of the glenoid component by using the glenoid base plate. While the screws sent over the glenoid baseplate increase early period stability, it is aimed to increase the late-period stability with the osteointegration provided by the press-fit porous surface.

6.3.3 Controversies in Evolving Designs

6.3.3.1 Lateralization of the CoR

A 155° humerosocket was used in the Grammont design, aiming to increase stability. However, scapular notching, which is quite common, revealed the need for new humerosocket designs with reduced inclination. In a retrospective study comparing 143° inclination humerosocket (Zimmer-Biomet Comprehensive Reverse Shoulder System®), and 155° inclination humerosockets (Aequalis Perform RSA® Tornier), scapular notching was reported to be 16.2% and 60.7%, respectively, at a mean follow-up of 20 months [34] (Fig. 6.6). In a meta-analysis involving 2222 shoulders by Erickson et al., a 155° inclination humerosocket and 135° inclination humerosocket and lateralized glenoid combination were compared, and scapular notching was reported as 16.80% and 2.83%, respectively [35]. It should also be kept in mind that decrease of humerosocket inclination promises reduction of scapular notching risk in the expense of instability [36].

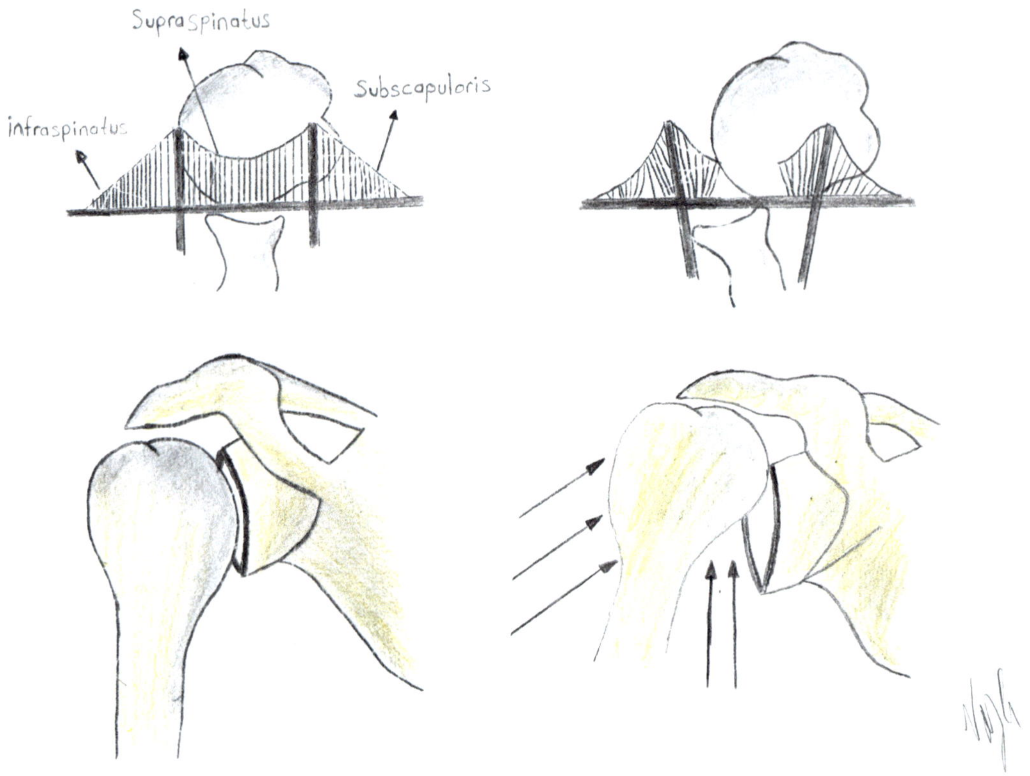

Fig. 6.5 Breaking of the cable of the suspension bridge, the loss of functionality of the bridge, and the superior migration of the humeral head by the effective force couple

While lateralization on the humeral side is achieved by decreasing the humerosocket inclination angle (Fig. 6.6), lateralization on the glenoid side can be achieved in two ways: (1) by placing autologous bone between the glenoid component and the glenoid bone surface (bony increased-offset RSA: BIO-RSA) (Fig. 6.7), (2) by increasing the glenosphere diameters or thickness [37]. While the glenosphere is a complete hemisphere in the Grammont prosthesis, the glenosphere that is 2/3 of a sphere is preferred in new designs (DJO Surgical Reverse Shoulder Prosthesis RSP™). Although the lateralization of the CoR enables the effective use of more deltoid fibers that enable abduction and external rotation, its disadvantage is that it increases the risk of loosening in the glenoid component due to the increase in torque at the bone and glenoid interface [38]. However, Virani et al. modeled seven glenoid designs with different radii, offset, and sphericity with the finite element method, and they reported that the baseplate motion was below 150 μm and that it was within acceptable limits [39]. The findings of Virani et al.'s study support that the increased micro-motion emerged by the lateralized CoR obtained by the glenoid component design at the glenoid baseplate–bone interface causes a clinically negligible disadvantage [39].

6.3.3.2 Inlay-Onlay Designs

Inlay (Zimmer-Biomet Comprehensive Reverse Shoulder System®, the DePuy Synthes DELTA XTEND™, the SMR® Shoulder System of Lima Corporate, DJO Surgical Reverse Shoulder Prosthesis RSP™) and onlay (the Stryker ReUnion® Reverse Shoulder Arthroplasty) glenoid baseplate designs were created to prevent loosening at the interface of the bone and glenoid baseplate. Designs in which the baseplate is embedded in the glenoid bone by opening a peg in the middle of the glenoid is called an inlay. In

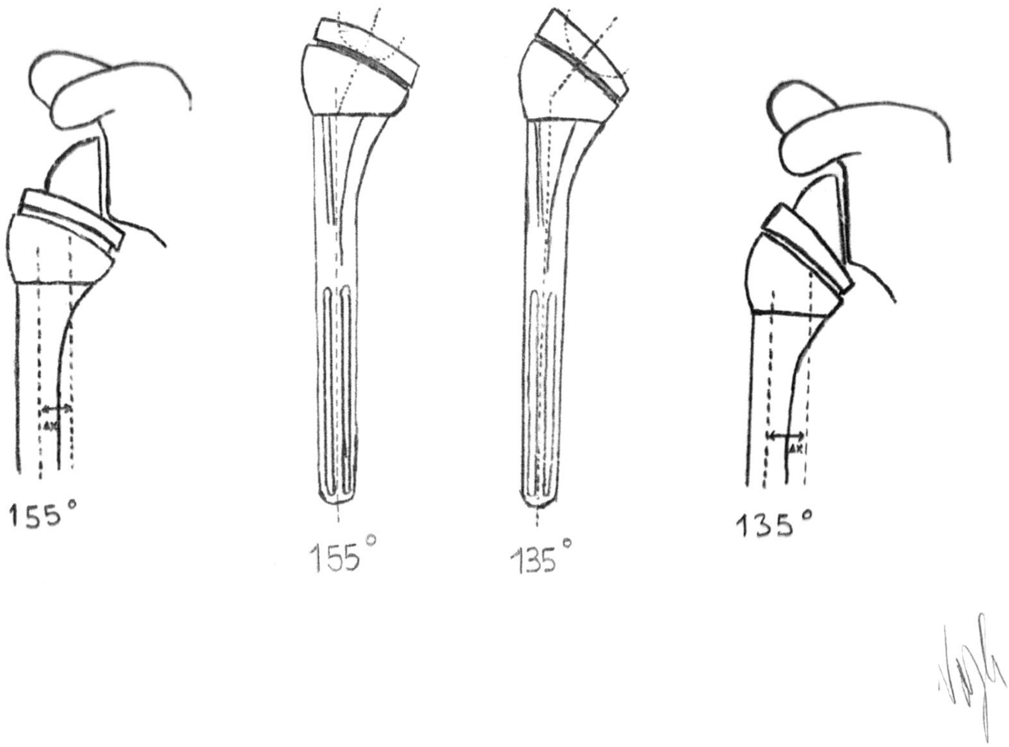

Fig. 6.6 Long deltoid lever arm (Δx) in the varus 135° inclination humerosocket design provides increased deltoid muscle moment

onlay designs, any extension of the baseplate is not embedded in the glenoid. Instead, fixing is provided by screws sent through the baseplate. In their cadaveric study comparing inlay and onlay glenoid components by Gagliano et al., it was reported that inlay designs provided more stable fixation and had fewer loosening rates [40]. For the humeral side, inlay and onlay designs have been developed to aiming to reduce scapular notching and better fixation. In DePuy DELTA XTEND™ Shoulder Prosthesis, which is an inlay design, the metaphyseal component is located inside the humeral metaphysis, while in the Zimmer-Biomet Comprehensive Reverse Shoulder System®, The Ascend Flex® Tornier Inc., or Equinoxe System® Exactech Inc., which are onlay design, it is located on the metaphysis (Fig. 6.8). Less scapular notching but more scapular stress fractures were reported with humeral lateralization achieved in humeral onlay designed prostheses [41]. Another advantage of onlay systems is that they facilitate modularity and convertibility owing to having more bone-preserving design [42]. Since it is very difficult to convert inlay system prostheses from anatomical design to reverse design, most of the convertible systems are designed as onlay.

6.3.3.3 Short-Stem or Stemless Options

Another popular topic with the design updates on the humeral side is humeral stem length. The short-stem or stemless options of almost every system have come into the market or continue to do so. If the bone quality is good, stemless or short stems are usually preferred. However, since the patient group who are encountered with is often osteoporotic, stems which are placed press-fitly in the metaphysis, are more preferred even if they are short. Similar favorable outcomes have been reported in long-term follow-up with short-stem and conventional stemmed RTSA [43–46].

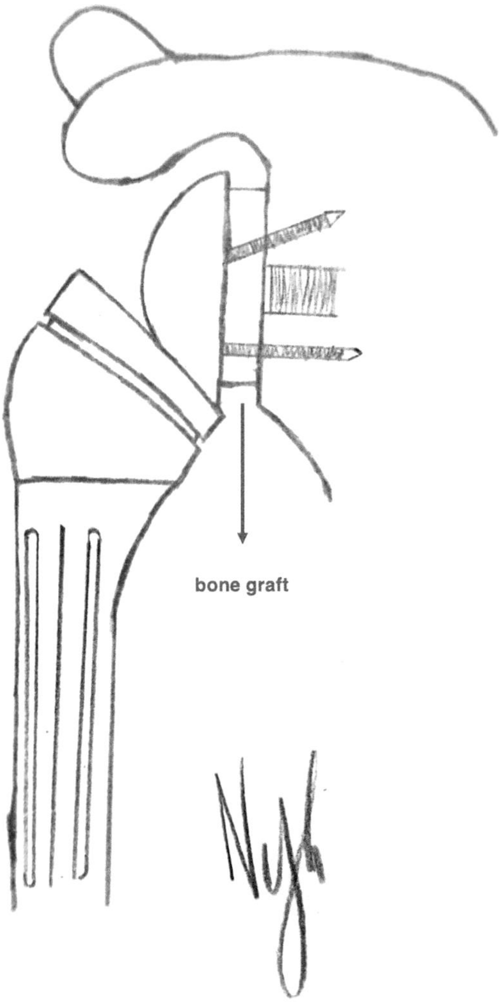

Fig. 6.7 BIO-RSA: Bony increased-offset revers shoulder arthroplasty

Additionally, in a study by Ballas and Béguin, in the 58-month follow-up of 56 patients who underwent stemless RTSA, satisfactory functional outcomes similar to conventional stemmed RTSA were reported [47]. Actually, reported poor outcomes with RTSA are actually associated with loosening of the glenoid component and scapular notching regardless of humeral stem design [48].

6.3.3.4 Polyethylene Glenospheres

The idea of replacing the "polyethylene liner" with a metal liner came up to prevent the osteolytic process which is thought to occur with debris formed as a result of the scapular notching coming into contact with the scapular pillar. With this idea, 40 mm and 44 mm diameter "ultra-high molecular-weight polyethylene (UHMWPE) glenospheres" were preferred in the SMR® Shoulder System of Lima design (Fig. 6.9a). In the AGILON® Omarthrosis prosthesis, the glenospheres are designed to be made of UHMWPE material of 33 mm, 40 mm, and 44 mm diameters (Fig. 6.9b). In a study of 133 patients with 24 months of follow-up in which 36 mm standard chromium-cobalt and 36 mm eccentric chromium-cobalt glenospheres were compared with 44 mm UHMWPE glenospheres, it is noteworthy that scapular notching rates were reported as 28.2%, 66.7%, and 14.3%, respectively [49]. However, it should be kept in mind that the prostheses with UHMWPE glenosphere design do not have long-term outcomes.

6.3.4 Solutions to Limitations

Modern RTSA designs have been providing favorable outcomes in the treatment of pseudo paralysis related to RC tears and cuff tear arthropathy, which has been difficult to deal with in shoulder surgery for a long period. However, some issues accompany this unnatural GH joint biomechanics. The most important criteria to minimize these issues are proper indication/patient selection and thorough preop planning.

6.3.4.1 Instability

It is the most common complication of RTSA. It almost always occurs in the anterior direction. Dislocation risk is minimized by lateralizing the COR (still medially compared to that in normal GH joint biomechanics), avoiding excessive humeral retroversion, and using a large glenosphere provided that high humeral base plate depth/glenosphere diameter is preferred.

6.3.4.2 Large Dead Space

It is the non-anatomical space formed by medialization and distalization of the COR. The risks of hematoma and associated infection minimized with the use of drains.

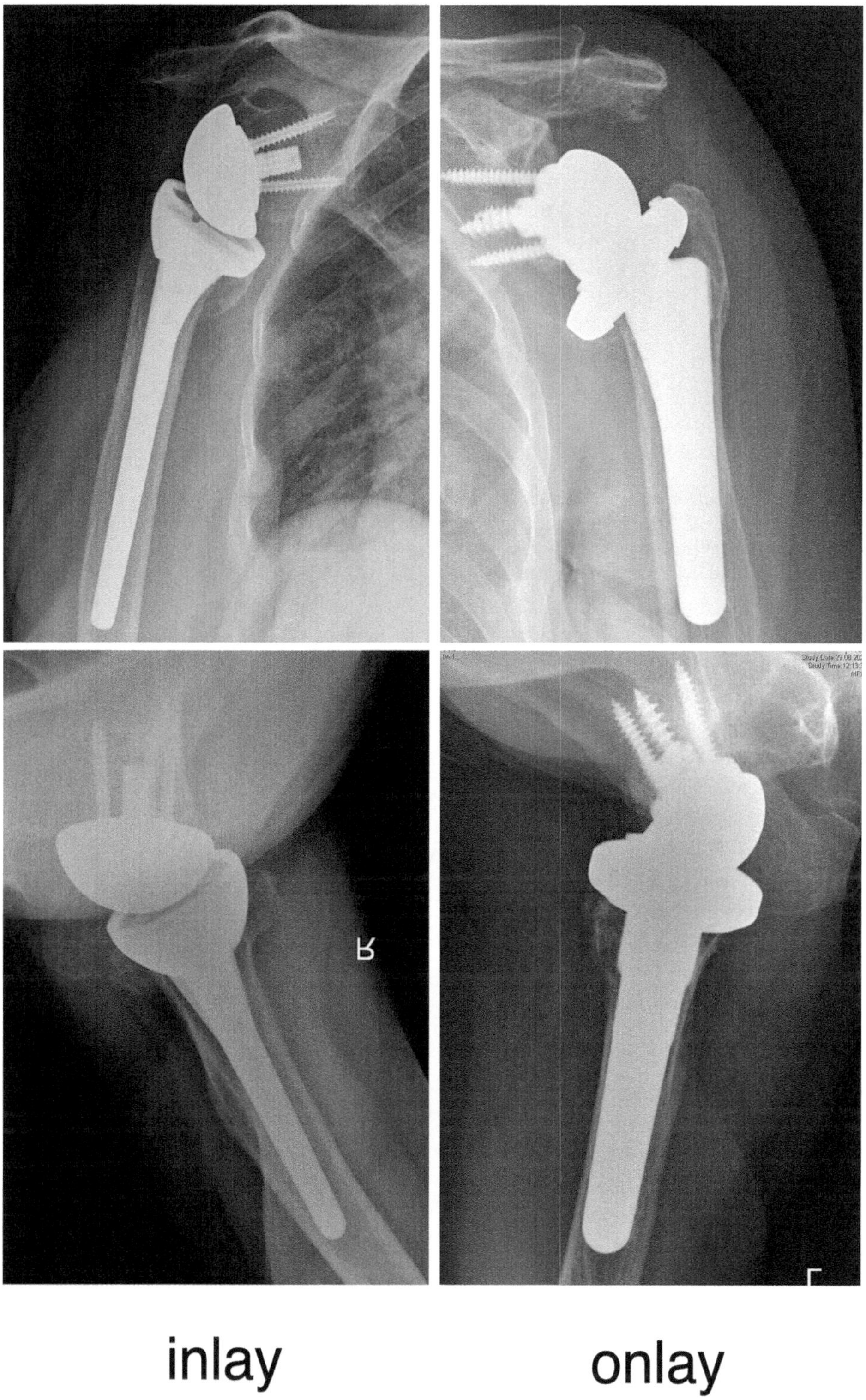

Fig. 6.8 A sample of DELTA XTEND™ DePuy Synthes reverse total shoulder arthroplasty which is an inlay design, and a sample of Zimmer-Biomet Comprehensive Reverse Shoulder System® which is an onlay design of reverse shoulder arthroplasty

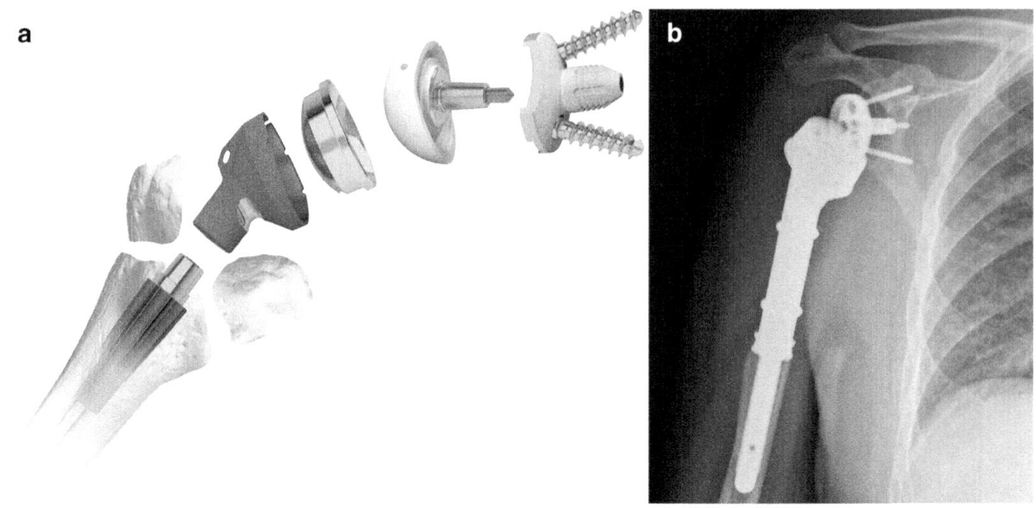

Fig. 6.9 (**a**) SMR® Shoulder System of Lima design with polyethylene glenosphere. Courtesy of Limacorporate S.p.A.—Italy. (**b**) X-ray radiographs of a 79-year-old male patient diagnosed with carcinoma metastasis of proximal humerus treated with MUTARS® Proximal and Inverse Humeral Replacement with polyethylene glenospheres

6.3.4.3 Decrease in External Rotation

It occurs due to the fact that the posterior deltoid muscle fibers, which normally function as external rotators, remain relatively lateral when the COR is medialized. Another substantial reason for the decrease in external rotation is that the transport of the COR to medial results in slackening in the infraspinatus and teres minor muscle fibers. Preop detection of fatty infiltration in teres minor and infraspinatus is crucial. Lateralized glenoid components can improve external rotation, and latissimus dorsi tendon transfer may need to be added.

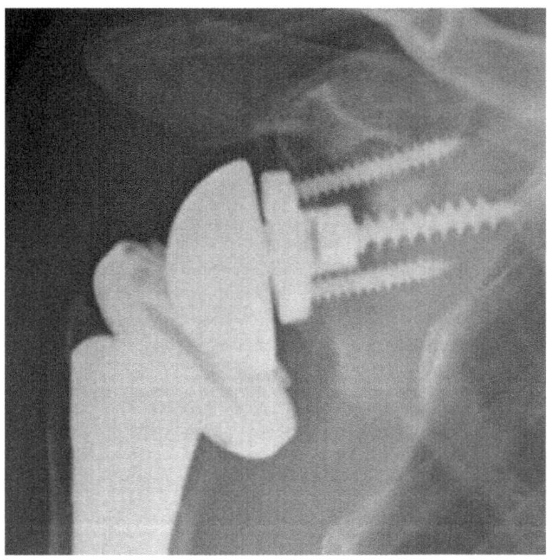

Fig. 6.10 Scapular notching

6.3.4.4 Scapular Notching

It occurs because of using a humeral component in 155° valgus and the medialization of the COR (Fig. 6.10). Another mechanism may be the induction of osteolysis by the eroding polyethylene particles [50]. Overall, although the opinion that scapular notching has no significant effects on clinical outcomes is common, its clinical significance is still controversial. Some authors claim that it may lead to impairment of functional outcomes and glenoid component loosening [51]. It has been reported that scapular neck length leads to increased rates of notching [52]. The scapular notching risk can be minimized by using larger diameter glenosphere or positioning the glenoid baseplate with inferior overhang, inferior tilt or using glenoid bone augmentation [51]. Scapular notching can also be reduced by opting for prosthetic designs with less humeral inclination that provides lateral offset [35]. Using all these technical details simultaneously to reduce scapular notching risk may result in decreased deltoid moment arm, increased risk of a stress fracture, and increased risk of instability. A medialized glenoid component with a lateralized humerus center of rotation can provide a longer deltoid moment arm while maintaining tension on the rotator cuff and reduces the risk of notching at the same time. However, it seems a more reasonable option to consider the use of combinations of aforementioned options individually for each patient.

6.3.4.5 Acromial Fractures and Displacement of the Os Acromiale

It occurs due to the increased deltoid tension caused by distalization of the COR (Fig. 6.11). Excessive stiffness should be avoided preoperatively. The presence of os acromiale should not be overlooked. If any, it should be fixed primarily.

6.3.4.6 Glenoid Bone Erosion

One of another current issue in RSA is glenoid bone erosion. This erosion can endanger the secure placement of the glenoid component. Aiming to surpass this, posterior and/or superior standardized augmented baseplate designs have also been developed to fill the region with excessive erosion (Equinoxe RTSA Platform Shoulder System® Exactech, Aequalis Perform RSA® Tornier). Standardized augmented baseplates may not be sufficient in cases with severe erosion. Metal-filled baseplates which are commercially available as The Ascend Flex™ Wright Medical that can be produced individually can be useful.

6.3.5 Future

6.3.5.1 Patient-Specific Instrumentation

In order to place optimally the glenoid component, it was aimed to use computed navigation more effectively. Although favorable outcomes have been reported, there is not enough evidence yet to say that computed navigation is superior to the classical method [53]. However, computed navigation has some disadvantages such as the prolongation of the surgical time and the need for technically demanding additional steps [54]. It

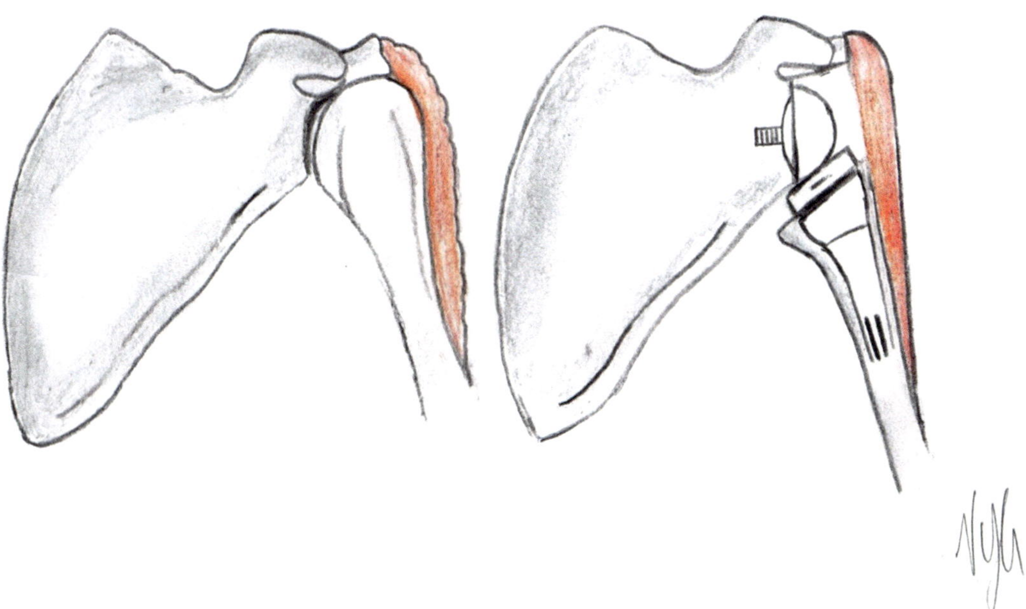

Fig. 6.11 Increase of deltoid tension caused by distalization of the COR

has been shown that the patient-specific instrumentation and patient-specific drill guides can provide correct placement of the glenoid baseplate as equivalent to the computed navigation [54]. Although these guidelines also allow better humeral cut, the main difference is that they ensure that the glenoid components are placed securely in the ideal position [54, 55]. It has been reported that the PSI decreased the mean deviation in the glenoid version from 6.9° to 4.3° and from 11.6° to 2.9° in the glenoid inclination in performing RSTA [56]. In the RSTA industry, it is expected that the use of patient-specific guides will become widespread in order to place the glenoid in the most appropriate position ensuring ideal version, inclination, depth and secure primary fixation according to the specific deformity.

6.3.5.2 3D Planning Softwares

At the present time, preop three-dimensional (3D) planning softwares (Blueprint™—Tournier, Virtual Implant Positioning™ System—Arthrex, GPS™—Exactech, Materialise—DJO) are being used with increasing frequency in TSA surgery as well as RTSA surgery. In these softwares, 3D reconstruction obtained from the patient's shoulder computed tomography scan is utilized. The size of the components, their placement positions to be used before the surgery and bone cuts to be performed can be determined on these 3D reconstructions. Although 3D planning software is a promising development, the error rates of the glenoid version, inclination, and subluxation measurements obtained automatically with such software are outstanding today. Erickson et al. compared the version, inclination, and subluxation measurements carried out by five fellowship-trained sports medicine/shoulder surgeons with the measurements obtained by four different 3D planning software. The authors reported lower reliability in the measurements obtained with 3D planning softwares and warned surgeons to be cautious in preoperative use of these softwares [57]. Today, there is not enough evidence to support that 3D planning softwares can enable arthroplasties clinically better outcomes by providing accurate and reliable measurements. Nevertheless, the authors of this chapter believe that improvements to these softwares can contribute to favorable shoulder arthroplasty surgeries in the future.

6.3.5.3 Augmented Reality

Augmented reality is one of the new technologies that may be used to place the glenoid component in the optimal position which is a key factor in RTSA success [58]. Basically, in the augmented reality, preoperative planning information is made use of during live surgery via a head-mounted display (Fig. 6.12). Although no clinical trials have yet been conducted, the result of Kriechling et al.'s study on cadavers seems encouraging [59]. The authors determined the entry point and trajection of the central guide wire by preoperative planning on 10 cadaveric scapulae. Central guide wires were sent using the head-mounted display to which this information is transferred. And high accuracies were reported with a mean deviation of 2.3 mm ± 1.1 mm at the entrance site and a mean deviation of 2.7° ± 1.3° in trajection. Although it is at earlies of its development, augmented reality technology seems to be another promising technology that can contribute to providing favorable outcomes in RTSA due to its nature that is open to evolution and does not lead to prolonged operation and also being cost-effective.

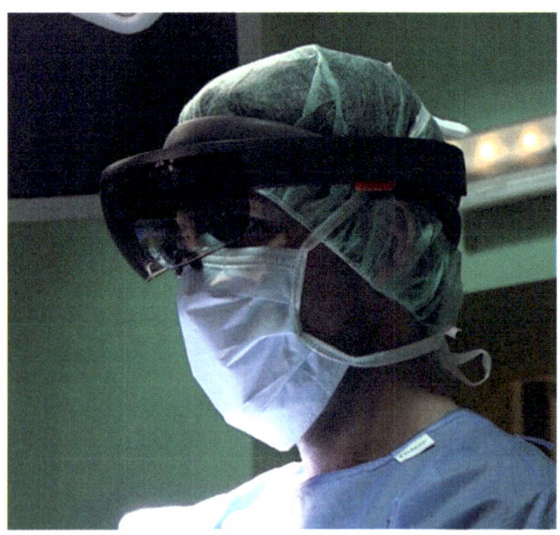

Fig. 6.12 A head-mounted display which is utilized in the augmented reality. Reproduced with permission from [60]

6.4 Conclusion

Cuff tear arthropathy (CTA) is known as an unsolvable disabling problem for many years. Lately, after the 2000s, evolution and worldwide dissemination of reverse shoulder arthroplasty have helped orthopedic surgeons and their patients gaining pain-free functional shoulder motion. Like every novel technique, there are limitations, and some problems were seen during the first years. Solutions for certain problems helped a better range of motion and longer durability of the implants. In future, more technology will help us to improve our results.

Acknowledgments The authors would like to thank Vedat Yakupçebioğlu for medical illustrations.

References

1. McCarty DJ, Lehr JR, Halverson PB. Crystal populations in human synovial fluid. Identification of apatite, octacalcium phosphate, and tricalcium phosphate. Arthritis Rheum. 1983;26(10):1220–4. https://doi.org/10.1002/art.1780261008.
2. Neer CS 2nd, Craig EV, Fukuda H. Cuff-tear arthropathy. J Bone Joint Surg Am. 1983;65(9):1232–44.
3. Nam D, Maak TG, Raphael BS, Kepler CK, Cross MB, Warren RF. Rotator cuff tear arthropathy: evaluation, diagnosis, and treatment: AAOS exhibit selection. J Bone Joint Surg Am. 2012;94(6):e34. https://doi.org/10.2106/JBJS.K.00746.
4. Adams R. Illustrations of the effects of rheumatic gout, or chronic rheumatic arthritis, on all the articulations. Lond: Churchill; 1857.
5. Smith RW. Observations upon chronic rheumatic arthritis of the shoulder. Dublin Q J Med Sci. 1853;15(1):1–16. https://doi.org/10.1007/BF02943957.
6. Codman EA. The shoulder: rupture of the supraspinatus tendon and other lesions in or about the subacromial Bursa. T Todd Company, Printers; 1934.
7. DeSeze SRA, Rampon S. L'epaule senile hemorragique. L'actualite Rhumatologique. 1968;1:107–14.
8. McCarty DJ, Halverson PB, Carrera GF, Brewer BJ, Kozin F. "Milwaukee shoulder"--association of microspheroids containing hydroxyapatite crystals, active collagenase, and neutral protease with rotator cuff defects. I. Clinical aspects. Arthritis Rheum. 1981;24(3):464–73. https://doi.org/10.1002/art.1780240303.
9. Visotsky JL, Basamania C, Seebauer L, Rockwood CA, Jensen KL. Cuff tear arthropathy: pathogenesis, classification, and algorithm for treatment. J Bone Joint Surg Am. 2004;86-A(Suppl 2):35–40.
10. Hurov J. Anatomy and mechanics of the shoulder: review of current concepts. J Hand Ther. 2009;22(4):328–42.; quiz 43. https://doi.org/10.1016/j.jht.2009.05.002.
11. Labriola JE, Lee TQ, Debski RE, McMahon PJ. Stability and instability of the glenohumeral joint: the role of shoulder muscles. J Shoulder Elb Surg. 2005;14(1 Suppl S):32S–8S. https://doi.org/10.1016/j.jse.2004.09.014.
12. Ecklund KJ, Lee TQ, Tibone J, Gupta R. Rotator cuff tear arthropathy. J Am Acad Orthop Surg. 2007;15(6):340–9. https://doi.org/10.5435/00124635-200706000-00003.
13. Lucas DB. Biomechanics of the shoulder joint. Arch Surg. 1973;107(3):425–32. https://doi.org/10.1001/archsurg.1973.01350210061018.
14. Burkhart SS. Fluoroscopic comparison of kinematic patterns in massive rotator cuff tears. A suspension bridge model. Clin Orthop Relat Res. 1992;284:144–52.
15. Zeman CA, Arcand MA, Cantrell JS, Skedros JG, Burkhead WZ Jr. The rotator cuff-deficient arthritic shoulder: diagnosis and surgical management. J Am Acad Orthop Surg. 1998;6(6):337–48. https://doi.org/10.5435/00124635-199811000-00002.
16. Peach CA, Zhang Y, Dunford JE, Brown MA, Carr AJ. Cuff tear arthropathy: evidence of functional variation in pyrophosphate metabolism genes. Clin Orthop Relat Res. 2007;462:67–72. https://doi.org/10.1097/BLO.0b013e31811f39de.
17. Jensen KL, Williams GR Jr, Russell IJ, Rockwood CA Jr. Rotator cuff tear arthropathy. J Bone Joint Surg Am. 1999;81(9):1312–24. https://doi.org/10.2106/00004623-199909000-00013.
18. Collins DN, Harryman DT 2nd. Arthroplasty for arthritis and rotator cuff deficiency. Orthop Clin North Am. 1997;28(2):225–39. https://doi.org/10.1016/s0030-5898(05)70282-0.
19. Toma T, Suenaga N, Taniguchi N, Oizumi N, Yamaguchi H, Tome Y, et al. Humeral head histopathological changes in cuff tear arthropathy. J Orthop Surg (Hong Kong). 2019;27(1):2309499018816428. https://doi.org/10.1177/2309499018816428.
20. Zingman A, Li H, Sundem L, DeHority B, Geary M, Fussel T, et al. Shoulder arthritis secondary to rotator cuff tear: a reproducible murine model and histopathologic scoring system. J Orthop Res. 2017;35(3):506–14. https://doi.org/10.1002/jor.23383.
21. Kramer EJ, Bodendorfer BM, Laron D, Wong J, Kim HT, Liu X, et al. Evaluation of cartilage degeneration in a rat model of rotator cuff tear arthropathy. J Shoulder Elb Surg. 2013;22(12):1702–9. https://doi.org/10.1016/j.jse.2013.03.014.
22. Lugli T. Artificial shoulder joint by Pean (1893): the facts of an exceptional intervention and the prosthetic method. Clin Orthop Relat Res. 1978;133:215–8.
23. Neer CS. Articular replacement for the humeral head. J Bone Joint Surg Am. 1955;37-A(2):215–28.

24. Neer CS 2nd, Watson KC, Stanton FJ. Recent experience in total shoulder replacement. J Bone Joint Surg Am. 1982;64(3):319–37.
25. Neer CS 2nd. Replacement arthroplasty for glenohumeral osteoarthritis. J Bone Joint Surg Am. 1974;56(1):1–13.
26. Lorenzetti AJ, Stone GP, Simon P, Frankle MA. Biomechanics of reverse shoulder arthroplasty: current concepts. Instr Course Lect. 2016;65:127–43.
27. Marmor L. Hemiarthroplasty for the rheumatoid shoulder joint. Clin Orthop Relat Res. 1977;122:201–3.
28. Reeves B, Jobbins B, Flowers F, Dowson D, Wright V. Some problems in the development of a total shoulder endo-prosthesis. Ann Rheum Dis. 1972;31(5):425–6. https://doi.org/10.1136/ard.31.5.425-b.
29. Neer C 2nd, Craig E, Fukuda H. Cuff-tear arthropathy. J Bone Joint Surg Am. 1983;65(9):1232–44.
30. Flatow EL, Harrison AK. A history of reverse total shoulder arthroplasty. Clin Orthop Relat Res. 2011;469(9):2432–9. https://doi.org/10.1007/s11999-010-1733-6.
31. Poncet D, Van Tongel A, De Wilde L. Depuy Synthes DELTA XTEND™. In: Reverse shoulder arthroplasty. Springer; 2016. p. 337–42.
32. Grammont PM, Baulot E. Delta shoulder prosthesis for rotator cuff rupture. Orthopedics. 1993;16(1):65–8.
33. Boileau P, Watkinson DJ, Hatzidakis AM, Balg F. Grammont reverse prosthesis: design, rationale, and biomechanics. J Shoulder Elb Surg. 2005;14(1 Suppl S):147S–61S. https://doi.org/10.1016/j.jse.2004.10.006.
34. Kempton LB, Balasubramaniam M, Ankerson E, Wiater JM. A radiographic analysis of the effects of prosthesis design on scapular notching following reverse total shoulder arthroplasty. J Shoulder Elb Surg. 2011;20(4):571–6. https://doi.org/10.1016/j.jse.2010.08.024.
35. Erickson BJ, Frank RM, Harris JD, Mall N, Romeo AA. The influence of humeral head inclination in reverse total shoulder arthroplasty: a systematic review. J Shoulder Elb Surg. 2015;24(6):988–93. https://doi.org/10.1016/j.jse.2015.01.001.
36. Oh JH, Shin SJ, McGarry MH, Scott JH, Heckmann N, Lee TQ. Biomechanical effects of humeral neck-shaft angle and subscapularis integrity in reverse total shoulder arthroplasty. J Shoulder Elb Surg. 2014;23(8):1091–8. https://doi.org/10.1016/j.jse.2013.11.003.
37. Boileau P, Moineau G, Roussanne Y, O'Shea K. Bony increased-offset reversed shoulder arthroplasty: minimizing scapular impingement while maximizing glenoid fixation. Clin Orthop Relat Res. 2011;469(9):2558–67. https://doi.org/10.1007/s11999-011-1775-4.
38. Henninger HB, Barg A, Anderson AE, Bachus KN, Burks RT, Tashjian RZ. Effect of lateral offset center of rotation in reverse total shoulder arthroplasty: a biomechanical study. J Shoulder Elb Surg. 2012;21(9):1128–35. https://doi.org/10.1016/j.jse.2011.07.034.
39. Virani NA, Harman M, Li K, Levy J, Pupello DR, Frankle MA. In vitro and finite element analysis of glenoid bone/baseplate interaction in the reverse shoulder design. J Shoulder Elb Surg. 2008;17(3):509–21. https://doi.org/10.1016/j.jse.2007.11.003.
40. Gagliano JR, Helms SM, Colbath GP, Przestrzelski BT, Hawkins RJ, DesJardins JD. A comparison of onlay versus inlay glenoid component loosening in total shoulder arthroplasty. J Shoulder Elb Surg. 2017;26(7):1113–20. https://doi.org/10.1016/j.jse.2017.01.018.
41. Kennon JC, Lu C, McGee-Lawrence ME, Crosby LA. Scapula fracture incidence in reverse total shoulder arthroplasty using screws above or below metaglene central cage: clinical and biomechanical outcomes. J Shoulder Elb Surg. 2017;26(6):1023–30. https://doi.org/10.1016/j.jse.2016.10.018.
42. Polisetty TS, Baessler AM, Levy JC, Badman BL. Onlay versus inlay reverse Total shoulder arthroplasty: a retrospective comparison of radiographic and clinical outcomes. In: Seminars in Arthroplasty : JSES; 2020. https://doi.org/10.1053/j.sart.2020.11.013.
43. Giuseffi SA, Streubel P, Sperling J, Sanchez-Sotelo J. Short-stem uncemented primary reverse shoulder arthroplasty: clinical and radiological outcomes. Bone Joint J. 2014;96-B(4):526–9. https://doi.org/10.1302/0301-620X.96B3.32702.
44. Atoun E, Van Tongel A, Hous N, Narvani A, Relwani J, Abraham R, et al. Reverse shoulder arthroplasty with a short metaphyseal humeral stem. Int Orthop. 2014;38(6):1213–8. https://doi.org/10.1007/s00264-014-2328-8.
45. Levy O, Narvani A, Hous N, Abraham R, Relwani J, Pradhan R, et al. Reverse shoulder arthroplasty with a cementless short metaphyseal humeral implant without a stem: clinical and radiologic outcomes in prospective 2- to 7-year follow-up study. J Shoulder Elb Surg. 2016;25(8):1362–70. https://doi.org/10.1016/j.jse.2015.12.017.
46. Dukan R, Bahman M, Rousseau MA, Boyer P. Outcomes of reverse shoulder arthroplasty using a short stem through a superolateral approach. J Shoulder Elb Surg. 2020;29(6):1197–205. https://doi.org/10.1016/j.jse.2019.09.025.
47. Ballas R, Beguin L. Results of a stemless reverse shoulder prosthesis at more than 58 months mean without loosening. J Shoulder Elb Surg. 2013;22(9):e1–6. https://doi.org/10.1016/j.jse.2012.12.005.
48. Werner BC, Wong AC, Mahony GT, Craig EV, Dines DM, Warren RF, et al. Causes of poor postoperative improvement after reverse total shoulder arthroplasty. J Shoulder Elb Surg. 2016;25(8):e217–22. https://doi.org/10.1016/j.jse.2016.01.002.
49. Bloch HR, Budassi P, Bischof A, Agneskirchner J, Domenghini C, Frattini M, et al. Influence of glenosphere design and material on clinical outcomes of reverse total shoulder arthroplasty. Shoulder Elbow. 2014;6(3):156–64. https://doi.org/10.1177/1758573214535574.

50. Nyffeler RW, Werner CM, Simmen BR, Gerber C. Analysis of a retrieved delta III total shoulder prosthesis. J Bone Joint Surg Br. 2004;86(8):1187–91. https://doi.org/10.1302/0301-620x.86b8.15228.
51. Friedman RJ, Barcel DA, Eichinger JK. Scapular notching in reverse Total shoulder arthroplasty. J Am Acad Orthop Surg. 2019;27(6):200–9. https://doi.org/10.5435/JAAOS-D-17-00026.
52. Fortun CM, Streit JJ, Horton SA, Muh SJ, Gillespie RJ, Gobezie R. Scapular neck length and implications for reverse total shoulder arthroplasty: an anatomic study of 442 cadaveric specimens. Int J Shoulder Surg. 2015;9(2):38–42. https://doi.org/10.4103/0973-6042.154754.
53. Verborgt O, De Smedt T, Vanhees M, Clockaerts S, Parizel PM, Van Glabbeek F. Accuracy of placement of the glenoid component in reversed shoulder arthroplasty with and without navigation. J Shoulder Elb Surg. 2011;20(1):21–6. https://doi.org/10.1016/j.jse.2010.07.014.
54. Levy JC, Everding NG, Frankle MA, Keppler LJ. Accuracy of patient-specific guided glenoid baseplate positioning for reverse shoulder arthroplasty. J Shoulder Elb Surg. 2014;23(10):1563–7. https://doi.org/10.1016/j.jse.2014.01.051.
55. Villatte G, Muller AS, Pereira B, Mulliez A, Reilly P, Emery R. Use of patient-specific instrumentation (PSI) for glenoid component positioning in shoulder arthroplasty. A systematic review and meta-analysis. PLoS One. 2018;13(8):e0201759. https://doi.org/10.1371/journal.pone.0201759.
56. Hendel MD, Bryan JA, Barsoum WK, Rodriguez EJ, Brems JJ, Evans PJ, et al. Comparison of patient-specific instruments with standard surgical instruments in determining glenoid component position: a randomized prospective clinical trial. J Bone Joint Surg Am. 2012;94(23):2167–75. https://doi.org/10.2106/JBJS.K.01209.
57. Erickson BJ, Chalmers PN, Denard P, Lederman E, Horneff G, Werner BC, et al. Does commercially available shoulder arthroplasty preoperative planning software agree with surgeon measurements of version, inclination, and subluxation? J Shoulder Elb Surg. 2020; https://doi.org/10.1016/j.jse.2020.05.027.
58. Jud L, Fotouhi J, Andronic O, Aichmair A, Osgood G, Navab N, et al. Applicability of augmented reality in orthopedic surgery - a systematic review. BMC Musculoskelet Disord. 2020;21(1):103. https://doi.org/10.1186/s12891-020-3110-2.
59. Kriechling P, Roner S, Liebmann F, Casari F, Furnstahl P, Wieser K. Augmented reality for base plate component placement in reverse total shoulder arthroplasty: a feasibility study. Arch Orthop Trauma Surg. 2020; https://doi.org/10.1007/s00402-020-03542-z.
60. McKnight RR, Pean CA, Buck JS, Hwang JS, Hsu JR, Pierrie SN. Virtual reality and augmented reality-translating surgical training into surgical technique. Curr Rev Musculoskelet Med. 2020;13(6):663–74. https://doi.org/10.1007/s12178-020-09667-3.

Rationale of Tendon-to-Bone Healing

7

Giuseppe Milano, Niccolò Vaisitti, Giuseppe Frizziero, and Maristella F. Saccomanno

7.1 Introduction

Rotator cuff tears (RCT) are a growing problem affecting 30–50% of people older than 50 years [1–4]. Given the increasing functional demands of the population and constant development of surgery, recent trends show an increase in the number of patients electing to undergo surgical treatment of RCT, leading to a 500% increase in rotator cuff repair (RCR) since 2001 [5]. Unfortunately, regardless of technologic advances and development of surgical techniques and despite the high number of surgeries performed yearly, re-tear after RCR remains a significant problem. It appears that re-tear rates after RCR may be as high as 20–40% for small-to-medium tears and go up till 94% for large or chronic tears [6–9].

G. Milano (✉)
Department of Medical and Surgical Specialties, Radiological Sciences, and Public Health, University of Brescia, Brescia, Italy

Department of Bone and Joint Surgery, Spedali Civili, Brescia, Italy

N. Vaisitti · G. Frizziero
Department of Medical and Surgical Specialties, Radiological Sciences, and Public Health, University of Brescia, Brescia, Italy

M. F. Saccomanno
Department of Bone and Joint Surgery, Spedali Civili, Brescia, Italy

Risk factors for failure of surgical treatment include tendon quality, patient's age, tear size and chronicity, and muscle atrophy/fatty infiltration. All these factors are predictors for an inferior ability of tendons to heal and for a higher chance of re-tear after repair [10, 11].

7.2 Tendon-to-Bone Healing

Tendon-to-bone healing is a multifactorial, complex, biological process that is poorly defined and not fully understood. One of the key aspects is the tendon-to-bone interface, called *enthesis*, where most of RCTs occur.

The enthesis develops postnatally and is considered as a four-zone structure at maturity [12]. The first zone is tendinous and composed primarily of type-I collagen, with the presence of the proteins *decorin* and *biglycan* and populated by tendon fibroblasts [13]. The second zone is formed by uncalcified fibrocartilage and presents type-I and -II collagen with aggrecan. The resident cell population mainly consists of fibrochondrocytes. The third zone is constituted by calcified fibrocartilage and consists of hypertrophic fibrochondrocytes producing type-II collagen, type-X collagen, aggrecan, and alkaline phosphatase [14–16]. The fourth zone consists of bone containing dense type-I collagen [13]. This complex histologic structure allows forces to be

G. Huri et al. (eds.), *Fundamentals of the Shoulder*, https://doi.org/10.1007/978-3-030-94702-6_7

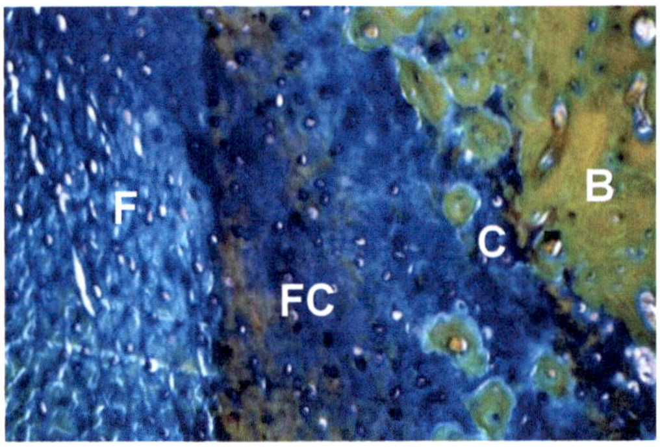

Fig. 7.1 Histologic appearance of the four-zone structure of native bone–tendon junction. This complex histologic structure allows forces to be transferred from the soft tendon tissue to the relatively stiff bone. Gomori-Halmi trichrome staining. Magnification 20× (*F* fibrous tissue, *FC* fibrocartilage, *C* calcified fibrocartilage, *B* bone)

transferred from the soft tendon tissue to the relatively stiff bone (Fig. 7.1).

After surgery, the healing process can be roughly divided into three phases. First, there is an inflammatory phase, with the injury site being infiltrated by numerous inflammatory cells. Macrophages and monocytes eliminate the necrotic tissue, promote cell migration, proliferation, and differentiation, and increase vascularization through the release of cytokines. The phenomenon starts in less than 24 h from the injury and lasts for several days. Second, there is a proliferation phase with cells, including tenocytes and fibroblasts, recruited to the repair site. This phase lasts several weeks, and it is during this period that the type-III collagen is deposited. Finally, there is a remodeling phase, with the type-III collagen being replaced by type-I collagen, and a gradual decrease in cellularity of the site. The result of this process is the repair of the tendon–bone interface by reactive scar formation without the restoration of the native structure of the enthesis in its four zones. This is extremely relevant given that the newly formed scar tissue has inferior mechanical properties compared to the native insertion site, thus making it more susceptible to failure [17].

Developing new approaches to re-establish the four zones that connect the bone to the tendon could result in improved tendon healing with better long-term outcome.

7.3 Biologic Augmentation of Tendon-to-Bone Healing

Interest in biologic augmentation of RCR, intended to enhance the healing response and to provide a mechanical aid for a tension-free repair, has been constantly rising in recent years.

Within the concept of biologic augmentation, two main development areas can be identified: "*biochemical augmentation*" and "*biomechanical augmentation*." The first regards the addition of stem cells and macromolecules like growth factors (GFs) that will exert a predominantly biochemical effect on the healing process. On the other hand, the second refers to grafts (such as auto-, xeno-, and allografts, or synthetics) that exercise a primarily mechanical effect of reinforcement on the repair, while still present some degree of indirect biological healing effects [18].

7.3.1 Biochemical Augmentation

7.3.1.1 Stem Cell Therapies

The idea behind the use of stem cell therapies is to improve the success of RCR by augmenting the local healing using multi- or pluripotent mesenchymal stem cells (MSCs). These cells can differentiate into varied mesodermal tissues mirroring normal healing. Autogenous MSCs are generally preferred over fetal stem cells given they are easier to obtain [19].

So far, stem cells for RC augmentation have shown promising results in a large number of animal models from both histological and biomechanical points of view [20–22].

7.3.1.2 Bone Marrow-Derived Mesenchymal Stem Cells

Clinical studies on the topic are scarce although encouraging. Hernigou et al. performed a case-controlled study using autogenous bone marrow-derived mesenchymal stem cells (BM-MSCs) harvested from the iliac crest and implanted into repaired tendon at the insertion site. At 6 months, a magnetic resonance imaging (MRI) was performed on all patients, and the BM-MSC-treated patients showed a 100% of healed RCs to the

67% of control patients. At 10 years, 87% of the BM-MSC-treated group presented intact RC on MRI to the 44% of control patients [23].

The iliac crest is not the only BM-MSC source, and studies have shown that a similar quantity of cells can be harvested from the proximal humerus via intraoperative aspiration

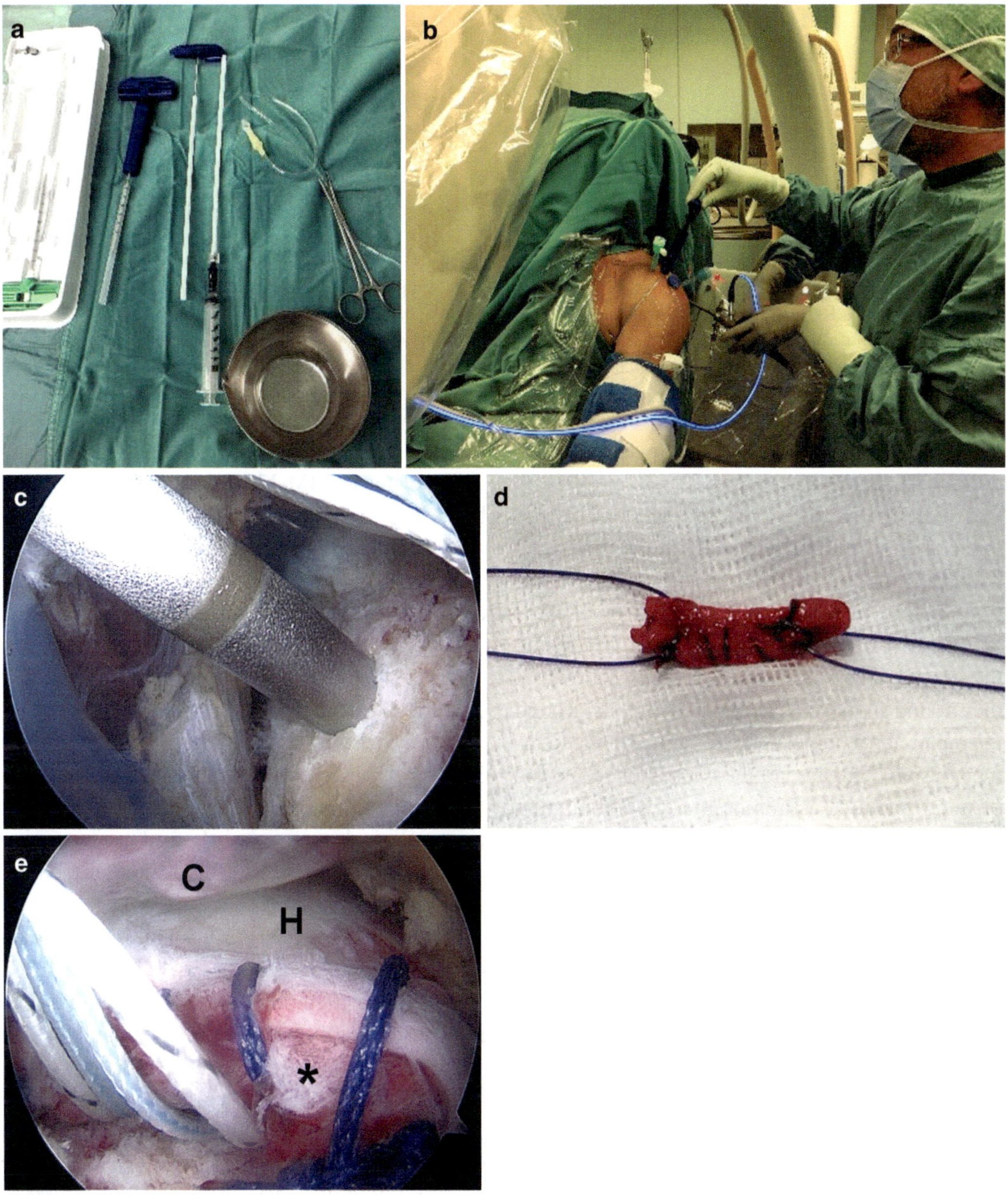

Fig. 7.2 Bone marrow mesenchymal stem cells (BM-MSC) can be harvested from the proximal humerus via intraoperative aspiration. (**a**) Disposable kit for bone marrow aspiration (Marrow Cellution™, Aspire Medical Innovation, Prinzregentenhof, Germany). (**b**) The cannula is introduced into the greater tuberosity to aspirate bone marrow through a percutaneous portal. (**c**) Correct placement of the cannula is achieved under simultaneous arthroscopic view and fluoroscopy. (**d**) Cells are embedded in a collagen scaffold to be applied in the repair site. (**e**) Scaffold with cells embedded (*asterisk*) is fixed at tendon–bone interface (*H* humeral head, *C* rotator cuff)

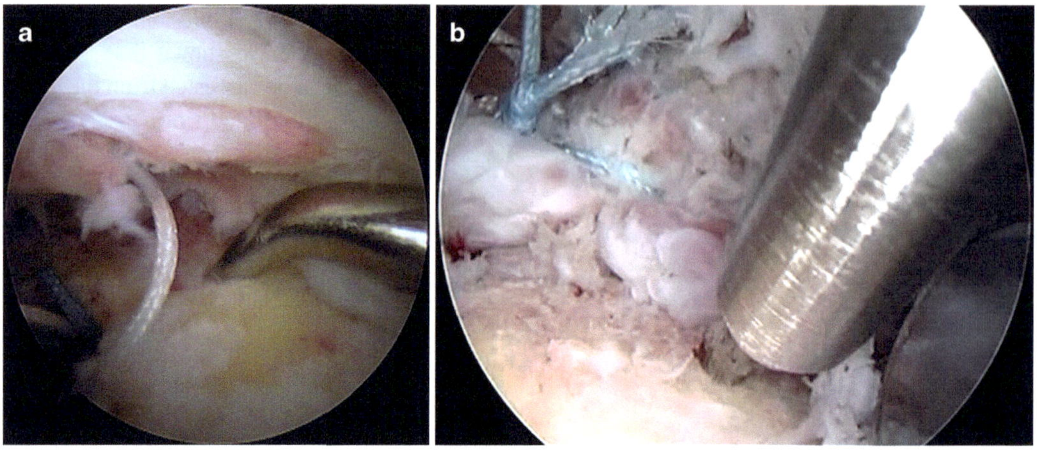

Fig. 7.3 (**a**) Microfracture of the greater tuberosity. Arthroscopic view. Right shoulder. (**b**) Bone vents on the greater tuberosity can be performed with a small and deep arthroscopic awl (nanofractures). Arthroscopic view. Right shoulder

solving the problem of additional invasive procedures [24] (Fig. 7.2).

An alternative to harvesting the stem cells is the idea of recruiting such cells in the repair site by performing microfractures at the tendon footprint before repair (Fig. 7.3). Kida et al. [25] labeled BM-MSCs in a rat model before drilling the greater tuberosity. At a 2-months follow-up, they found higher levels of labeled cells around the repair and a higher force to failure in the treatment group compared to the control one. The use of microfractures as a bone marrow-stimulating (BMS) technique for RC healing was also investigated by Bilsel et al. [26], showing that microfractures promoted tendon healing achieving a significantly increased force to failure, thicker collagen bundles, and more fibrocartilage histologically at 8 weeks post-repair in a rabbit model.

Several clinical studies confirmed that recruiting BM-MSCs via microfractures could be a safe and viable option to augment RCR. Although no superiority in terms of functional improvement was proven, bone marrow stimulation with microfractures during RCR seems to reduce retear rate compared to the standard repair [27–30].

Lastly, some experimental studies on RCR investigated the effect of BM-MSCs modified to overexpress specific factors known to be overrepresented at tendon–bone insertion sites during embryogenesis. BM-MSCs overexpressing membrane type-1 MMP (MT1-MMP) and scleraxis were compared with BM-MSCs, the former showing higher amount of fibrocartilage formation, greater load to failure, and higher stiffness at 4 weeks after RCR [22, 31].

7.3.1.3 Other Sources of Mesenchymal Stem Cells

Studies on several kinds of MSCs have been performed on animal models and, despite being far from clinical practice, showed encouraging results promising a fruitful avenue for future researches.

Valencia Mora et al. [32] found in a study on rats that the use of *adipose tissue–derived MSCs* (A-MSCs) to augment RCR led to less inflammation postoperatively, which could result in less scarring and therefore better elastic properties (Fig. 7.4).

Oh et al. [33] used allogenic A-MSCs to augment subscapularis repair in adult male rabbits. At 6-week follow-up, the augmented group showed a muscle potential area almost equal to that of the native subscapularis and a decrease in fatty infiltration.

Animal models for the use of *muscle-derived MSCs* (M-MSCs) in augmenting RCR showed

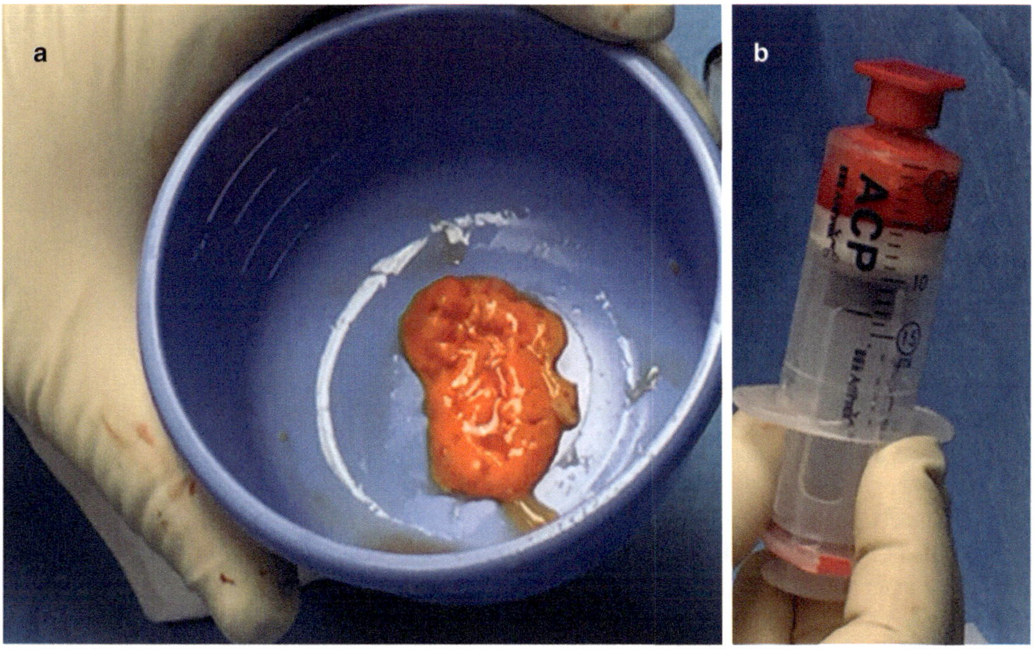

Fig. 7.4 (**a**) Fat tissue is a potential source of MSC. (**b**) Fat tissue can be processed to concentrate cells that can be injected or applied at the repair site

the cells ability to differentiate into a fibroblastic phenotype [34] and significantly lessened fatty infiltration compared to a non-M-MSC-receiving group. However, M-MSCs revealed inferior results as load to failure and contraction force compared to BM-MSCs [35].

Shen et al. [36] used allogenic *tendon-derived MSCs* (T-MSCs) to augment RCR in rabbits. At 12-week follow-up, improved biomechanical and histological findings were observed compared to a group of non-augmented repairs. However, T-MSCs are still poorly characterized, and further studies are necessary for a better understanding of their mechanism and use.

Bursa-derived MSCs (B-MSCs) have demonstrated significant adipogenesis and osteogenesis, making them an easily accessible and multipotent source of MSCs [37], albeit their application and functionality remain to be studied.

MSCs for RC surgery are one of the most promising branches of biologic augmentation, having shown, in vitro and animal models, excellent results. Clinical trials in humans, although limited in numbers, appear to support the earlier studies with positive results [11]. However, the use of MSCs for RCR presents some controversial issues. First, cell death after implant occurs at a high rate, therefore demanding a large number of MSCs to reach therapeutic levels [38, 39]. Second, the differentiation of MSCs into the required cell lineages needs both angiogenesis and a specific balance of growth factors and cytokines that has not fully understood yet [39]. Lastly, MSCs represent a high-cost therapy, for both the provider and the patient, making it potentially problematic to use on a large scale [11].

Overall MSCs' use for augmentation of RCR showed promising results and, despite its drawbacks, worth of further researches.

7.3.1.4 Growth Factors

The healing of RCT is a process dependent on an organized succession of GF release. These factors are responsible for cell migration into the defect, a fundamental step toward healing. Growth factor levels raise first in the proximal part of the myotendinous insertion and only later in the distal part of tendon defect [40]. Growth factor levels remain sustained on average for 3–8 weeks from injury [41].

Cytokines involved in tendon healing include transforming growth factor-β1 (TGF-β1), fibroblast growth factor (FGF), BMP, interleukins (ILs), platelet-derived growth factor (PDGF), and vascular endothelial growth factor (VEGF).

7.3.1.5 Platelet-Rich Plasma

Platelet-rich plasma (PRP) is a centrifuged of autologous blood, with a platelet concentration 3–5 times higher than in whole blood, that is used locally on damaged tissue to achieve better healing [42]. A higher concentration of platelets grants higher levels of those GFs contained within platelets, such as insulin-like growth factor-1 (IGF-1), PDGF, VEGF, and TGF-β, all implicated in tendon healing [43] (Fig. 7.5).

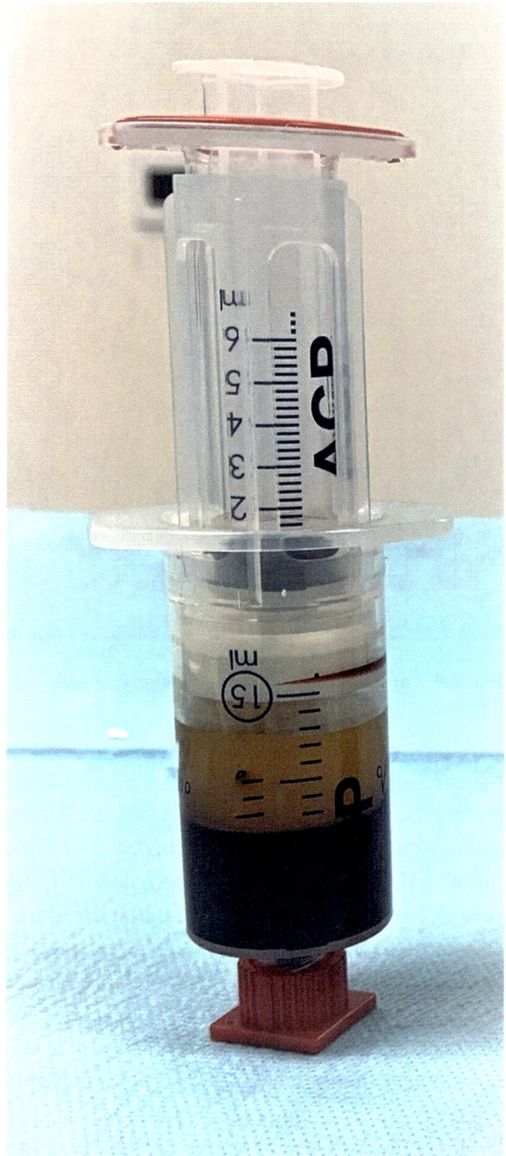

Fig. 7.5 Platelet concentrates enhance soft tissue healing by promoting proliferation of fibroblasts and differentiation of undifferentiated cells into fibrous cell lineage (Autologous Platelet Concentrate; Arthrex, Naples, FL, USA)

Early studies showed promising results for the use of PRP as an augment for RCR. In vitro analysis showed that PRP stimulated fibroblasts and tenocytes proliferation in chronically torn RCs [44]. Furthermore, animal studies showed improved histological and mechanical features associated with intraoperative administration of autologous PRP [45]. Unfortunately, these encouraging results were not confirmed in clinical studies, with only a minority of clinical trials showing evidence of a positive effect of PRP on RCR. Most studies on the effect of PRP on RCR found no difference between PRP-augmented and control group. Charousset et al. [46] evaluated the outcome of massive RCTs repaired arthroscopically with or without PRP and found no significant differences in functional or radiological outcome, nor in re-tear rate at 2-year follow-up.

Ruiz-Moneo et al. [47], in a double-blinded randomized controlled trial on the use of PRP on all-sized arthroscopic RCR, did not highlight any significant difference in functional outcomes, patient satisfaction, or MRI-based analysis of tendon healing between the treatment and the control groups.

In another double-blinded randomized controlled trial, Weber et al. [48] found no significant difference in perioperative pain, functional outcome, or structural integrity between the PRP-treated group and the control group in the repair of all-sized RCTs.

To this day, significant evidence to support the clinical use of PRP to augment RCR is lacking. Future researches should focus on standardizing PRP application forms and types of platelet concentrates, as the difference in protocols has so far limited the chances for proper comparisons.

7.3.1.6 Other Growth Factors

Transforming growth factor-β (TGF-β) is a family of factors synthetized by every cell involved in the healing process. On the other hand, normal tendons show low concentrations of TGF-β [49, 50]. It has been discovered that higher levels of the TGF-β1 isoform are associated with a significantly higher the formation of scar tissue, while the TGF-β3 is linked to lower levels of scarring tissue and better healing [41, 49]. The potential negative effect of TGF-β1 was investigated in an animal study in which TGF-β1 was delivered in a rat supraspinatus tendon tear model leading to an increase in type-III collagen and a scar-mediated response [51]. A study by Kovacevic et al. [52] showed that RCR augmented with a matrix containing TGF-β3 resulted in a significantly improved strength of the repair tissue and in a more favorable collagen I/collagen III ratio when compared to the control group.

Within the TGF-β family, the effects of bone morphogenetic protein (BMP) have been studied as well, given that the formation of the enthesis bears some similarities to the enchondral ossification [53] and that ingrowth of bone into the tendon is part of the tendon–bone healing process [54]. For example, a BMP-12-augmented RCR in a sheep model showed, at 8 weeks postoperatively, tendon load to failure and stiffness two times better as well as histologically accelerated healing compared to controls [55]. Another study using BMP-13 in rats demonstrated improved mechanical strength and better histological tissue organization at 6 weeks after RCR [56].

Fibroblast growth factors are another family of GFs involved in tendon healing, remaining highly upregulated during the entire healing process, at least in some of their isoforms. These factors stimulate in vitro the proliferation of RC tendon cells and suppress the secretion of collagen [57]. Although definitive evidence is lacking, several authors have reported improved tendon healing after the addition of bFGF [58, 59], with a significant increase in tendon strength [60, 61].

VEGF-augmented RCR was studied by Zumstein et al. [62] in a small, randomized controlled trial in chronic RCTs. Doppler analysis was performed and showed improved vascularization in VEGF-treated patients in the early phase after repair, but no significant difference at 12-week follow-up as well as no clinical difference between the treated group and the control one.

Early studies on GFs used, as a delivery mechanism, a single injection, usually intraoperatively, in such a way that it does not mimic the physiological expression of GFs during healing. To address that, more sustained delivery methods have been studied.

Vesicular phospholipid gels (VPGs) are lecithin and aqueous solutions able to grant a sustained, adjustable local delivery of GFs, while being non-toxic and easy to produce [63, 64].

Another delivery mechanism that has been studied is the use of sutures coated with GFs, as they ensure local delivery of GFs directly to the site. Moreover, the coating of sutures with GFs requires no additional surgical steps and does not affect the mechanical properties of the sutures, making it a safe way to deliver GFs locally and in a sustainable way [65, 66].

7.3.1.7 Matrix Metalloproteinase Inhibitors

Matrix Metalloproteinases (MMP) are enzymes responsible for the degradation of the extracellular matrix (ECM) and are overexpressed both in degenerated tendons and in repaired tendon tissue [67] Specifically, MMP-13 seems to play a role in this tendinopathy [68].

The effects of inhibiting the action of MMP both locally (with alpha-2-macroglobulin) and systemically (with oral doxycycline) have been studied to augment RCR in animal models [69, 70]. These studies confirmed the reduction in levels of local MMP-13, but yielded disparate results as far as biomechanical outcomes were concerned, emphasizing the need for further researches on the topic.

7.3.2 Biomechanical Augmentation

By protecting the repair in the early postoperative period, enhancing the rate and quality of the healing [71], scaffold augmentation has led to promising outcomes and a low failure rate [72]. These materials provide a collagen-based structure, that over time hosts cell populations that gradually remodel the scaffold, thus improving the quality of the resulting tissue [73]. At the same time, they also influence the mechanical fixation bearing up to 45% of the total load after repair [74].

Scaffolds can be divided into two main categories: *biological* and *synthetic*.

Biological scaffolds include autografts, xenografts, and allografts, while synthetics comprises bioengineered polymeric matrices. Efficacy of biological scaffolds is supported by a great body of work, showing that the addition of such grafts reduces gap formation at the repair site, increases load to failure and protects the repair by supporting a mean of 35% of load locally applied [75]. Synthetic scaffolds are on the rise thanks to good clinical outcomes, but not without some concerns, such as the limited growth potential and possible inflammatory responses, both acute and chronic, to foreign materials [76].

7.3.2.1 Autografts

The first autograft used for biological augmentation in RCR was the biceps tendon [77], given its ready availability and limited donor site morbidity, while providing more collagen to the repair and, therefore, more healing potential [78] (Fig. 7.6).

Another studied autograft was fascia lata, due to its properties similar to those of the RC tendons [79, 80]. In recent years, it was investigated the possibility to reinforce these grafts with poly-L-lactic acid (PLLA) or PLLA/polyglycolic acid polymer braids to provide mechanical augmentation and minimize tendon retraction, obtaining a reduction in gap formation and better mean postoperative scores with a lower re-tear rate on MRI [81].

Another autograft described was the coracoacromial ligament [82], which produced excellent outcomes in terms of subjective and functional scores and ROM, normal MRI tendon signals, and no complications, while iliotibial band [83] and periosteal flap [84] showed higher rates of complications.

7.3.2.2 Allografts

These grafts are allogenic matrices produced by decellularization of cadaveric material to reduce

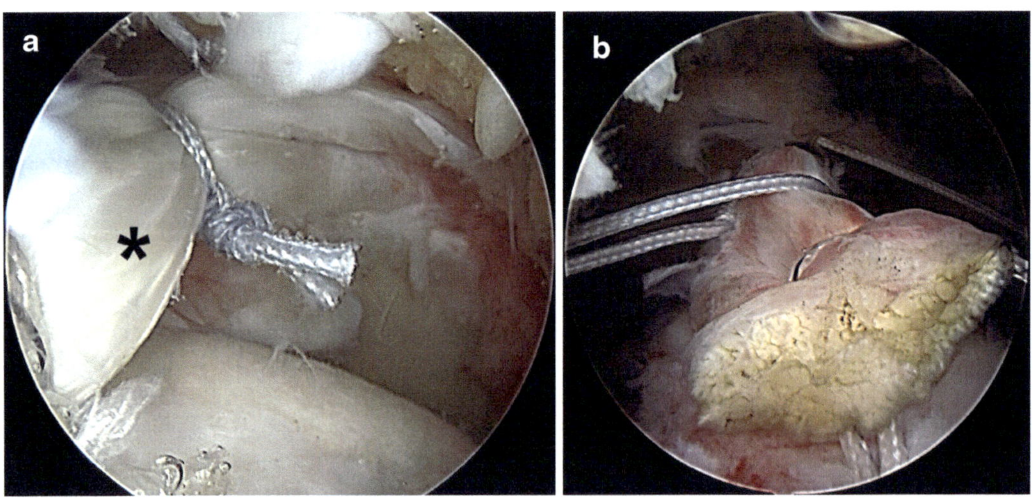

Fig. 7.6 (**a**) Arthroscopic view. Right shoulder. Long head of biceps tendon (LHBT) (arrow) acts as biological augmentation to enhance healing in rotator cuff repair. (**b**) Arthroscopic view. Right shoulder. Proximal biceps stump can be used as biomechanical augmentation to reinforce rotator cuff repair

the risk of graft rejection [3]. They are used for their high failure load and their ability to increase the strength of the repaired tendon [85]. Although a human dermal matrix was demonstrated to bring a significant strengthening to repaired tendons, complete healing on MRIs, and excellent postoperative scores [86, 87], there have been concerns about the possible inflammatory response and degeneration due to the presence of residual DNA [88].

7.3.2.3 Xenografts

Xenografts were among the first studied, and the two main options are porcine dermal matrix and submucosal grafts. Despite good results in other surgical uses, the porcine submucosal grafts proved disappointing for RCR.

Sclamberg et al. [89] found that most patients with large retracted tears at a 6-month follow-up showed no improvement in functional scores and half of those actually had worse scores. Other studies on porcine submucosal grafts had similar results with only a minority of patients presenting healed tears [90, 91].

Iannotti et al. [90] reported increased pain in patients treated with the graft, and Walton et al. [91] reported four patients with severe reactions needing revision surgery. This could be explained by the strong inflammatory response most likely due to the DNA remained in these grafts [92], making submucosal graft not a viable option anymore.

Porcine dermal grafts, on the other hand, have yielded good outcomes [93–95] (Fig. 7.7). Avanzi et al. [95] reported that RCR augmented with a porcine dermal patch is a safe and effective technique leading to excellent clinical outcomes with a high healing rate and close-to-normal MRI findings. These findings might overcome the idea that xenografts, in general, appear to be just inferior to allografts due to the hardly avoidable inflammatory response they cause, with consequent inferior outcomes.

Up to date, there is no consensus whether to prefer one type of scaffold over the other, as each of them shows advantages and disadvantages, in the absence of long-term outcome studies and paucity of randomized clinical trials.

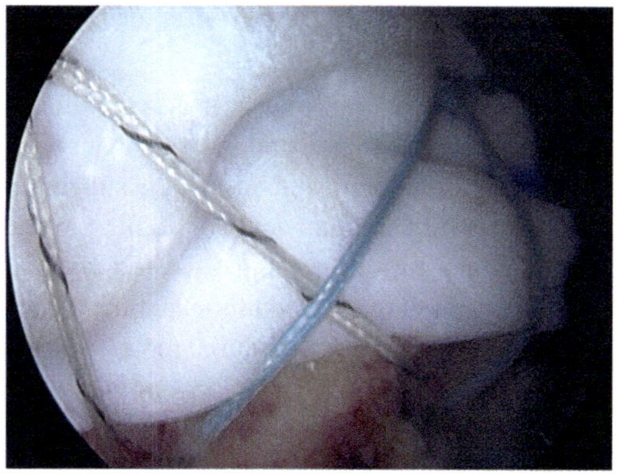

Fig. 7.7 Arthroscopic view. Right shoulder. Porcine dermal graft has been used as augmentation in rotator cuff repair and fixed with a suture-bridge technique

7.3.2.4 Synthetic Scaffolds

The interest in developing synthetic, biodegradable biomaterials to augment the repair of soft-to-hard tissue interfaces has been constantly on the rise in recent years [96, 97]. Synthetic grafts hold such interest as they offer mechanical strengthening of the repair while replicating the extracellular microenvironment and at the same time delivering biological factors and cells to enhance healing [98, 99]. For example, Gore-Tex patches showed high elongation values [100], while a woven poly-L-lactide graft led to a significant increase in ultimate load [87].

Currently, some interest raised for nanofiber scaffolds. These synthetic scaffolds have a high surface area/volume ratio, and they are very adaptable: surface and dimensions can be easily customized, they present adjustable degradation rate and might be modified with nanotopography-mediated cell response with stem cells or with GFs [101].

Recent clinical studies showed encouraging results in healing rates, biocompatibility, regenerated tendon area, load to failure, ROM, and retear rates [86, 102], although some concerns arose over the degradation products of the polymers used to produce synthetic scaffolds [103].

7.4 Summary

Biologic augmentation for RCR is a significant area of research for its potential to enhance the integration of injured soft-to-hard tissue interfaces improving the outcomes of RCR. Several strategies are currently being explored such as GFs, stem cell therapies, and biomaterials. These strategies are used to augment the repair site and to enhance the regeneration and integration of the tendon–bone interface after surgery.

Despite the general enthusiasm surrounding this new development area and early promising results, caution should be used when drawing conclusions on the role of biologic augmentation in RC tears. While some scientific and clinical data are compelling, the overall body of work is very mixed, and results still inconclusive. For these reasons, biologic agents merit further investigation with more rigorous testing to hopefully define formulations and approaches that can become gold standards in the future.

References

1. Yamaguchi K, Ditsios K, Middleton WD, Hildebolt CF, Galatz LM, Teefey SA. The demographic and morphological features of rotator cuff disease: a comparison of asymptomatic and symptomatic shoulders. J Bone Joint Surg Am. 2006;88(8):1699–704.
2. Isaac C, Gharaibeh B, Witt M, Wright VJ, Huard J. Biologic approaches to enhance rotator cuff healing after injury. J Shoulder Elbow Surg. 2012;21(2):181–90.
3. Thangarajah T, Pendegrass CJ, Shahbazi S, Lambert S, Alexander S, Blunn GW. Augmentation of rotator cuff repair with soft tissue scaffolds. Orthop J Sports Med. 2015;3(6):2325967115587 49.
4. Zhao J, Zhao S, Zhao J, Dong S, Huangfu X, Li B, et al. Biological augmentation of rotator cuff repair using bFGF-loaded electrospun poly(lactide-co-glycolide) fibrous membranes. Int J Nanomedicine. 2014;9:2373–85.
5. Hakimi O, Mouthuy P-A, Carr A. Synthetic and degradable patches: an emerging solution for rotator cuff repair. Int J Exp Pathol. 2013;94(4):287–92.
6. Boileau P. Arthroscopic repair of full-thickness tears of the supraspinatus: does the tendon really heal? J Bone Joint Surg Am. 2005;87(6):1229.
7. Bishop J, Klepps S, Lo IK, Bird J, Gladstone JN, Flatow EL. Cuff integrity after arthroscopic versus open rotator cuff repair: a prospective study. J Shoulder Elbow Surg. 2006;15(3):290–9.
8. Sears BW, Choo A, Yu A, Greis A, Lazarus M. Clinical outcomes in patients undergoing revision rotator cuff repair with extracellular matrix augmentation. Orthopedics. 2015;38(4):e292–6.
9. Galatz LM, Ball CM, Teefey SA, Middleton WD, Yamaguchi K. The outcome and repair integrity of completely arthroscopically repaired large and massive rotator cuff tears. J Bone Joint Surg Am. 2004;86(2):219–24.
10. Björnsson HC, Norlin R, Johansson K, Adolfsson LE. The influence of age, delay of repair, and tendon involvement in acute rotator cuff tears: structural and clinical outcomes after repair of 42 shoulders. Acta Orthop. 2011;82(2):187–92.
11. Patel S, Gualtieri AP, Lu HH, Levine WN. Advances in biologic augmentation for rotator cuff repair: biologic augmentation for rotator cuff repair. Ann N Y Acad Sci. 2016;1383(1):97–114.
12. Jensen PT, Lambertsen KL, Frich LH. Assembly, maturation, and degradation of the supraspinatus enthesis. J Shoulder Elbow Surg. 2018;27(4):739–50.
13. Galatz L, Rothermich S, VanderPloeg K, Petersen B, Sandell L, Thomopoulos S. Development of the supraspinatus tendon-to-bone insertion: localized expression of extracellular matrix and growth factor genes. J Orthop Res. 2007;25(12):1621–8.
14. Shen G. The role of type X collagen in facilitating and regulating endochondral ossification of articular cartilage. Orthod Craniofac Res. 2005;8(1):11–7.
15. Hettrich CM, Gasinu S, Beamer BS, Stasiak M, Fox A, Birmingham P, et al. The effect of mechanical load on tendon-to-bone healing in a rat model. Am J Sports Med. 2014;42(5):1233–41.
16. Dyment NA, Breidenbach AP, Schwartz AG, Russell RP, Aschbacher-Smith L, Liu H, et al. Gdf5 progenitors give rise to fibrocartilage cells that mineralize via hedgehog signaling to form the zonal enthesis. Dev Biol. 2015;405(1):96–107.
17. Schär MO, Rodeo SA. Biology of injury and repair of soft tissues of the shoulder. In: Milano G, Grasso A, editors. Shoulder arthroscopy. London: Springer; 2014. p. 59–72. [cited 2020 Dec 18]. https://doi.org/10.1007/978-1-4471-5427-3_5.
18. Avanzi P, Dei Giudici L, Gigante A, Zorzi C. How to manage failed rotator cuff repair: biologic augmentation. In: Milano G, Grasso A, Calvo A, Brzóska R, editors. Management of failed shoulder surgery. Berlin: Springer; 2018. p. 219–28. [cited 2020 Dec 18]. https://doi.org/10.1007/978-3-662-56504-9_29.
19. Valencia Mora M, Ruiz Ibán MA, Díaz Heredia J, Barco Laakso R, Cuéllar R, Arranz MG, M. Stem cell therapy in the management of shoulder rotator cuff disorders. World J Stem Cells. 2015;7(4):691.
20. Gulotta LV, Kovacevic D, Ehteshami JR, Dagher E, Packer JD, Rodeo SA. Application of bone marrow-derived mesenchymal stem cells in a rotator cuff repair model. Am J Sports Med. 2009;37(11):2126–33.

21. Valencia Mora M, Ruiz Ibán MA, Díaz Heredia J, Barco Laakso R, Cuéllar R, García AM. Stem cell therapy in the management of shoulder rotator cuff disorders. World J Stem Cells. 2015;7(4):691–9.
22. Gulotta LV, Kovacevic D, Montgomery S, Ehteshami JR, Packer JD, Rodeo SA. Stem cells genetically modified with the developmental gene MT1-MMP improve regeneration of the supraspinatus tendon-to-bone insertion site. Am J Sports Med. 2010;38(7):1429–37.
23. Hernigou P, Flouzat Lachaniette CH, Delambre J, Zilber S, Duffiet P, Chevallier N, et al. Biologic augmentation of rotator cuff repair with mesenchymal stem cells during arthroscopy improves healing and prevents further tears: a case-controlled study. Int Orthop. 2014;38(9):1811–8.
24. Mazzocca AD, McCarthy MBR, Chowaniec DM, Cote MP, Arciero RA, Drissi H. Rapid isolation of human stem cells (connective tissue progenitor cells) from the proximal humerus during arthroscopic rotator cuff surgery. Am J Sports Med. 2010;38(7):1438–47.
25. Kida Y, Morihara T, Matsuda K-I, Kajikawa Y, Tachiiri H, Iwata Y, et al. Bone marrow-derived cells from the footprint infiltrate into the repaired rotator cuff. J Shoulder Elbow Surg. 2013;22(2):197–205.
26. Bilsel K, Yildiz F, Kapicioglu M, Uzer G, Elmadag M, Pulatkan A, et al. Efficacy of bone marrow-stimulating technique in rotator cuff repair. J Shoulder Elbow Surg. 2017;26(8):1360–6.
27. Taniguchi N, Suenaga N, Oizumi N, Miyoshi N, Yamaguchi H, Inoue K, et al. Bone marrow stimulation at the footprint of arthroscopic surface-holding repair advances cuff repair integrity. J Shoulder Elbow Surg. 2015;24(6):860–6.
28. Jo CH, Shin JS, Park IW, Kim H, Lee SY. Multiple channeling improves the structural integrity of rotator cuff repair. Am J Sports Med. 2013;41(11):2650–7.
29. Li Z, Zhang Y. Efficacy of bone marrow stimulation in arthroscopic repair of full thickness rotator cuff tears: a meta-analysis. J Orthop Surg Res. 2019;14(1):36. https://doi.org/10.1186/s13018-019-1072-6. Jan 29 [cited 2020 Dec 18];14. Available from: https://www.ncbi.nlm.nih.gov/pmc/articles/PMC6352336/
30. Milano G, Saccomanno MF, Careri S, Taccardo G, De Vitis R, Fabbriciani C. Efficacy of marrow-stimulating technique in arthroscopic rotator cuff repair: a prospective randomized study. Arthroscopy. 2013;29(5):802–10.
31. Gulotta LV, Kovacevic D, Packer JD, Deng XH, Rodeo SA. Bone marrow–derived mesenchymal stem cells transduced with scleraxis improve rotator cuff healing in a rat model. Am J Sports Med. 2011;39(6):1282–9.
32. Valencia Mora M, Antuña Antuña S, Arranz García M, Carrascal MT, Barco R. Application of adipose tissue-derived stem cells in a rat rotator cuff repair model. Injury. 2014;45:S22–7.
33. Oh JH, Chung SW, Kim SH, Chung JY, Kim JY. 2013 Neer award: effect of the adipose-derived stem cell for the improvement of fatty degeneration and rotator cuff healing in rabbit model. J Shoulder Elbow Surg. 2014;23(4):445–55.
34. Pelinkovic D, Lee J-Y, Engelhardt M, Rodosky M, Cummins J, Fu FH, et al. Muscle cell-mediated gene delivery to the rotator cuff. Tissue Eng. 2003;9(1):143–51.
35. Coleman S, Ehteshami J, Kisiday J, Altchek D, Warren R, Turner A. The effects of mesenchymal stem cells on rotator cuff muscle in a chronic injury model in sheep. Paper No 167. In: American Shoulder and Elbow Surgeons 2009 Closed Meeting. October 24-27, 2009 New York.
36. Shen W, Chen J, Yin Z, Chen X, Liu H, Heng BC, et al. Allogenous tendon stem/progenitor cells in silk scaffold for functional shoulder repair. Cell Transplant. 2012;21(5):943–58.
37. Utsunomiya H, Uchida S, Sekiya I, Sakai A, Moridera K, Nakamura T. Isolation and characterization of human mesenchymal stem cells derived from shoulder tissues involved in rotator cuff tears. Am J Sports Med. 2013;41(3):657–68.
38. Kim Y-S, Lee H-J, Ok J-H, Park J-S, Kim D-W. Survivorship of implanted bone marrow-derived mesenchymal stem cells in acute rotator cuff tear. J Shoulder Elbow Surg. 2013;22(8):1037–45.
39. Otarodifard K, Bruce Canham R, Galatz LM. Biologic augmentation of rotator cuff repair. Semin Arthroplasty. 2014;25(4):220–5.
40. Kobayashi M, Itoi E, Minagawa H, Miyakoshi N, Takahashi S, Tuoheti Y, et al. Expression of growth factors in the early phase of supraspinatus tendon healing in rabbits. J Shoulder Elbow Surg. 2006;15(3):371–7.
41. Würgler-Hauri CC, Dourte LM, Baradet TC, Williams GR, Soslowsky LJ. Temporal expression of 8 growth factors in tendon-to-bone healing in a rat supraspinatus model. J Shoulder Elbow Surg. 2007;16(5):S198–203.
42. Dohan Ehrenfest DM, Rasmusson L, Albrektsson T. Classification of platelet concentrates: from pure platelet-rich plasma (P-PRP) to leucocyte- and platelet-rich fibrin (L-PRF). Trends Biotechnol. 2009;27(3):158–67.
43. Lu HH, Vo JM, Chin HS, Lin J, Cozin M, Tsay R, et al. Controlled delivery of platelet-rich plasma-derived growth factors for bone formation. J Biomed Mater Res A. 2008;86A(4):1128–36.
44. Hoppe S, Alini M, Benneker LM, Milz S, Boileau P, Zumstein MA. Tenocytes of chronic rotator cuff tendon tears can be stimulated by platelet-released growth factors. J Shoulder Elbow Surg. 2013;22(3):340–9.
45. Dolkart O, Chechik O, Zarfati Y, Brosh T, Alhajajra F, Maman E. A single dose of platelet-rich plasma improves the organization and strength of a surgically repaired rotator cuff tendon in rats. Arch Orthop Trauma Surg. 2014;134(9):1271–7.

46. Charousset C, Zaoui A, Bellaïche L, Piterman M. Does autologous leukocyte-platelet–rich plasma improve tendon healing in arthroscopic repair of large or massive rotator cuff tears? Arthroscopy. 2014;30(4):428–35.
47. Ruiz-Moneo P, Molano-Muñoz J, Prieto E, Algorta J. Plasma rich in growth factors in arthroscopic rotator cuff repair: a randomized, double-blind, controlled clinical trial. Arthroscopy. 2013;29(1):2–9.
48. Weber SC, Kauffman JI, Parise C, Weber SJ, Katz SD. Platelet-rich fibrin matrix in the management of arthroscopic repair of the rotator cuff: a prospective, randomized. Double-blinded study. Am J Sports Med. 2013;41(2):263–70.
49. Molloy T, Wang Y, Murrell GAC. The roles of growth factors in tendon and ligament healing. Sports Med. 2003;33(5):381–94.
50. Hsu C, Chang J. Clinical implications of growth factors in flexor tendon wound healing. J Hand Surg Am. 2004;29(4):551–63.
51. Kim HM, Galatz LM, Das R, Havlioglu N, Rothermich SY, Thomopoulos S. The role of transforming growth factor beta isoforms in tendon-to-bone healing. Connect Tissue Res. 2011;52(2):87–98.
52. Kovacevic D, Fox AJ, Bedi A, Ying L, Deng X-H, Warren RF, et al. Calcium-phosphate matrix with or without TGF-β_3 improves tendon-bone healing after rotator cuff repair. Am J Sports Med. 2011;39(4):811–9.
53. Benjamin M, Ralphs JR. Fibrocartilage in tendons and ligaments—an adaptation to compressive load. J Anat. 1998;193(4):481–94.
54. Rodeo SA, Arnoczky SP, Torzilli PA, Hidaka C, Warren RF. Tendon-healing in a bone tunnel. A biomechanical and histological study in the dog. J Bone Joint Surg Am. 1993;75(12):1795–803.
55. Seeherman HJ, Archambault JM, Rodeo SA, Turner AS, Zekas L, D'Augusta D, et al. rhBMP-12 accelerates healing of rotator cuff repairs in a sheep model. J Bone Joint Surg Am. 2008;90(10):2206–19.
56. Murray DH, Kubiak EN, Jazrawi LM, Araghi A, Kummer F, Loebenberg MI, et al. The effect of cartilage-derived morphogenetic protein 2 on initial healing of a rotator cuff defect in a rat model. J Shoulder Elbow Surg. 2007;16(2):251–4.
57. Takahashi S, Nakajima M, Kobayashi M, Wakabayashi I, Miyakoshi N, Hirosh M, et al. Effect of recombinant basic fibroblast growth factor (bFGF) on fibroblast-like cells from human rotator cuff tendon. Tohoku J Exp Med. 2002;198(4):207–14.
58. Chan BP, Fu S, Qin L, Lee K, Rolf CG, Chan K. Effects of basic fibroblast growth factor (bFGF) on early stages of tendon healing: a rat patellar tendon model. Acta Orthop Scand. 2000;71(5):513–8.
59. Tang JB, Cao Y, Zhu B, Xin K-Q, Wang XT, Liu PY. Adeno-associated virus-2-mediated bFGF gene transfer to digital flexor tendons significantly increases healing strength: an in vivo study. J Bone Joint Surg Am. 2008;90(5):1078–89.
60. Ide J, Kikukawa K, Hirose J, Iyama K, Sakamoto H, Fujimoto T, et al. The effect of a local application of fibroblast growth factor-2 on tendon-to-bone remodeling in rats with acute injury and repair of the supraspinatus tendon. J Shoulder Elbow Surg. 2009;18(3):391–8.
61. Ide J, Kikukawa K, Hirose J, Iyama K, Sakamoto H, Mizuta H. The effects of fibroblast growth factor-2 on rotator cuff reconstruction with acellular dermal matrix grafts. Arthroscopy. 2009;25(6):608–16.
62. Zumstein MA, Berger S, Schober M, Boileau P, Nyffeler RW, Horn M, et al. Leukocyte- and platelet-rich fibrin (L-PRF) for long-term delivery of growth factor in rotator cuff repair: review, preliminary results and future directions. Curr Pharm Biotechnol. 2012;13(7):1196–206.
63. Tian W, Schulze S, Brandl M, Winter G. Vesicular phospholipid gel-based depot formulations for pharmaceutical proteins: development and in vitro evaluation. J Control Release. 2010;142(3):319–25.
64. Buchmann S, Sandmann GH, Walz L, Reichel T, Beitzel K, Wexel G, et al. Growth factor release by vesicular phospholipid gels: in-vitro results and application for rotator cuff repair in a rat model. BMC Musculoskelet Disord. 2015;16(1):82.
65. Uggen C, Dines J, McGarry M, Grande D, Lee T, Limpisvasti O. The effect of recombinant human platelet-derived growth factor BB–coated sutures on rotator cuff healing in a sheep model. Arthroscopy. 2010;26(11):1456–62.
66. Dines JS, Grande DA, ElAttrache N, Dines DM. Biologics in shoulder surgery: suture augmentation and coating to enhance tendon repair. Tech Orthop. 2007;22(1):20–5.
67. Del Buono A, Oliva F, Longo UG, Rodeo SA, Orchard J, Denaro V, et al. Metalloproteases and rotator cuff disease. J Shoulder Elbow Surg. 2012;21(2):200–8.
68. Lo IKY, Marchuk LL, Hollinshead R, Hart DA, Frank CB. Matrix metalloproteinase and tissue inhibitor of matrix metalloproteinase mRNA levels are specifically altered in torn rotator cuff tendons. Am J Sports Med. 2004;32(5):1223–9.
69. Bedi A, Kovacevic D, Hettrich C, Gulotta LV, Ehteshami JR, Warren RF, et al. The effect of matrix metalloproteinase inhibition on tendon-to-bone healing in a rotator cuff repair model. J Shoulder Elbow Surg. 2010;19(3):384–91.
70. Bedi A, Fox AJS, Kovacevic D, Deng X-H, Warren RF, Rodeo SA. Doxycycline-mediated inhibition of matrix metalloproteinases improves healing after rotator cuff repair. Am J Sports Med. 2010;38(2):308–17.
71. Ricchetti ET, Aurora A, Iannotti JP, Derwin KA. Scaffold devices for rotator cuff repair. J Shoulder Elbow Surg. 2012;21(2):251–65.
72. Barber FA, Burns JP, Deutsch A, Labbé MR, Litchfield RB. A prospective, randomized evaluation of acellular human dermal matrix augmenta-

tion for arthroscopic rotator cuff repair. Arthroscopy. 2012;28(1):8–15.
73. DeFranco MJ, Derwin K, Iannotti JP. New therapies in tendon reconstruction. J Am Acad Orthop Surg. 2004;12(5):298–304.
74. Aurora A, McCarron JA, van den Bogert AJ, Gatica JE, Iannotti JP, Derwin KA. The biomechanical role of scaffolds in augmented rotator cuff tendon repairs. J Shoulder Elbow Surg. 2012;21(8):1064–71.
75. Shea KP, Obopilwe E, Sperling JW, Iannotti JP. A biomechanical analysis of gap formation and failure mechanics of a xenograft-reinforced rotator cuff repair in a cadaveric model. J Shoulder Elbow Surg. 2012;21(8):1072–9.
76. Babensee JE, Anderson JM, McIntire LV, Mikos AG. Host response to tissue engineered devices. Adv Drug Deliv Rev. 1998;33(1–2):111–39.
77. Neviaser JS. Ruptures of the rotator cuff of the shoulder. New concepts in the diagnosis and operative treatment of chronic ruptures. Arch Surg. 1971;102(5):483–5.
78. Nassos JT, Chudik SC. Arthroscopic rotator cuff repair with biceps tendon augmentation. Am J Orthop (Belle Mead NJ). 2009;38(6):279–81.
79. Fuchs B, Weishaupt D, Zanetti M, Hodler J, Gerber C. Fatty degeneration of the muscles of the rotator cuff: assessment by computed tomography versus magnetic resonance imaging. J Shoulder Elbow Surg. 1999;8(6):599–605.
80. Goutallier D, Postel JM, Bernageau J, Lavau L, Voisin MC. Fatty muscle degeneration in cuff ruptures. Pre- and postoperative evaluation by CT scan. Clin Orthop. 1994;304:78–83.
81. Mori D, Funakoshi N, Yamashita F. Arthroscopic surgery of irreparable large or massive rotator cuff tears with low-grade fatty degeneration of the infraspinatus: patch autograft procedure versus partial repair procedure. Arthroscopy. 2013;29(12):1911–21.
82. Bektaşer B, Oçgüder A, Solak S, Gönen E, Yalçın N, Kılıçarslan K. Free coracoacromial ligament graft for augmentation of massive rotator cuff tears treated with mini-open repair. Acta Orthop Traumatol Turc. 2010;44(6):426–30.
83. Mihara S, Ono T, Inoue H, Kisimoto T. A new technique for patch augmentation of rotator cuff repairs. Arthrosc Tech. 2014;3(3):e367–71.
84. Scheibel M, Brown A, Woertler K, Imhoff AB. Preliminary results after rotator cuff reconstruction augmented with an autologous periosteal flap. Knee Surg Sports Traumatol Arthrosc. 2007;15(3):305–14.
85. Barber FA, Herbert MA, Boothby MH. Ultimate tensile failure loads of a human dermal allograft rotator cuff augmentation. Arthroscopy. 2008;24(1):20–4.
86. Proctor CS. Long-term successful arthroscopic repair of large and massive rotator cuff tears with a functional and degradable reinforcement device. J Shoulder Elbow Surg. 2014;23(10):1508–13.
87. McCarron JA, Milks RA, Chen X, Iannotti JP, Derwin KA. Improved time-zero biomechanical properties using poly-L-lactic acid graft augmentation in a cadaveric rotator cuff repair model. J Shoulder Elbow Surg. 2010;19(5):688–96.
88. Derwin KA, Baker AR, Spragg RK, Leigh DR, Iannotti JP. Commercial extracellular matrix scaffolds for rotator cuff tendon repair: biomechanical, biochemical, and cellular properties. J Bone Joint Surg Am. 2006;88(12):2665–72.
89. Sclamberg SG, Tibone JE, Itamura JM, Kasraeian S. Six-month magnetic resonance imaging follow-up of large and massive rotator cuff repairs reinforced with porcine small intestinal submucosa. J Shoulder Elbow Surg. 2004;13(5):538–41.
90. Iannotti JP, Codsi MJ, Kwon YW, Derwin K, Ciccone J, Brems JJ. Porcine small intestine submucosa augmentation of surgical repair of chronic two-tendon rotator cuff tears: a randomized, controlled trial. J Bone Joint Surg Am. 2006;88(6):1238–44.
91. Walton JR, Bowman NK, Khatib Y, Linklater J, Murrell GAC. Restore orthobiologic implant: not recommended for augmentation of rotator cuff repairs. J Bone Joint Surg Am. 2007;89(4):786–91.
92. Gilbert TW, Freund JM, Badylak SF. Quantification of DNA in biologic scaffold materials. J Surg Res. 2009;152(1):135–9.
93. Badhe SP, Lawrence TM, Smith FD, Lunn PG. An assessment of porcine dermal xenograft as an augmentation graft in the treatment of extensive rotator cuff tears. J Shoulder Elbow Surg. 2008;17(1 Suppl):35S–9S.
94. Cho C-H, Lee S-M, Lee Y-K, Shin H-K. Mini-open suture bridge repair with porcine dermal patch augmentation for massive rotator cuff tear: surgical technique and preliminary results. Clin Orthop Surg. 2014;6(3):329.
95. Avanzi P, Giudici LD, Capone A, Cardoni G, Lunardi G, Foti G, et al. Prospective randomized controlled trial for patch augmentation in rotator cuff repair: 24-month outcomes. J Shoulder Elbow Surg. 2019;28(10):1918–27.
96. O'Brien FJ. Biomaterials & scaffolds for tissue engineering. Mater Today. 2011;14(3):88–95.
97. Place ES, Evans ND, Stevens MM. Complexity in biomaterials for tissue engineering. Nat Mater. 2009;8(6):457–70.
98. Atala A, Kasper FK, Mikos AG. Engineering complex tissues. Sci Transl Med. 2012;4(160):160rv12.
99. Gardin C, Ferroni L, Favero L, Stellini E, Stomaci D, Sivolella S, et al. Nanostructured biomaterials for tissue engineered bone tissue reconstruction. Int J Mol Sci. 2012;13(1):737–57.
100. McKeown AD, Beattie RF, Murrell GA, Lam PH. Biomechanical comparison of expanded polytet-

rafluoroethylene (ePTFE) and PTFE interpositional patches and direct tendon-to-bone repair for massive rotator cuff tears in an ovine model. Shoulder Elbow. 2016;8(1):22–31.
101. Ma B, Xie J, Jiang J, Shuler FD, Bartlett DE. Rational design of nanofiber scaffolds for orthopedic tissue repair and regeneration. Nanomedicine (Lond). 2013;8(9):1459–81.
102. Ciampi P, Scotti C, Nonis A, Vitali M, Di Serio C, Peretti GM, et al. The benefit of synthetic versus biological patch augmentation in the repair of posterosuperior massive rotator cuff tears: a 3-year follow-up study. Am J Sports Med. 2014;42(5):1169–75.
103. Meyer F, Wardale J, Best S, Cameron R, Rushton N, Brooks R. Effects of lactic acid and glycolic acid on human osteoblasts: a way to understand PLGA involvement in PLGA/calcium phosphate composite failure. J Orthop Res. 2012;30(6):864–71.

Part II

Diagnosis of Shoulder Conditions

Physical Examination for Glenohumeral Joint Pathologies

8

H. Çağdaş Basat and Mehmet Armangil

8.1 Introduction

The glenohumeral joint examination is a part of the shoulder examination. Whole shoulder should be examined before the glenohumeral joint examination. However, in this chapter, the glenohumeral joint examination is focused, and tests of the glenohumeral joints are only described. Other parts of the shoulder examination are described in the following chapters.

It should be taken into account that precise history is the first step to achieve true diagnosis, like other system examinations. Thus, the examiner could choose special tests to diagnosis for the pathology, because all tests are not essential for every patient [1, 2]. Main complaint should be ascertained such as onset, location of the pain, aggravating and alleviating factors, etc. Besides, demographic features—age, gender, work, activities, hand dominance, etc.—general health condition, and background of the patient should be evaluated. Identification of features of pain is helpful for true diagnosis [3]. Pain around the deltoid could be especially related with glenohumeral joint. Bicipital grove pain could be associated with lesion of tendon of biceps or superior labrum. It is also important to differentiate whether symptoms are acute or chronic. Acute trauma forcing the shoulder external rotation and abduction is related to dislocation or subluxation of the glenohumeral joint or tear of the superior labrum. However, adessive capsulitis or arthrosis should be considered that insidious onset or chronicity of the symptom and loss of the range of motion is detected from the history of patient [4].

Inspection is the first step of glenohumeral examination. Patients should be undressed for inspection, and upper extremity, neck, thorax should be checked symmetrically for atrophy of deltoid or supra-infraspinatus, trauma (ecchymosis, abrasion), infection, incision of prior surgery, effusion (more than 15 mL fluid in glenohumeral joint could be detected by inspection), deformity, tumor, and scapular dyskinesia [3, 5, 6].

Palpation is the following step of the glenohumeral examination. All bony and soft tissue structures around the glenohumeral joint should be palpated for sensitivity, atrophy, swelling, crepitus, increased temperature. But also, the long head of the biceps tendon should be checked for tenderness. Crepitus could be noticed by palpation during the movement of the shoulder. It could be sign of arthrosis, tendinitis, adhesive capsulitis of glenohumeral joint [4, 7].

Range of motion (ROM) of the glenohumeral joint should be evaluated both actively and passively. These measurements should be compared

H. Ç. Basat (✉)
Orthopaedics and Traumatology Department, Kırşehir Ahi Evran University School of Medicine, Kırşehir, Turkey

M. Armangil
Orthopaedics and Traumatology Department, Ankara University School of Medicine, Ankara, Turkey

G. Huri et al. (eds.), *Fundamentals of the Shoulder*, https://doi.org/10.1007/978-3-030-94702-6_8

with the opposite shoulder. It is suggested that four ROM should be evaluated; flexion, abduction, internal rotation, and external rotation both at the side and 90° of abduction [8]. Pain arises midrange of motion for glenohumeral arthrosis and end range of motion for adhesive capsulitis. Patients having rotator cuff disease have limited active ROM but near normal passive ROM [4, 5, 8].

8.2 Special Tests

8.2.1 Anterior Instability Tests

8.2.1.1 Apprehension Test

Apprehension test was described first by Row et al. for anterior shoulder instability in 1981 [9]. This test could be performed while patient is either in standing or lying in supine position. The shoulder is placed at 90° of abduction and elbow is placed at 90° flexion. While in this position, examiner brings patients shoulder at 90° of external rotation using one hand and performs anterior pressure using the other hand. This maneuver continues until patient feels apprehension or pain (Fig. 8.1). Apprehension is much sensitive and specific than pain [10]. Hegedus reviewed meta-analysis of shoulder physical examination in literature. Sensitivity and specificity of the apprehension test were reported as 65.6% and 95.4%, respectively [11].

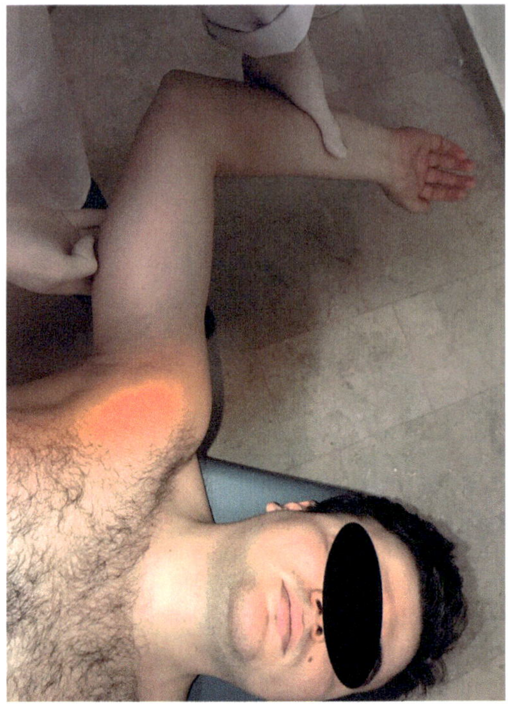

Fig. 8.1 Apprehension test is presented

8.2.1.2 Jobe Relocation Test

This test was defined by F. Jobe to identify secondary impingement [12]. But now, this test is implemented to detect anterior instability with apprehension test. Relocation test is performed with apprehension test. First, the examiner brings about the patient shoulder to 90° abduction and external rotation until patient feels pain or apprehension. Afterwards, the examiner applies posterior directed force on the anterior aspect of the shoulder (Fig. 8.2). If patient symptom disappears, this test is a positive. Relief of the pain is more related with rotator cuff disease, posterior SLAP lesion, and AC joint disease. Relief of apprehension is more related with anterior insta-

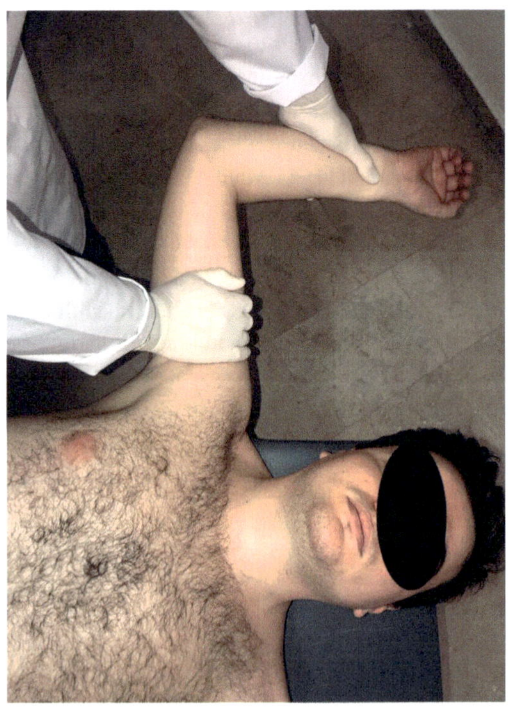

Fig. 8.2 Jobe relocation test is presented

bility [13, 14]. Relocation test improves sensitivity and specificity of the apprehension test. A meta-analysis report showed that sensitivity and specificity of the relocation test were 64.6% and 90.2%, respectively [11]. But if relocation test was performed with apprehension test, sensitivity and specificity were reported as 81% and 98%, respectively [10].

8.2.1.3 Anterior Drawer Test

Gerber and Ganz presented this test in 1984 [15]. The position of the patient is supine, and the shoulder is over the edge of the table. The examiner should stand in front of the shoulder of the patient. Afterwards, the examiner places his/her one hand's index and middle finger to the spine of scapula of the patient and his thumb to the coracoid process of the scapula of the patient to stabilize the scapula. The shoulder of the patient should be abducted 80°–120°, flexed 0°–20° and, placed in 0°–30° of external rotation. The shoulder of the patient is forced to anterior direction with other hand of the examiner (Fig. 8.3), while axial load is applied to the arm [15]. The amount of the displacement is measured by method of the McFarland [16]. Grades are described as follows: grade 1, if the humeral head is moved to the glenoid rim; grade 2, if the humeral head is displaced over the glenoid rim but reduces spontaneously; grade 3, if the humeral head is displaced and does not reduce spontaneously. Sensitivity and specificity of the anterior drawer test were reported by Farber et al. [10] as 53% and 85%, respectively. However, they reported that if anterior drawer test was combined with apprehension, the sensitivity decreased, and specificity increased.

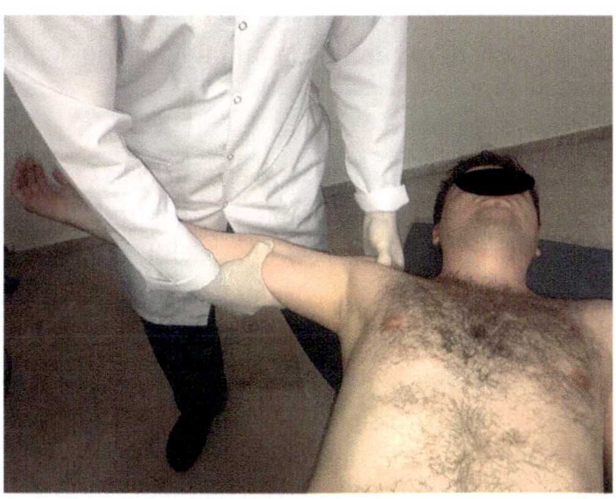

Fig. 8.3 Anterior drawer test is presented

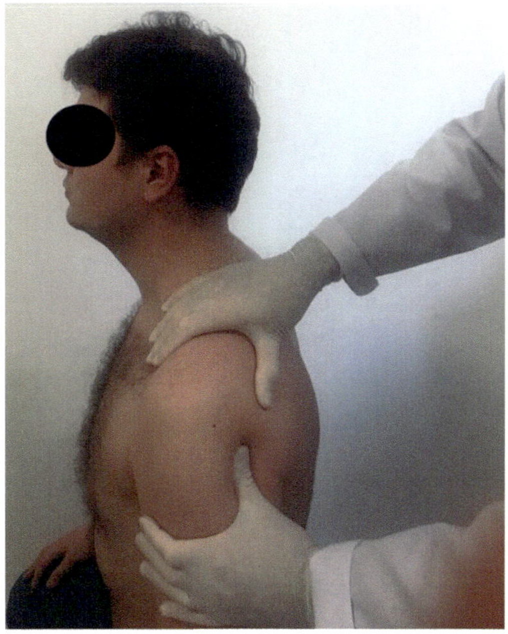

Fig. 8.4 Load and shift test is presented

8.2.1.4 Load and Shift Test

Load and shift test was presented by Hawkins and Silliman [17]. This test is similar to anterior drawer test. The examiner should be at back of the patient while patient is sitting. The examiner holds the scapula of the patient using his/her one hand to stabilize the scapula. The shoulder of the patient is forced in anterior direction using the other hand of the examiner (Fig. 8.4). The amount of the displacement is measured by the method of McFarland [16], as anterior drawer test. Tzannes et al. reported that sensitivity and specificity of the load and shift test were 50% and 100%, respectively.

8.2.1.5 Hyper Extension–Internal Rotation (HERI) Test

This test was introduced by Lafosse et al. in 2015. Lafosse emphasized that the patient does not feel apprehension during movements and has no risk of dislocation throughout this test. Aim of

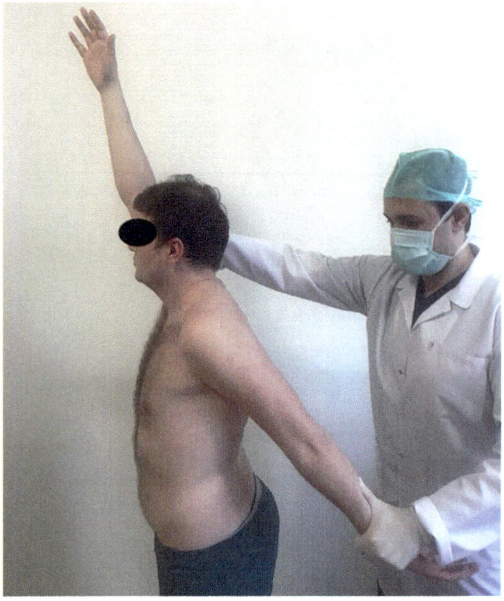

Fig. 8.5 HERI test is presented

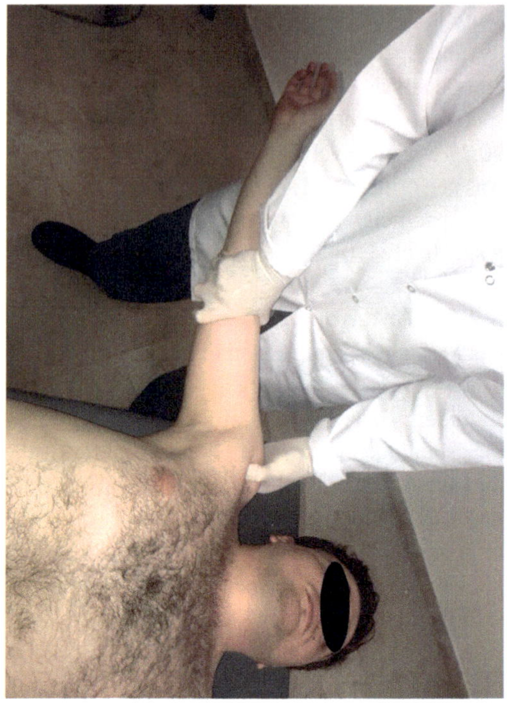

Fig. 8.6 Posterior drawer test is presented

the HERI test is to measure of the glenohumeral extension angle. Description of the HERI test is performed for right shoulder. The position of the patient is standing as examiner is behind of the patient. The examiner forces the patient's left scapula with his/her left elbow and rises left upper limb of the patient to stabilize the thoracic region. The examiner holds the patient right forearm with his/her right hand as the patient forearm is pronated and elbow is extended position. The examiner forces the shoulder of the patient to hyperextension until feeling end point (Fig. 8.5). This extension angle is calculated, and this test is repeated opposite shoulder and both results is compared. There is no study about sensitivity and specificity of HERI test.

8.2.2 Posterior Instability Tests

8.2.2.1 Posterior Drawer Test

Gerber and Ganz presented this test in1984 [15]. The position of the patient is supine, and the shoulder is over the edge of the table. The examiner should stand in front of shoulder of the patient. Afterwards, examiner place his/her one hand's index and middle finger to the spine of scapula of the patient and his thumb to the coracoid process of the scapula of the patient to stabilize the scapula. The shoulder of the patient should be abducted 80°–120°, flexed 0°–20° and placed in 0°–30° of external rotation. The shoulder of the patient is forced to posterior direction with other hand of the examiner (Fig. 8.6), while axial load is applied to the arm [15]. The amount of the displacement is measured by the method of McFarland [16]. There is no enough data about sensitivity and specificity of posterior drawer test.

8.2.2.2 Posterior Apprehension Test

The posterior apprehension test was introduced by Kessell [6]. This test is performed while the patient is in sitting position. The shoulder of the patient is brought 90° of forward flexion, adduction, and internal rotation. Afterwards, the examiner forces humerus to the posterior direction. If the patient feels apprehension, test is a positive (Fig. 8.7). There is no data to show test reliability.

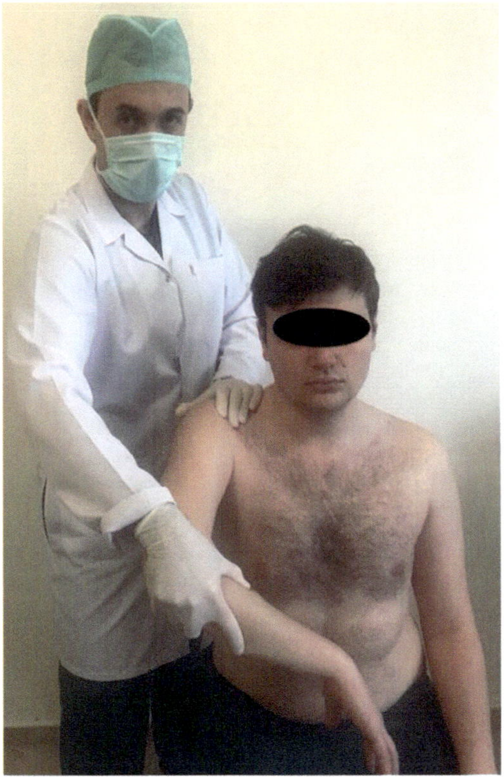

Fig. 8.7 Posterior apprehension test is presented

8.2.2.3 Jerk Test

The jerk test was introduced by Matsen et al. This test is performed while the patient is in sitting position [18]. Examiner is standing near the shoulder and grasps the patient's scapula with one hand. Afterwards, the shoulder is brought 90° of forward flexion, internal rotation and elbow is flexed to 90°. Examiner grasps the patient's elbow and applies force to the posterior direction (Fig. 8.8a). If a sudden jerk is detected, this test is a positive. If the shoulder is brought to 90° abduction, the second jerk could be detected (Fig. 8.8b). This jerk occurs resulting from the movement of the instable humeral head. Kim et al. [19] reported that sensitivity and specificity of jerk test were 73% and 98%, respectively.

8.2.3 Multidirectional Instability Tests

8.2.3.1 Sulcus Sign

Sulcus sign test was performed by Neer and Foster to evaluate multidirectional shoulder instability [20]. Superior glenohumeral and rota-

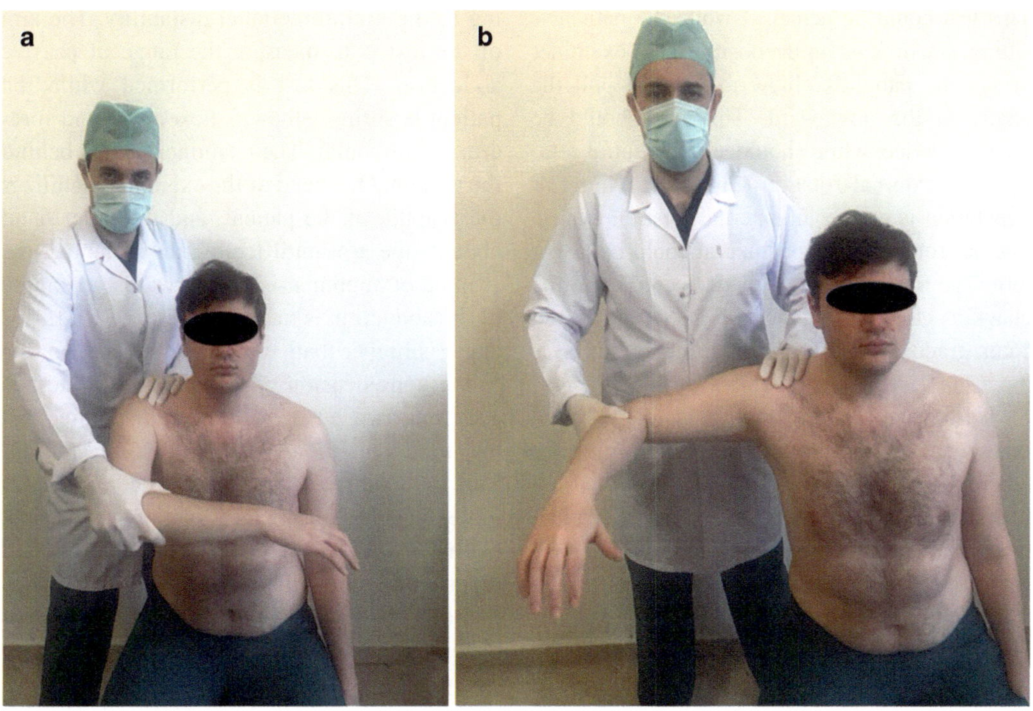

Fig. 8.8 Jerk test is presented. (**a**) Shoulder in internal rotation, 90° of forward flexion, and elbow is flexed to 90° while being forced to posterior direction. (**b**) Shoulder is brought 90° of abduction

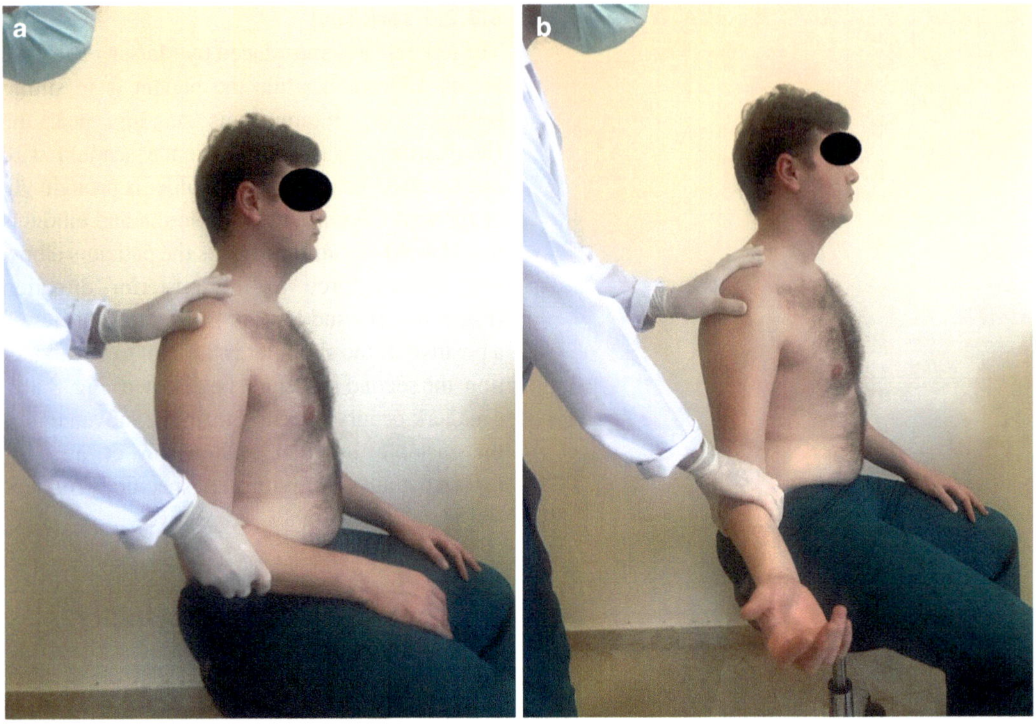

Fig. 8.9 Sulcus sign is presented in neutral rotation (**a**) and external rotation (**b**)

tor interval could be evaluated by sulcus sign. This test could be achieved while the patient is sitting, standing or supine position. The examiner brings the patient's elbow to 90° and pull the elbow to the downward. This test should be repeated twice, while shoulder is in neutral rotation and external rotation (Fig. 8.9). If inferior translation is greater in external rotation than in internal rotation, rotator interval could be lax or torn. The measurement is made according to the Hawkins et al. [21]. Grade I is between 0.5 and 1 cm, grade II is between 1 and 2 cm, grade III is more than 2 cm. The reliability of this measurement is poor [22]. Nonetheless, it was reported that specificity of this test was 97%. However, the sensitivity of this test was poor (28%). So, using this test alone can cause to missing the patient having the multidirectional instability [23].

8.2.3.2 Hyperabduction Test

The hyperabduction test was presented by Gagey in 2000 [24]. This test evaluates inferior glenohumeral ligament insufficiency, contributing to the multidirectional instability. The aim of the test is to measure the range of passive abduction. This test is performed while the patient is sitting, elbow is flexed 90°, and forearm is horizontal. The examiner stands behind the patient. One hand of the examiner stabilizes the scapula of the patient, and the other hand abducts the arm until feeling movement of the scapula or apprehension. Afterwards, the angle of the abduction is measured (Fig. 8.10). If this angle is higher than 105° or test is stopped due to the apprehension of the patient, this test is accepted as a positive. Ren et al. pointed out that three conditions should be existed to confirm the results as positive: (1) deep pain, (2) more than 20° differences with opposite side, (3) no or soft end point. Jouve et al. reported that if having more than 15° differences between the two sides, sensitivity and specificity of hyperabduction test were 77% and 91%, respectively [25].

8.3 Intraarticular Biceps Tests (SLAP)

8.3.1 Active Compression (O'Brien) Test

The Active Compression Test was presented by O'Brien in 1998 [26]. This test provides to detect SLAP (superior labral anterior and posterior) lesion or disease of acromioclavicular (AC) joint. The positions of the patient and the examiner are standing, and examiner is behind the patient during this test. The patient flexes shoulder to the 90° while elbow is fully extended. Afterward, the shoulder is moved to 10–15° adduction position and internal rotation with thumb down. The patient is asked to resist inferior force by performed the examiner. This test is repeated while the shoulder is external rotation with thumb up. If pain is detected on AC joint with first maneuver and disappeared with second maneuver, the test is considered as a positive for the AC joint pathology (Fig. 8.11). If the patient feels pain inside the joint with first maneuver and pain is disappeared with the second maneuver, the test is considered as a positive for SLAP lesion. O'Brien et al. reported that the sensitivity and specificity of the active compression test were 100% and 98.5%, respectively [26]. However, in the literature, the sensitivity and specificity of the active compression test were variable [27–29].

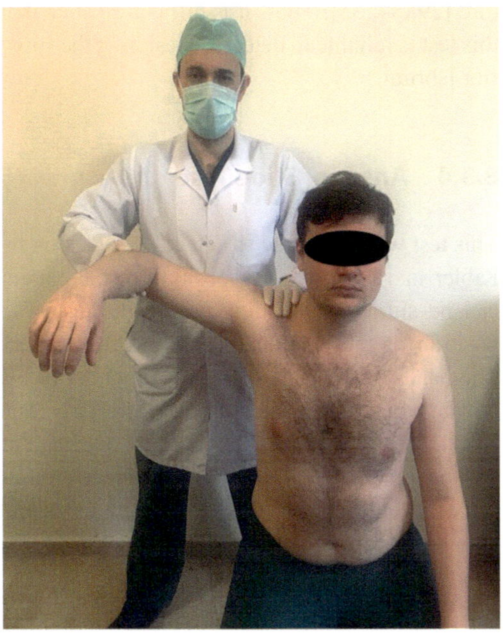

Fig. 8.10 Hyperabduction test is presented

8.3.2 Biceps Load Test I and II

The biceps load tests I and II were introduced to evaluate superior labrum by Kim et al. in 1999 and 2001, respectively [30, 31]. These tests are performed while patient is supine position. Patient shoulder is brought to 90° abduction for test I and 120° for test II and external rotation while elbow is flexed and supinated, as apprehension test. The external rotation is continued until feeling apprehension. The patient is asked to flex the elbow against performed opposite force by examiner (Fig. 8.12). If the patient feels apprehension or painful, test is considered as a positive. If the patient feels more comfortable or does not feel apprehension, then the test is a negative. Kim et al. [30, 31] stated that the sensitivity and

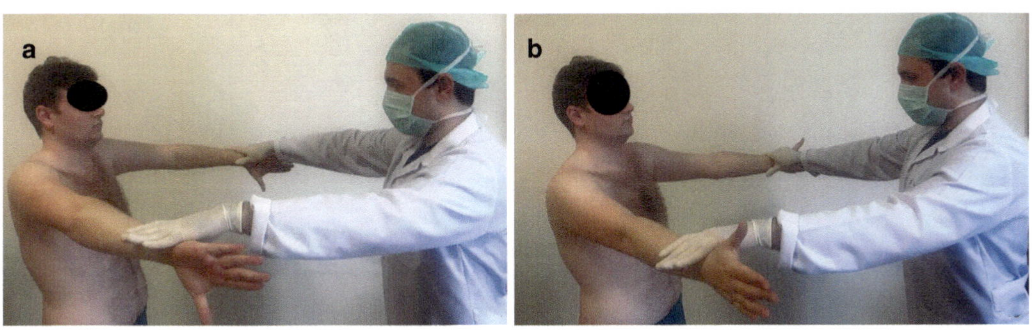

Fig. 8.11 Active compression test is presented in internal rotation (**a**) and external rotation (**b**)

specificity of biceps load test I were 90.9% and 96.9%, respectively and the sensitivity and specificity of the biceps load test II were 89.7% and 96.9%, respectively. However, Cook et al. [32] reported that the sensitivity and specificity of biceps load test II were 55% and 53%, respectively. Hereby, there is not enough data to mention the power of these tests, yet.

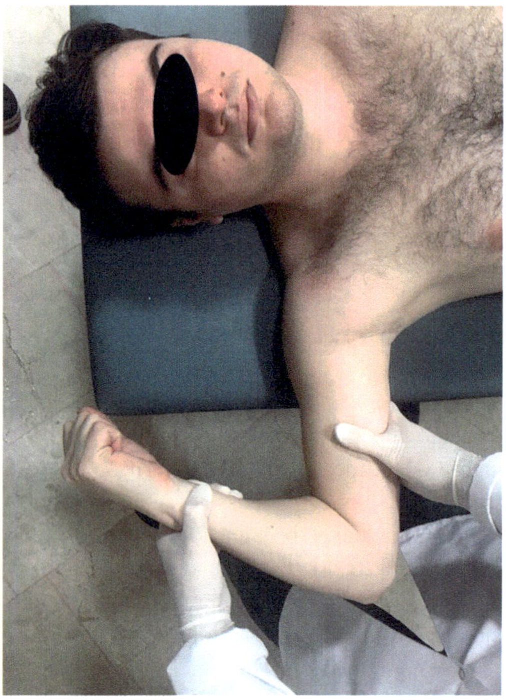

Fig. 8.12 Biceps load test is presented

8.3.3 Crank Test

The crank test was performed to evaluate the superior labrum by Liu et al. in 1996 [33]. This test can be performed while the patient is sitting or standing. The examiner takes position behind the patient. The shoulder is brought about 160° abduction and is applied axial load along the humerus. While applying axial force, the shoulder rotates internally and externally (Fig. 8.13). If the patient feels pain or symptoms reproduce, this test is a positive. Liu et al. shared their results that the sensitivity and specificity of biceps load test II were 91% and 93%, respectively. However, variable results have been reported in the literature [29, 34, 35]. Thus, it is difficult to state that this test is reliable to detect the lesion of the superior labrum.

8.3.4 Anterior Slide Test

This test was presented to detect SLAP lesion by Kibler in 1995 [36]. The position of the patient can be sitting or standing during the test. The position of the examiner is behind the patient. The hands of the patient are placed on her or his pelvis, while the thumb pointing posteriorly. The scapula of the patient is stabilized by one hand of the examiner. Afterwards, the examiner applies force in anterior and superior directions of the elbow, and the patient is asked to resist this force

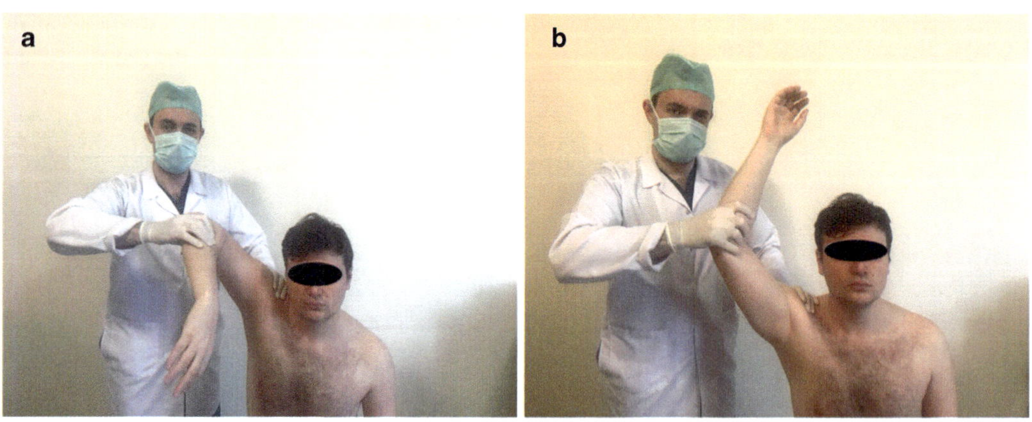

Fig. 8.13 Crank test is presented in internal rotation (**a**) and external rotation (**b**)

(Fig. 8.14). If the patient feels pain or sensation of the click or pop, the test is considered as a positive. Kibler et al. (36)reported that the sensitivity and specificity of anterior slide test were 78.4% and 91.5%, respectively. However, poor results are seen in literature [34, 37]. Parentis et al. [34] detected that the sensitivity and specificity of anterior slide test for SLAP type I and II were 10% and 81.5%, respectively.

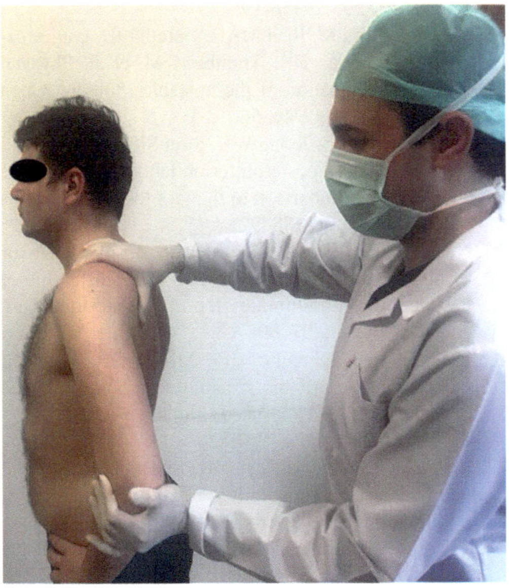

Fig. 8.14 Anterior slide test is presented

8.3.5 Compression Rotation Test

Compression rotation test was defined to detect the SLAP lesion by Snyder et al. [38]. This test is performed while the patient is supine, and the examiner is standing beside shoulder. The shoulder is abducted between 20° and 90°, elbow is flexed at 90°. While the axial force is applied from the elbow to the shaft of humerus, rotational movement is performed to catch the torn labrum in the joint (Fig. 8.15). If the patient feels pain or sensation of click, the test is considered as a positive. McFarland et al. reported their results that the sensitivity and specificity of compression rotation test for SLAP were 24% and 76%, respectively [39].

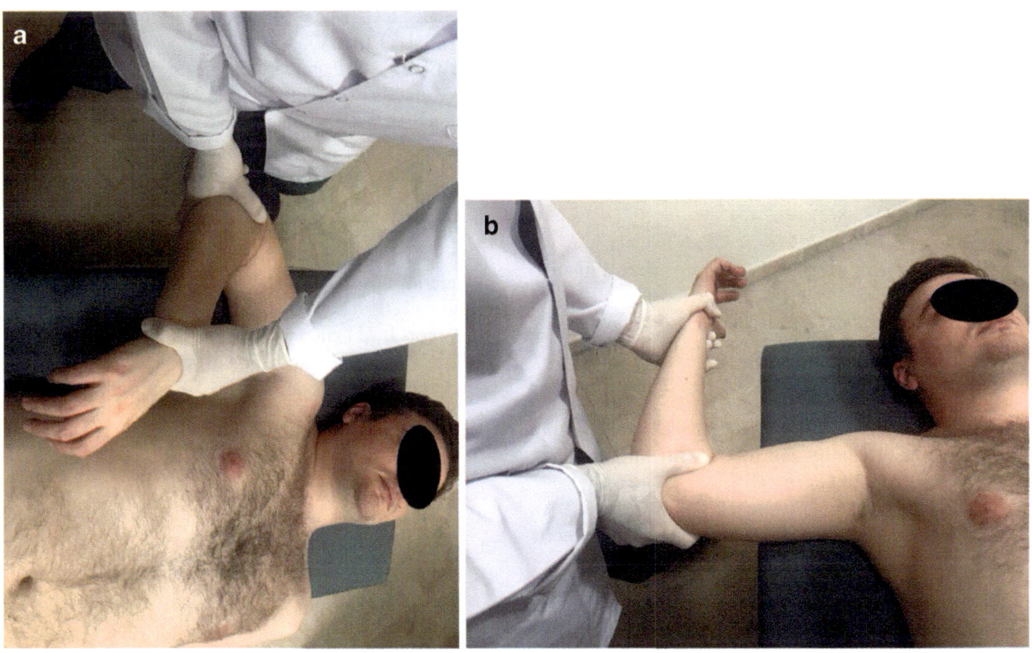

Fig. 8.15 Compression rotation test is presented in internal rotation (**a**) and external rotation (**b**)

References

1. King JJ, Wright TW. Physical examination of the shoulder. J Hand Surg Am. 2014;39(10):2103–12.
2. Woodward TW, Best TM. The painful shoulder: part I. Clinical evaluation. Am Fam Physician. 2000;61(10):3079–88.
3. McFarland EG, Tanaka MJ, Papp DF. Examination of the shoulder in the overhead and throwing athlete. Clin Sports Med. 2008;27(4):553–78.
4. Jonathan Barlow ACM, Jones GL. Stiff shoulder. In: MMS T, editor. DeLee, Drez and Miller's orthopaedic sports medicine, vol. 2. 5th ed. Amsterdam: Elsevier; 2019. p. 579–91.
5. Ombregt L. Clinical examination of the shoulder. In: Ombregt L, editor. A system of orthopaedic medicine. 3rd ed. Edinburgh: Churchill Livingstone; 2013. p. 207–17.e2.
6. McFarland JEB EG, Borade A, Malanga GA. Tutankhamen Pappoe. Physical examination of the shoulder. In: Malanga GA, Mautner K, editors. Musculoskeletal physical examination. 2nd ed. Philadelphia: Elsevier; 2017.
7. Waldman SD. Palpation of the shoulder. In: Waldman SD, editor. Physical diagnosis of pain: an atlas of signs and symptoms. 4th ed. Philadelphia: Elsevier; 2021. p. 40.
8. Jason S, Klein BGL, Namdari S, Williams GR. Rotator cuff tendinopathies and tears surgery and therapy. In: Skirven TM, Osterman AL, Fedorczyk JM, Amadio PC, Feldscher SB, Shin EK, editors. Rehabilitation of the hand and upper extremity. Philadelphia: Elsevier; 2021. p. 540–58.
9. Rowe CR, Zarins B. Recurrent transient subluxation of the shoulder. J Bone Joint Surg Am. 1981;63(6):863–72.
10. Farber AJ, Castillo R, Clough M, Bahk M, McFarland EG. Clinical assessment of three common tests for traumatic anterior shoulder instability. J Bone Joint Surg Am. 2006;88(7):1467–74.
11. Hegedus EJ, Goode AP, Cook CE, Michener L, Myer CA, Myer DM, et al. Which physical examination tests provide clinicians with the most value when examining the shoulder? Update of a systematic review with meta-analysis of individual tests. Br J Sports Med. 2012;46(14):964–78.
12. Jobe FW, Kvitne RS, Giangarra CE. Shoulder pain in the overhand or throwing athlete. The relationship of anterior instability and rotator cuff impingement. Orthop Rev. 1989;18(9):963–75.
13. Speer KP, Hannafin JA, Altchek DW, Warren RF. An evaluation of the shoulder relocation test. Am J Sports Med. 1994;22(2):177–83.
14. Burkhart SS, Morgan CD. The peel-back mechanism: its role in producing and extending posterior type II SLAP lesions and its effect on SLAP repair rehabilitation. Arthroscopy. 1998;14(6):637–40.
15. Gerber C, Ganz R. Clinical assessment of instability of the shoulder. With special reference to anterior and posterior drawer tests. J Bone Joint Surg Br. 1984;66(4):551–6.
16. McFarland EG, Torpey BM, Curl LA. Evaluation of shoulder laxity. Sports Med. 1996;22(4):264–72.
17. Silliman JF, Hawkins RJ. Classification and physical diagnosis of instability of the shoulder. Clin Orthop Relat Res. 1993;291:7–19.
18. Valencia Mora M, Iban MAR, Heredia JD, Gutierrez-Gomez JC, Diaz RR, Aramberri M, et al. Physical exam and evaluation of the unstable shoulder. Open Orthop J. 2017;11:946–56.
19. Kim SH, Park JS, Jeong WK, Shin SK. The Kim test: a novel test for posteroinferior labral lesion of the shoulder—a comparison to the jerk test. Am J Sports Med. 2005;33(8):1188–92.
20. Neer CS 2nd, Foster CR. Inferior capsular shift for involuntary inferior and multidirectional instability of the shoulder. A preliminary report. J Bone Joint Surg Am. 1980;62(6):897–908.
21. Hawkins RJ, Schutte JP, Janda DH, Huckell GH. Translation of the glenohumeral joint with the patient under anesthesia. J Shoulder Elbow Surg. 1996;5(4):286–92.
22. Levy AS, Lintner S, Kenter K, Speer KP. Intra- and interobserver reproducibility of the shoulder laxity examination. Am J Sports Med. 1999;27(4):460–3.
23. Tzannes A, Murrell GA. Clinical examination of the unstable shoulder. Sports Med. 2002;32(7):447–57.
24. Gagey OJ, Gagey N. The hyperabduction test. J Bone Joint Surg Br. 2001;83(1):69–74.
25. Jouve F, Schlur C, Klouche S, Bauer T, Rousselin B, Hardy P. SHART: shoulder hyperabduction radiological test. Knee Surg Sports Traumatol Arthrosc. 2011;19(12):2080–4.
26. O'Brien SJ, Pagnani MJ, Fealy S, McGlynn SR, Wilson JB. The active compression test: a new and effective test for diagnosing labral tears and acromioclavicular joint abnormality. Am J Sports Med. 1998;26(5):610–3.
27. Davis C, Immormino J, Higgins BM, Clark K, Engebose S, Garcia AN, et al. Diagnostic utility of the active compression test for the superior labrum anterior posterior tear: a systematic review. Shoulder Elbow. 2019;11(5):321–31.
28. Walton J, Mahajan S, Paxinos A, Marshall J, Bryant C, Shnier R, et al. Diagnostic values of tests for acromioclavicular joint pain. J Bone Joint Surg Am. 2004;86(4):807–12.
29. Stetson WB, Templin K. The crank test, the O'Brien test, and routine magnetic resonance imaging scans in the diagnosis of labral tears. Am J Sports Med. 2002;30(6):806–9.
30. Kim SH, Ha KI, Han KY. Biceps load test: a clinical test for superior labrum anterior and posterior lesions in shoulders with recurrent anterior dislocations. Am J Sports Med. 1999;27(3):300–3.
31. Kim SH, Ha KI, Ahn JH, Kim SH, Choi HJ. Biceps load test II: a clinical test for SLAP lesions of the shoulder. Arthroscopy. 2001;17(2):160–4.

32. Cook C, Beaty S, Kissenberth MJ, Siffri P, Pill SG, Hawkins RJ. Diagnostic accuracy of five orthopedic clinical tests for diagnosis of superior labrum anterior posterior (SLAP) lesions. J Shoulder Elbow Surg. 2012;21(1):13–22.
33. Liu SH, Henry MH, Nuccion SL. A prospective evaluation of a new physical examination in predicting glenoid labral tears. Am J Sports Med. 1996;24(6):721–5.
34. Parentis MA, Glousman RE, Mohr KS, Yocum LA. An evaluation of the provocative tests for superior labral anterior posterior lesions. Am J Sports Med. 2006;34(2):265–8.
35. Myers TH, Zemanovic JR, Andrews JR. The resisted supination external rotation test: a new test for the diagnosis of superior labral anterior posterior lesions. Am J Sports Med. 2005;33(9):1315–20.
36. Kibler WB. Specificity and sensitivity of the anterior slide test in throwing athletes with superior glenoid labral tears. Arthroscopy. 1995;11(3):296–300.
37. Jia X, Petersen SA, Khosravi AH, Almareddi V, Pannirselvam V, McFarland EG. Examination of the shoulder: the past, the present, and the future. J Bone Joint Surg Am. 2009;91(Suppl 6):10–8.
38. Snyder SJ, Karzel RP, Del Pizzo W, Ferkel RD, Friedman MJ. SLAP lesions of the shoulder. Arthroscopy. 1990;6(4):274–9.
39. McFarland EG, Kim TK, Savino RM. Clinical assessment of three common tests for superior labral anterior-posterior lesions. Am J Sports Med. 2002;30(6):810–5.

Physical Examination for Subacromial and Acromioclavicular Pathologies

9

Emre Anıl Özbek
and Ahmet Mehmet Demirtaş

9.1 History

Prior to starting physical examination, a thorough history must be taken in patients presenting with pain in the shoulder girdle. While taking the history, the age and profession of the patient, the mechanism of injury if it is of trauma origin, the localization and severity of the pain, and the onset of symptoms should be questioned. In addition to these, the presence of activities that increase or decrease the patient's pain, history of previous surgery around the shoulder, and the presence of numbness in the fingers of the same limb should be questioned. The results obtained from the history will make up a pre-diagnosis of the patient's pathology, allowing the use of more specific physical examination methods. For example, in a 75-year-old patient with pain in the anterior shoulder for 2 years, pre-diagnoses such as chronic rotator cuff rupture and subacromial impingement are considered, and specific tests are performed for these pre-diagnoses. However, in a 40-year-old patient with pain in the shoulder girdle for 6 months and complaints of numbness in the fingers of the same limb, specific physical examination tests should be selected for cervical pathologies.

9.2 Physical Examination

After a detailed history, physical examination should be started. As with all orthopedic examinations, physical examination should begin with inspection, palpation, evaluation of the shoulder range of motion (ROM), and completed with specific tests.

9.2.1 Inspection

Physical examination of the shoulder girdle should start with inspection, as with any orthopedic physical examination. The patient is prepared with both shoulders fully open and is inspected by comparison with the contralateral shoulder. While standing in front of the patient, the presence of deformity, swelling or dislocation is evaluated in the sternoclavicular (SC) joint, clavicle, and acromioclavicular (AC) joints. Then, the findings of ecchymosis, edema, previous surgical scars, contour, and volume difference between the pectoralis major muscle and its contralateral, and auto-rupture (Popeye sign) of the biceps muscle are assessed. When assessing the lateral side of the patient, atrophy of the deltoid muscle, which

E. A. Özbek (✉)
Orthopedic Surgery Department, Ankara University School of Medicine, Ankara, Turkey

A. M. Demirtaş
Hand Surgery Division, Orthopedic Surgery Department, Ankara Memorial Hospital, Ankara, Turkey

G. Huri et al. (eds.), *Fundamentals of the Shoulder*, https://doi.org/10.1007/978-3-030-94702-6_9

may be caused by C5 spinal nerve or axillary neuropathy, is evaluated. On the back of the patient, scapular malposition and dyskinesia are assessed. In the meantime, atrophy of the supraspinatus and infraspinatus muscles is noted. When considering that chronic rupture may be present in isolated atrophy of the supraspinatus muscle, the atrophic appearance of the supraspinatus and infraspinatus muscles together may indicate that the suprascapular nerve is compressed in the suprascapular notch or there may be chronic rupture of both muscles [1]. If it is isolated atrophy of the infraspinatus muscle, it may be due to compression caused by a possible paralabral cyst in the spinoglenoid notch of the suprascapular nerve [1].

9.2.2 Palpation and ROM Assessment

Palpation is started from the cervical area, since the pain in the shoulder and its surroundings can also be caused by cervical pathologies. The bony landmarks of the cervical vertebrae and the paraspinal region are palpated for sensitivity. Positive results should suggest degenerative cervical pathologies, muscle strain/sprain, or cervical vertebral fractures. Then, palpation of the shoulder area is started. Sternoclavicular (SC) and acromioclavicular (AC) joints are palpated for dislocation and tenderness. The bicipital groove is then palpated for tenderness that may be caused by pathologies in the long head tendon of the biceps muscle or in the tendonous part of the subscapularis muscle.

Range of motion assessment should also be started from the cervical area. Flexion, extension, lateral flexion, and rotation movements of the cervical region are evaluated. The physician should be alert to the cervical vertebra pathologies in the case of the limitation of these movements [2]. Then, bilateral comparative ROM assessment of the shoulder joint should be started. The anterior flexion, external rotation, internal rotation, and only the shoulder abduction movements of the glenohumeral joint by manually determining the scapula of the shoulder joint by the physician should be assessed actively and passively. The normal range of motion of these movements are 180°, 80°, 90°, and 90°, respectively. While assessing the shoulder ROMs of the patients, the coexistence of active and passive ROM limitation, especially in external rotation, suggests adhesive capsulitis, while intact passive ROMs but limited active ROMs suggest large rotator cuff ruptures [2].

9.2.3 Specific Tests

Many specific tests have been described to detect the origin of pain or other symptoms around the shoulder girdle. These specific tests are used to evaluate the pathologies originating from the cervical region, subacromial impingement, and the pathologies of the muscles that make up the rotator cuff.

9.2.3.1 Cervical Spine Tests

Cervical pathologies should be considered in cases of pain in the form of radiculopathy radiating from the shoulder joint to the distal side and pain around the neck. Due to the high false-positive rates of electrodiagnostic and imaging techniques used for the evaluation of cervical pathologies, it is essential to assess the patient with specific tests [3]. Due to the high specificity of the tests specific to the cervical region, they are extremely useful for physicians to detect cervical pathologies [2].

1. *Spurling Test*

   The physician is positioned on the back of the patient in the sitting position and makes lateral flexion and rotation movements in the extended neck toward the side of the pain. Increased radicular pain and paresthesia are considered a positive finding [4].
2. *Axial Manual Traction*

   A traction of 15 kg is applied in the axial direction from the head of the patient lying in the supine position. A reduction in pain and other symptoms is considered a positive finding [5].
3. *Shoulder Abduction Test*

   The hand of the symptomatic upper limb of the patient who is positioned in the sitting position is abducted over the head. A decrease in symptoms is interpreted as a positive test [6].
4. *Valsalva Test*

   The patient positioned in the sitting position breathes deeply holds his/her breath and exhales for 2–3 s. An increase in radicular symptoms during the test is interpreted as a positive test [6].

9.2.3.2 Subacromial Impingement Tests

The tuberculum major, rotator cuff muscles, subacromial bursa are located inferior to the AC joint, and if contact occurs between these structures and the AC joint during shoulder movements, it is referred to as subacromial impingement [2]. Since the specific tests described for subacromial impingement have high specificity rates, it is possible to consider that the symptoms are not caused by subacromial impingement when these tests are negative [7].

1. *Neer Test*

   The examiner is positioned on the lateral side of the standing patient. The examiner stabilizes the patient's scapula with one hand while passively placing the affected upper limb in forward elevation. Anterior or lateral shoulder pain is interpreted as a positive test [8]. The Neer test alone has very high specificity for subacromial bursitis and partial rotator cuff rupture [7].
2. *Hawkins Test*

   The patient positioned in the standing position is asked to place the affected upper limb in 90° forward elevation with the arm internally rotated. Anterior and lateral shoulder pain is evaluated as a positive test [9].
3. *Painful Arch*

   The patient in standing position is asked to place the affected limb in 180° forward elevation. Anterior shoulder pain during active joint movement between 70° and 120° is evaluated as a positive test [10].

9.2.3.3 Specific Tests for Rotator Cuff

Shoulder pain is among the most common symptoms of musculoskeletal problems as it occurs in more than 60% of the entire population [11]. The most common causes of shoulder pain include subacromial impingement, rotator cuff rupture, and frozen shoulder [12]. The rotator cuff consists of subacromial muscle and tendon group surrounding the shoulder joint. The subscapularis, supraspinatus, infraspinatus, and teres minor muscle, which is the lowest reported tear rate (0.9%), are consisting of the rotator cuff [13].

A correct diagnosis is required to plan the most accurate treatment or rehabilitation for shoulder problems. A correct diagnosis depends on taking a good patient history, specific clinical tests, and imaging techniques [14]. In this section, specific clinical tests for each muscle that makes up the rotator cuff will be reviewed.

Supraspinatus Tests

1. *Jobe Test*

   The examiner stands behind the patient in standing position. After the affected shoulder is placed in 90° abduction, 30° horizontal adduction, and the arm in full internal rotation, force is applied to the limb from top to bottom (Fig. 9.1). Weakness or pain in the shoulder is considered a positive test result [8].
2. *Full Can Test*

   The only difference of this test, which is performed in the same way as the Jobe test, is that the arm, that is the humerus, is positioned in 45° is external rotation and force is applied to the limb from top to bottom (Fig. 9.2). Weakness or pain in the shoulder is reported as a positive test result [15]. The shoulder is positioned in the impingement position, which is in internal rotation, during the Jobe test, while the shoulder is positioned in the external rotation to reduce false-positive results caused by impingement during full can test [2].
3. *Drop Arm Test*

   The patient in standing position is asked to actively place the affect limb in 90° abduction. Afterwards, the patient is asked to place the same limb in neutral position by slowly lowering (Fig. 9.3). It is reported as a positive test result if the patient is unable to lower the limb slowly and the limb suddenly drops [2].

Infraspinatus Tests

1. *External Rotation Test*

   While the patient is standing, the upper limbs in bilateral neutral position are positioned in 90° elbow flexion and the forearms in 90° pronation (Fig. 9.4). While applying counter resistance, the patient is asked to externally rotate the shoulders. Muscle weakness or shoulder pain is reported as a positive test result [16].

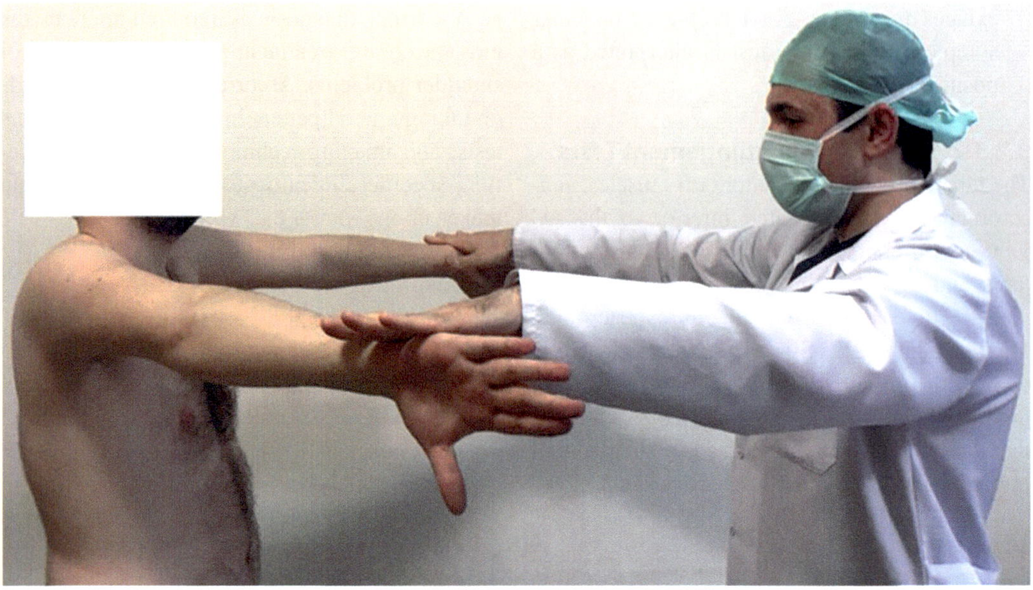

Fig. 9.1 Jobe test

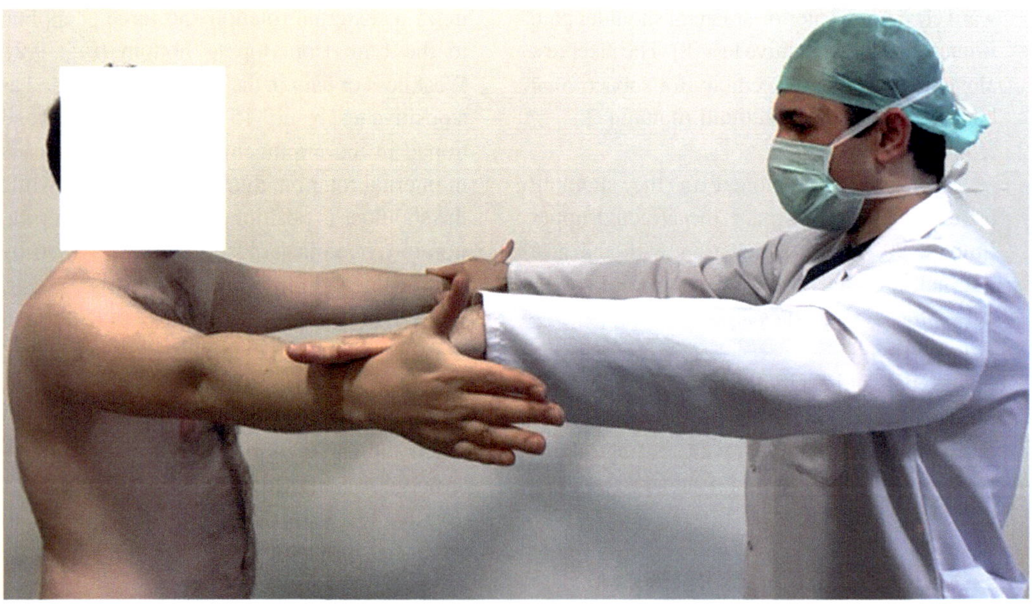

Fig. 9.2 Full can test

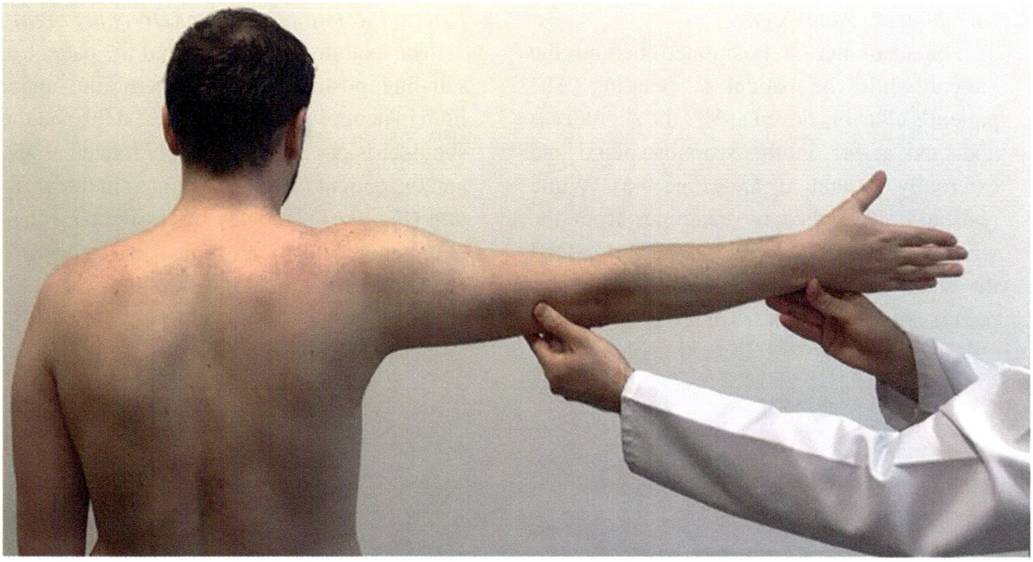

Fig. 9.3 Drop arm test

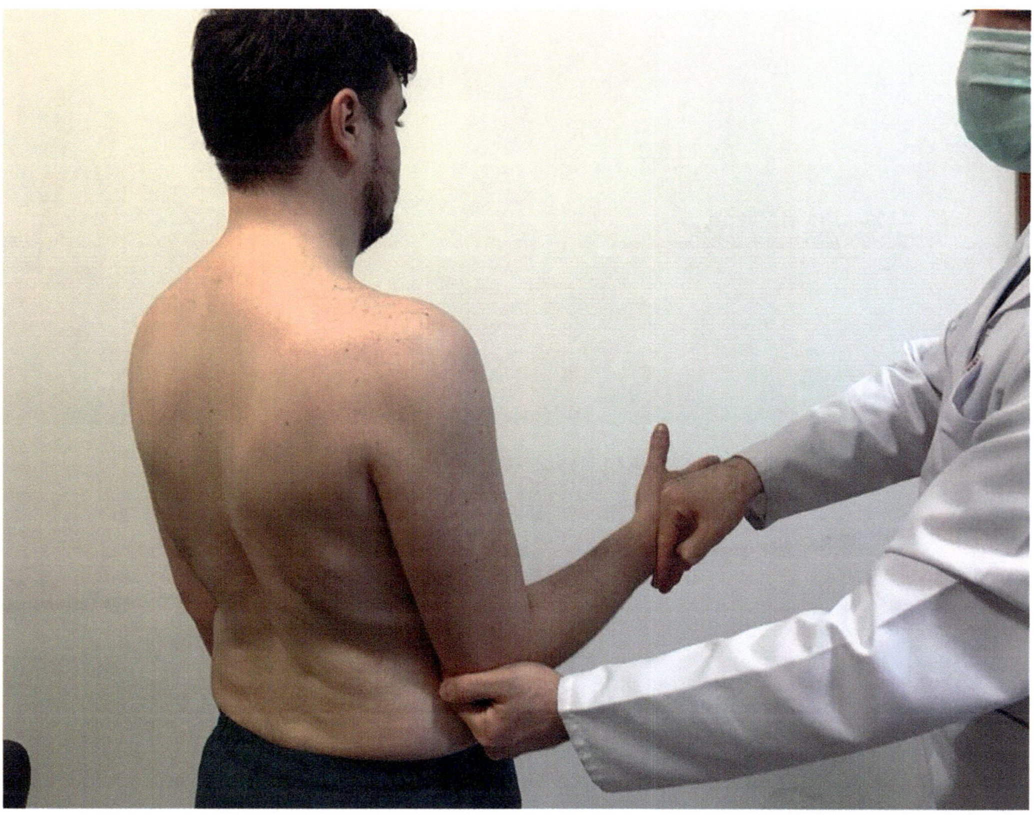

Fig. 9.4 External rotation test

2. *Hornblower (Patte) Sign*

   The examiner is positioned behind the patient while the patient is standing. The patient's elbow is flexed at 90°, the shoulder is abducted at 90° in the scapular plane and externally rotated at 45° (Fig. 9.5). While resistance is applied by examiner from the wrist dorsum, the patient is asked to extend the shoulder. Muscle weakness or shoulder pain is reported as a positive test result [17].

3. *External Rotation Lag Sign (Dropping Sign)*

   The examiner stands behind the patient in standing position. The elbow of the upper limb in neutral position is flexed at 90° and the shoulder is passively externally rotated at 45°, and the patient is asked to remain in this position (Fig. 9.6). The inability of the patient to maintain this position is reported as a positive test result [17].

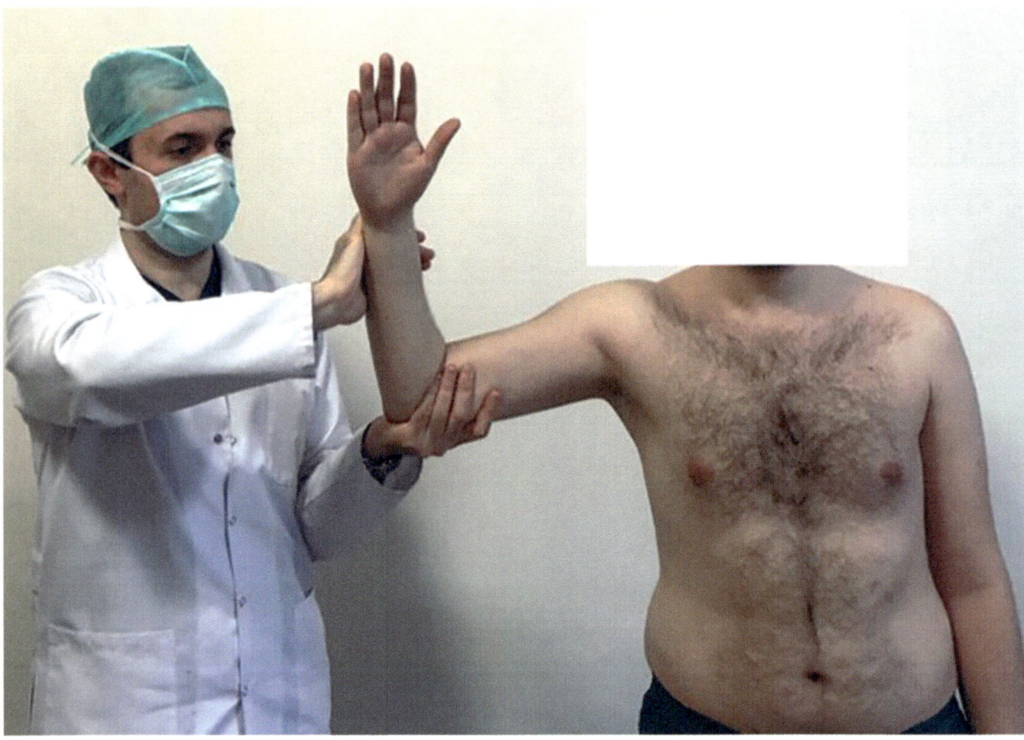

Fig. 9.5 Hornblower (Patte) test

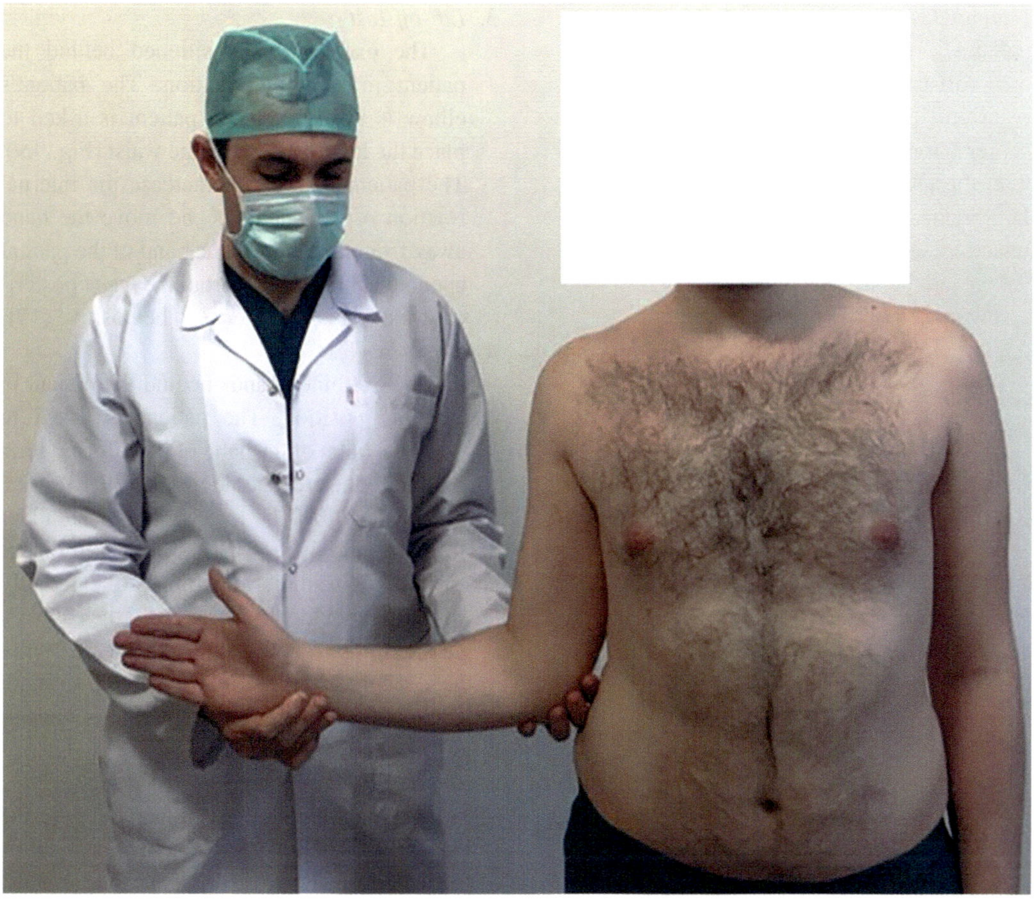

Fig. 9.6 External rotation lag test (Dropping Sign)

Subscapularis Tests

1. *Belly Press (Napoleon) Test*

   The examiner stands behind the patient in standing position. When the elbow of the patient's affected limb is in 90° flexion, the patient's palm is placed on the abdomen (Fig. 9.7). The patient is asked to press the palm on the abdomen with the elbow not going behind the body. If the patient can press the wrist on the abdomen without flexing it, the test is evaluated as negative. However, if the patient can press it on the abdomen by flexing the wrist, a positive test result is reported considering the deficiency of the subscapularis muscle and the compensatory effect of the posterior deltoid muscle [18]. Its superiority to the lift-off test is that it can be easily applied to patients with limited internal rotation such as frozen shoulder [2].

2. *Belly-off Test*

   The patient, patient's upper limb, and the examiner are positioned in the same way as the belly press test (Fig. 9.8). While the examiner supports the patient's elbow with one hand, s/he presses on the patient's wrist of the patient using the palm to bring the limb into full internal rotation with the other hand. When the examiner stops applying force with his/her hands, the patient's inability to continue the movement or flexing the wrist is considered as a positive test result [2].

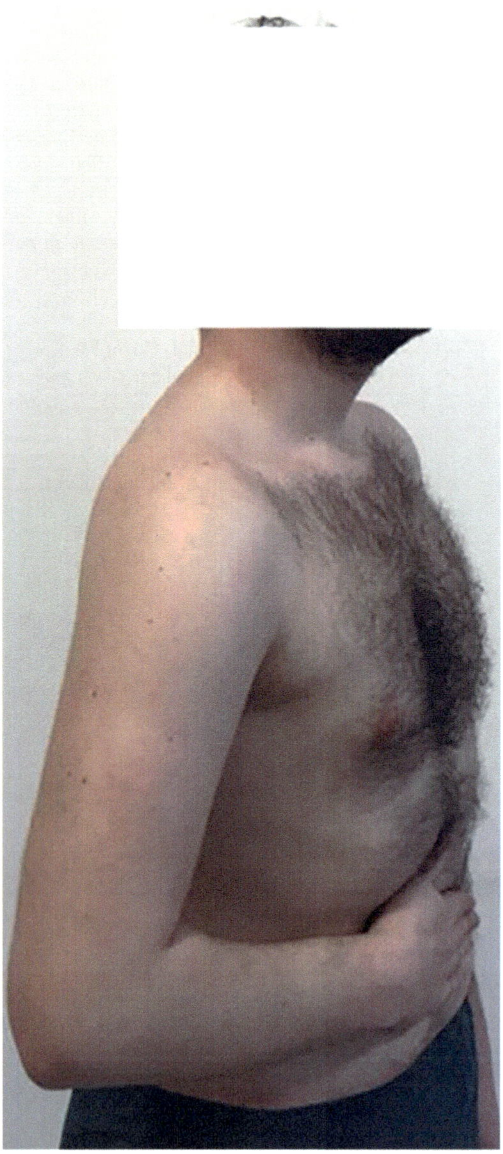

Fig. 9.7 Belly press (Napoleon) test

3. *Lift-off Test*

   The examiner is positioned behind the patient in standing position. The patient's elbow is flexed, and the patient is asked to place the hand dorsum on the waist (Fig. 9.9). The patient is asked to increase the internal rotation of the shoulder and move the hand away from the waist. If the hand of the patient cannot be moved away from the waist posteriorly, the test is considered to be positive [19].

4. *Bear Hug Test*

   The examiner stands behind the patient in standing position. The patient is asked to place the palm of the hand of the affected limb on the opposite shoulder (Fig. 9.10). The elbow should not be lower than the shoulder level and the fingers should be in full extension. The examiner attempts to externally rotate the shoulder by trying to lift the patient's affected hand from the opposite shoulder head. More than 20% loss of muscle strength compared to the contralateral side is considered a positive test result [18].

5. *Internal Rotation Lag Sign*

   The examiner is positioned behind the patient in standing position. The patient is asked to place the hand dorsum of the affected limb on the waist. The examiner places the patient's shoulder in full internal rotation with one hand holding the patients elbow and the other hand holding the patient's hand on the waist. Afterwards, the examiner only leaves the patient's hand and asks the patient to maintain the same position. If the patient cannot maintain the position, it is considered as a positive test result [20].

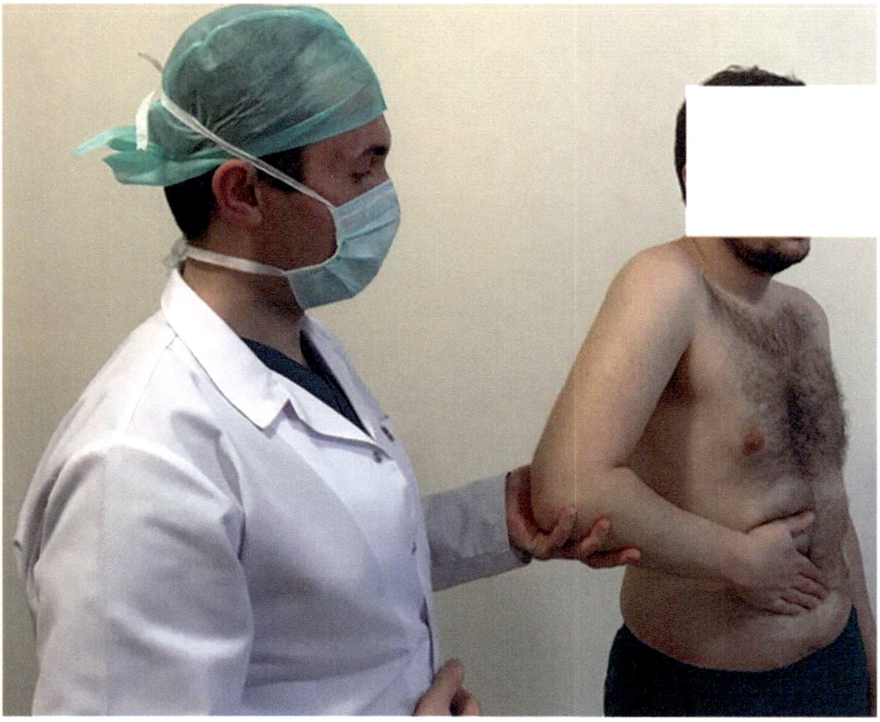

Fig. 9.8 Belly off test

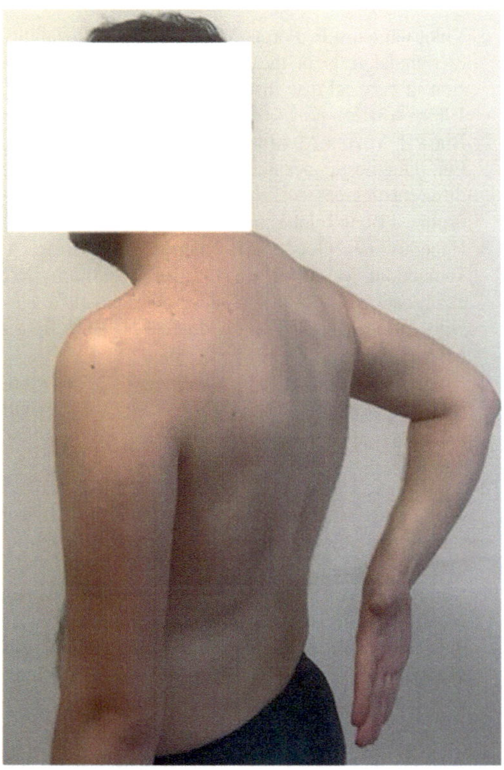

Fig. 9.9 Lift off test

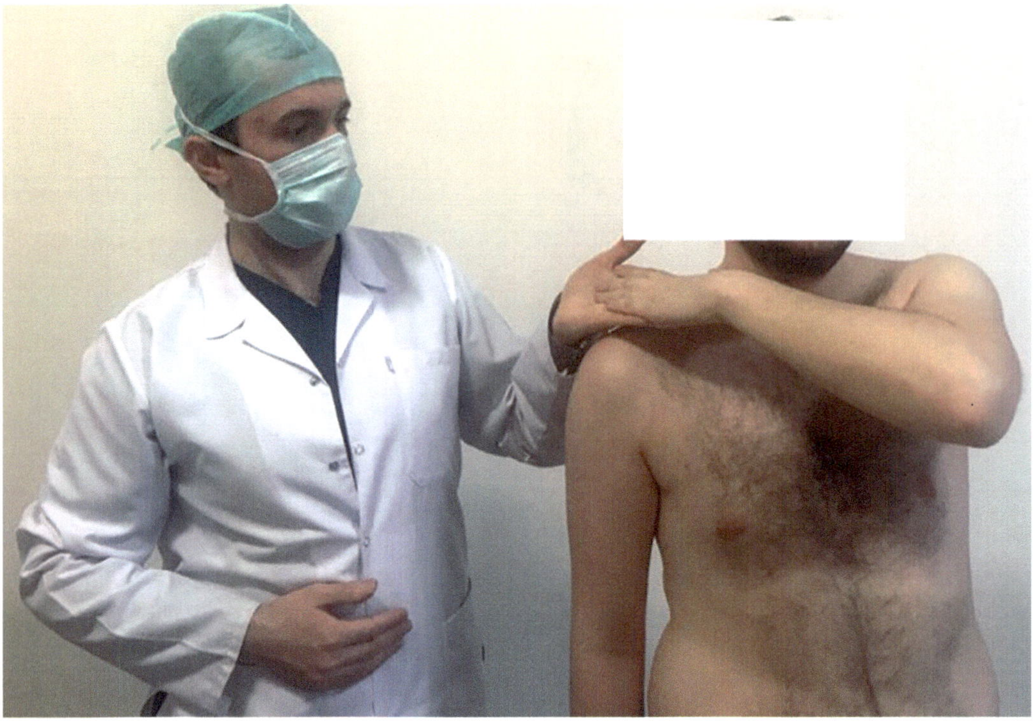

Fig. 9.10 Bear hug test

References

1. Gill TJ. Shoulder—diagnosis and management. In: MMS T, editor. DeLee, Drez and Miller's orthopaedic sports medicine, vol. 2. 5th ed. Amsterdam: Elsevier; 2020. p. 402–8. Available from: https://www.clinicalkey.com/#!/content/book/3-s2.0-B9780323544733000376.
2. Hippensteel KJ, Brophy R, Smith MV, Wright RW. A comprehensive review of physical examination tests of the cervical spine, scapula, and rotator cuff. J Am Acad Orthop Surg. 2019;27(11):385–94.
3. Rubinstein SM, Pool JJM, van Tulder MW, Riphagen II, de Vet HCW. A systematic review of the diagnostic accuracy of provocative tests of the neck for diagnosing cervical radiculopathy. Eur Spine J. 2007;16(3):307–19.
4. Tong HC, Haig AJ, Yamakawa K. The Spurling test and cervical radiculopathy. Spine (Phila Pa 1976). 2002;27(2):156–9.
5. Wainner RS, Fritz JM, Irrgang JJ, Boninger ML, Delitto A, Allison S. Reliability and diagnostic accuracy of the clinical examination and patient self-report measures for cervical radiculopathy. Spine (Phila Pa 1976). 2003;28(1):52–62.
6. Viikari-Juntura E, Porras M, Laasonen EM. Validity of clinical tests in the diagnosis of root compression in cervical disc disease. Spine (Phila Pa 1976). 1989;14(3):253–7.
7. Bin PH, Yokota A, Gill HS, El Rassi G, McFarland EG. Diagnostic accuracy of clinical tests for the different degrees of subacromial impingement syndrome. J Bone Joint Surg Am. 2005;87(7):1446–55.
8. Hegedus EJ, Goode A, Campbell S, Morin A, Tamaddoni M, Moorman CT 3rd, et al. Physical examination tests of the shoulder: a systematic review with meta-analysis of individual tests. Br J Sports Med. 2008;42(2):80–92. discussion 92
9. Hughes PC, Taylor NF, Green RA. Most clinical tests cannot accurately diagnose rotator cuff pathology: a systematic review. Aust J Physiother. 2008;54(3):159–70.
10. MacDonald PB, Clark P, Sutherland K. An analysis of the diagnostic accuracy of the Hawkins and Neer subacromial impingement signs. J Shoulder Elb Surg. 2000;9(4):299–301.
11. Miranda H, Viikari-Juntura E, Heistaro S, Heliövaara M, Riihimäki H. A population study on differences in the determinants of a specific shoulder disorder versus nonspecific shoulder pain without clinical findings. Am J Epidemiol. 2005;161(9):847–55.

12. Bron C, Franssen J, Wensing M, Oostendorp RAB. Interrater reliability of palpation of myofascial trigger points in three shoulder muscles. J Man Manip Ther. 2007;15(4):203–15.
13. Williams MD, Edwards TB, Walch G. Understanding the importance of the teres minor for shoulder function: functional anatomy and pathology. J Am Acad Orthop Surg. 2018;26(5):1–12.
14. Uysal Ö, Demirci S, Kara D, Yıldız Tİ, Sevinç C, Eraslan LS, et al. Tender point examination with palpation in different shoulder pathologies: a retrospective study. J Orthop Sci. 2021;S0949-2658(20):30383–3. https://doi.org/10.1016/j.jos.2020.12.015.
15. Itoi E, Minagawa H, Yamamoto N, Seki N, Abe H. Are pain location and physical examinations useful in locating a tear site of the rotator cuff? Am J Sports Med. 2006;34(2):256–64.
16. Beaudreuil J, Nizard R, Thomas T, Peyre M, Liotard JP, Boileau P, et al. Contribution of clinical tests to the diagnosis of rotator cuff disease: a systematic literature review. Joint Bone Spine. 2009;76(1):15–9.
17. Walch G, Boulahia A, Calderone S, Robinson AH. The "dropping" and "hornblower's" signs in evaluation of rotator-cuff tears. J Bone Joint Surg Br. 1998;80(4):624–8.
18. Barth JRH, Burkhart SS, De Beer JF. The bear-hug test: a new and sensitive test for diagnosing a subscapularis tear. Arthroscopy. 2006;22(10):1076–84.
19. Gerber C, Hersche O, Farron A. Isolated rupture of the subscapularis tendon. J Bone Joint Surg Am. 1996;78(7):1015–23.
20. Lin L, Yan H, Xiao J, Ao Y, Cui G. Internal rotation resistance test at abduction and external rotation: a new clinical test for diagnosing subscapularis lesions. Knee Surg Sports Traumatol Arthrosc. 2015;23(4):1247–52.

Radiological Assessment of the Shoulder

10

Üstün Aydıngöz

10.1 Introduction

Radiology has several roles in the assessment and management of shoulder disorders. First and foremost, it gives important clues about the underlying conditions or the exact cause of shoulder pain, which, frequently along with restricted range of motion, is the most common symptom of this joint. Second, radiological imaging provides guidance in selecting treatment choices as well as determining the type and extent of surgery to be performed in problems such as rotator cuff tears and bony Bankart lesions. Third, imaging is an important tool in the assessment of the glenohumeral and/or acromioclavicular joints following treatment or surgery. Last but not the least, radiology can also be used directly in the treatment of some conditions (e.g., ultrasonography-guided lavage of calcific deposits in calcific tendinitis) and for the intra-articular injection of anesthetics and/or corticosteroids.

10.2 Radiological Modalities and Techniques of Shoulder Imaging

Radiological armamentarium available for the shoulder joint includes radiography, arthrography, ultrasonography, computed tomography, and magnetic resonance imaging. Each of these modalities has special roles and techniques employed in the assessment of shoulder conditions ranging from traumatic to degenerative and inflammatory to neoplastic. Selection of the imaging modality to be used and determination of the technical aspects of the radiological examination depend on the nature of the underlying condition and/or presenting symptom of the patient. It is therefore of crucial importance to provide the radiologist with correct, relevant, and sufficient clinical information at the time of requesting these procedures. Sometimes, the radiologist might change or fine-tune the modality or technique in view of the patient characteristics or presumptive diagnosis.

10.2.1 Radiography

Similar to its use elsewhere in the musculoskeletal system, radiography is usually the first-line imaging tool for the shoulder. Overall, a set of internal and external rotation anteroposterior (AP) views and a scapular Y view is a commonly

Ü. Aydıngöz (✉)
Department of Radiology, Hacettepe University School of Medicine, Ankara, Turkey
e-mail: uaydingo@hacettepe.edu.tr

G. Huri et al. (eds.), *Fundamentals of the Shoulder*, https://doi.org/10.1007/978-3-030-94702-6_10

used combination [1]. Internal and external rotation AP views profile the middle and superior facets of the greater humeral tubercle, and thereby the infraspinatus and supraspinatus tendon insertions, respectively. Scapular Y view shows the glenoid joint surface *en face,* displaying the relationship of the humeral head with respect to the glenoid fossa. Axillary view, which also shows this relationship and profiles the glenohumeral joint, is not easy to obtain in the posttraumatic painful setting. If the arm cannot be sufficiently abducted for the classic axillary view, modifications to the axillary view, including the Velpeau view, can be made. Another variant of AP views is the Grashey view, which profiles the glenohumeral joint and is obtained while the patients rotate their body 45° with respect to the X-ray detector by distancing their contralateral shoulder away from the detector surface. I recommend replacing external rotation AP view with Grashey view in routine work-up of patients with shoulder pain as the latter profiles not only the glenohumeral joint but also the supraspinatus tendon insertion, and displays the critical shoulder angle (CSA), which was shown by multiple studies to be a relevant predictor for the development of either osteoarthritis or rotator cuff tears (Fig. 10.1) [2] (see Sect. 10.3.1).

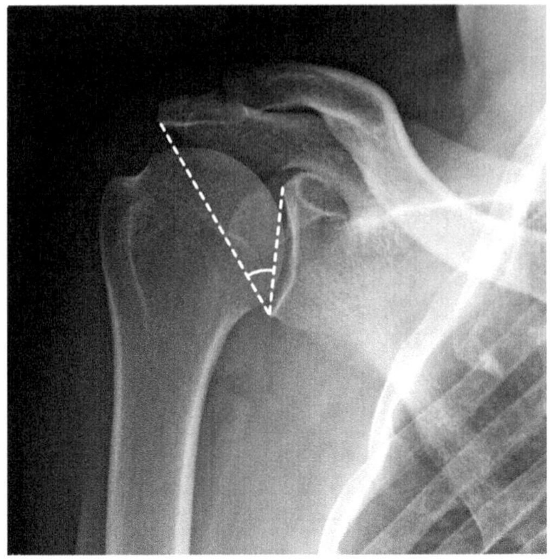

Fig. 10.1 Radiographic Grashey view is a 45° oblique anteroposterior projection that profiles the glenohumeral joint. Critical shoulder angle (CSA) forms between a line through the superior and inferior margins of the glenoid cavity and a line through the inferior glenoid margin and lateral acromial margin. Some studies suggest that a CSA >35° correlates with rotator cuff tears and a CSA <30° correlates with osteoarthritis of the glenohumeral joint [2]

10.2.2 Arthrography

X-ray arthrography has become practically obsolete in musculoskeletal imaging. However, imaging guidance for intraarticular or parafascial/peritendinous/intrabursal therapeutic injections entails the use of either X-rays or ultrasonography. For the wrist, it is also important to follow during arthroscopic fluoroscopy the path of the injectate through joint compartments during stress maneuvers. For the shoulder joint, however, MR- and CT-arthrography has essentially replaced conventional X-ray arthrography.

10.2.3 Ultrasonography (US)

Ultrasonography is widely used in the assessment of rotator cuff tendons and long head of the biceps tendon (LHBT). An experienced operator is essential for the best use of US in the shoulder. Awareness of the imaging and positioning pitfalls, use of an appropriate transducer and rotatable chairs for the examiner and the patient, employment of specific positions of the patient's arm and forearm for different structures, and examination of the contralateral side as needed are among the key points in shoulder US. Major downsides of shoulder US are its limited ability to show intraarticular structures such as the glenoid labrum, the glenohumeral ligaments and the joint cartilage, and its inability to display bone marrow.

10.2.4 Computed Tomography (CT)

Computed tomography, which provides detailed information about the bony structures of the shoulder in a few minutes, is mostly used in the shoulder joint to characterize and classify acute occult or complex fractures, and to identify a bony Bankart (or reverse Bankart) lesion as well as to assess the glenoid bony stock [3].

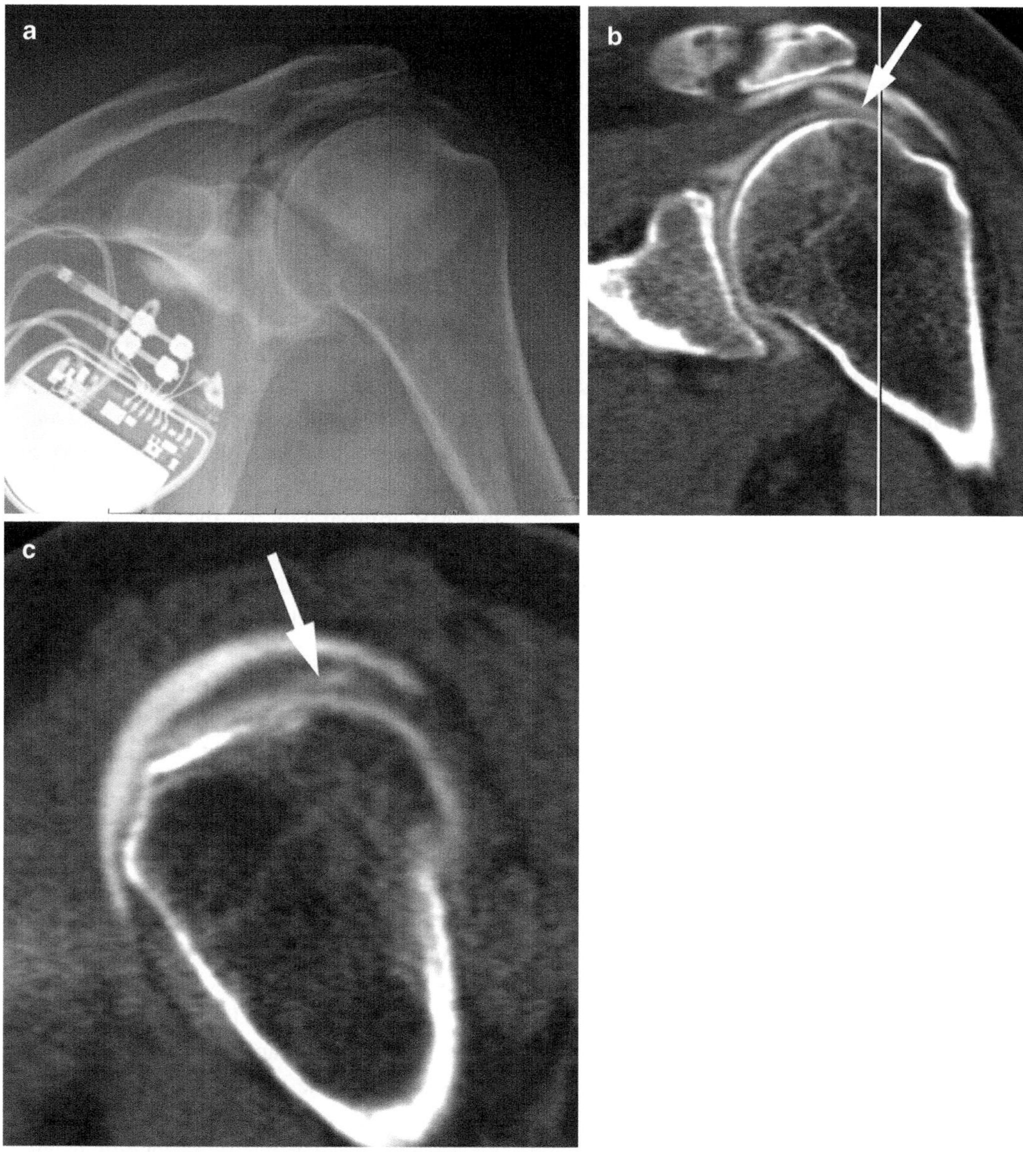

Fig. 10.2 CT-arthrography in a 59-year-old man, who could not undergo an MRI examination because of a cardiac defibrillator (**a**, fluoroscopy image obtained during intraarticular injection). Coronal oblique (**b**) and sagittal oblique (**c**) reformatted CT-arthrography images show a delamination tear at the supraspinatus tendon (*arrows*) (perpendicular line on **b** refers to the section on **c**)

The latter is facilitated by the possibility of excluding the humeral head from three-dimensional CT images [1]. When radiography is inconclusive for the assessment of shoulder arthroplasty complications, CT with metal artifact reduction software can show conditions such as loosening, scapular notching, and heterotopic ossification [4].

10.2.4.1 CT-Arthrography

After MR-arthrography began to be extensively used in the early years of this century, CT arthrography gained popularity, following its initial use in patients with MR-incompatible devices and claustrophobia (Fig. 10.2). CT arthrography might indeed be performed as a salvage procedure when what started as an MR-arthrography procedure needs to be modified due to unanticipated claustrophobia or significant patient motion. In fact, cartilage lesions in the glenohumeral joint are detected on CT arthrography even better than on direct MR-arthrography [5]. CT-arthrography is also accurate in the detection of full-thickness and articular surface partial tears of the supraspinatus and infraspinatus, shows similar sensitivities and specificities with 3 T MR-arthrography in the assessment of lesions of the proximal LHBT, effectively detects superior labrum anterior-to-posterior (SLAP) lesions, and distinguishes between normal variants affecting the anterosuperior labrum and labral-bicipital complex [6].

10.2.5 Magnetic Resonance Imaging (MRI)

Magnetic resonance imaging is by far the most common cross-sectional imaging method used for the evaluation of shoulder joint. It superbly depicts all structures of the shoulder region relevant to the practice of clinicians, be it an orthopedic surgeon, a rheumatologist, or a physical therapy and rehabilitation expert. Its ability to exquisitely show osseous structures (in conditions like bone contusions, edema-like changes, and focal or infiltrative lesions) as well as soft tissues such as tendons, capsulolabral, and ligamentous structures makes MRI an excellent imaging tool for the assessment of the musculoskeletal system in general and the shoulder joint in particular. A novel imaging technique, called zero echo-time (ZTE) MRI, allows obtaining radiograph- or CT-like images during an MRI examination, making it easier in some cases to identify bony Bankart lesions and assess the glenoid track as well as bone stock [7, 8].

Many caveats related to MRI involve variant anatomy (see Sects. 10.3.1 and 10.4.1), imaging pitfalls [9, 10], and the possibility that some shoulder conditions might be encountered in asymptomatic persons (see Sect. 10.3.1). In other words, what appears to be a positive finding on MRI is not necessarily pathological or clinically significant.

The best use of MRI in the shoulder joint necessitates the use of a dedicated surface coil, optimal positioning of the arm, and forearm (usually lying alongside the patient's body with the palm of the hand facing either the patient's body or the ceiling—and not the examination table) with liberal padding to ensure avoidance of motion artifacts, and a tailored examination with imaging planes that best depict normal structures and their pathological conditions. Internal rotation of the shoulder during MRI results in anterior capsulolabral and ligamentous redundancy and can conceal tears at this region. Adding a sequence obtained in ABduction and External Rotation (ABER) would tension this region during MR-arthrography of the shoulder, which already helps overcome such redundancy with intraarticular contrast distention of the glenohumeral joint (see Sect. 10.2.5.1).

10.2.5.1 MR-Arthrography

MR-arthrography can be performed in two ways: indirect or direct. Indirect MR-arthrography employs the use of an intravenously administered contrast material, which diffuses into the glenohumeral joint in a few minutes during which the patients move their shoulder joint to facilitate the diffusion; magnetic resonance imaging then follows. Although it entails no need for the set-up of a fluoroscopy- or US-guided injection into the glenohumeral joint, indirect MR-arthrography suffers from the lack of distention of the joint in addition to the contrast enhancement of surrounding structures, which might mask some of the conditions investigated. Therefore, direct MR-arthrography, in which an injectate usually containing a gadolinium-based contrast agent is injected into the shoulder joint under imaging guidance, is much more commonly used.

Direct MR-arthrography is the best imaging method for the overall assessment of shoulder joint including the rotator cuff tendons, capsulolabral and ligamentous structures, LHBT, and articular cartilage. Imaging sequences built into direct MR-arthrography also allow visualization of the muscles and bone marrow. Although recent studies have demonstrated gadolinium

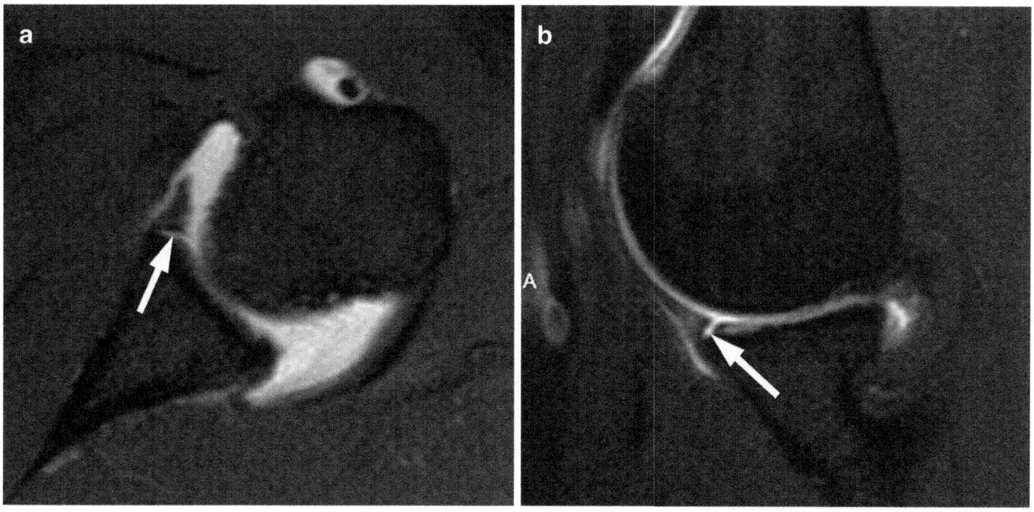

Fig. 10.3 Perthes lesion (*arrows*) denoting a non-displaced tear of the anteroinferior labrum on a transverse fat-saturated T1-weighted MR-arthrography image (**a**) in a 29-year-old man is better depicted on the oblique fat-saturated T1-weighted image (**b**) obtained at the abduction and external rotation (ABER) position

deposition in the brain in patients with normal renal function following intravenous gadolinium-based contrast agent administration, investigators of a recent study found no MRI evidence of intracranial gadolinium deposition following MR-arthrography [11].

From a technical standpoint, MR-arthrography of the shoulder can be made using an anterior or preferably a posterior approach. Anterior approach runs the risk of contaminating the subacromial-subdeltoid bursa by way of inadvertently injecting into the subcoracoid bursa (these two bursae are connected in some patients): The subacromial-subdeltoid bursa could have been otherwise filled by a full-thickness supraspinatus and/or infraspinatus tendon tear and anterior injection might compromise the diagnosis of such a tear, especially when it is small.

Addition of the ABER sequence to MR-arthrography of the shoulder helps depict better some glenohumeral joint structures and surrounding tissues under a position relevant to several pathologic conditions. The abduction and external rotation position during this sequence tensions the anteroinferior glenohumeral ligament and labrum and releases tension on the supraspinatus and infraspinatus relative to the normal coronal view obtained with the arm in adduction. Among the lesions better shown in this position are subtle partial-thickness articular sided tears of the supraspinatus and infraspinatus tendons (Fig. 10.3), subtle tears of the anteroinferior portion of the glenoid labrum (such as the Perthes lesion) and anterior band of the inferior glenohumeral band [12].

10.3 Imaging in Rotator Cuff Abnormalities

Rotator cuff muscles and tendons are best displayed on MRI or MR-arthrography. Radiographs are a useful adjunct in this regard and should be the first-line imaging tool. Ultrasonography is also being increasingly used in the assessment—and sometimes during treatment—of rotator cuff abnormalities.

10.3.1 Rotator Cuff Tendonosis and Tendon Tears

Rotator cuff disorders are a major source of shoulder pain. Rotator cuff tendinopathy is an umbrella term that encompasses tendon tears and tendonosis (the latter covers both tendon inflammation and degeneration). Tendonosis on MRI refers to the mild signal increase within a tendon which does not amount to fluid intensity, which usually means a tear. Although US has been validated for the assessment of rotator cuff tears with reported sensitivities and specificities that rival that of conventional MRI, it is with MRI that a more comprehensive evaluation of the shoulder, including more accurate appraisal of articular cartilage and labroligamentous structures as well as depiction of the bones, is possible [13]. A partial-thickness tear involves either the substance or the articular or, less commonly, bursal side of a rotator cuff tendon. Considering an expected thickness of 10–12 mm for a tendon, small and medium tears are 3 mm deep and 3–6 mm deep, respectively, both involving less than 50% of tendon thickness [13]. Large partial-thickness tears, still non-communicating as the small and medium tears, involve greater than 50% of tendon thickness. Delaminating tears are longitudinally oriented and involve the tendon substance, sometimes reaching the articular or bursal surface at one (commonly the distal) end (Fig. 10.4). Easily identified on MRI, such tears are difficult to visualize at arthroscopy. Although full-thickness tears by definition communicate between the articular and bursal sides of the cuff, a small percentage of them fail to show the characteristic fluid or gadolinium-based contrast filled "communicating gap" appearance on MRI or MR-arthrography, respectively. Instead, there might be a heterogeneous T2 signal, likely due to inflammation with subsequent granulation tissue formation within the region of tear, volume averaging pitfall of small tears, or coaptation of torn tendon edges [13]. The best practice on reporting MRI is to describe the location and three-dimensional extent of all rotator cuff tendon tears (partial or full thickness) (Fig. 10.5). As important full-thickness rotator cuff tear descriptors on MRI, tear size, degree of tendon retraction, and degrees of atrophy and fatty infiltration of muscles help guide surgical management [13]. Intramuscular "sentinel cysts" and humeral head cysts at or near the footprints of rotator cuff tendons can help in the MRI diagnosis or suggestion of rotator cuff tears in cases with equivocal findings. Anterior humeral cysts at the supraspinatus and subscapularis tendon insertions show a high correlation with rotator cuff disorders, whereas cysts at the infraspinatus tendon insertion and posterolateral "bare area" at the anatomic neck have little such correlation, are asymptomatic and likely related to vascular intrusions [14].

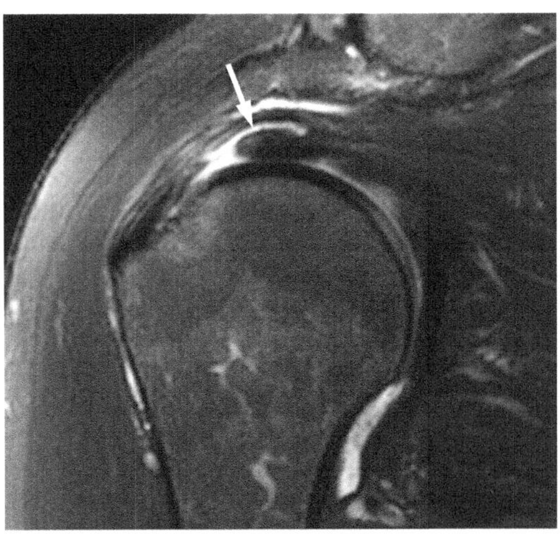

Fig. 10.4 Coronal oblique fat-saturated T2-weighted MR image shows a Partial thickness Articular-sided tear with INTratendinous extension (i.e., delamination, *arrow;* the so-called PAINT lesion) in the background of infraspinatus tendonosis

MRI findings of rotator cuff tendonosis and tears do not necessarily correspond to shoulder problems. In other words, asymptomatic persons can have rotator cuff tendonosis and tears on MRI. Up to 46% of entire population and nearly 40% of persons over 60 years of age may have rotator cuff tears [15]. Up to 40% of elite overhead athletes have partial or full thickness rotator cuff tears on MRI with no reported problems [16].

Subscapularis tendon evaluation on routine MRI is challenging. Although MR-arthrography is quite sensitive and specific for subscapularis tears, routine MRI has a lower sensitivity [13]. A four-step approach to subscapularis evaluation may improve the sensitivity of MRI: Start with transverse fluid-sensitive images, evaluate LHBT for evidence of subluxation, assess subscapularis fatty infiltration or atrophy on T1-weighted

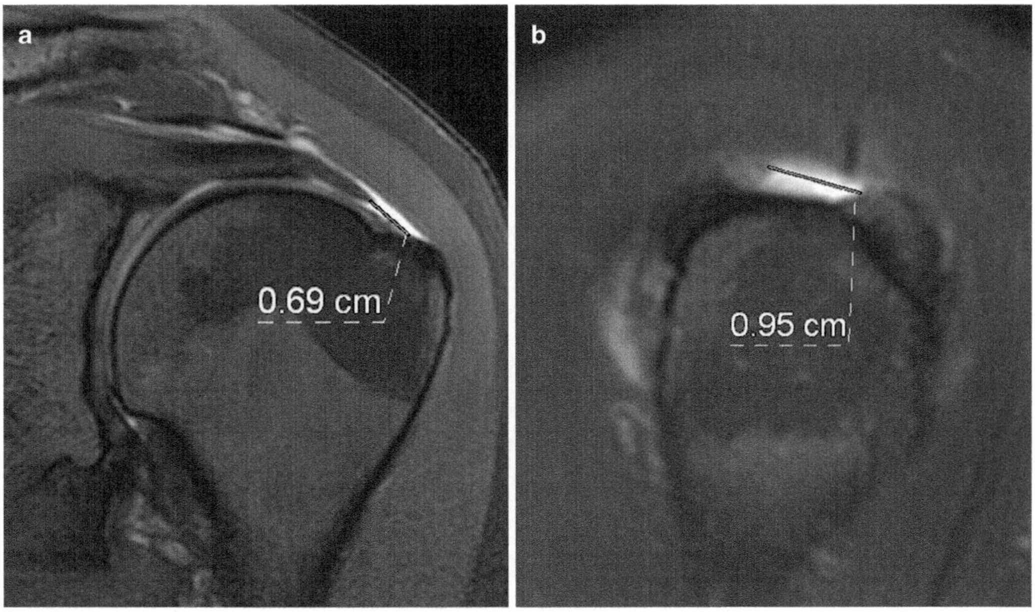

Fig. 10.5 Coronal oblique (**a**) and sagittal oblique (**b**) fat-saturated T2-weighted MR images show a full-thickness tear of the supraspinatus tendon with size measurements

images, and look for a tear on fluid-sensitive sagittal oblique images [17].

MRI plays an important role in identifying the "novel lesion" of the infraspinatus, which is an isolated atraumatic rupture of the interstitial infraspinatus (Fig. 10.6) [18]. Visualization of such tears is challenging at arthroscopy, they may progress to severe atrophy, and MRI guidance is essential in surgical intervention.

Assessment of the muscle stock associated with rotator cuff tears is an important contribution of radiological imaging. Goutallier et al.'s staging, which was described on CT and later modified by Fuchs et al. for MRI, and the tangent method on MRI are routinely used in the evaluation of rotator cuff muscle quality and quantity, respectively [19, 20].

Degenerative rotator cuff disease can manifest itself on radiographs with entheseal changes at the tendon insertions such as osteopenia, bony sclerosis, surface irregularity, and cyst formation [14]. Acromion morphology based on T1-weighted sagittal oblique MRI may be inferiorly flat, curved (most common), hooked (associated with increased incidence of impingement), or convex (upturned). Reporting on MRI of an os acromiale is important; when unreported and if unstable, this unfused ossification center, which normally appears at around 15 years of age and fuses by the age of 25 years [9], might compromise rotator cuff surgery. Critical shoulder angle (CSA) integrates two risk factors of rotator cuff tears by quantifying the extent of acromial coverage of the humerus and the inclination of the glenoid [2]. As a simple and highly reproducible parameter, CSA is a promising—yet controversial—tool for discriminating between rotator cuff tears and osteoarthritis (Fig. 10.1). Pre- and postoperative CSA measurements also appear to be useful for assessing the retear risk [2]. Unfortunately, routine MRI (without an isotropic 3D gradient echo or the recently introduced ZTE sequence) is not convenient for the measurement of CSA, which is readily displayed on an appropriately obtained radiographic Grashey view (see Sect. 10.2.1).

Postoperative evaluation of the shoulder for a retear is mostly made with MRI or, better, MR-arthrography. A 1.5 Tesla (T) system should be preferred over 3 T MRI equipment as artifacts created by surgical implants are more problem-

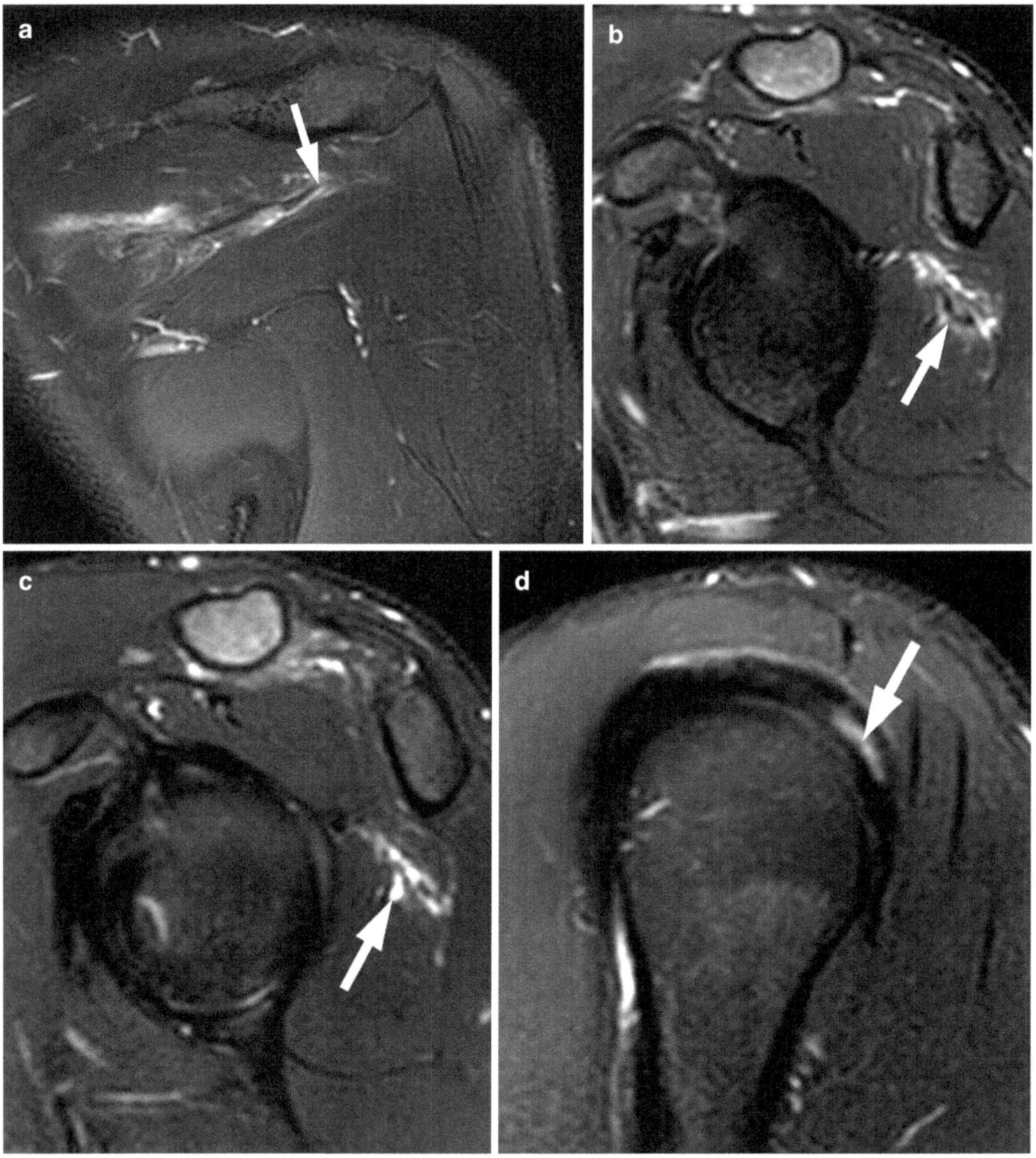

Fig. 10.6 Coronal oblique (**a**) and sagittal oblique (**b–d**) fat-saturated T2-weighted MR images show a retracted (*arrow*, **a**) delaminated intrasubstance (*arrows*, **c**, **d**) tear of the infraspinatus tendon (the "novel" lesion), associated with myotendinous junction edema (**a**, **b**) in a 24-year-old man who felt sudden pain during weightlifting. Such tears might not be seen on arthroscopy

atic on the latter. Metal artifact reduction sequences need to be employed. CT-arthrography or US may need to be substituted for MRI in the postoperative setting. Repair does not provide a "watertight" cuff and subacromial–subdeltoid bursal fluid, which might even communicate into this bursa from the glenohumeral joint during MR-arthrography, is not necessarily abnormal [21]. Repaired tendon can appear heterogeneous and thin on MRI (Fig. 10.7). Edema-like signal can be seen at the humeral head. Fluid-filled defect within the tendon suggests "retear."

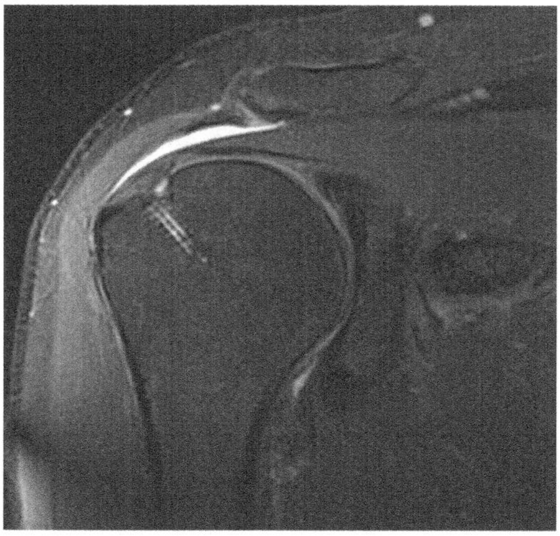

Fig. 10.7 Coronal oblique fat-saturated T2-weighted MR image shows a repaired supraspinatus tendon without retear. Repaired tendon can appear heterogeneous and thin on MRI

10.3.2 Rotator Cuff Tear Arthropathy

Massive tears of the rotator cuff can result in rotator cuff tear arthropathy, which has characteristic imaging findings on radiography and MRI. The acromiohumeral interval is narrowed (<7 mm) with superior migration of the humeral head, proximal humerus is "femoralized" because of the repetitive hitting of the greater tubercle to the acromion in the absence of a buffering rotator cuff, and "acetabularization" (rounding) of the coracoacromial arch occurs as an adaptation (Fig. 10.8) [22]. The resulting secondary osteoarthritis of the glenohumeral joint, with a more superiorly conspicuous joint cartilage loss, is usually different from primary shoulder osteoarthritis.

Although technically challenging to repair, massive rotator cuff tears are not necessarily irreparable. However, static superior migration of the humeral head, along with a narrowed or absent acromiohumeral interval and fatty infiltration affecting 50% or more of the rotator cuff muscles, is among the signs of irreparability and currently rotator cuff tear arthropathy is a major indication for reverse total shoulder arthroplasty [22].

With a massive rotator cuff tear, glenohumeral articular and subacromial-subdeltoid bursal fluid may gain access to the acromiohumeral (AC) joint through an eroded inferior AC joint capsule. This fluid can then traverse the AC joint to present as a cystic mass overlying the AC joint under the skin. This finding on MRI is called "the geyser sign" and suggests rotator cuff tear arthropathy.

10.3.3 Milwaukee Shoulder

The very rare unique combination of glenohumeral joint effusion with hydroxyapatite crystal deposition, rotator cuff tear(s), and rapidly destructive arthropathy is known as "Milwaukee shoulder" (Fig. 10.9). This condition is classically seen in the dominant hand side over 60 years of age and predominantly females. When bilateral, this condition is almost always more advanced on the dominant side. Although shoulder involvement is more common, the knees and the hips can also be affected [23, 24].

10.3.4 Adhesive Capsulitis (Frozen Shoulder)

Although adhesive capsulitis is typically a clinical diagnosis made on the basis of patient's history and physical examination, rotator cuff abnormalities and osteoarthritis can sometimes cause similar symptoms and signs, therefore necessitating imaging to rule it in or out. On MRI, findings suggestive of adhesive capsulitis include pericapsular fibroinflammatory changes with thickening of the joint capsule at the axillary pouch (≥4 mm) or the rotator interval, along with thickening of the coracohumeral ligament (≥4 mm on MR arthrography) and obliteration of the subcoracoid fat triangle (Fig. 10.10) [25, 26]. According to a recent study, capsular thickness on MRI in the humeral portion of the axillary recess correlates with pain intensity and is greatest at the first of four clinical stages; obliteration of the subcoracoid fat triangle is also more frequent in the earlier stages [27].

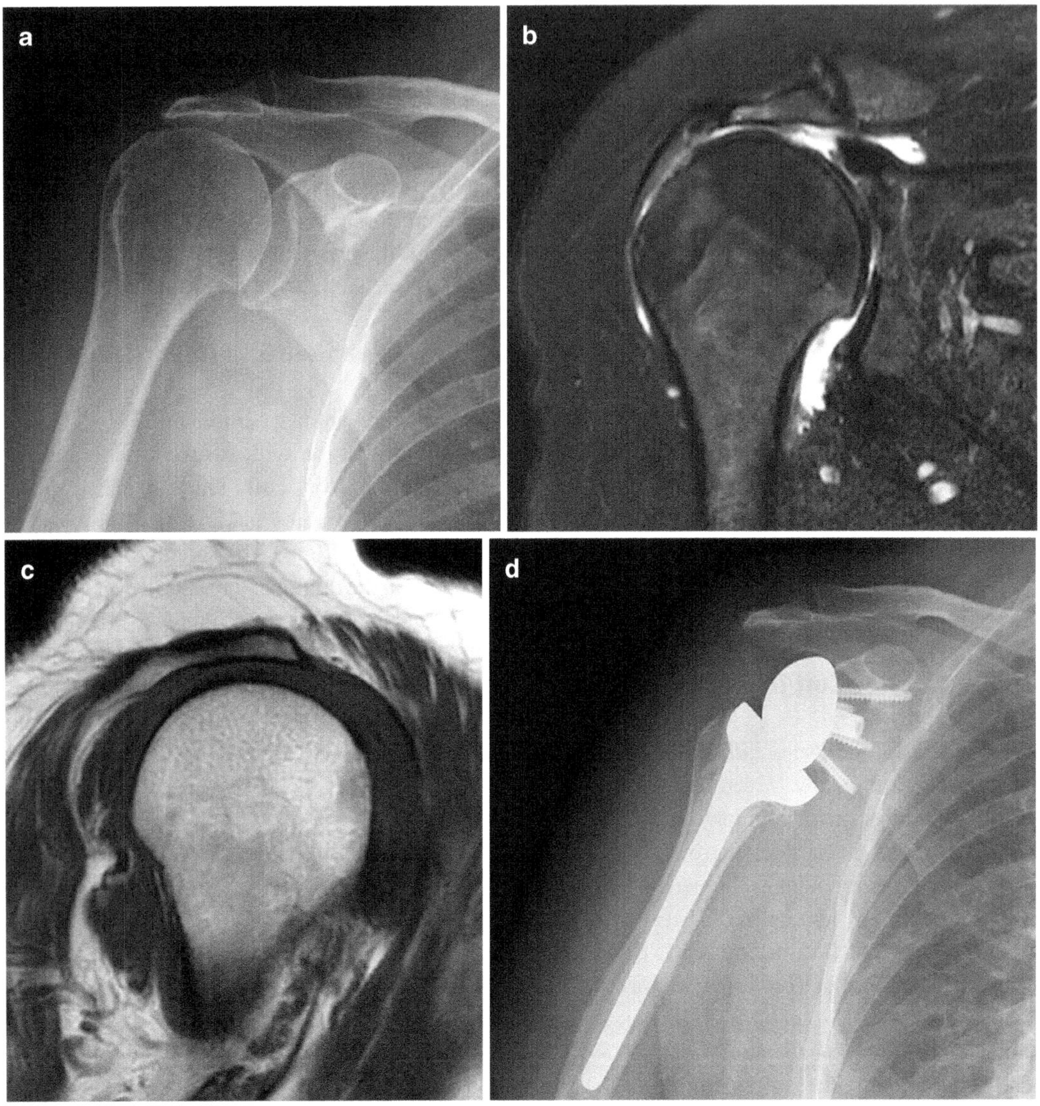

Fig. 10.8 Anteroposterior radiograph in external rotation (**a**), coronal oblique fat-saturated T2-weighted (**b**) and sagittal oblique T1-weighted (**c**) MR images of a 77-year-old woman show a decreased acromiohumeral distance due to a chronic massive rotator cuff tear, "femoralization" of the humeral head (**a**, **b**), and "acetabularization" of the coracoacromial arch (**c**), findings characteristic of rotator cuff tear arthropathy. Later, the patient underwent reverse total shoulder arthroplasty (**d**)

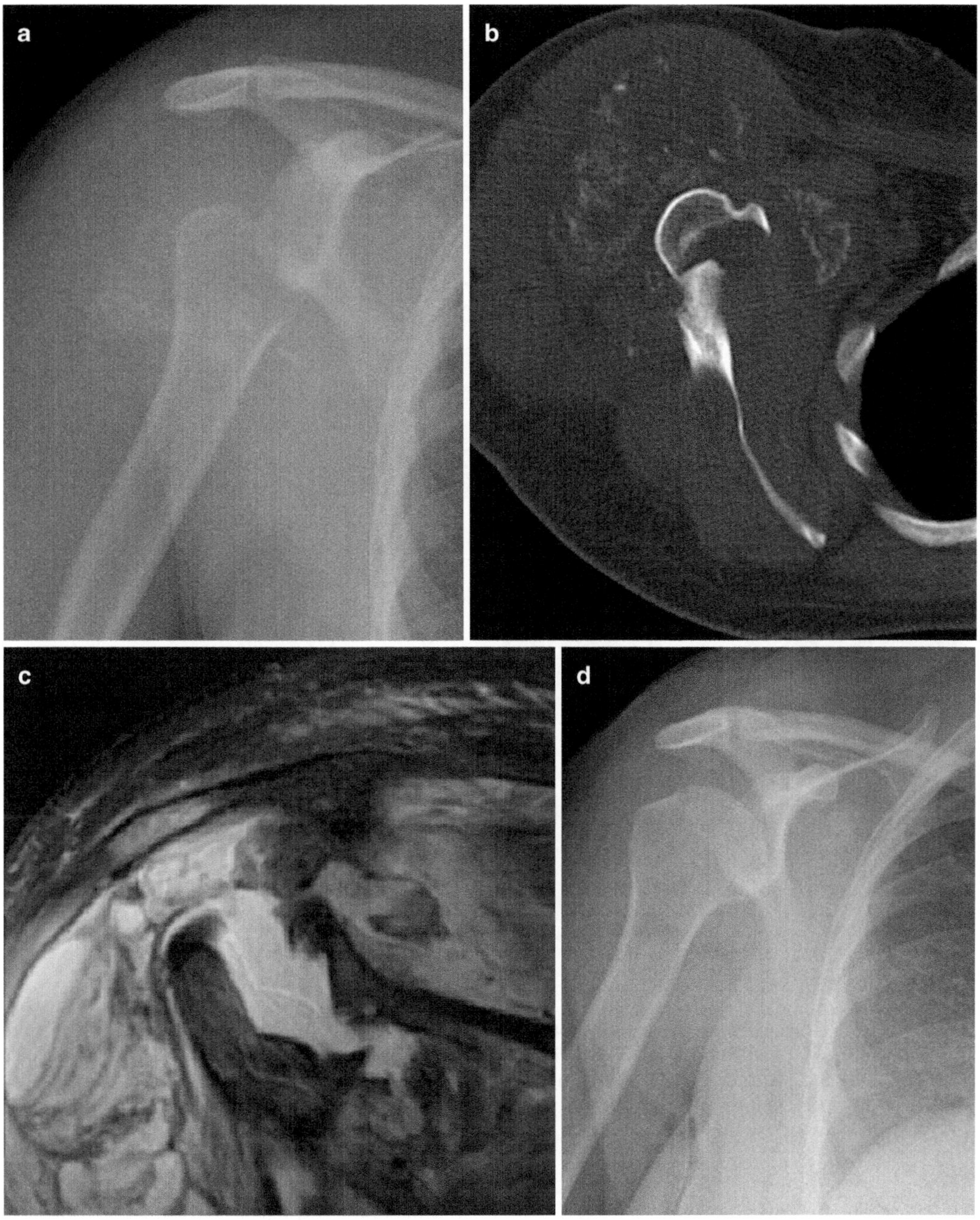

Fig. 10.9 Anteroposterior (AP) radiograph (**a**), transverse CT (**b**), and coronal (**c**) fat-saturated T2-weighted MR images show destruction of the glenohumeral joint and a large effusion with calcifications (**a**, **b**) in a 53-year-old woman, who presented with right shoulder pain and swelling. Active range of motion was severely limited. She had had no recent trauma to the shoulder area. Joint aspirate revealed hydroxyapatite crystals. An AP chest radiograph from 7 months earlier (**d**) was unremarkable for this region. Current condition of the patient is therefore consistent with a "Milwaukee shoulder." (Case courtesy of Zeynep Maraş Özdemir, MD, Malatya, Turkey)

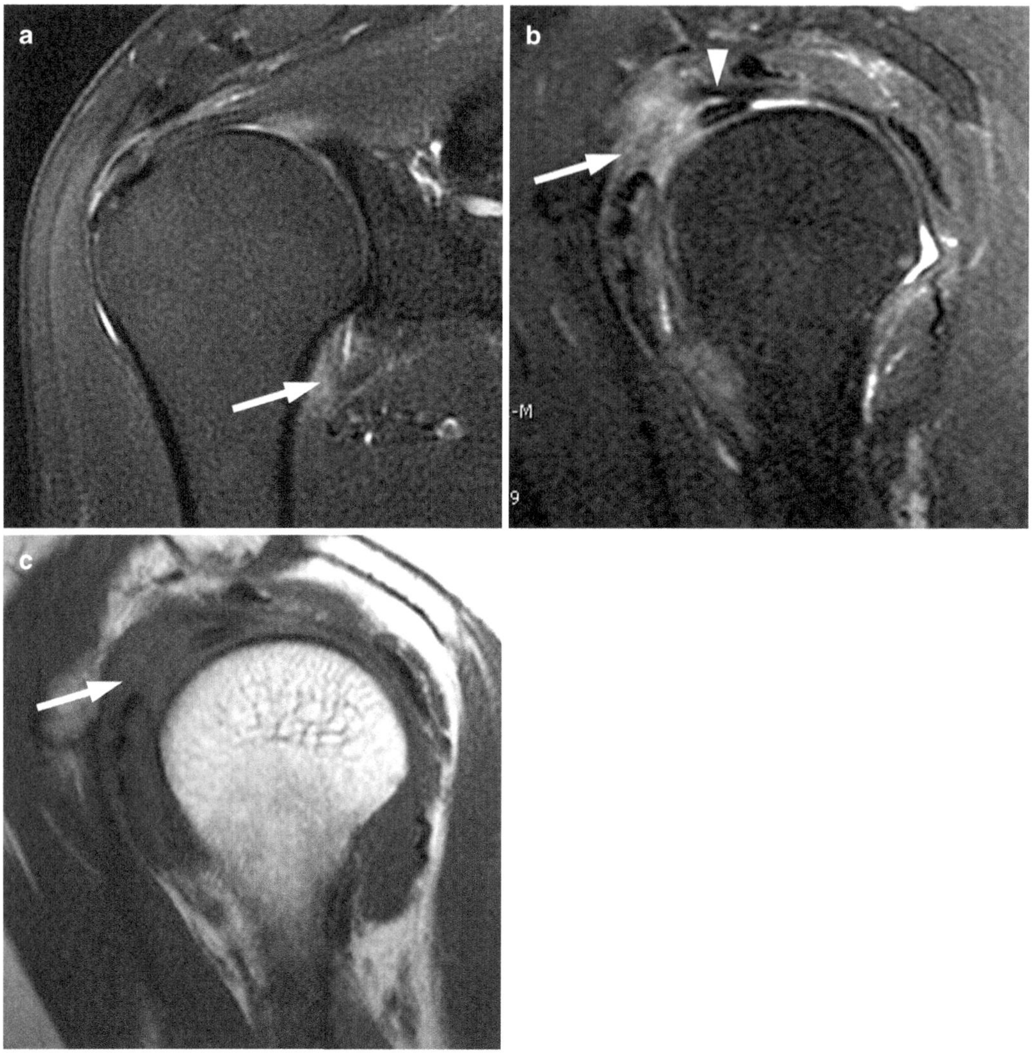

Fig. 10.10 Coronal oblique (**a**) and sagittal oblique (**b**) fat-saturated T2-weighted, and T1-weighted (**c**) MR images show thickening and edema of the joint capsule (inferior glenohumeral ligament) at the axillary pouch (**a**, *arrow*), and obliteration of the subcoracoid fat triangle by fibroinflammatory changes (*arrows*, **b**, **c**). The coracohumeral ligament is also mildly thickened (*arrowhead*, **b**). All of these findings are consistent with adhesive capsulitis

10.3.5 Calcific Tendinitis

Calcific tendinitis is characterized by the deposition of hydroxyapatite crystals within the tendons, most commonly of the rotator cuff. Patients are usually middle aged (between 30 and 60 years). Although this condition is usually self-healing with spontaneous resolution of the calcific deposits and surrounding inflammation over time, it may cause chronic moderate-to-marked pain with functional disability.

Magnetic resonance imaging exquisitely shows calcific deposits surrounded with soft tissue or bone inflammation. It is important to correlate the findings on MRI with radiography as small deposits might be subtle. For small deposits at the distal aspects of rotator cuff tendons, T1-weighted sagittal images are especially useful as tendons at this location are otherwise susceptible to the so-called magic angle phenomenon and would not usually display the hypointensity that could have masked such deposits. The novel

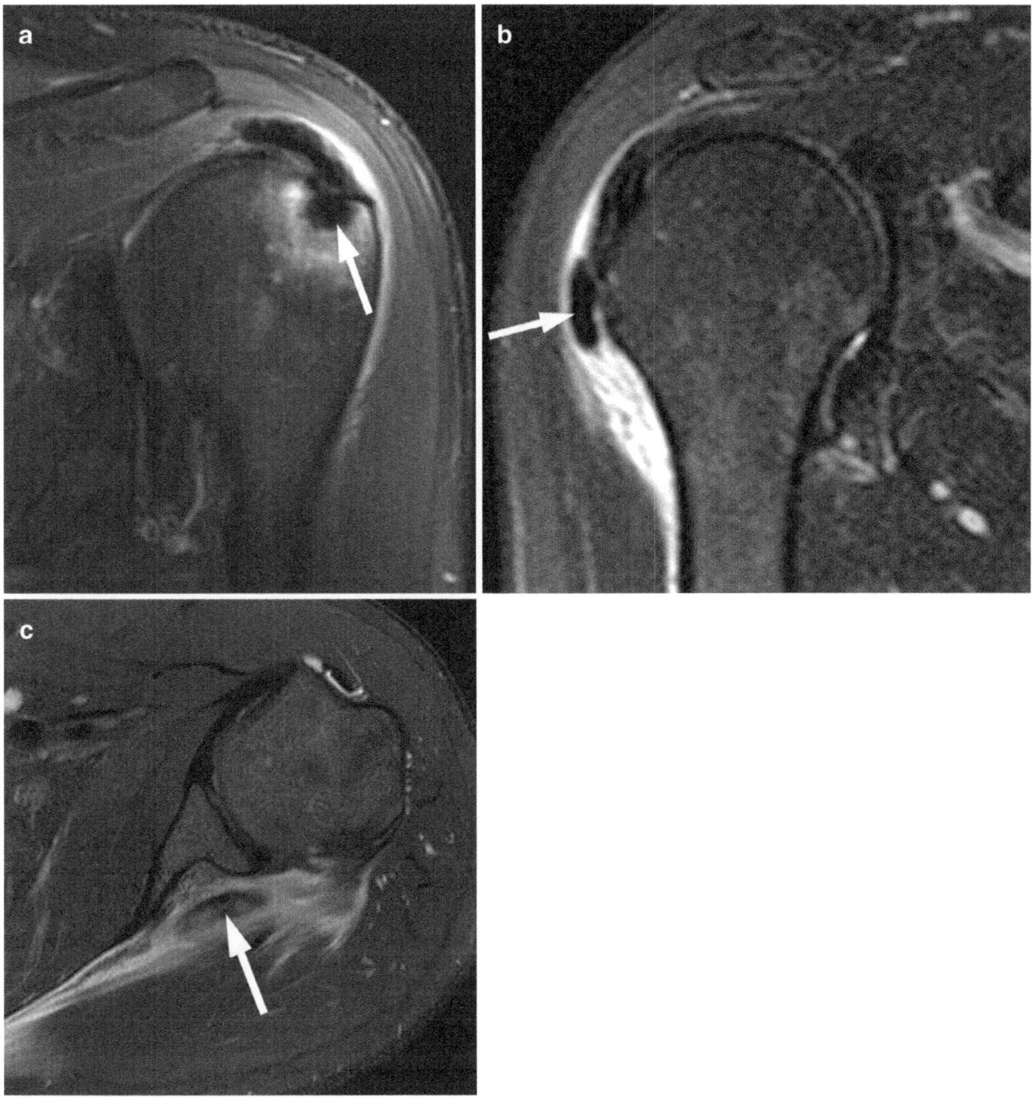

Fig. 10.11 Coronal oblique (**a**, **b**) and transverse (**c**) fat-saturated T2-weighted MR images show in three different patients migration of the calcific tendinitis deposits (*arrows*) into the humeral head, subdeltoid bursa, and infraspinatus muscle belly, respectively. Active inflammation surrounds the migrated deposits in each case

ZTE sequence also exquisitely displays such calcifications on MRI.

Migration of the calcific deposits into the bursae, humerus, and into the muscle belly can also occur and is easily depicted on MRI or CT (Fig. 10.11). Such migrated deposits and their surrounding inflammation can also subside over time.

Ultrasonographic appearance of calcific deposits are classified as hard (hyper-reflexive nodule with a well-circumscribed dorsal acoustic shadow), soft (well-circumscribed, homogeneous hyperechoic foci without posterior shadow), and fluid (hyperechoic peripheral rim with hypoechoic/anechoic center) [28]. Ultrasonography-guided percutaneous irrigation of calcific deposits is a valid treatment option as it is less invasive, quicker and with less post-procedural complication in comparison to arthroscopic removal (Fig. 10.12) [28]. Although different US-guided techniques and approaches have been reported using one or two needles of different sizes to remove calcium, no definite evidence exists in favor of using a specific size or number of needles [28].

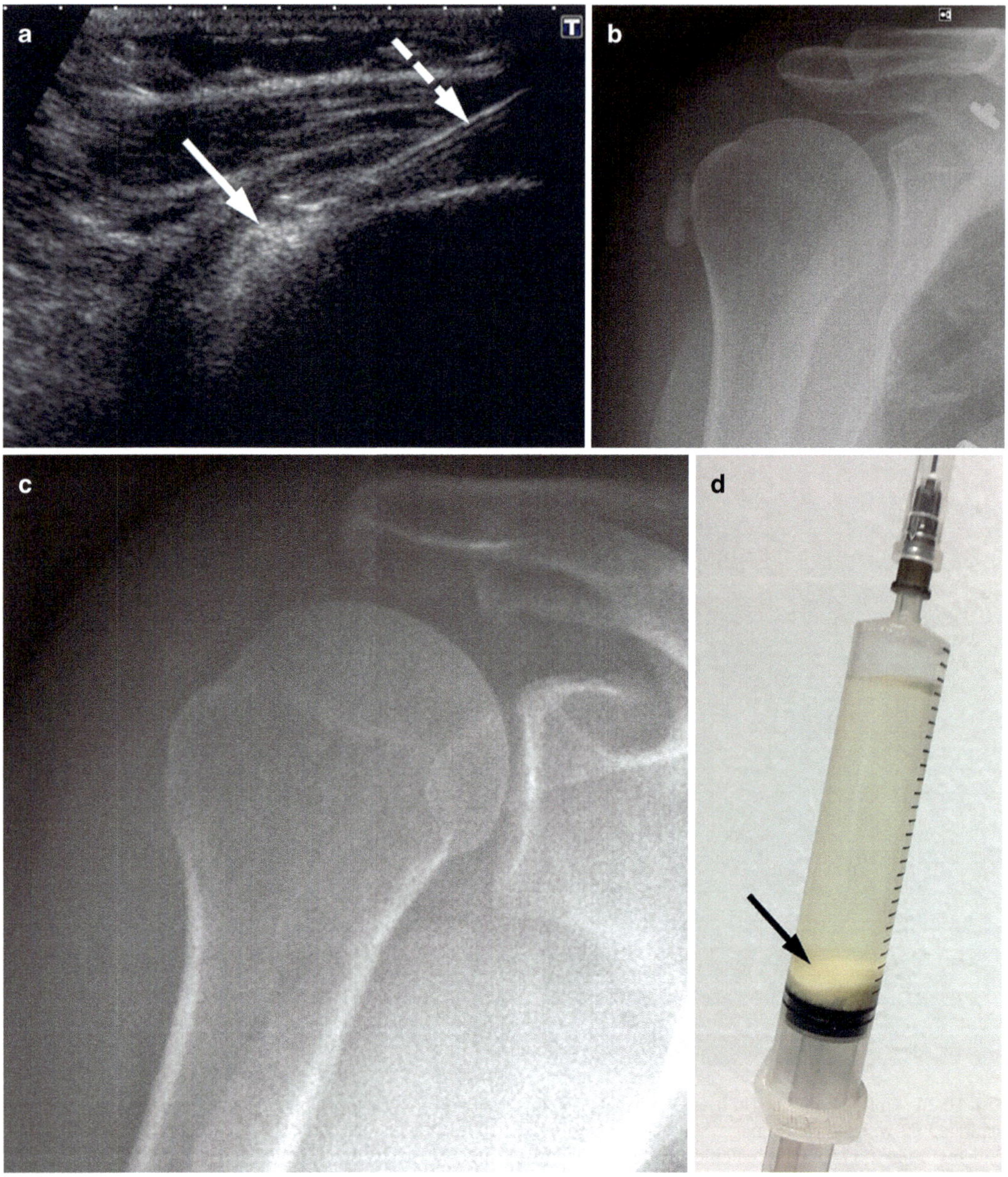

Fig. 10.12 Ultrasonography-guided percutaneous irrigation (**a**) of a calcific deposit (*arrow*) that migrated into the subdeltoid bursa in a 71-year-old woman (needle, *dashed arrow*). Grashey projection radiographs before (**b**) and after (**c**) the procedure show nearly complete removal of the calcific deposit. Calcium deposits layer (*arrow*, **d**) in the syringe filled with the fluid that was withdrawn after irrigation with physiologic saline

10.4 Imaging in Shoulder Instability

Radiography is the first-line imaging tool in the assessment of shoulder instability. In the acute setting, radiographs readily show the dislocation. Humeral head and glenoid fractures that are frequently associated with shoulder dislocations are also displayed on the standard trauma series, which usually comprises an AP view (neutral or with internal/external rotation), the scapular Y view, and the axillary view (or one of its modifications). Nevertheless, CT with 3D reconstructions is widely used to quantify bone stock in deciding and planning surgery. Magnetic resonance imaging is the preferred cross-sectional imaging modality for the comprehensive assessment of bone and soft tissue lesions. Although MR-arthrography has little or no role in the acute setting (because of the usual presence of a joint effusion already creating an arthrographic effect), it exquisitely demonstrates in later stages capsulolabral as well as bony injury and rotator cuff tendon tears that might be associated with shoulder dislocation. MR-arthrography has a sensitivity of 88%–96% and a specificity of 91%–98% in the evaluation of glenoid labrum lesions [29]. In the last decade, the widespread availability of 3 T with improved image quality has somewhat decreased the use of MR-arthrography.

10.4.1 Anterior Instability

Concomitant bone injuries of the glenoid and humeral head, which are detected in up to 80% of patients with anterior shoulder instability, predispose to recurrent episodes of anterior shoulder dislocation with a cumulative effect depending on size and location [29]. Both posterolateral humeral head impaction fracture (Hill-Sachs lesion) and, more importantly, anterior glenoid rim fracture are important to characterize and quantify preoperatively because of their importance in prognosis and surgical guidance.

Soft tissue lesions of anterior shoulder instability include Bankart, Perthes, anterior labroligamentous periosteal sleeve avulsion (ALPSA), glenoid labrum articular disruption (GLAD), and humeral avulsion of the (inferior) glenohumeral ligament (HAGL) lesions (Fig. 10.13). The most commonly torn rotator cuff tendon in anterior shoulder dislocation is the subscapularis [29]. In soft tissue Bankart lesions, which occur at the anteroinferior aspect of the glenoid (3–6 o'clock), both the labrum and its capsular insertion along with the glenoid periosteum are torn, and the labrum may be partially or completely detached from the glenoid rim. In a recent study, Bankart tears demonstrated on MR-arthrography fluid signal more often (92%) on T2-weighted images than gadolinium-based contrast signal (76%) on T1-weighted images, suggesting that resynovialization could cause joint fluid to be trapped beneath the tear [29, 30]. Perthes lesion denotes a non-displaced tear of the anteroinferior labrum and the inferior glenohumeral ligament (IGHL), whereby the glenoid periosteum remains intact. Since the labrum remains anatomically positioned, this lesion may be overlooked at arthroscopy. With the addition of ABER sequence to MR-arthrography, Perthes lesions become often more conspicuous (Fig. 10.3) [29]. GLAD lesion refers to the presence of articular cartilage injury (in the form of fibrillation, erosion, impaction, detachment, or delamination) adjacent to a non-displaced tear of the anteroinferior labrum. The periosteum and the IGHL usually remain intact. Both the HAGL lesion, characterized by humeral avulsion of the IGHL, and glenoid avulsion of the IGHL are exquisitely shown on MRI or MR-arthrography. Either type of these lesions can sometimes present with a detached bony fragment at the humeral or glenoid side (radiography correlation is particularly useful since MRI might not readily show small bony fragments). A "floating" anterior IGHL results from the rare occurrence of a HAGL lesion with a concomitant anteroinferior capsulolabral tear [29].

It is more important and relevant for the radiologist to describe the injured structures and the extent of injury along with the presence of any displacement rather than to furnish a pinpoint term-based diagnosis. Although anatomic variants such as a sublabral foramen (a gap between the anterosuperior labrum and the glenoid) and the Buford complex (cordlike thickening of the middle glenohumeral ligament in the absence of the anterosuperior labrum) can be seen on shoulder MRI, it is important to realize that there is an association between these conditions and a superior labral tear likely due to increased stress on the bicipitolabral complex [9].

In anterior and anteroinferior dislocations, the Pico method, which utilizes a best-fit circle drawn on the glenoid joint surface on CT reconstructions showing the glenoid *en face*, accurately quantifies glenoid bone loss and is associated with recurrent dislocation when above 20% (Fig. 10.14) [29].

With the better visibility of rotator cuff insertion than on CT and the possibility of producing CT-like images with the recently introduced ZTE sequence [7, 8], MRI affords a better feasibility of determining whether the bipolar (i.e., humeral and glenoid) bone lesions (Hill-Sachs and bony

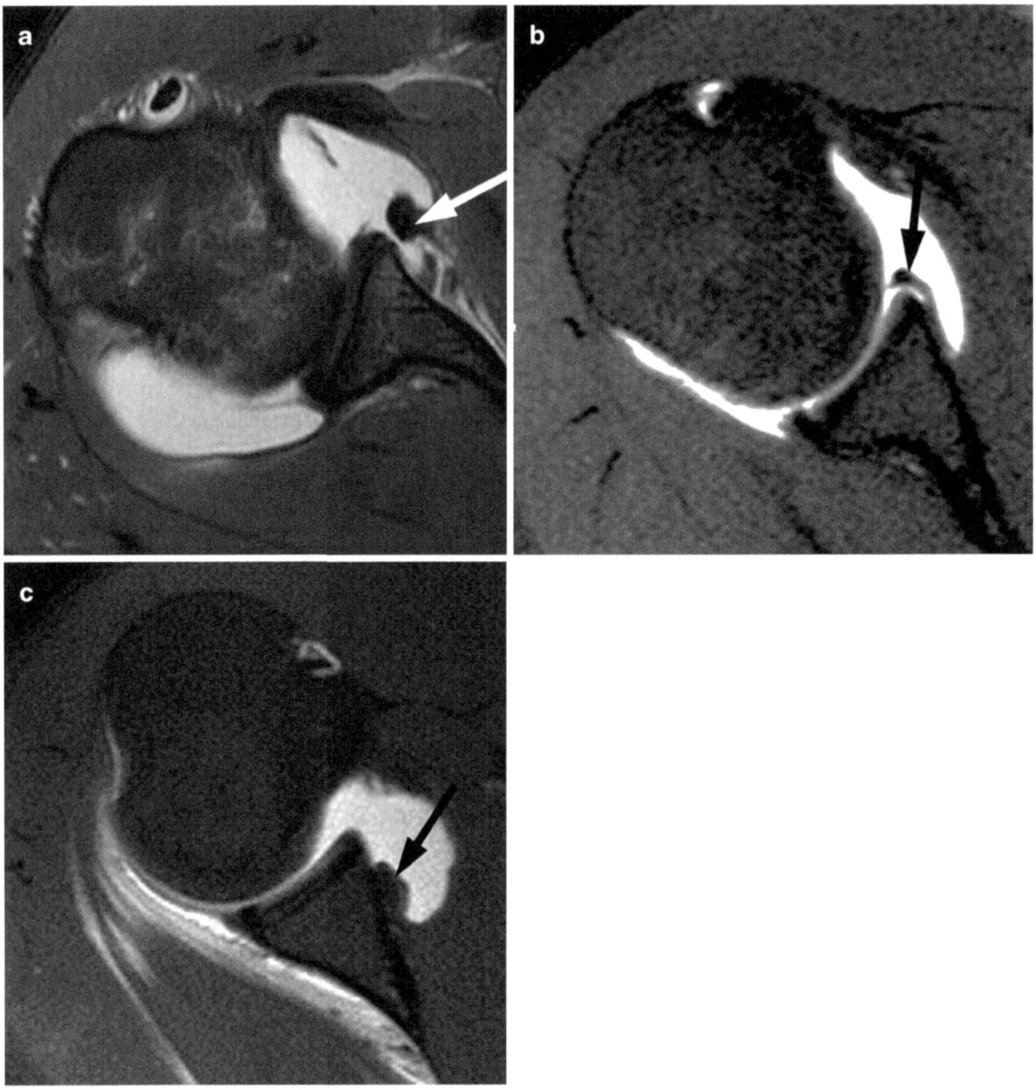

Fig. 10.13 Transverse fat-saturated T2-weighted MR (**a**) and fat-saturated T1-weighted MR-arthrography (**b**, **c**) images show Bankart, Perthes, and anterior labroligamentous periosteal sleeve avulsion (ALPSA) lesions (*arrows*) in three different patients. Conventional MRI in the patient with Bankart lesion (**a**) was obtained shortly after trauma and shows moderate joint effusion, obviating the need for an MR-arthrography (**b**, **c**)

Bankart lesions, respectively) are "on-track" or "off-track" [29]. Bipolar bone lesions are considered "off-track" and at risk for engagement (i.e., recurrent dislocation and the need for revision surgery) if the Hill-Sachs interval (HSI) is greater than the glenoid track. HSI is composed of the Hill-Sachs impaction fracture width plus the intact bone bridge width between the Hill-Sachs lesion and the medial edge of the rotator cuff tendon insertion on transverse MR images. The glenoid track is given by the formula: $0.83 \times (D–d)$, where D represents the diameter of the intact glenoid as outlined in a circle *en face* as in the Pico method, and d represents the amount of anterior glenoid bone loss (both in millimeters). Bone lesions associated with anterior instability are considered on-track and not at risk for engagement (and recurrent dislocation) if the HSI is less than the glenoid track [29, 31]. The recommended treatment strategy for on-track and off-track lesions depending on the glenoid bone loss being <25% or ≥ 25% is different [31].

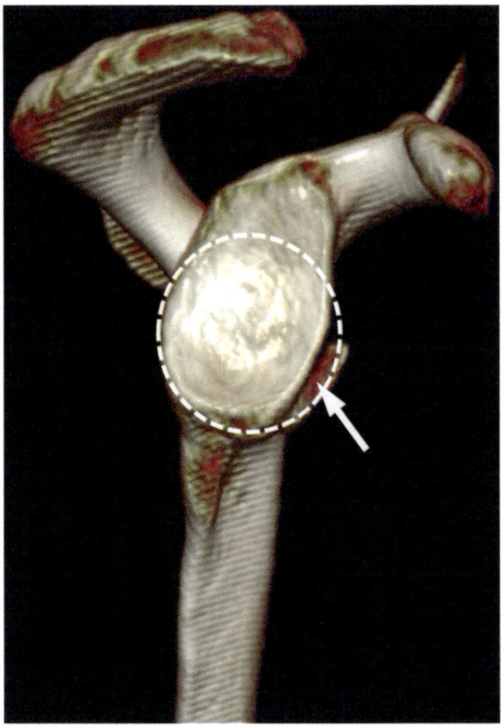

Fig. 10.14 Imaging-based pre-operative assessment of the glenoid stock utilizes a best-fit circle drawn on the glenoid joint surface on a three-dimensional CT reconstruction showing the glenoid *en face*, which in this patient also features a bony Bankart lesion (*arrow*)

10.4.2 Posterior Instability

The much less common condition of posterior shoulder instability also has imaging findings. Radiographs can show an engaged posterior shoulder dislocation or its sequelae. Radiographic findings on AP projection include the absence of normal overlap of the humeral head and glenoid, the "lightbulb sign," which denotes the fixed internal rotation of the humerus, and the "trough sign," which refers to the double contour of the medial humeral head representing the long axis of a reverse Hill-Sachs fracture parallel to the normal medial cortex (Fig. 10.15) [14]. Reverse Hill-Sachs lesion (also known as McLaughlin lesion), which is an impaction fracture at the anteromedial aspect of the humeral head and reverse bony Bankart lesion at the posteroinferior glenoid rim are usually better seen on CT and MRI. If more than 30% of the articular surface is involved in a reverse Hill-Sachs lesion or when there are associated injuries of the posterior capsulolabral ligamentous structures, the risk of instability increases [32]. Soft tissue lesions associated with posterior dislocation include reverse Bankart lesion, posterior labrocapsular

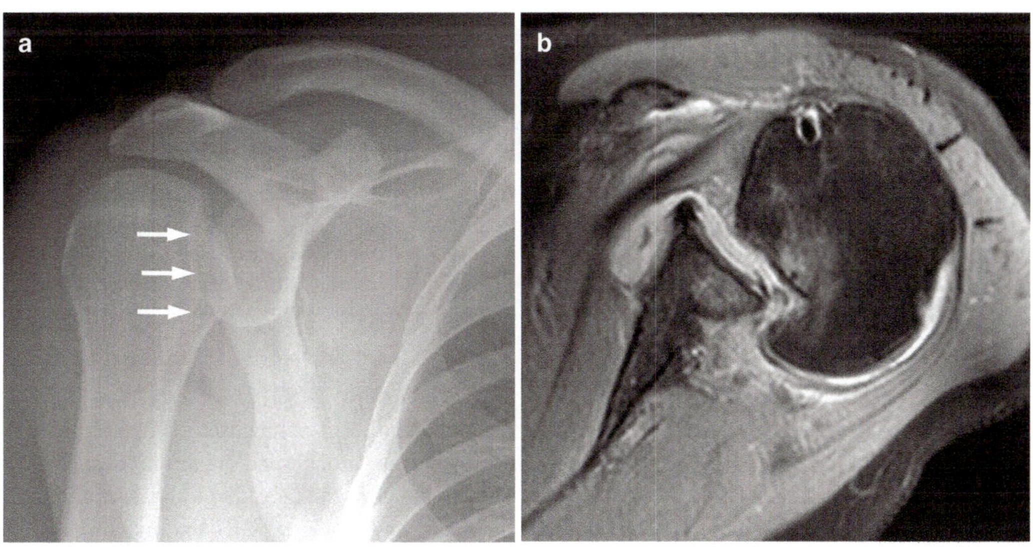

Fig. 10.15 A neutral anteroposterior radiograph of the shoulder (**a**) displays the "trough sign," created by the double contour of the medial humeral head representing the long axis of a reverse Hill-Sachs fracture parallel to the medial cortex (*arrows*). Transverse MR image (**b**) in another patient, who likewise sustained a posterior shoulder dislocation, shows an engaging reverse Hill-Sachs lesion

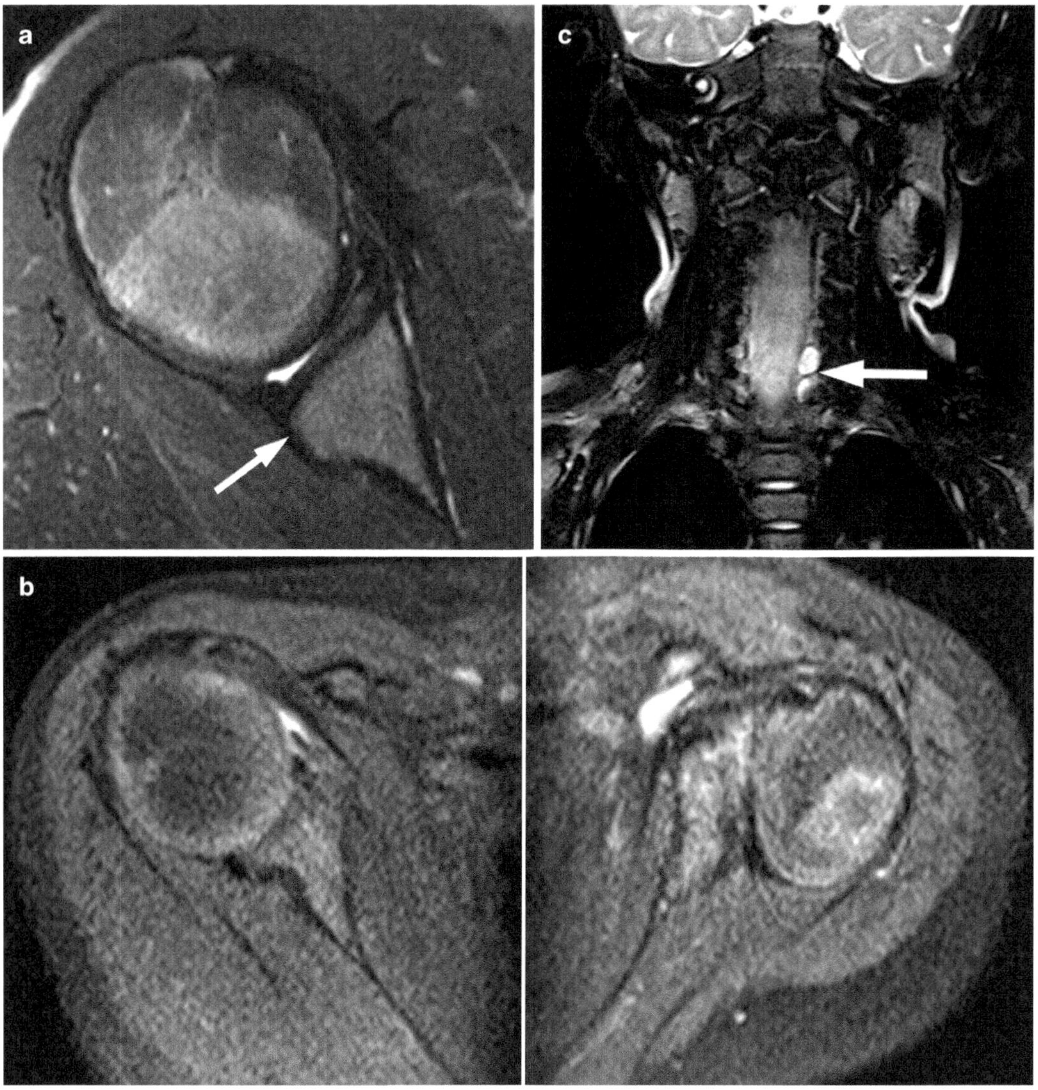

Fig. 10.16 Transverse fat-saturated T2-weighted MR image (**a**) shows a dysplastic posterior glenoid (*arrow*) in a young adult. A composite fat-saturated proton-density MR image showing both shoulders (**b**) and a coronal T2-weighted MR image (**c**) show left-sided glenohumeral dysplasia due to obstetric brachial plexus palsy in an infant with lower cervical spinal nerve root avulsions that resulted in meningeal cysts/diverticula (*arrow*, **c**). Note retroversion of the left glenoid cavity with subluxation of the left humeral head and atrophy of the rotator cuff muscles (**b**)

periosteal sleeve avulsion (POLPSA), and Kim lesion, which is a superficial tear between the posterior glenoid labrum and glenoid articular cartilage without labral detachment.

Glenoid dysplasia, which is a congenital abnormality likely resulting from underdevelopment of the inferior glenoid ossification center, predisposes to posterior instability. In this condition, the scapular neck and the humeral head can be hypoplastic and humeral neck can display varus deformity. MRI shows a dysplastic posterior glenoid with compensatory hypertrophy of the posterior cartilage and/or labrum (Fig. 10.16a). Glenohumeral dysplasia due to obstetric brachial plexus palsy produces MRI findings that can be observed in infants as young as 3 months [33] and include retroversion of the glenoid cavity, developmental delay, posterior translation/subluxation of the humeral head, and atrophy of the rotator cuff muscles (Fig. 10.16b, c).

10.5 Imaging in Some Other Shoulder Conditions

10.5.1 Long Head Biceps Tendon Lesions

Long head of the biceps tendon (LHBT) lesions occur most commonly at the subacromial part of this tendon (i.e., at the rotator interval) and they are an important pain generator. Since clinical tests are often equivocal, imaging plays an important role in ascertaining these lesions. One commonly encountered lesion on shoulder MRI is LHBT tendonosis either by itself or in combination with rotator cuff conditions. A systematic review of the literature showed that chronic LHBT tendinopathy is associated with chronic supraspinatus tendinopathy [34]. It is imperative to follow the LHBT on all three MRI planes as it courses around the humeral head to insert on the supraglenoid tubercle (Fig. 10.17).

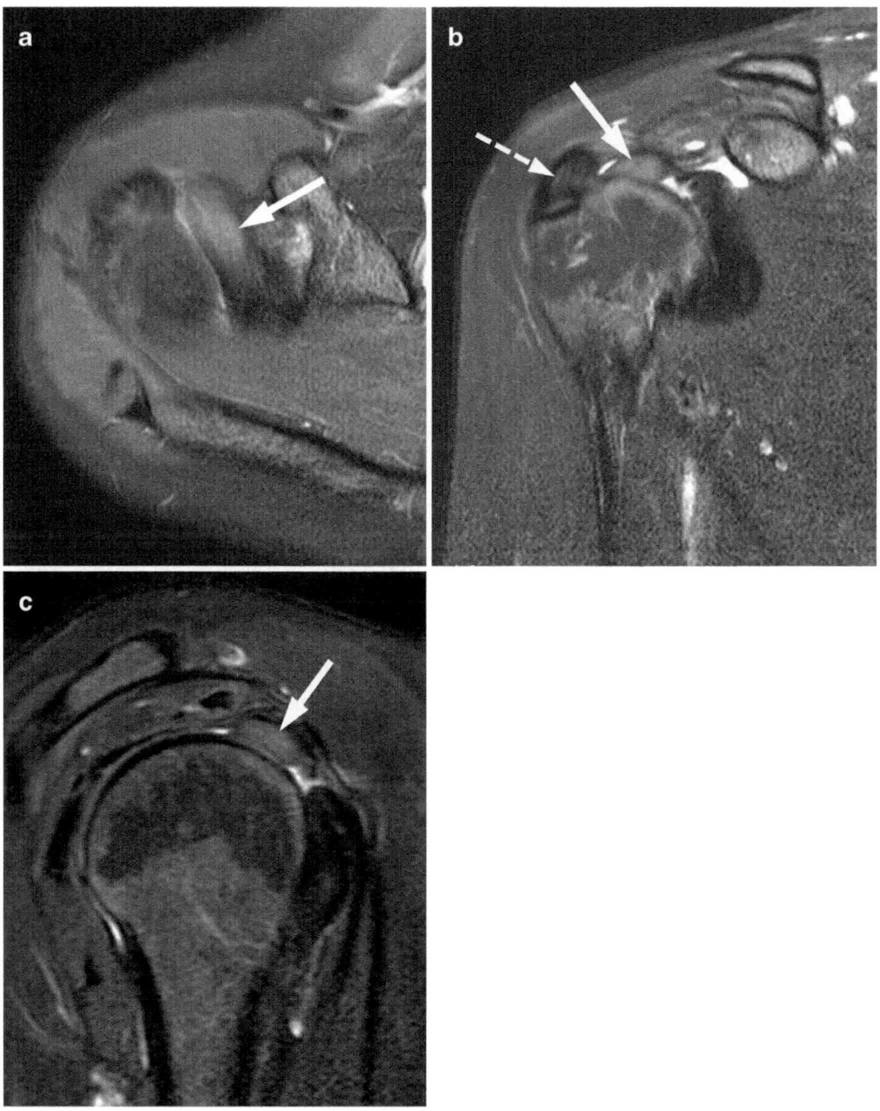

Fig. 10.17 Transverse (**a**), coronal oblique (**b**), and sagittal oblique (**c**) fat-saturated T2-weighted MR images show tendonosis of the long head biceps tendon (LHBT, *arrows*). Note also tendonosis of the supraspinatus tendon (*dashed arrow*, **b**)

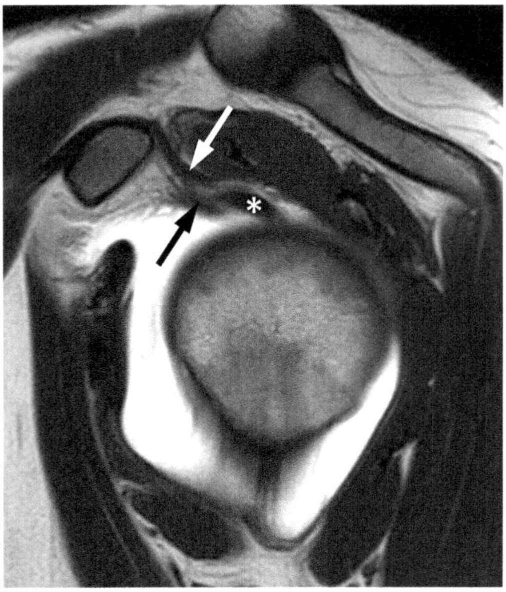

Fig. 10.18 Biceps pulley and the intraarticular portion of the long head biceps tendon (LHBT). Sagittal T1-weighted MR-arthrography image shows the coracohumeral ligament (CHL, *white arrow*) and the superior glenohumeral ligament (SGHL, *black arrow*) anteriorly enveloping the biceps tendon long head (*asterisk*)

MR-arthrography superbly shows the LHBT and its supporting structures (mainly the biceps pulley) [35]. Biceps pulley is defined either somewhat narrowly as the ligamentous sling consisting of the CHL and SGHL (Fig. 10.18) or more broadly as the entire stabilizing structure for LHBT consisting of the CHL, SGHL, and upper border of the subscapularis tendon as well as the anterior border of the supraspinatus tendon. Biceps pulley guides and stabilizes the LHBT during its deflection from the intraarticular horizontal segment to the extraarticular vertical portion. It prevents medial subluxation of the LHBT, along with which it also stabilizes the shoulder against superior translation. The latter is especially important in patients with supraspinatus tendon tears.

It is important to realize that shape variations and slightly eccentric position of LHBT within the humeral groove on MRI can also be seen in healthy volunteers [36]. Medial subluxation of the LHBT usually implies a subscapularis tendon lesion and biceps pulley tear. This finding on MR-arthrography is highly specific but not very sensitive [35]. The signs of nonvisibility or discontinuity of the SGHL and caudad and/or anterior displacement of the LHBT relative to the subscapularis tendon on MR-arthrography have higher sensitivity (with quite high specificity) for biceps pulley tears.

Biceps pulley lesions are often associated with supraspinatus and/or subscapularis tendon tears and SLAP tears. Therefore, all relevant structures need to be carefully assessed on MR-arthrography images. LHBT can dislocate superficial or deep to the subscapularis tendon, depending on which pulley components are torn (Fig. 10.19) [37].

10.5.2 Distal Clavicular Osteolysis

As a recognized cause of shoulder pain, distal clavicular osteolysis can mimic physical examination findings—as well as symptoms—of rotator cuff injury and labral tears, with which it can coexist [38]. It typically follows an acute traumatic incident or, more commonly, chronic repetitive stress to the acromioclavicular joint (usually related to weightlifting and overhead sports). Initial radiographs may be normal; later periostitis, resorption, and/or osteopenia at the distal clavicle can be seen. This condition is seen on MRI as a stress reaction in the form of bone marrow edema/contusion sometimes accompanied with a frank stress fracture (Fig. 10.20a), before characteristic osteolysis of distal clavicle (Fig. 10.20b) is visible on conventional radiographs.

10.5.3 Nerve Compression or Entrapment

Suprascapular nerve compression or entrapment at the levels of suprascapular or spinoglenoid notches can result in denervation changes at the supraspina-

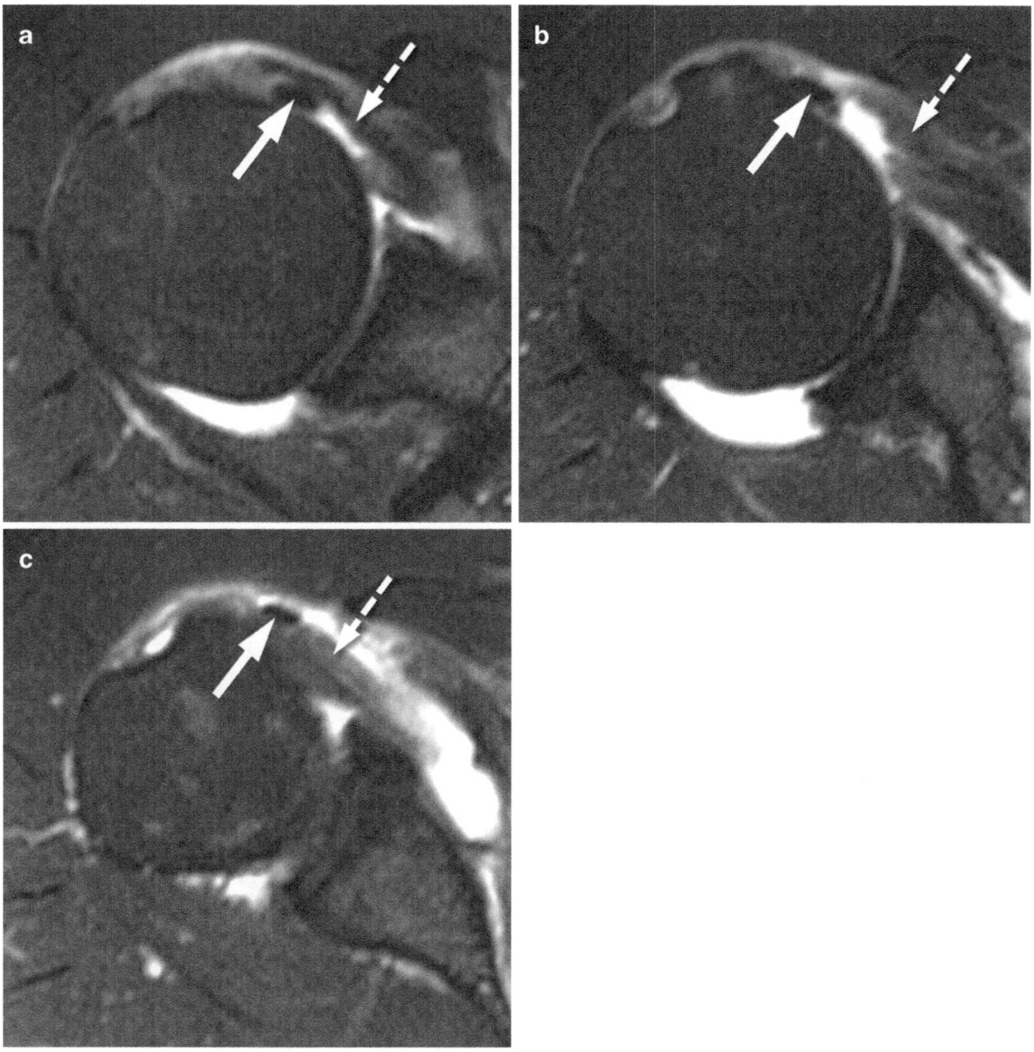

Fig. 10.19 Craniocaudally consecutive transverse fat-saturated proton-density MR images (**a–c**) show a medially dislocated long head biceps tendon (LHBT, *arrows*) crossing the torn distal portion of the subscapularis tendon (*dashed arrows*)

tus and infraspinatus muscle bellies or only infraspinatus muscle belly, respectively (Fig. 10.21).

Teres minor denervation changes on MRI have an incidence of 3% (Fig. 10.22) [39]. Quadrilateral space syndrome results from the compression of axillary nerve or one of its major branches in the quadrilateral space, affecting the deltoid and/or teres minor. Axillary nerve is susceptible to injury not only in its segment at the quadrilateral space but also as it courses around the glenoid and posterior capsule and can be injured by repetitive microtrauma associated with humeral head instability [13]. Preservation of teres minor muscle bulk and function is an important prognostic factor in reverse shoulder arthroplasty and tendon transfers [40].

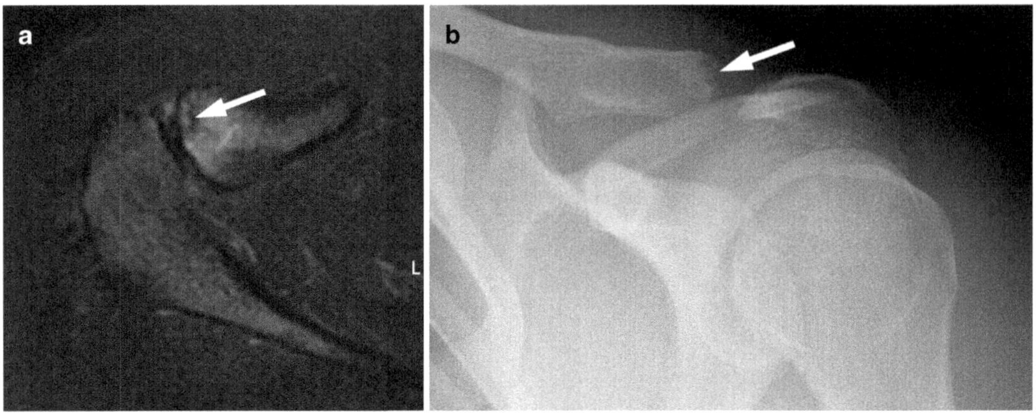

Fig. 10.20 Transverse fat-saturated T2-weighted MR image (**a**) displays a subchondral insufficiency fracture (*arrow*) surrounded with bone marrow edema-like signal. Anteroposterior radiograph in internal rotation (**b**) shows resorptive changes (*arrow*) characteristic of distal clavicular osteolysis in another patient who displayed distal clavicular bone marrow edema-like changes on MRI

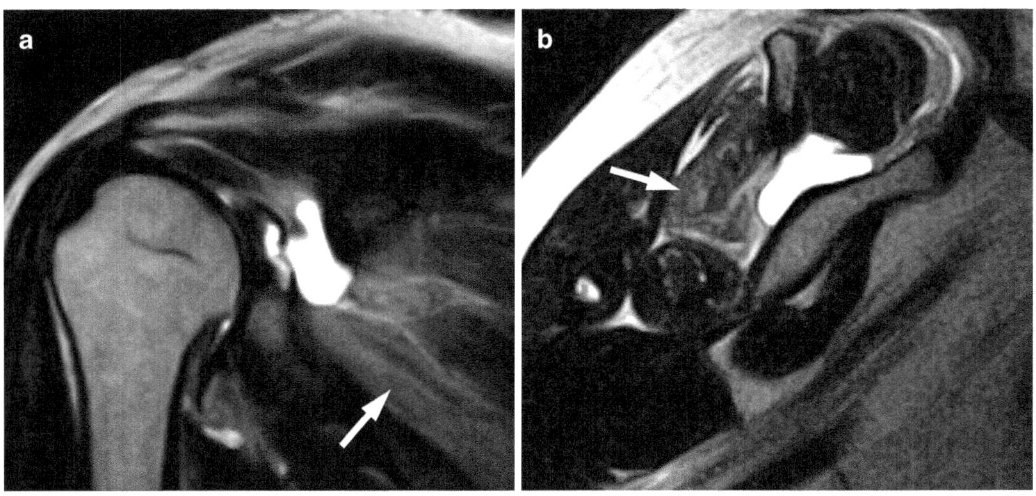

Fig. 10.21 Coronal oblique (**a**) and sagittal oblique T2-weighted (**b**) MR images show a paralabral cyst centered at the level of the spinoglenoid notch causing suprascapular nerve compression that resulted in denervation changes at the infraspinatus muscle (*arrows*), sparing the supraspinatus

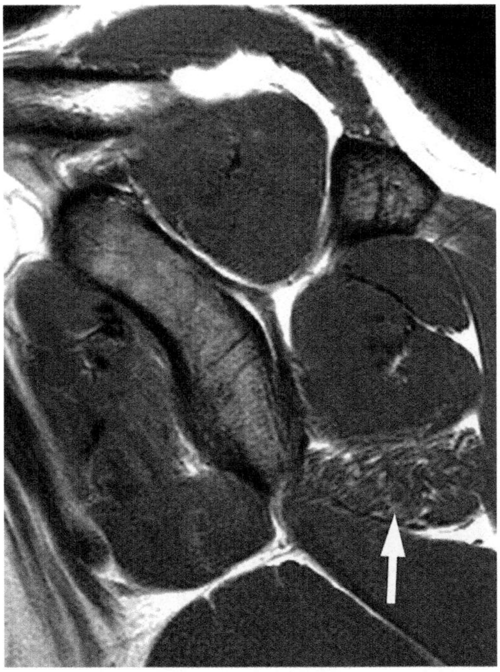

Fig. 10.22 Sagittal oblique T1-weighted MR image shows isolated atrophy of the teres minor (*arrow*) with Goutallier stage 2 (or Fuchs grade 2) fatty infiltration in a 43-year-old man who sustained a fall several months earlier. A branch of the axillary nerve might have been injured

10.5.4 Enchondroma

Differentiating an enchondroma from a low-grade chondrosarcoma affecting the appendicular skeleton (proximal to the metacarpals and metatarsals) is challenging even for pathologists, who take into account parameters such as cellularity, number of mitoses, and cellular atypia. Some radiological features can be helpful in these instances. Deep endosteal scalloping more than two-thirds of the cortex is associated with an increased risk of malignancy (Fig. 10.23) [41]. Other clinical and imaging features that favor chondrosarcoma include pain related to the lesion, cortical destruction and soft-tissue mass, periosteal reaction, and marked uptake of radionuclide (greater than the anterior iliac crest) at bone scintigraphy. MRI is the best imaging modality to follow up cartilage matrix lesions after treatment with curettage and cementing.

10.6 Conclusion

Radiological assessment is essential in addressing a wide range of shoulder problems. The combination of radiography with MRI and/or MR-arthrography is the most commonly used set of imaging tools. Computed tomography is useful for further assessment of fractures and dislocations and helps in surgical planning. Zero echo-time (ZTE) imaging, which is a recently introduced MRI sequence producing CT-like images, may obviate the need to obtain additional or complementary CT for glenoid stock estimation and glenoid track assessment. CT-arthrography needs to be considered whenever there is a contraindication for MR-arthrography. Ultrasonography is used in the shoulder mostly for the evaluation of bursitis, rotator cuff and LHBT abnormalities, as well as for guidance during MR- and/or CT-arthrography, and percutaneous irrigation of calcific tendinitis.

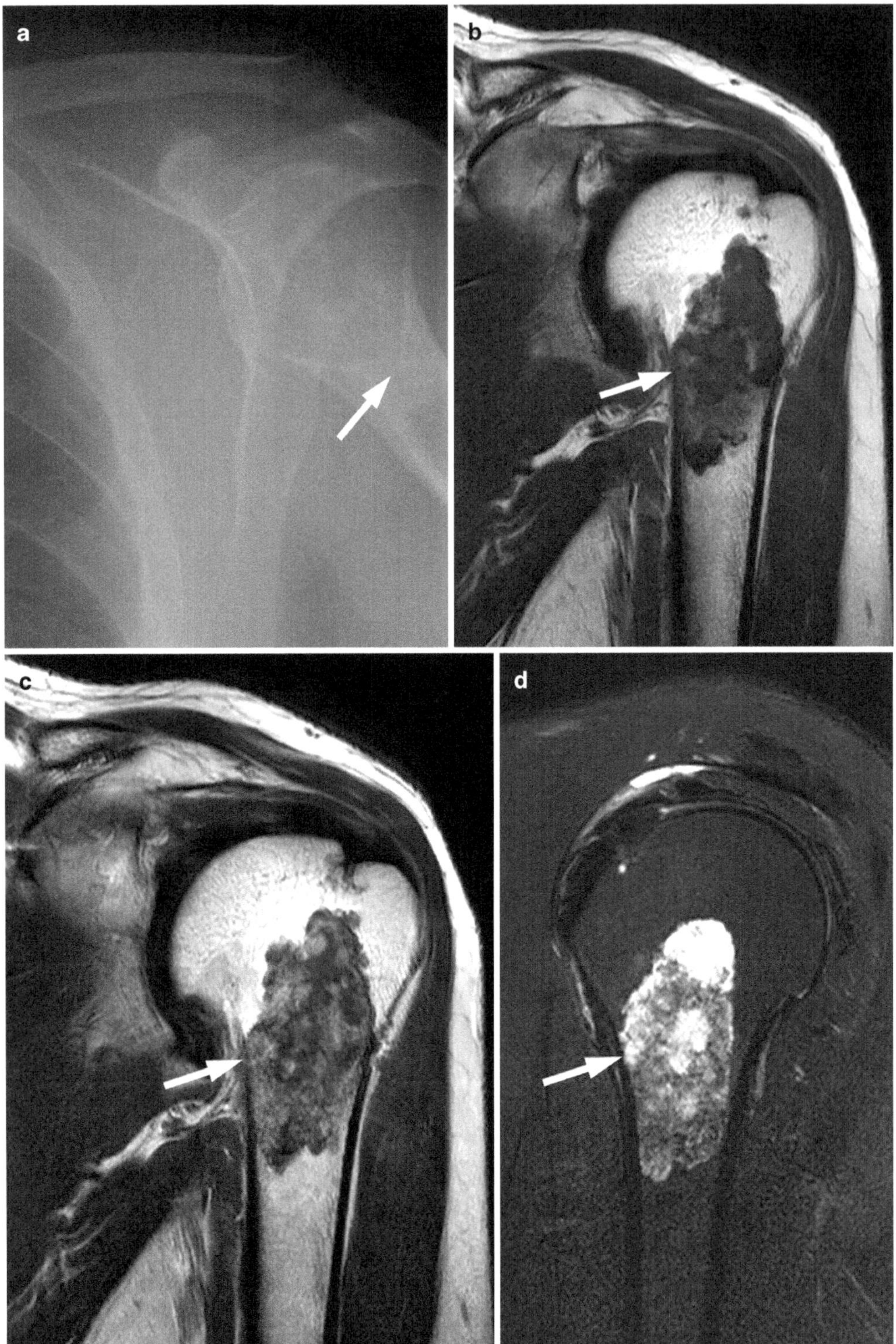

Fig. 10.23 An incidental lesion with stippled calcification in the proximal left humerus metaphysis was encountered on a posteroanterior chest radiograph (**a**). Coronal T1-weighted pre- (**b**) and post-contrast (**c**), and sagittal oblique fat-saturated T2-weighted (**d**) MR images show a lesion with chondroid matrix that features ring and arc type enhancement (**c**) and internal calcifications. The thinning of the cortex medially (*arrows,* **b**–**d**) suggests malignancy (i.e., low-grade chondrosarcoma)

References

1. Aydıngöz Ü, Canbulat N, Demirhan M. Radiological assessment of the shoulder region. Turk J Phys Med Rehabil. 2014;60:S68–77. https://doi.org/10.5152/tftrd.2014.36744.
2. Loriaud A, Bise S, Meyer P, Billaud A, Dallaudiere B, Silvestre A, Pesquer L. Critical shoulder angle: what do radiologists need to know? Skeletal Radiol. 2020;49(4):515–20. https://doi.org/10.1007/s00256-019-03337-3.
3. Magarelli N, Milano G, Sergio P, Santagada DA, Fabbriciani C, Bonomo L. Intra-observer and interobserver reliability of the 'Pico' computed tomography method for quantification of glenoid bone defect in anterior shoulder instability. Skeletal Radiol. 2009;38:1071–5.
4. Combes D, Lancigu R, Desbordes de Cepoy P, Caporilli-Razza F, Hubert L, Rony L, Aubé C. Imaging of shoulder arthroplasties and their complications: a pictorial review. Insights Imaging. 2019;10(1):90. https://doi.org/10.1186/s13244-019-0788-5.
5. Omoumi P, Rubini A, Dubuc JE, Vande Berg BC, Lecouvet FE. Diagnostic performance of CT-arthrography and 1.5T MR-arthrography for the assessment of glenohumeral joint cartilage: a comparative study with arthroscopic correlation. Eur Radiol. 2015;25(4):961–9. https://doi.org/10.1007/s00330-014-3469-2.
6. Rhee RB, Chan KK, Lieu JG, Kim BS, Steinbach LS. MR and CT arthrography of the shoulder. Semin Musculoskelet Radiol. 2012;16(1):3–14. https://doi.org/10.1055/s-0032-1304297.
7. Breighner RE, Endo Y, Konin GP, Gulotta LV, Koff MF, Potter HG. Zero echo time imaging of the shoulder: enhanced osseous detail by using MR imaging. Radiology. 2018;286(3):960–6. https://doi.org/10.1148/radiol.2017170906.
8. de Mello RAF, Ma YJ, Ashir A, et al. Three-dimensional zero echo time magnetic resonance imaging versus 3-dimensional computed tomography for glenoid bone assessment. Arthroscopy 2020;36(9):2391–400. https://doi.org/10.1016/j.arthro.2020.05.042.
9. Motamedi D, Everist BM, Mahanty SR, Steinbach LS. Pitfalls in shoulder MRI: Part 1—normal anatomy and anatomic variants. AJR Am J Roentgenol. 2014;203(3):501–7. https://doi.org/10.2214/AJR.14.12848.
10. Motamedi D, Everist BM, Mahanty SR, Steinbach LS. Pitfalls in shoulder MRI: Part 2—biceps tendon, bursae and cysts, incidental and postsurgical findings, and artifacts. AJR Am J Roentgenol. 2014;203(3):508–15. https://doi.org/10.2214/AJR.14.12849.
11. Bunnell KM, Hemke R, Husseini JS, Torriani M, Huang SY, Bredella MA. Does MR arthrography cause intracranial gadolinium deposition? Skeletal Radiol. 2020;49(7):1051–6. https://doi.org/10.1007/s00256-020-03380-5.
12. Aydıngöz Ü, Maraş Özdemir Z, Ergen FB. Demystifying ABER (ABduction and External Rotation) sequence in shoulder MR arthrography. Diagn Interv Radiol. 2014;20(6):507–10. https://doi.org/10.5152/dir.2014.14117.
13. McCrum E. MR imaging of the rotator cuff. Magn Reson Imaging Clin N Am. 2020;28(2):165–79. https://doi.org/10.1016/j.mric.2019.12.002.
14. Goes PCK, Pathria MN. Radiographic/MR imaging correlation of the shoulder. Magn Reson Imaging Clin N Am. 2019;27(4):575–85. https://doi.org/10.1016/j.mric.2019.07.005.
15. Lawrence RL, Moutzouros V, Bey MJ. Asymptomatic rotator cuff tears. JBJS Rev. 2019;7(6):e9. https://doi.org/10.2106/JBJS.RVW.18.00149.
16. Connor PM, Banks DM, Tyson AB, Coumas JS, D'Alessandro DF. Magnetic resonance imaging of the asymptomatic shoulder of overhead athletes: a 5-year follow-up study. Am J Sports Med. 2003;31(5):724–7.
17. Adams CR, Brady PC, Koo SS, Narbona P, Arrigoni P, Karnes GJ, Burkhart SS. A systematic approach for diagnosing subscapularis tendon tears with preoperative magnetic resonance imaging scans. Arthroscopy. 2012;28(11):1592–600. https://doi.org/10.1016/j.arthro.2012.04.142.
18. Lunn JV, Castellanos-Rosas J, Tavernier T, Barthélémy R, Walch G. A novel lesion of the infraspinatus characterized by musculotendinous disruption, edema, and late fatty infiltration. J Shoulder Elbow Surg. 2008;17(4):546–53. https://doi.org/10.1016/j.jse.2007.11.016.
19. Somerson JS, Hsu JE, Gorbaty JD, Gee AO. Classifications in brief: Goutallier classification of fatty infiltration of the rotator cuff musculature. Clin Orthop Relat Res. 2016;474(5):1328–32. https://doi.org/10.1007/s11999-015-4630-1.
20. Zanetti M, Gerber C, Hodler J. Quantitative assessment of the muscles of the rotator cuff with magnetic resonance imaging. Invest Radiol. 1998;33(3):163–70.
21. Pierce JL, Nacey NC, Jones S, Rierson D, Etier B, Brockmeier S, Anderson MW. Postoperative shoulder imaging: rotator cuff, labrum, and biceps tendon. Radiographics. 2016;36(6):1648–71.
22. Eajazi A, Kussman S, LeBedis C, Guermazi A, Kompel A, Jawa A, Murakami AM. Rotator cuff tear arthropathy: pathophysiology, imaging characteristics, and treatment options. AJR Am J Roentgenol. 2015;205(5):W502–11. https://doi.org/10.2214/AJR.14.13815.
23. Ersoy H, Pomeranz SJ. Milwaukee shoulder syndrome. J Surg Orthop Adv. 2017;26(1):54–7.
24. Nguyen VD. Rapid destructive arthritis of the shoulder. Skeletal Radiol. 1996;25(2):107–12.
25. Husseini JS, Levin M, Chang CY. Capsular injury and inflammation. Magn Reson Imaging Clin N Am. 2020;28(2):257–67. https://doi.org/10.1016/j.mric.2019.12.008.

26. Mengiardi B, Pfirrmann CW, Gerber C, Hodler J, Zanetti M. Frozen shoulder: MR arthrographic findings. Radiology. 2004;233(2):486–92. https://doi.org/10.1148/radiol.2332031219.
27. Park S, Lee DH, Yoon SH, Lee HY, Kwack KS. Evaluation of adhesive capsulitis of the shoulder with fat-suppressed T2-weighted MRI: association between clinical features and MRI findings. AJR Am J Roentgenol. 2016;207(1):135–41. https://doi.org/10.2214/AJR.15.15200.
28. Messina C, Banfi G, Orlandi D, Lacelli F, Serafini G, Mauri G, Secchi F, Silvestri E, Sconfienza LM. Ultrasound-guided interventional procedures around the shoulder. Br J Radiol. 2016;89(1057):20150372. https://doi.org/10.1259/bjr.20150372.
29. Burke CJ, Rodrigues TC, Gyftopoulos S. Anterior instability: what to look for. Magn Reson Imaging Clin N Am. 2020;28(2):195–209. https://doi.org/10.1016/j.mric.2019.12.004.
30. Nacey NC, Fox MG, Bertozzi CJ, Pierce JL, Said N, Diduch DR. Incidence of gadolinium or fluid signal within surgically proven glenoid labral tears at MR arthrography. Skeletal Radiol. 2019;48(8):1185–91. https://doi.org/10.1007/s00256-018-3143-x.
31. Itoi E. 'On-track' and 'off-track' shoulder lesions. EFORT Open Rev. 2017;2(8):343–51. https://doi.org/10.1302/2058-5241.2.170007.
32. Albano D, Messina C, Sconfienza LM. Posterior shoulder instability: what to look for. Magn Reson Imaging Clin N Am. 2020;28(2):211–21. https://doi.org/10.1016/j.mric.2019.12.005.
33. Menashe SJ, Tse R, Nixon JN, Ishak GE, Thapa MM, McBroom JA, Iyer RS. Brachial plexus birth palsy: multimodality imaging of spine and shoulder abnormalities in children. AJR Am J Roentgenol. 2015;204(2):W199–206. https://doi.org/10.2214/AJR.14.12862.
34. Redondo-Alonso L, Chamorro-Moriana G, Jiménez-Rejano JJ, López-Tarrida P, Ridao-Fernández C. Relationship between chronic pathologies of the supraspinatus tendon and the long head of the biceps tendon: systematic review. BMC Musculoskelet Disord. 2014;15:377. https://doi.org/10.1186/1471-2474-15-377.
35. Schaeffeler C, Waldt S, Holzapfel K, Kirchhoff C, Jungmann PM, Wolf P, Stat D, Schröder M, Rummeny EJ, Imhoff AB, Woertler K. Lesions of the biceps pulley: diagnostic accuracy of MR arthrography of the shoulder and evaluation of previously described and new diagnostic signs. Radiology. 2012;264(2):504–13. https://doi.org/10.1148/radiol.12112007.
36. Buck FM, Dietrich TJ, Resnick D, Jost B, Pfirrmann CW. Long biceps tendon: normal position, shape, and orientation in its groove in neutral position and external and internal rotation. Radiology. 2011;261(3):872–81. https://doi.org/10.1148/radiol.11110914.
37. Petchprapa CN, Beltran LS, Jazrawi LM, Kwon YW, Babb JS, Recht MP. The rotator interval: a review of anatomy, function, and normal and abnormal MRI appearance. AJR Am J Roentgenol. 2010;195(3):567–76. https://doi.org/10.2214/AJR.10.4406.
38. Kassarjian A, Llopis E, Palmer WE. Distal clavicular osteolysis: MR evidence for subchondral fracture. Skeletal Radiol. 2007;36(1):17–22.
39. Sofka CM, Lin J, Feinberg J, Potter HG. Teres minor denervation on routine magnetic resonance imaging of the shoulder. Skeletal Radiol. 2004;33(9):514–8.
40. Williams MD, Edwards TB, Walch G. Understanding the importance of the teres minor for shoulder function: functional anatomy and pathology. J Am Acad Orthop Surg. 2018;26(5):150–61. https://doi.org/10.5435/JAAOS-D-15-00258.
41. Murphey MD, Flemming DJ, Boyea SR, Bojescul JA, Sweet DE, Temple HT. Enchondroma versus chondrosarcoma in the appendicular skeleton: differentiating features. Radiographics. 1998;18(5):1213–37; quiz 1244–1245

Part III

Shoulder Pathologic Conditions

11 Shoulder Cartilage and Osteoarthritis

Sercan Akpinar and Bülent Özdemir

11.1 Introduction

The shoulder joint, also known as the glenohumeral joint, is the most mobile joint in the body. It is resistant to compression and distraction forces due to the harmony between the glenoid and humerus [1]. The glenoid is concave and pyriform, with a narrow upper side and wide lower side. The glenoid cavity is covered with hyaline cartilage that tapers in thickness from the outer part to the center. Cartilage thickness is low in a small area in the center, and although it appears like a defect, this is a normal finding. The glenoid labrum, which is a fibrocartilaginous structure, surrounds the joint surface like a ring and contributes to joint stability by increasing the depth of the glenoid cavity. The glenoid cavity, together with the glenoid labrum, is approximately 9 mm deep in the superoinferior (vertical) direction and approximately 5 mm deep in the anteroposterior (horizontal) direction [2]. Whereas the glenoid labrum is firmly attached to the joint cartilage at the inferior region, it is loose and mobile at the superior region. In contrast, the variations in the thickness of the humeral articular cartilage exhibit the opposite pattern to that of the glenoid articular cartilage; the surrounding cartilage thickness is approximately 1 mm, which increases in the range of 1.2–1.3 mm at the center of the humeral head [3].

Glenohumeral joint degeneration is similar to osteoarthritis of other joints. It is the third most common joint affected by degenerative diseases, after the knee and hip joints [4, 5]. However, primary osteoarthritis of this joint is less common since it is not a weight-bearing joint like the knee and hip joints. Glenohumeral joint osteoarthritis (GHJ-OA) is a progressive, degenerative disease of the capsular and bone structures of the joint, particularly of the glenohumeral joint articular cartilage. This disease develops after numerous mechanical and biochemical processes [6] and leads to osteophyte formation that results in subchondral sclerosis, labral lesions, loose bodies, articular cartilage defects, and capsular thickening. Cartilage lesions are not uncommon even in young patients and are often detected during arthroscopic procedures performed due to various pathological conditions [7–9]. Small cartilage lesions associated with rotator cuff or glenohumeral ligament damage cause different topographic stresses in various regions of the joint surface [10]. If left untreated, these lesions can progress to degenerative osteoarthritis. Since the glenohumeral joint cartilage cannot self-renew, GHJ-OA can occur rapidly even after minor lesions such as instability or rotator cuff injury [10]. GHJ-OA is a disabling condition

S. Akpinar (✉) · B. Özdemir
Department of Orthopedics and Traumatology, Medline Hospital, Adana, Turkey

G. Huri et al. (eds.), *Fundamentals of the Shoulder*, https://doi.org/10.1007/978-3-030-94702-6_11

characterized predominantly by pain, stiffness, and subsequent decreased functional activity. It is often complicated by secondary adhesive capsulitis due to incompatible joint surfaces, osteophytes, and capsular scarring. Moreover, it is frequently concurrent with subacromial bursitis, acromioclavicular joint arthritis, labral tears, tendinopathy of the long head of the biceps tendon, and adhesive capsulitis [11]. Hence, treatment of these conditions is as important as chondral injury treatment.

Several surgical options are available to manage primary GHJ-OA, including simple arthroscopic joint debridement [12] and more complex techniques such as resurfacing using fascia lata or meniscus transplantation [13], osteochondral autologous transplantation [14], resurfacing arthroplasty [15], and total arthroplasty [16]. If detected at an early stage, minor joint injuries can be treated non-operatively. More severe joint cartilage damage that causes pain or leads to limitation or loss of function needs to be evaluated carefully. The extent of joint damage and life expectancy of the patient should also be considered in the treatment plan. For example, total joint arthroplasty is not suitable in young patients due to higher incidence of component failure and worse outcome scores.

11.2 Epidemiology

Although it is less common than osteoarthritis of large joints such as the hip and knee, GHJ-OA is important because it causes disability at least as much as these joints. The glenohumeral joint is the second most common site of chronic pain after the knee joint; however, GHJ-OA is rare [17]. Symptomatic chondral lesions of the glenohumeral joint have been reported to have an incidence of 5–17% [18, 19]. The prevalence of this condition increases gradually over the age of 45 in men and above the age of 55 in women, whereas its prevalence is 32.8% in both sexes above the age of 60 [20, 21]. The incidence of GHJ-OA is high in women and increases between the ages of 60 and 70 [22, 23]. It may also occur in patients with shoulder instability in the age group of 30–40 years [21].

Etiologically, GHJ-OA should be evaluated in two categories as primary and secondary. Primary GHJ-OA is less common than secondary GHJ-OA and is more common in women, whereas secondary GHJ-OA is more common in men [21]. Whereas incomplete rotator cuff lesion, traumatic/atraumatic instability, and frozen shoulder are the three most common causes of GHJ-OA under the age of 45, full-thickness rotator cuff tears, degenerative joint disease, and frozen shoulder are among the most common causes above this age [24].

11.3 Classification Based on Etiology

Similar to knee joint lesions, shoulder cartilage lesions are classified according to the Outerbridge classification system [25]. Grade 0 is normal cartilage structure; Grade I refers to softened and swollen cartilage; Grade II refers to fragmentation and fissure formation that do not reach subchondral bone or exceed 1.5 cm in diameter; Grade III refers to fragmentation and fissure formation extending to the level of subchondral bone in an area with a diameter >1.5 cm; Grade IV refers to loss of cartilage approaching or transitioning to the subchondral bone.

Walch classification system of glenoid morphology is also used to classify glenoid pathology in primary osteoarthritis as follows: Type A is central or symmetrical arthritis without posterior subluxation of the humeral head. Type A1 has little central wear or erosion, whereas Type A2 has severe or major central wear or erosion. Type B is characterized by asymmetric arthritis with posterior subluxation of the humeral head. Type B1 is characterized by narrowing of the posterior joint space, subchondral sclerosis, and osteophytes without pronounced glenoid erosion. Type B2 has pronounced or obvious wear of the posterior glenoid that creates a biconcave appearance of the glenoid. Type C is characterized by more than 25° of glenoid retroversion, regardless of glenoid erosion or the position of the humeral head relative to the glenoid [26, 27].

11.4 Primary GHJ-OA

Primary GHJ-OA is less common than osteoarthritis in the knee or hip. In primary GHJ-OA, the glenoid cavity is mainly affected, whereas the humeral head is affected less [24]. The cause of primary GHJ-OA is largely unknown. It typically results in posterior glenoid wear, whereas posterior humeral head subluxation occurs in up to 50% of affected shoulders [28]. Rotator cuff tear occurs in less than 5–10% of the cases of primary GHJ-OA [28]. Joint space narrowing occurs most often in the lower humeral head with the formation of periarticular osteophytes. Symptoms develop gradually, and in the long term; patients have shoulder pain that lasts for many years and increases in severity [21]. Further, positional shoulder pain could occur, which gets worse when lying on the side of the affected shoulder [21]. Joint movement is gradually restricted. After the anterior structures are stretched due to degeneration, the external rotatory movement is restricted first, and patients complain of friction and crepitation with movement of the arm [24, 29]. Despite marked degeneration of the joint observed by radiography, patients may present with only limitation of movement rather than pain [29]. There are many etiologies, including idiopathy, inflammation, autoimmune disorders, post-infectious arthritis, shoulder instability, and even iatrogenesis.

11.5 Secondary GHJ-OA

Secondary GHJ-OA is often caused by trauma, inflammatory diseases, acute or recurrent dislocation, or previous surgery. Rotator cuff lesion is observed in 76–92% of patients with GHJ-OA [30]. Full-thickness rotator cuff tears contribute to the development of GHJ-OA due to joint instability and deterioration of perfusion due to the leakage of joint fluid into the subacromial space [31]. In a cadaver study, it was found that articular cartilage degeneration was two times higher in patients with rotator cuff tears [32]. Further, shoulder instability is an important cause of secondary GHJ-OA in young people [11]. Arthritis is frequently seen in these patients. It is believed to be caused by repetitive trauma to the articular surfaces of the joint, abnormal loading of the joint surfaces and "over-tightening" of the joint with instability operations. Whereas iatrogenic chondrolysis has been reported in patients with shoulder injuries treated by thermal capsulorrhaphy [33].

Another cause of GH arthritis in the young patient is inflammatory arthropathy, especially juvenile rheumatoid arthritis [34]. Therefore, a significant proportion of these young patients will need shoulder arthroplasty [35, 36]. Avascular necrosis of the shoulder is rare; it is most common in adults of the age group 20–50. Non-traumatic osteonecrosis occurs in the humeral head, the second most common region affected, after the femoral head [37]. Subsequent to various known risk factors such as hip osteonecrosis, interosseous pressure increases, and bone death occurs. Corticosteroid use is reported as the most common reason for this [37]. Although sickle cell disease, alcohol abuse, and coagulation disorders are also among the causes of osteonecrosis, many cases are idiopathic [38, 39]. In addition, adhesive capsulitis is a factor; GHJ-OA and adhesive capsulitis may occur simultaneously and may be difficult to distinguish clinically. Moreover, secondary GHJ-OA can occur due to less common conditions such as glenoid dysplasia and osteochondritis dissecans [40]. Glenoid dysplasia is rare in the general population (−14%), it has been demonstrated to predispose patients to early osteoarthritis [41].

11.6 Clinical Evaluation

Patients rarely present with isolated cartilage damage in the shoulder. Such conditions are often accompanied by secondary injuries that cause pain. Determining the occupation and physical activities of the patient is highly effective in determining the etiology, patient's expectations of treatment, and the treatment strategy of the surgeon. Prior history of any trauma or shoulder surgery should be assessed. Patients generally present with symptoms of pain, joint swelling, stiffness, and crepitation.

In cases of GHJ-OA, physical examination of joint movement reveals crepitation, pain, and limitation in passive movement, whereas muscular strength is unaffected. Range of motion should be evaluated both passively and actively using parameters such as passive abduction, forward flexion, and internal and external rotation (adduction and 90° of abduction). Palpable or audible clicks with shoulder motion may indicate bursitis, biceps tendon pathology, or osteophytes. Comparative evaluation of the ranges of motion of the affected and non-affected shoulders should be performed. Pain is often felt in the front and sides of the shoulder and deep along the deltoid, which may be accompanied by locking and freezing of the glenohumeral joint. Whereas pain during full abduction and flexion suggests impingement, pain during mid-abduction or mid-flexion of the shoulder mostly suggests a chondral lesion [23].

11.7 Imaging

The diagnosis of GHJ-OA is made by direct radiography. However, magnetic resonance imaging (MRI) is used to evaluate additional pathologies. However, direct radiographic findings related to osteoarthritis, such as joint space narrowing and subchondral cysts, emerge in the advanced stage of the disease [42]. True anteroposterior radiography (Grashey radiography: arm in 45° internal rotation) as well as lateral scapular (scapular outlet) and axillary (arm in abduction, cassette in superior) radiography are performed (Fig. 11.1a–c). Osteophytes in the glenoid and humeral head at the inferior position are the classic radiological finding on direct radiography. Subchondral sclerosis and cystic changes occur in the superolateral part of the humeral head and in the central part of the glenoid. Weinstein et al. defined a four-stage radiographic classification of GHJ-OA. Stage I includes normal radiographs; however, there is arthroscopic evidence of joint cartilage changes. Stage II includes minimal joint narrowing, both in the humeral head and glenoid, whereas Stage III includes moderate joint narrowing with early inferior osteophyte formation (Fig. 11.1d). Finally, Stage IV includes severe loss of joint space with osteophyte formation and loss of concentricity between the humeral head and glenoid [12].

In the early stages of the disease, MRI may be useful for evaluating chondral lesions (Fig. 11.2). Thus, both chondral and soft tissue pathology can

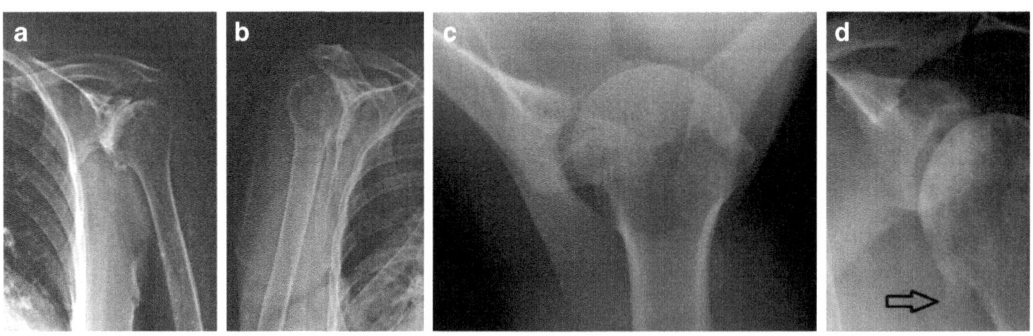

Fig. 11.1 (**a**) Anteroposterior radiography, (**b**) scapular outlet radiography, (**c**) axillary radiography, (**d**) radiographic image of early inferior osteophyte formation

be evaluated using this technique. Three-dimensional computed tomography (CT) may be useful to evaluate underlying bone pathologies, as well as the degree of arthritis deformity, osteophyte formation, and glenoid version (Fig. 11.3). However, the gold standard for revealing chondral lesions nowadays is arthroscopic imaging (Fig. 11.4).

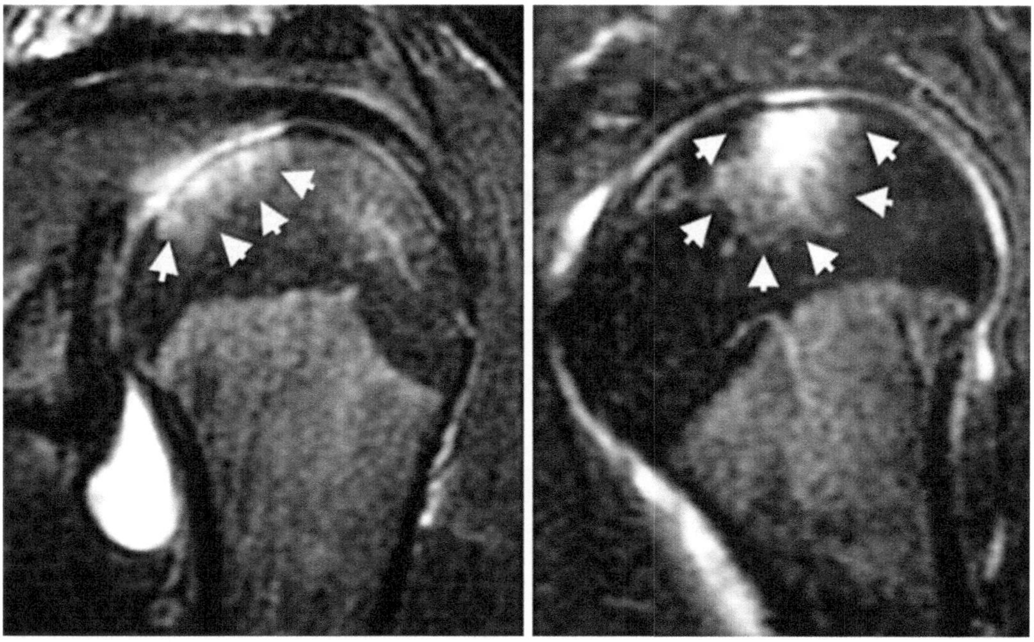

Fig. 11.2 Images with MRG in the early stages of the disease

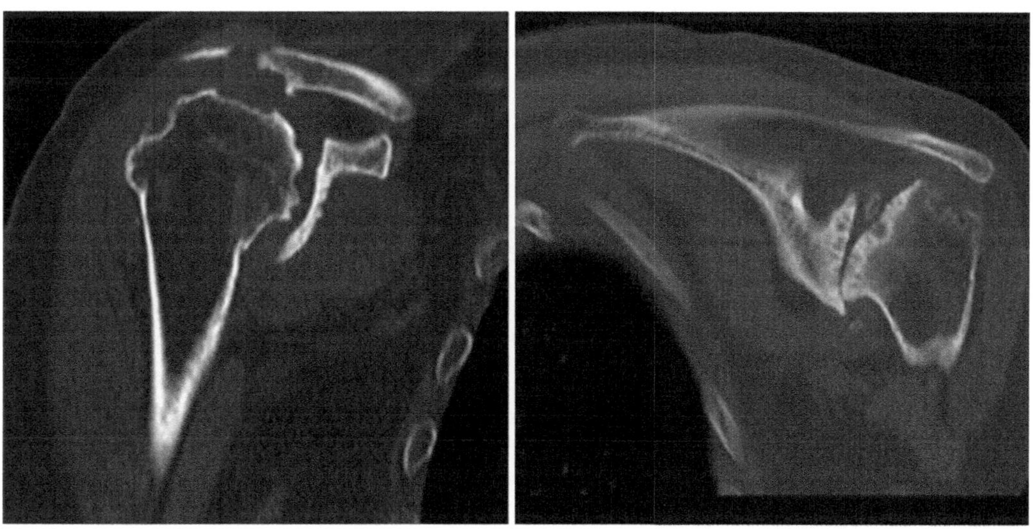

Fig. 11.3 CT images

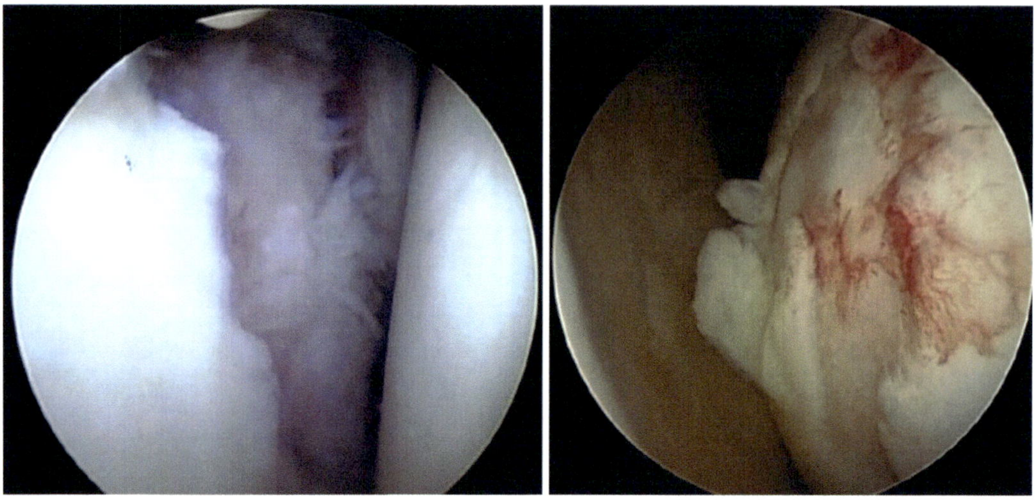

Fig. 11.4 Arthroscopic view of the cartilage lesion

11.8 Treatment

The treatment of symptomatic glenohumeral chondral lesions depends on various factors such as the patient's age, occupation, comorbidities, activity level, degree of injury, and accompanying shoulder pathology. In addition, the extent, depth, and location of symptomatic cartilage damage should be carefully evaluated. There are many conservative and surgical methods for treatment of GHJ-OA. Although non-operative management of GHJ-OA will not alter disease progression, it may be effective in relieving symptoms. In the early stages and in cases where the joint is affected to a lesser extent, treatment is done conservatively. Conservative treatment options include lifestyle and activity changes, occupational changes, physiotherapy, and administration of non-steroidal anti-inflammatory drugs, intra-articular corticosteroid injection, platelet-rich plasma, and viscosupplementation. Surgical treatment modalities include arthroscopic debridement and capsulotomy, microfracture, autologous chondrocyte implantation, osteochondral auto and allograft transplantation, interposition arthroplasty, partial resurfacing arthroplasty, and total shoulder arthroplasty (TSA). Ideal candidates for osteochondral grafting and autologous chondrocyte implantation are of young middle age, physically active, and those with isolated focal chondral defects [28]. Age, activity level, occupation, and current pathologies are important in choosing surgical treatment. Non-arthroplastic treatment strategies for GHJ-OA have been developed to delay the need for TSA and to prevent premature glenoid loosing. These techniques are an integral part of the management of younger patients with painful chondral lesions [28]. However, the results of these treatments are highly variable in terms of pain relief and restoration of function.

11.8.1 Arthroscopic Debridement

The aim of this treatment is to reduce pain, increase range of motion, and delay the need for joint replacement rather than to repair cartilage lesions. The most commonly used joint preserving operations option is GHE debridement. Cartilage fragments, degenerative soft tissues, and loose body are removed in arthroscopic debridement. Generally, although this treatment modality is not a permanent solution, it is preferred for delaying arthroplasty and in elderly patients with low activity demands. In this technique, a stable transition zone between degenerative and intact joint cartilage is created using arthroscopic shaver machines, baskets, and curettes. Care is taken to leave the healthy carti-

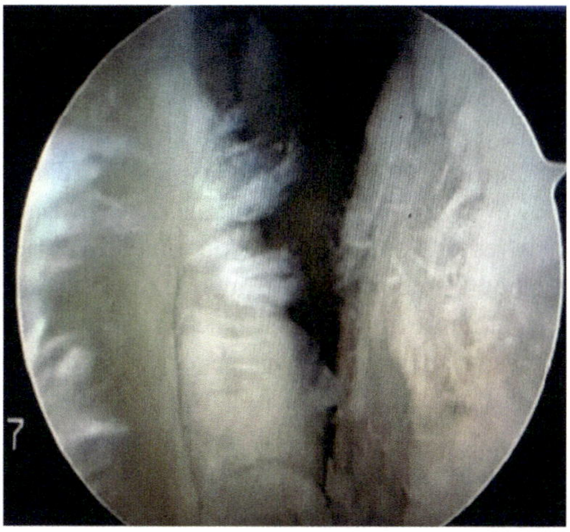

Fig. 11.5 Arthroscopic debridement

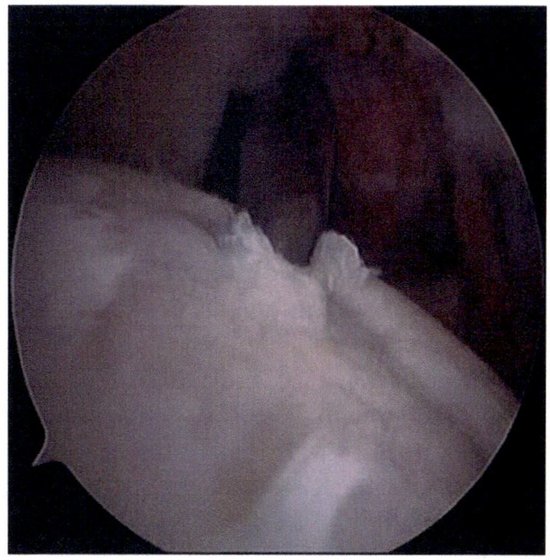

Fig. 11.6 Arthroscopic microfracture

lage tissue intact while removing loose tissue (Fig. 11.5). Small osteophytes can be removed; however, debridement of large inferior osteophytes is not recommended due to the risk of neurovascular injury [43]. Weinstein et al. [12] performed arthroscopic debridement, subacromial bursectomy, and loose body excision in 25 patients with Stage II–III osteoarthritis and obtained excellent results in 80% of the patients after 34 months of follow-up [12]. Guyette et al. reported good results in patients with Stage I–III disease and poor results in those with Stage IV disease [44]. Satisfactory results can be obtained with arthroscopic treatment in cases of low-severity osteoarthritis. Arthroscopic correction of other accompanying pathologies (such as subacromial decompression, capsule release, labral repair, biceps tenodesis, or tenotomy) increases the chance of success.

11.8.2 Microfracture

As a bone marrow stimulation procedure, the microfracture technique was first described as a repair option for a full-thickness focal cartilage defect in the knee with good clinical results and low surgical morbidity [45]. It is frequently used in knee joints due to the high success rate and low morbidity. Cartilages do not have vascularity that allows recovery after injury. The ideal candidate for this treatment should have a focal, Outerbridge stage IV lesion with the cartilage roof intact, and with no response to conservative treatment [46]. The purpose of microfracture is to allow the passage of pluripotent stem cells and to generate new cartilage tissue by causing a mesenchymal clot [47]. After the lesion is exposed, holes are drilled at 3–4 mm intervals 2–3 mm deep, starting from the periphery. Care should be taken to avoid fractures between the holes. A vertical cartilaginous wall surrounding the lesion is created with curettes to contain the clot (Fig. 11.6). Millett et al. obtained good results in the 47-month follow-up of their patients with Stage IV cartilage lesions [48]. Since the glenohumeral joint is a non-weight-bearing joint, direct pendulum exercises can be started in patients after this treatment to aid in healing.

11.8.3 Osteochondral Allograft Transfer

Osteochondral allograft transplantation is a treatment method used in many joint diseases. It is a useful technique for treating major injuries of the joint cartilage. Donor site morbidity does not occur in this technique, and it can be applied to larger defects. Although osteochondral allograft transplantation is widely used for the treatment of osteochondral defects in the knee, it is less commonly used in the treatment of bone defects of the humeral head. A major factor that can reduce the effectiveness of this procedure is chondrocyte cell death that occurs during the storage, preparation, and implantation of osteochondral grafts. Limited chondrocyte viability, loss of the matrix structure, and risk of disease transmission are possible disadvantages [49]. In the glenohumeral joint, allograft is most commonly used for tumor resections followed by the treatment of Hill–Sachs lesions caused by bone deficiencies and glenohumeral instability [50, 51]. Allografts are available in fresh, fresh-frozen, or frozen-stored forms. Fresh osteochondral allografts have the highest percentage of viable chondrocytes; however, they have a shelf life of approximately 30 days and require precise preoperative sizing. Fresh-frozen allografts have a small number of viable chondrocytes [52]. It has been reported that cryopreserved osteochondral grafts have a shelf life of up to 2 years and can protect 70.5% of viable chondrocytes [52]. In a study on patients with lesions in the humeral head (33 Hill–Sacs lesion, one osteochondritis dissecans lesion, and one iatrogenic injury), the rates of complications and reoperations after osteochondral allograft transfer were 20%–30% and 26.67%, respectively [53]. Although fresh allografts were transplanted in only three patients, graft resorption, necrosis, or arthritic changes were not reported in these patients. In many studies in the literature, allografts are reported regarding their use in Hill–Sacs lesions rather than in shoulder osteoarthritis. An excellent result was reported in a 2-year follow-up of a case report using fresh glenoid allografts [54].

11.8.4 Autologous Chondrocyte Implantation (ACI)

Patients with large lesions who are not suitable for other repair techniques and who do not respond to conservative treatment may be suitable for this technique. It is a two-stage procedure in which healthy articular cartilage is curetted, produced in vitro, and implanted after 3–4 weeks. ACI techniques have been successful in the treatment of knee cartilage lesions [55]. In this technique, there is no donor site morbidity. However, ACI is time consuming, expensive, and requires multiple surgeries to complete. This method is not routinely applied for shoulder lesions at present. Young and active patients with high level of shoulder function and isolated focal lesions of the humeral cartilage are suitable candidates for this type of treatment [49]. Two previous studies in the literature have reported good results with this technique [56, 57].

11.8.5 Osteochondral Autograft Transfer

This technique is mostly used in small lesions and is usually transferred from the femoral condyle. This treatment can treat bone defects in addition to full-thickness cartilage lesions (Fig. 11.7). Little is currently known about the results of this procedure in the glenohumeral joint. However, osteochondral transplants also carry the risk of donor site morbidity, which although unusual is a serious complication that the surgeon should consider. Scheibel et al. conducted a study with a small series and found an improvement in shoulder scores and 20% morbidity in the donor site after 32 months of follow-up [58]. In a case report, good results were reported when this technique was used in the treatment of cartilage defects arising due to osteochondritis dissecans in the humeral head [59].

11.8.6 Interposition Arthroplasty

Although mainly for research purposes, the treatment of glenoid focal chondral defects using biological glenoid resurfacing was first described in the late 1980s [60]. Arthroscopic resurfacing with meniscal allograft was first described by Pennington and Bartz [61]. Arthroscopic interposition arthroplasty uses biological and synthetic

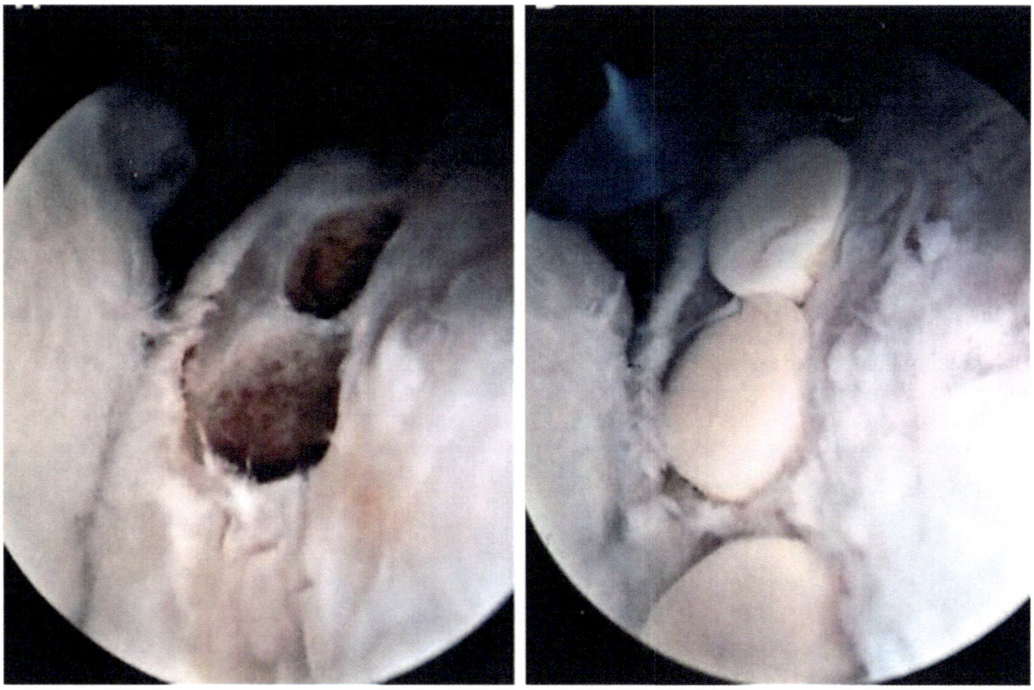

Fig. 11.7 Arthroscopic osteochondral autograft transplantation

acellular materials to resurface the glenoid. The goal of this treatment is to install a synthetic or biological scaffold with sufficiently high tensile strength to allow repopulation by host cells [62]. Autogenous fascia lata, anterior shoulder capsule, meniscus allografts, and regenerative tissue matrices are used in open reconstructive procedures [63, 64]. Biological glenoid resurfacing procedures, such as interposition of fascial, capsular, or meniscal allograft tissue, may be considered to address the problem of glenoid erosion [65]. Many authors conclude that this procedure is not a reliable method of treatment in young patients with shoulder OA, based on both the poor clinical outcome and the absence of the graft acting as a durable glenoid surface [49]. Gobezie et al. published devastating results in 13 patients with a mean age of 34 and 18–49, treated with soft tissue resurfacing of the glenoid, and 77% of patients in the study underwent revision total shoulder prosthesis [66].

11.8.7 Resurfacing Arthroplasty

Typically, resurfacing arthroplasty is a final attempt before considering TSA or glenohumeral arthrodesis. Shoulder hemiarthroplasty is performed when the rotator cuff and glenoid are intact. Instead of cutting the humerus head, only the cartilage of the humeral side is removed, and a new metal articular surface is used as replacement. Resurfacing arthroplasty has advantages such as minimal bone resection, short operative time, no humeral canal penetration, and low risk of periprosthetic fracture. A potential benefit of this technique is that it facilitates arthroplasty in patients with altered anatomy (for example, after malunion) (Fig. 11.8). In their study, Rai et al. achieved successful results in the 12-year follow-up of 40 patients (46 cases of shoulders, 40 osteoarthritis, 2 rheumatoid arthritis, 2 rotator cuff injuries, 1 instability, 1 avascular necrosis). TSA had to be performed after periprosthetic fracture in three patients [67].

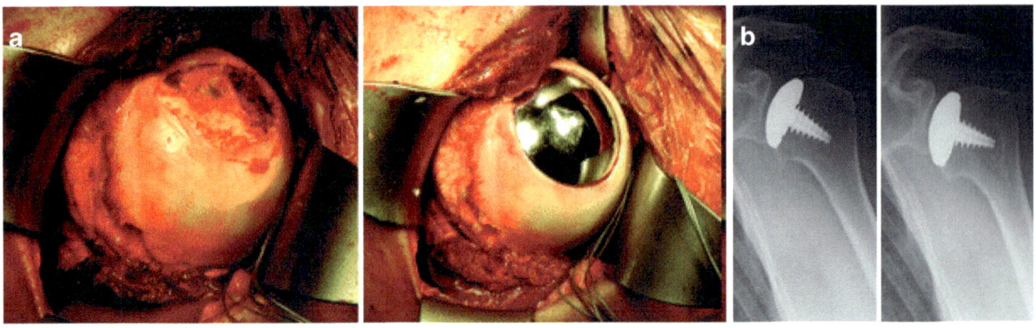

Fig. 11.8 (**a**) Intraoperative views of resurfacing arthroplasty, (**b**) postoperative X-ray images of resurfacing arthroplasty

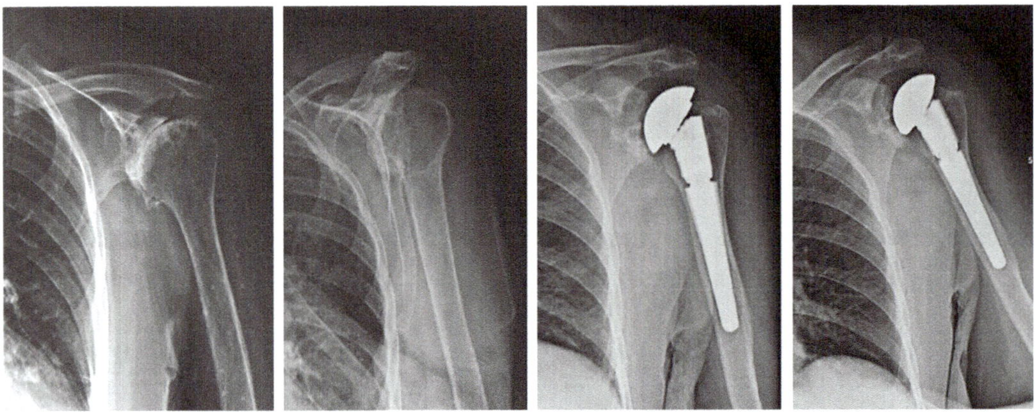

Fig. 11.9 Radiographic view of TSA in shoulder osteoarthritis

11.8.8 Total Shoulder Arthroplasty

TSA can be considered the gold standard for the treatment of shoulder osteoarthritis. Although operations in which the glenoid bone stock can be preserved are more preferred, total shoulder arthroplasty, remains the most reliable operation in terms of pain relief and ROM [68]. Indications for this technique are primary or secondary osteoarthritis causing instability, osteonecrosis, inflammatory joint diseases, and in some cases, complex proximal humerus fractures. Although TSA provides excellent symptomatic relief in GHJ-OA, it should be the last resort in younger and more active patients due to the loosing of the glenoid component and low satisfaction rates [69]. TSA is preferred for many patients with end-stage glenohumeral arthrosis. The presence of a working rotator cuff is imperative to achieve a good result (Fig. 11.9). In 1974, Neer introduced first generation TSA [70], and this prosthesis included a monoblock-cemented humeral stem and an all-polyethylene glenoid prosthesis. TSA has been shown to be a good long-term solution for degenerative shoulder joints and some post-traumatic proximal humeral fracture sequelae [71]. Several studies have reported that loosening and failure of the glenoid component is the most common long-term complication of TSA, which accounts for approximately 24% of all TSA complications [72]. The cause of cemented polyethylene-glenoid loosening is multifactorial, including implant design, surgical technique, cement used, patient characteristics, rotator cuff function, and infection [71]. Proper placement and fixation technique of the glenoid implant are crucial to maximizing the long-term success of TSA [71]. In general, the aim of glenoid component placement should be to achieve <10°of retroversion and ≥89% glenoid component placement, as well as to provide high-quality bone stock for support. Preservation of the subchondral bone of the glenoid is crucial for the implant to provide a rigid structural support and to prepare the glenoid bone surface perfectly [73]. Ho et al. showed that the insertion of the glenoid component with >15° of retroversion was associated with a five-fold increase in osteolysis around the central peg [74]. In severe cases of glenoid deformity, complete correction may not be possible to restore normal anatomical structure. If reaming proceeds medially, the glenoid dome narrows, thus reducing the amount of bone stock available for implantation [71], whereas excessive reaming results in small glenoid component placement, thus causing severe incompatibility between the glenoid and humeral head [71]. Additionally, excessive medialization of the glenoid can reduce tension within the rotator cuff and result in possible harmful functional consequences. Excessive glenoid reaming to correct the glenoid version may increase the risk of medial collapse [75]. For the bone stock to be considered sufficient for the implant, the glenoid depth should be at least 15 mm [76]. If osteoarthritis is associated with recurrent instability or excessive capsular laxity, it may be possible to increase joint stability and capsular tension with proper soft tissue balance or the use of a large head [77]. Anatomical replacement or resurfacing arthroplasty is possible in patients younger than 50 years of age with rotator cuff insufficiency, a solid coracoacromial arch, and sufficient abduction strength provided by the deltoid muscle [77, 78]. In cases with glenoid dysplasia or insufficient bone stock, grafting of the deficient area with the placement of the glenoid component, glenoid osteotomy or grafting with hemiarthroplasty, or metal reinforcements with the glenoid component are used for the deficient bone area [65]. In difficult-to-reconstruct cases like biconcave glenoid, the use of TSA is associated with high failure rates due to early glenoid loosening or recurrent posterior instability [26, 79]. Implant failure is often caused by bone resorption and nonunion of the tuberosity and rotator cuff deficiency [80]. Major complications in shoulder prostheses include periprosthetic fractures, infections, instability, rotator cuff lesions, loosening of the glenoid component, and neurological injuries. Sperling et al. reported a 38% incidence of glenoid component failure requiring revision surgery in a recent series of 33 patients with a mean age of 46 at the time of TSA [81].

11.8.9 Reverse TSA

Reverse shoulder prosthesis changes the orientation of the shoulder. The joint surface acting as a socket is replaced by a ball. Conversely, the humeral head, which is normally a ball, is turned into a socket. The center of rotation is medialized and the humeral head is lateralized. Thus, it was predicted to increase the shoulder functions by extending the deltoid muscle strength arm. Consequently, it increases the strength exerted by the deltoid muscle (Fig. 11.10).

The first cases of GHJ-OA caused by rupture of the rotator cuff were described by Adams and Smith in the 1850s [33]. The term "cuff tear arthropathy" was first defined by Neer in 1983 as a complex condition of shoulder dysfunction due to rotator cuff deficiency [82]. Paul Grammont and colleagues modernized the reverse shoulder arthroplasty implant in 1987 to treat this condition [83]. The RTSA was originally designed to treat a massive irreparable rotator cuff with superior migration of the humeral head combined with glenohumeral arthritis (cuff tear arthropathy) [31, 82, 84, 85]. Reverse TSA (RTSA) was introduced to treat rotator cuff tear arthropathy; however, it is now also used to treat various conditions such as fracture sequel, acute fractures, fixed glenohumeral dislocation, chronic pseudoparalysis without arthritis, revision arthroplasty, tumors, and instability [86]. Reverse total shoulder arthroplasty should be reserved as a salvage operation for the young

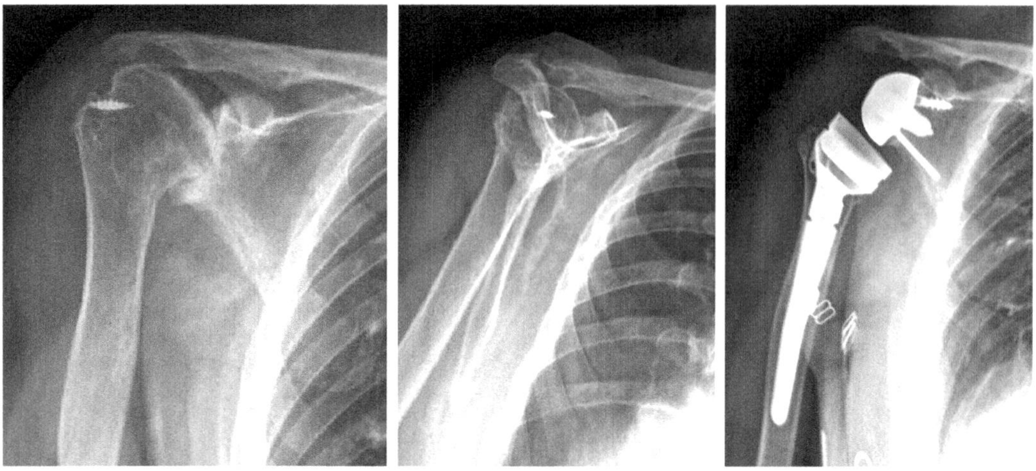

Fig. 11.10 Radiographic view of RTSA in shoulder osteoarthritis

patient with GHEOA. Severely impaired deltoid function, an isolated supraspinatus tear, a massive rotator cuff tear and fully active shoulder elevation with arthritis are contraindications for RTSA [86]. For appropriately selected candidates with symptomatic and rotator cuff dysfunction, RTSA can result in life-changing improvements in pain, movement, function, and patient satisfaction. Rotator cuff tear arthropathy is the single most common indication for RTSA [87]. The RTSA effectively relieves symptoms and restores function for patients with cuff tear arthropathy and irreparable rotator cuff tears. Clinical manifestations are variable and largely depend on the degree of arthritis and the specific rotator cuff tendons ruptured. Loosening of the glenoid component is the most common cause of TSA failure, especially when used on the shoulder with rotator cuff deficiency [88]. Early failure due to glenoid loosening with anatomical TSA in patients with severe rotator cuff deficiency led to the development of RTSA [89]. The advent of RTSA was revolutionary in the treatment of cuff tear arthropathy. The reverse prosthesis was able to address both glenohumeral joint pathology and instability caused by the deficient rotator cuff [90]. Since the development of RTSA, many clinical studies have proved that RTSA is an effective treatment for cuff tear arthropathy with reduced range of motion in the shoulder joint [91]. Frankle et al. evaluated 60 patients who underwent RTSA for rotator cuff tear arthropathy and GHJ-OA and found that 95% of the patients were satisfied with the procedure [92]. Satisfaction with the postoperative outcome in young patients than in elderly patients is lower who have been treated with reverse total shoulder arthroplasty [93, 94].

11.9 Conclusion

Further research is needed to determine treatment modalities that provide the best long-term results for shoulder cartilage pathologies. Although numerous techniques used in the treatment GHJ-OA have been thoroughly investigated in the literature, there is a shortage of long-term comparative studies. Therefore, no consensus exists on the selection of the appropriate surgical procedure for treatment. With regard to GHJ-OA, the aims of treatment are to relieve pain, restore function, and delay the need for arthroplasty. Therefore, long-term randomized prospective studies are necessary to determine the effectiveness of various treatment modalities.

References

1. Quran B, Pioneer J. Glenohumeral osteoarthritis: primary and secondary causes. Turk J Phys Med Rehabil. 2012;58:236–42.
2. Kanatli U, Bolukbasi S, Ekin O, Simsek A. Anatomy, biomechanics, and pathophysiology of instability of the glenohumeral joint. Acta Orthop Traumatol Turc. 2005;39(1):4–13.
3. Fox JA, Cole BJ, Romeo AA, et al. Articular cartilage thickness of the humeral head: an anatomic study. Orthopedics. 2008;31(3):216.
4. Sankaye P, Ostlere S. Arthritis at the shoulder joint. Semin Musculoskelet Radiol. 2015;19(3):307–18.
5. Matsen FA III, Rockwood CA, Wirth MA, Lippitt SB, Parsons M. Glenohumeral arthritis and its management. In: Rockwood C, Matsen III F, Wirth M, Lippitt S, editors. The shoulder, vol. 2. 3rd ed. Philadelphia: Saunders; 2004. p. 879–1009.
6. Snyder SJ, Banas MP, Karzel RP. An analysis of 140 injuries to the superior glenoid labrum. J Shoulder Elbow Surg. 1995;4:243–8.
7. Cameron ML, Kocher MS, Briggs KK, Horan MP, Hawkins RJ. The prevalence of glenohumeral osteoarthrosis in unstable shoulders. Am J Sports Med. 2003;31(1):53–5.
8. Gartsman GM, Taverna E. The incidence of glenohumeral joint abnormalities associated with full-thickness, reparable rotator cuff tears. Arthroscopy. 1997;13(4):450–5.
9. Warner JJ, Bowen MK, Deng XH, Hannafin JA, Arnoczky SP, Warren RF. Articular contact patterns of the normal glenohumeral joint. J Shoulder Elbow Surg. 1998;7(4):381–8.
10. Porcellini G, Merolla G, Campi F, Pellegrini A, Bodanki CS, PaladiniP. Arthroscopic treatment of early glenohumeral arthritis. J Orthop Traumatol. 2013;14(1):23–9.
11. George MS. Arthroscopic management of shoulder osteoarthritis. Open Orthop J. 2008;2:23–6.
12. Weinstein DM, Bucchieri JS, Pollock RG, Flatow EL, Bigliani LU. Arthroscopic debridement of the shoulder for osteoarthritis. Arthroscopy. 2000;16(5):471–6.
13. Ho JY, Miller SL. Allografts in the treatment of athletic injuries of the shoulder. Sports Med Arthrosc Rev. 2007;15(3):149–57.
14. Scheibel M, Bartl C, Magosch P, Lichtenberg S, Habermeyer P. Osteochondral autologous transplantation for the treatment of full-thickness articular cartilage defects of the shoulder. J Bone Joint Surg Br. 2004;86:991–7.
15. Lollino N, Pellegrini A, PaladiniP CF, Porcellini G. Gleno-humeral arthritis in young patients: clinical and radiographic analysis of humerus resurfacing prosthesis and meniscus interposition. Musculoskelet Surg. 2011;95(1):59–63.
16. Sperling JW, Steinmann SP, Cordasco FA, Henshaw DR, Coons DA, Burkhead WZ. Shoulder arthritis in the young adult: arthroscopy to arthroplasty. Instr Course Lect. 2006;55:67–74.
17. Andersson G, American Academy of Orthopaedic Surgeons. The burden of musculoskeletal disease in the United States: prevalence, societal and economic cost. Rosemont: American Academy of Orthopedic Surgeons; 2008.
18. Frank RM, Van Thiel GS, Slabaugh MA, et al. Clinical outcomes after microfracture of the glenohumeral joint. Am J Sports Med. 2010;38:772–81.
19. Elser F, Braun S, Dewing CB, Millett PJ. Glenohumeral joint preservation: current options for managing cartilage lesions in young, active patients. Arthroscopy. 2010;26:685–96.
20. Millett PJ, Gobezie R, Boykin RE. Shoulder osteoarthritis: diagnosis and management. Am Fam Physician. 2008;78:605–11.
21. Farid N, Bruce D, Chung CB. Miscellaneous conditions of the shoulder: anatomical, clinical, and pictorial review emphasizing potential pitfalls in imaging diagnosis. Eur J Radiol. 2008;68:88–105.
22. Paley KJ, Jobe FW, Pink MM, Kvitne RS, ElAttrache NS. Arthroscopic findings in the overhand throwing athlete: evidence for posterior internal impingement of the rotator cuff. Arthroscopy. 2000;16(1):35–40.
23. Cole BJ, Yanke A, Provencher MT. Nonarthroplasty alternatives for the treatment of glenohumeral arthritis. J Shoulder Elbow Surg. 2007;16(5):231–40.
24. Gerber A, Lehtinen JT, Warner JJ. Glenohumeral osteoarthritis in active patients: diagnostic tips and complete management options. Phys Sportsmed. 2003;31:33–40.
25. Outerbridge RE. The etiology of chondromalacia patella. J Bone Joint Surg Br. 1961;43:752–7.
26. Walch G, Badet R, Boulahia A, Khoury A. Morphologic study of the glenoid in primary glenohumeral osteoarthritis. J Arthroplasty. 1999;14:756–60.
27. Kidder JF, Rouleau DM, DeFranco MJ, Pons-Villanueva J, Dynamidis S. Revisited: Walch classification of the glenoid in glenohumeral osteoarthritis. Shoulder Elbow. 2012;4:11–5.
28. Bhatia S, Hsu A, Lin EC, Chalmers P, Ellman M, Cole BJ, Verma NN. Surgical treatment options for the young and active middle-aged patient with glenohumeral arthritis. Adv Orthop. 2012:846843.
29. Gomoll AH, Katz JN, Warner JJ. Millett PJ rotator cuff disorders: recognition and management among patients with shoulder pain. Arthritis Rheum. 2004;50:3751–61.
30. Nové-Josserand L, Walch G, Adlerine P, Courpron P. Effect of age on the natural history of the shoulder: a clinical and radiological study in the elderly. Rev Chir Orthop Reparatrice Appar Mot. 2005;91:508–14.
31. Feeley BT, Gallo RA, Craig EV. Cuff tear arthropathy: current trends in diagnosis and surgical management. J Shoulder Elbow Surg. 2009;18:484–94.
32. Feeney M, O'Dowd J, Kay EW, Colville J. Glenohumeral articular cartilage changes in rotator cuff disease. J Shoulder Elbow Surg. 2003;12:20–3.

33. Petty DH, Jazrawi LM, Estrada LS, Andrews JR. Glenohumeral chondrolysis after shoulder arthroscopy: case reports and review of the literature. Am J Sports Med. 2004;32(2):509–15.
34. Glueck D, Gellman H. Management of the upper extremity in juvenile rheumatoid arthritis. J Am Acad Orthop Surg. 2005;13:254–66.
35. Barlow JD, Yuan BJ, Schleck CD, Harmsen WS, Cofield RH, Sperling JW. Shoulder arthroplasty for rheumatoid arthritis: 303 consecutive cases with minimum 5-year follow-up. J Shoulder Elbow Surg. 2014;23:791–9.
36. Betts HM, Abu-Rajab R, Nunn T, Brooksbank AJ. Total shoulder replacement in rheumatoid disease: a 16- to 23-year follow-up. J Bone Joint Surg Br. 2009;91:1197–200.
37. Harreld KL, Marker DR, Wiesler ER, Shafig B, Mont MA. Osteonecrosis of the humeral head. J Am Acad Orthop Surg. 2009;17(6):345–55.
38. Lau MW, Blinder MA, Williams K, Galatz LM. Shoulder arthroplasty in sickle cell patients with humeral head avascular necrosis. J Shoulder Elbow Surg. 2007;16:129–34.
39. Orfaly RM, Rockwood CA Jr, Esenyel CZ, Wirth MA. Shoulder arthroplasty in cases with avascular necrosis of the humeral head. J Shoulder Elbow Surg. 2007;16(3):27–32.
40. Meyer DC, Fucentese SF, Koller B, Gerber C. Association of osteopenia of the humeral head with full-thickness rotator cuff tears. J Shoulder Elbow Surg. 2004;13(3):333–7.
41. Allen B, Schoch B, Sperling JW, Cofield RH. Shoulder arthroplasty for osteoarthritis secondary to glenoid dysplasia: an update. J Shoulder Elbow Surg. 2014;23:214–20.
42. Cameron BD, Galatz LM, Ramsey ML, Williams GR, Lannotti JP. Non-prosthetic management of grade IV osteochondral lesions of the glenohumeral joint. J Shoulder Elbow Surg. 2002;11:25–32.
43. Ramsey JR, Savoie FH, Field LD. Arthroscopic treatment of the arthritic shoulder. In: Craig EV, editor. The shoulder. 2nd ed. Philadelphia: Lippincott Williams & Wilkins; 2004. p. 115–28.
44. Guyette TM, Bae H, Warren RF, Craig E, Wickiewicz TL. Results of arthroscopic subacromial decompression in patients with subacromial impingement and glenohumeral degenerative joint disease. J Shoulder Elbow Surg. 2002;11(4):299–304.
45. Steadman JR, Briggs KK, Rodrigo JJ, Kocher MS, Gill TJ, Rodkey WG. Outcomes of microfracture for traumatic chondral defects of the knee: average 11-year follow-up. Arthroscopy. 2003;19(5):477–84.
46. Bilsel K. Cartilage approaches in the shoulder joint. In: Bozkurt M, Tandoğan NR, Elmalı N, Aktekin CN, editors. Cartilage. Istanbul: Turkey; 2016. p. 177–85.
47. Mithoefer K, Williams RJ 3rd, Warren RF, et al. Chondral resurfacing of articular cartilage defects in the knee with the microfracture technique. Surgical technique. J Bone Joint Surg Am. 2006;88(Suppl 1 Pt 2):294–304.
48. Millett PJ, Huffard BH, Horan MP, Hawkins RJ, Steadman JR. Outcomes of full-thickness articular cartilage injuries of the shoulder treated with microfracture. Arthroscopy. 2009;25(8):856–63.
49. Meijden OA, Gaskill TR, Millett PJ. Glenohumeral joint preservation: a review of management options for young. Active patients with osteoarthritis. Adv Orthop. 2012;2012:160923.
50. Chapovsky F, Kelly JD IV. Osteochondral allograft transplantation for treatment of glenohumeral instability. Arthroscopy. 2005;21(8):1007.
51. Groot H, Donati D, Liddo MD, Gozzi E, Mercuri M. The use of cement in osteoarticular allografts for proximal humeral bone tumors. Clin Orthop Relat Res. 2004;427:190–7.
52. Geraghty S, Kuang JQ, Yoo D, LeRoux-Williams M, Vangsness CT Jr, Danilkovitch A. A novel, cryopreserved, viable osteochondral allograft designed to augment marrow stimulation for articular cartilage repair. J Orthop Surg Res. 2015;10:66.
53. Saltzman BM, Riboh JC, Cole BJ, Yanke AB. Humeral head reconstruction with osteochondral allograft transplantation. Arthroscopy. 2015;31(9):1827–34.
54. Petrera M, Veillette CJ, Taylor DW, Park SS, Theodoropoulos JS. Use of fresh osteochondral glenoid allograft to treat posteroinferior bone loss in chronic posterior shoulder instability. Am J Orthop (Belle Mead NJ). 2013;42(2):78–82.
55. Peterson L, Brittberg M, Kiviranta I, Akerlund EL, Lindahl A. Autologous chondrocyte transplantation: biomechanics and long-term durability. Am J Sports Med. 2002;30(1):2–12.
56. Romeo AA, Cole BJ, Mazzocca AD, Fox JA, Freeman KB, Joy E. Autologous chondrocyte repair of an articular defect in the humeral head. Arthroscopy. 2002;18(8):925–9.
57. Boehm E, Minkus M, Scheibel M. Autologous chondrocyte implantation for treatment of focal articular cartilage defects of the humeral head. J Shoulder Elbow Surg. 2020;29(1):2–11.
58. Scheibel M, Bartl C, Magosch P, Lichtenberg S, Habermeyer P. Osteochondral autologous transplantation for the treatment of full-thickness articular cartilage defects of the shoulder. J Bone Joint Surgery Br. 2004;86(7):991–7.
59. Pham TT, Kany J, Lakhal W, Gauzy JS, Accadbled F. Arthroscopic osteochondral autograft transfer for juvenile osteochondritis dissecans of the humeral head: a case report. JBJS Case Connect. 2017;7(3):63.
60. Nicholson GP, Goldstein JL, Romeo AA, et al. Lateral meniscus allograft biologic glenoid arthroplasty in total shoulder arthroplasty for young shoulders with degenerative joint disease. J Shoulder Elbow Surg. 2007;16(5):261–6.
61. Pennington WT, Bartz BA. Arthroscopic glenoid resurfacing with meniscal allograft: a minimally invasive alternative for treating glenohumeral arthritis. Arthroscopy. 2005;21:1517–20.
62. Bhatia DN, van Rooyen KS, du Toit DF, de Beer JF. Arthroscopic technique of interposition arthro-

plasty of the glenohumeral joint. Arthroscopy. 2006;22(5):570–5.
63. Burkhead WZ Jr, Hutton KS. Biologic resurfacing of the glenoid with hemiarthroplasty of the shoulder. J Shoulder Elbow Surg. 1995;4(4):263–70.
64. Ball CM, Galatz LM, Yamaguchi K. Meniscal allograft interposition arthroplasty for the arthritic shoulder: description of a new surgical technique. Tech Shoulder Elbow Surg. 2001;2:247–54.
65. Mattei L, Mortera S, Arrigoni C, Castoldi F. Anatomic shoulder arthroplasty: an update on indications, technique, results and complication rates. Joints. 2015;3(2):72–7.
66. Gobezie R, Lenarz CJ, Wanner JP, Streit JJ. All-arthroscopic biologic total shoulder resurfacing. Arthroscopy. 2011;27(11):1588–93.
67. Rai P, Davies O, Wand J, Bigsby E. Long-term follow-up of the Copeland mark III shoulder resurfacing hemi-arthroplasty. J Orthop. 2016;13(1):52–6.
68. Barlow JD, Abboud J. Surgical options for the young patient with glenohumeral arthritis. Int J Shoulder Surg. 2016;10(1):28–36.
69. Sperling JW, Cofield RH, Rowland CM. Neer hemi-arthroplasty and Neer total shoulder arthroplasty in patients fifty years old or less. Long-term results. J Bone Joint Surg Am. 1998;80:464–73.
70. Neer CS II. Replacement arthroplasty for glenohumeral osteoarthritis. J Bone Joint Surg Am. 1974;56:1–13.
71. Castagna A, Garofalo R. Journey of the glenoid in anatomic total shoulder replacement. Shoulder Elbow Surg. 2019;11(2):140–8.
72. Gonzalez JF, Alami GB, Baque F, et al. Complications of unconstrained shoulder prostheses. J Shoulder Elbow Surg. 2011;20:666–82.
73. Strauss EJ, Roche C, Flurin PH, et al. The glenoid in shoulder arthroplasty. J Shoulder Elbow Surg. 2009;18:819–33.
74. Ho JC, Sabesan VJ, Iannotti JP. Glenoid component retroversion is associated with osteolysis. J Bone Joint Surg Am. 2013;95:82.
75. Walch G, Mesiha M, Boileau P, et al. Three-dimensional assessment of the dimensions of the osteoarthritic glenoid. Bone Joint J. 2013;95-B: 1377–82.
76. Edwards TB, Boulahia A, Kempf JF, Boileau P, Némoz C, Walch G. Shoulder arthroplasty in patients with osteoarthritis and dysplastic glenoid morphology. J Shoulder Elbow Surg. 2004;13:1–4.
77. Matsen FA 3rd, Boileau P, Walch G, Gerber C, Bicknell RT. The reverse total shoulder arthroplasty. J Bone Joint Surg Am. 2007;89:660–7.
78. Lynch JR, Franta AK, Montgomery WH Jr, Lenters TR, Mounce D, Matsen FA 3rd. Self-assessed outcome at two to four years after shoulder hemiarthroplasty with concentric glenoid reaming. J Bone Joint Surg Am. 2007;89:1284–92.
79. Karelse A, Leuridan S, Van Tongel A, Piepers IM, Debeer P, De Wilde LF. A glenoid reaming study: how accurate are current reaming techniques? J Shoulder Elbow Surg. 2014;23:1120–7.
80. Boileau P, Krishnan SG, Tinsi L, Walch G, Coste JS, Molé D. Tuberosity malposition and migration: reasons for poor outcome after hemiarthroplasty for displaced fractures of the proximal humerus. J Shoulder Elbow Surg. 2002;11:401–12.
81. Sperling JW, Antuna SA, Sanchez-Sotelo J, Schleck C, Cofield RH. Shoulder arthroplasty for arthritis after instability surgery. JBJS Case Connect. 2002;84(10):1775–81.
82. Neer CS 2nd, Craig EV, Fukuda H. Cuff-tear arthropathy. J Bone Joint Surg Am. 1983;65(9):1232–44.
83. Grammont PM, Trouilloud P, Laffay JP, Deries X. Study and realization of a new shoulder prosthesis [in French]. Rhumatologie. 1987;39:17–22.
84. Boileau P, Sinnerton RJ, Chuinard C, Walch G. Arthroplasty of the shoulder. J Bone Joint Surg Br. 2006;88:562–75.
85. Ecklund KJ, Lee TQ, Tibone J, Gupta R. Rotator cuff tear arthropathy. J Am Acad Orthop Surg. 2007;15:340–9.
86. Drake GN, O'Connor DP, Edwards TB. Indications for reverse total shoulder arthroplasty in rotator cuff disease. Clin Orthop Relat Res. 2010;468(6):1526–33.
87. Wall B, Nove-Josserand L, O'Connor DP, Edwards TB, Walch G. Reverse total shoulder arthroplasty: a review of results according to etiology. J Bone Joint Surg Am. 2007;89:1476–85.
88. Hyun YS, Huri G, Garbis NG, McFarland EG. Uncommon indications for reverse total shoulder arthroplasty. Clin Orthop Surg. 2013;5(4):243–55.
89. Nolan BM, Ankerson E, Wiater JM. Reverse total shoulder arthroplasty improves function in cuff tear arthropathy. Clin Orthop Relat Res. 2011;469(9):2476–82.
90. Sellers TR, Abdelfattah A, Frankle MA. Massive rotator cuff tear: when to consider reverse shoulder arthroplasty. Curr Rev Musculoskelet Med. 2018;11(1):131–40.
91. Naveed MA, Kitson J, Bunker TD. The Delta III reverse shoulder replacement for cuff tear arthropathy: a single-centre study of 50 consecutive procedures. J Bone Joint Surg Br. 2011;93:57–61.
92. Frankle M, Siegal S, Pupello D, Saleem A, Mighell M, Vasey M. The reverse shoulder prosthesis for glenohumeral arthritis associated with severe rotator cuff deficiency. A minimum two-year followup study of sixty patients. J Bone Joint Surg Am. 2005;87(8):1697–705.
93. Ek ET, Neukom L, Catanzaro S, Gerber C. Reverse total shoulder arthroplasty for massive irreparable rotator cuff tears in patients younger than 65 years old: results after five to fifteen years. J Shoulder Elbow Surg. 2013;22:1199–208.
94. Muh SJ, Streit JJ, Wanner JP, Lenarz CJ, Shishani Y, Rowland DY, et al. Early follow-up of reverse total shoulder arthroplasty in patients sixty years of age or younger. J Bone Joint Surg Am. 2013;95:1877–83.

12 Inflammatory, Metabolic, and Neuromuscular Pathologies in Shoulder Joint

Hakan Turan Cift and Onur Kocadal

12.1 Polymyalgia Rheumatica

Polymyalgia rheumatica (PMR) is characterized pain in the shoulder and pelvic girdle included with prolonged early morning stiffness. The pain is typically inflammatory and worse at night. Morning stiffness eases up in the late morning or early afternoon. In the beginning of the disease shoulder pain can be unilateral and characteristically radiates distally toward the elbows. Shoulder pain may become bilateral over time. Bilateral synovial, tenosynovial, or bursal effusions occur in 76% of patients with polymyalgia rheumatica [1]. In these patients; bilateral rotator cuff tendinopathy, bursitis, biceps tenosynovitis, osteoarthritis of the acromioclavicular joint, and impingement syndrome can also accompany the clinical situation as low-grade fever and weight loss.

Simple statin myopathy, myositis, adhesive capsulitis, and fibromyalgia can be considered for the differential diagnosis of PMR. Myositis and statin myopathy can occasionally cause myalgia in addition to muscle weakness. However, myositis and statin myopathy are not associated with morning stiffness, while muscle strength is normal in PMR. In doubtful cases, measurements of muscle enzymes, electromyography, muscle MRI, or muscle biopsy can aid in securing the correct diagnosis.

For diagnosis, serum IL-6 is perhaps the most sensitive tool of disease activity [2]. However, autoimmune serology results, including antinuclear antibodies, rheumatoid factor, anti-cyclic citrullinated peptide antibodies (ACPA), and antineutrophil cytoplasmic antibodies (ANCA), are typically negative. Both the erythrocyte sedimentation rate (ESR) and C-reactive protein (CRP) levels are usually elevated in untreated patients with PMR [3, 4]. Scintigraphy demonstrates increased tracer accumulation in about 80% of patients with PMR, mainly around the shoulders. 18F-FDG PET has also been assessed as an imaging modality for patients with PMR. Increased FDG uptake in the shoulders, hips, and spinous processes of vertebrae has been observed in PMR patients.

For treatment, glucocorticoids are used at substantially lower doses in PMR. An initial dose of 12.5–25 mg/day of prednisone equivalent is suggested by EULAR/ACR recommendations. Calcium (1000–1500 mg/day) and vitamin D (800 IU/day) should be given to all patients who take glucocorticoids [5].

12.1.1 Rheumatoid Arthritis

Rheumatoid arthritis (RA) is a chronic, systemic autoimmune condition with variable manifestations. The primary expression of disease occurs

H. T. Cift (✉) · O. Kocadal
Department of Orthopaedics and Traumatology, Yeditepe University, İstanbul, Turkey

G. Huri et al. (eds.), *Fundamentals of the Shoulder*, https://doi.org/10.1007/978-3-030-94702-6_12

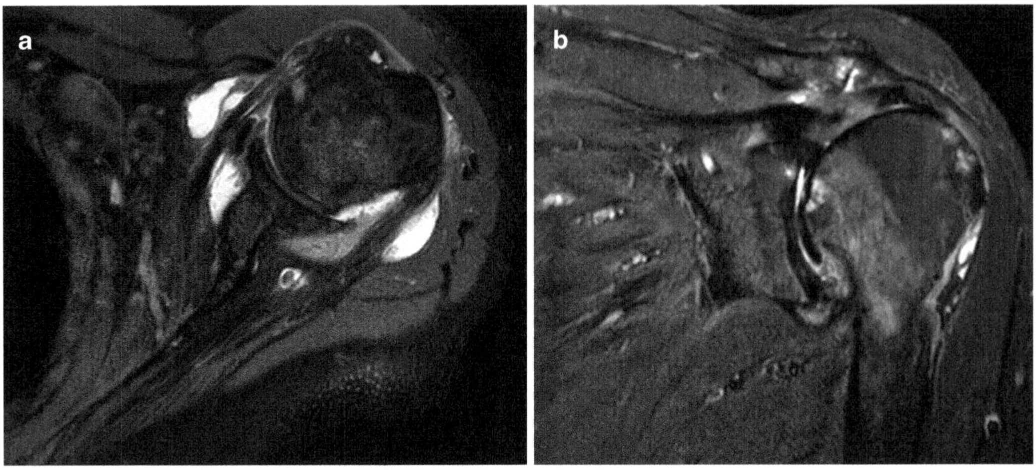

Fig. 12.1 GH MRI of a RA patient axial (**a**) and coronal (**b**) views

in the synovial tissues and is characterized clinically by symmetric polyarticular inflammation, which may lead to progressive joint damage. Rheumatoid arthritis (RA) can result in synovial and tenosynovial inflammation, as well as damage to bone and tendons.

The clinical course of RA follows an onset of disease that may be abrupt and acute, gradual and insidious, or subacute. A worse prognosis has been found in patients with disease that was gradual in onset and associated with involvement of large joints [6].

RA begins predominantly as an articular disease, and one or more joints such as shoulders, elbows, wrists, knees may be affected. Articular symptoms most often begin with stiffness, pain, and swelling involving the small joints of the hands and feet in a symmetric distribution, including the MCP, proximal interphalangeal (PIP), and MTP joints. The morning stiffness may last minutes and hours. Shoulder involvement typically produces significant limitation of motion in all planes. Acute monoarthritis of the shoulder can be mimic septic arthritis, pseudogout, or gouty process. It may also start as a systemic manifestation. For extraarticular involvement, systemic findings include diffuse polyarthralgia or polymyalgia, generalized weakness, fatigue and, less commonly, weight loss or low-grade fever.

For shoulder imaging early in disease, marginal erosions are usually seen in the rotator cuff attachments at the superolateral aspect of the humeral head. In the Grashey view, diffuse glenohumeral joint-space narrowing with loss of cartilage may be detected. In more chronic disease, rotator cuff atrophy and tearing due to superior migration of the humeral head, with narrowing of the acromiohumeral distance.

Magnetic resonance imaging (MRI) plays an important role in the diagnosis, prognosis, and monitoring of the effectiveness of therapies in clinical practice. MRI can excellently image ligaments, tendons, labrum, articular cartilage, synovium, and joint fluid. Shoulder MRI usually in axial gradient patient with rheumatoid arthritis a large synovial cyst can be detected along the long head of the biceps tendon. Pannus can be present in the glenohumeral joint (GH) along with irregular erosive changes of the glenoid and loose bodies can be seen in the GH joint (Fig. 12.1).

As shoulder and elbow surgeons, we can treat the complications of shoulder RA such as joint arthrosis. These patients use many drugs like corticosteroids so their bone quality may not be very so good, in other words osteoporotic. Hence, symptomatic treatment can be done with these patients. For severe joint arthrosis, anatomic-reverse shoulder arthroplasty (RSA) could also be performed for these patients. Intraoperative fractures, glenoid loosening, and infection rates are relatively high. However, RSA can be considered a reliable treatment option for patients with RA, resulting in significant pain relief and improvements in functional shoulder motion. The most common complication has been instability (4.7%), followed by intraoperative or postoperative fractures (4.6%), infection (3.8%), glenoid loosening (3.5%), and acromion or scapular spine fractures (1.5%). The revision rate has been reported as 10.1% [7].

12.1.2 Inflammatory Arthritis

Although the most common inflammatory arthritis involving the shoulder joint is RA, other systemic disorders such as psoriatic arthritis, systemic lupus erythematosus, ankylosing spondylitis, and systemic sclerosis (SSc) may cause glenohumeral degeneration. Motion is usually limited due to secondary soft tissue contractures.

In plain radiographies, narrowing of the glenohumeral joint space may occur, with erosion and cyst formation and without significant sclerosis or osteophytes. As the disease progresses, erosion of the superior and posterior portion of the glenoid with proximal subluxation of the humeral head may occur.

Treatment is initially conservative and is directed toward controlling pain, inducing a systemic remission, and maintaining joint motion through physical therapy. Anatomic shoulder arthroplasty cannot consider a treatment option for severe joint arthrosis.

Psoriatic arthritis (PsA): Psoriatic arthritis is a common inflammatory disease of the peripheral and axial skeleton. Psoriatic arthritis can develop at any age but it frequently appears between the ages of 30 and 50 years. Men and women are equally affected. Clinical features of the PsA are dactylitis, enthesitis, nail dystrophy, uveitis, and osteitis. In these patients, metabolic and cardiovascular disease may also be seen in 10–15% of the cases. PsA joint symptoms appear before signs of cutaneous psoriasis. Diagnosis is usually made by clinical findings; skin biopsy is rarely used to diagnose psoriasis. The CASPAR (Classification Criteria for Psoriatic Arthritis) criteria is used for diagnosis with evidence of psoriasis (current history and family history of psoriasis, nail dystrophy, negative RF, dactylitis and radiologically evidence of juxta-articular new bone formation) [8]. During shoulder roentgenography both bone erosion with new bone growth can be seen.

SLE (systemic lupus erythematosus): SLE is an autoimmune disorder characterized by antibodies to against nuclear and cytoplasmic antigens. It causes multisystem inflammation, with variable clinical manifestations, and a relapsing and remitting course. More than 90%of the cases of SLE occur in women. Muscle involvement can be either a typical inflammatory or a bland fibrotic myopathy with an elevation in the creatine phosphokinase level. The onset of muscle weakness is typically insidious, bilateral, symmetric, and painless, with involvement of proximal more than distal muscles. Patients with mild proximal weakness may develop shoulder contractures (Fig. 12.2).

The association of SLE with idiopathic polymyositis or dermatomyositis is reported to occur in the range of 4–6%. Dermatomyositis has a characteristic rash that may develop simultaneously with, or following the muscle symptoms. Gottron papules and a heliotrope rash are considered pathognomonic cutaneous features of dermatomyositis. Gottron papules are scaly; erythematous papules and plaques generally located over bony prominences, particularly the metacarpophalangeal and proximal and distal interphalangeal joints of the hands. Gottron sign is a macular erythema that occurs in the same distribution and over other extensor areas such as the elbows, knees, and ankles. Later in the course of disease, the affected skin lesions may become shiny, atrophic, and hypopigmented with telangiectasia. The "shawl" sign refers to a rash distributed over the nape of the neck, upper part of the back, and across both shoulders. Soft tissue calcification can occur in the chronic stage. Polyarthralgia and polyarthritis, if they occur,

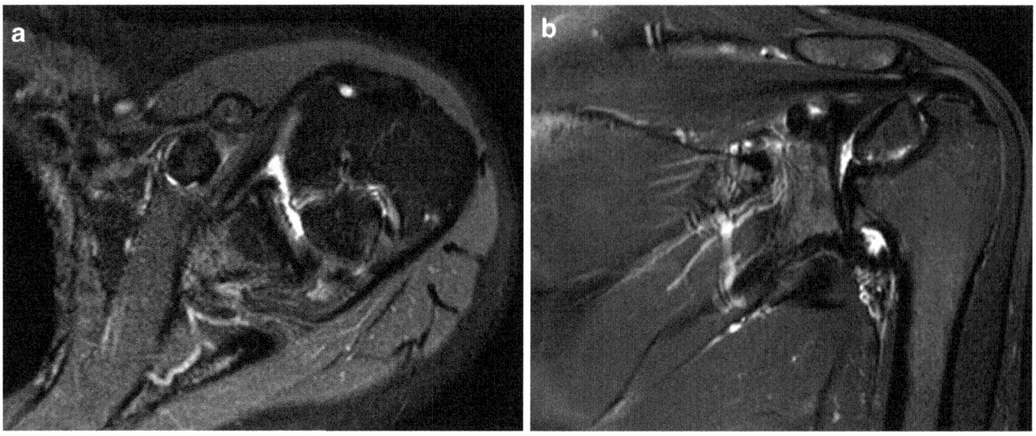

Fig. 12.2 Patient with SLE axial (**a**) and coronal (**b**) view of GH joint

present early in the course of disease. For clinical follow-up serum, CK is the most reliable enzyme test. During the flare of disease, serum CK levels usually increase weeks before the muscle weakness develops. Electromyography (EMG) is relatively sensitive but nonspecific for myositis. In myositis, typical EMG findings include irritability of myofibrils at rest and on needle insertion (with fibrillation potentials, complex repetitive discharges, and positive sharp waves), as well as short-duration, low-amplitude, complex (polyphasic) potentials on contraction. Open biopsy represents the current gold standard for confirmation of the diagnosis of inflammatory myopathy. Antinuclear or anticytoplasmic antibodies are detected in more than 90% of patients with myositis.

Ankylosing spondylitis (AS): Ankylosing spondylitis is a seronegative spondyloarthropathy and is associated with *Human Leukocyte Antigen* (HLA-B27). It is a chronic, progressive inflammatory rheumatic disease involving primarily the sacroiliac joints and the axial skeleton. The main clinical features are back pain and progressive stiffness of the spine. Oligoarthritis of the hips and shoulders, enthesopathy, and anterior uveitis are common; however, involvement of the heart and lungs is rare [9]. Typically presents itself between the ages of 20 and 40 years and is three times more likely to occur in men [10]. Clinical signs of AS include pain and stiffness of the neck and back, usually with insidious onset lasting longer than 3 months, beginning before the age of 40s. End-stage disease results in bony ankylosis of the sacroiliac joints and a "bamboo spine," characterized by ossified syndesmophytes bridging the spine disc spaces and osseous fusion of the facet joints [10].

Systemic sclerosis: SSc is characterized by fibrosis and microvascular alterations in the skin and internal organs, with an overproduction of collagen [11]. Early term of SSc often presents with a painful, symmetrical arthropathy. Musculoskeletal involvement (of the joints, tendons, and muscles) is a major cause of disability; however, the prognosis of the disease largely depends on the degree of visceral involvement [12]. The most common radiographic abnormalities in patients with SSc are subcutaneous calcinosis and digital tuft resorptions. Juxta-articular demineralization (osteoporosis/osteopenia), joint-space narrowing, intra-articular calcification, erosions, subluxation and—rarely—aseptic necrosis may also occur [13]. Subcutaneous calcinosis is more frequently found in limited SSc. Muscle weakness can be seen with the muscular involvement which may increase serum creatinine phosphokinase, aldolase and lactate dehydrogenase (LDH) levels. EMG, MRI, and muscle biopsy should also be performed. Proximal muscle weakness, mainly of the shoulder and hip girdles, is a common clinical feature in SSc.

12.1.3 Crystal-Associated Mono/Polyarthritis

Gout is the classic monoarthritis of the first metatarsophalangeal joint. Obese males, patients with hypertension, alcoholics, and postmenopausal females are more prone to gout. Patients are usually aged 45–60 year [14]. If left untreated, it can progress from recurrent monoarticular episodes to polyarticular phases. The initial involvements are usually lower extremities, but later with the polyarticular manifestations in the upper extremities may also occur. The polyarticular phase of gout can sometimes mimic RA. In acute settings, blood leukocytes counts, ESR, and CRP levels are elevated. In this phase, it should be distinguished from septic arthritis. Monosodium urate crystals (MSU) are diagnostic of gout. MSU crystals are needle-shaped, 5–30 μm in length, and highly birefringent [15]. They can be distinguished from other crystals by their properties in polarized light with an interference plate in the system. Routine radiographs frequently show no bony abnormalities (Fig. 12.3).

Pseudogout, more properly called acute calcium pyrophosphate (CPP) crystal arthritis, occurs mostly in elderly people. It can occur in relation to specific metabolic diseases, such as hemochromatosis and primary hyperparathyroidism. Calcinosis of cartilage and periarticular tissues can be seen on radiographs. Synovial fluid microscopy demonstrates rhomboid-shaped crystals at ×400 magnification, and this remains the diagnostic gold standard [16, 17]. It should also be distinguished from septic arthritis.

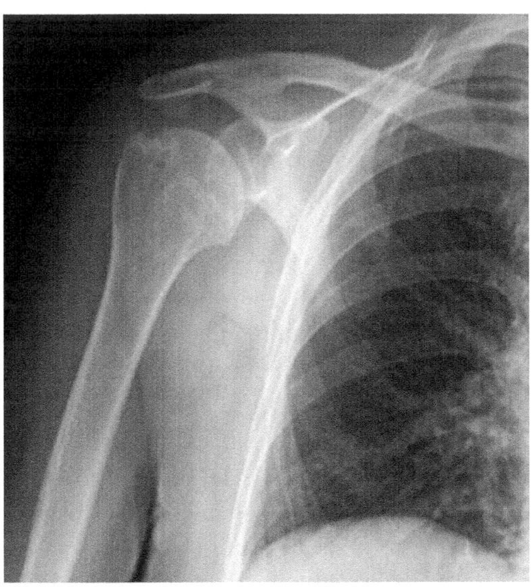

Fig. 12.3 Shoulder AP roentgenogram with Gout disease

Milwaukee shoulder (MSS): MSS is a destructive, calcium hydroxyapatite crystalline arthropathy which presents with recurrent bilateral shoulder effusions. The shoulder is most frequently involved joint. It is usually seen in elderly females. Calcium hydroxyapatite crystal disease is characterized by recurrent painful periarticular calcific deposits in tendons, soft tissues, or intra-articular surfaces. Radiographic changes on plain X-ray show joint-space narrowing, subchondral sclerosis with cyst formation, destruction of subchondral bone, soft tissue swelling, capsular calcifications, and intra-articular loose bodies. There are usually severe destructive changes of the glenohumeral joints on the MRI, massive tears of the rotator cuff can be seen [18, 19]. Unlike monosodium urate (MSU) and calcium pyrophosphate dehydrate crystals (CPPD), identifying calcium hydroxyapatite crystals has been elusive for many years, as they do not appear under plain and polarized microscopy [20, 21]. Synovial fluid is typically hemorrhagic, noninflammatory, and calcium apatite crystals are positive. Leukocyte count is typically within normal limits. Differential diagnosis of Milwaukee shoulder syndrome are rapidly destructive or progressive arthropathy, septic arthritis, neuropathic arthropathy, osteonecrosis, inflammatory arthritis, other crystal-associated arthropathy and arthropathy of late syphilis.

The most important part with crystal-associated mono/polyarthritis is to distinguish it from septic arthritis. After making the right diagnosis treatment should be done which is usually symptomatic. Steroid injection to the shoulder joint can be option for relieving pain and inflammation. If the diagnosis is gout, special treatment should be done, especially with diet, and medication with colchicine and uricolysis drugs. If crystal-associated mono/polyarthritis creates a severe GH joint arthrosis, then shoulder arthroplasty should be performed. In this point the most important part is the rotator cuff; if it is damaged then reverse shoulder arthroplasty may be an option.

12.2 Amyloidosis Associated with Immunocyte Dyscrasia (AL Amyloidosis)

Systemic AL amyloidosis, formerly known as primary amyloidosis, develops in about 2% of individuals with monoclonal gammopathies [22]. AL fibrils are derived from monoclonal immunoglobulin light chains. It can affect any organ in the body. The patients usually complain early loss of pain and temperature sensation. Later motor deficits can develop; additionally orthostatic hypotension, impotence, and GI disturbances may also occur with skin papules, nodules, and plaques. Amyloid arthropathy is rare entity but may superficially mimic acute polyarticular RA, or it may present as asymmetric arthritis affecting the hip or shoulder joints. Infiltration of the glenohumeral joint and surrounding soft tissues occasionally produces the characteristic "shoulder pad" sign. A rare but serious manifestation of AL amyloid is an acquired bleeding diathesis that may be associated with deficiency of factor X and sometimes factor IX or with increased fibrinolysis.

12.2.1 Neuropathic Arthropathy

Neuropathic arthropathy (NA) is a relatively painless, progressive, destructive arthropathy caused by a neurologic deficit. It is characterized by loss of nociception, pain sensation, or both [23]. A syrinx is believed to be the most common underlying etiology of the neuropathic shoulder [24, 25].

Arnold-Chiari malformations, post-traumatic syringomyelia, cervical spondylosis, infection, chronic alcoholism, and diabetes are other potential causes. Syphilitic myelopathy and tuberculosis spondylitis also have historical significance in etiology.

The resulting inflammation due to disturbance of sensation is considered essential to the pathophysiology of NA [26, 27]. The increase in proinflammatory cytokines leads to activation of the receptor activator of nuclear factor-κβ (NF-κβ) ligand (RANKL), which is essential for osteoclast activation, function, and survival [26]. The ligand's expression is induced by osteoblasts and binds to the RANK receptor on mononuclear precursor osteoclasts. This RANK-RANKL bond induces the formation of transcription factor NF-κβ, leading to the maturation of the osteoclasts and subsequently increasing osteoclastogenesis. A known inhibitor of the above processes is osteoprotegerin (OPG), which acts as a decoy receptor that also binds to the RANK ligand. OPGRANKL binding leads to inhibition of osteoclastogenesis and decreases survival of preexisting osteoclasts.

Symptoms are generalized pain and swelling about the shoulder region associated with limited range of motion (ROM) and weakness. Physical exam findings can be quite variable depending on the degree of GH arthritis. Swelling around the shoulder region with or without erythema should be distinguished from septic arthritis. Generalized weakness of the upper extremity with or without muscular atrophy is common.

In the differential diagnosis, diabetes mellitus, tabes dorsalis, tuberculosis, septic joint, chronic alcoholism, syringomyelia, spinal cord injury, and Gorham disease (vanishing bone disease) should be kept in mind. Radiographically, there are two patterns of disease: atrophic and hypertrophic [28]. Atrophic neuroarthropathy is characterized by bone resorption, while hypertrophic neuroarthropathy is frequently mistaken for hypertrophic osteoarthritis with findings of sclerosis, debris, fragmentation, and the presence of exuberant osteophytes [1] In addition to radiographs, MRI of the glenohumeral joint may be considered in order to confirm the diagnosis of NA (effusion, soft tissue inflammation, and glenohumeral joint destruction) or to examine for associated pathology (i.e. rotator cuff tears) [29] (Fig. 12.4).

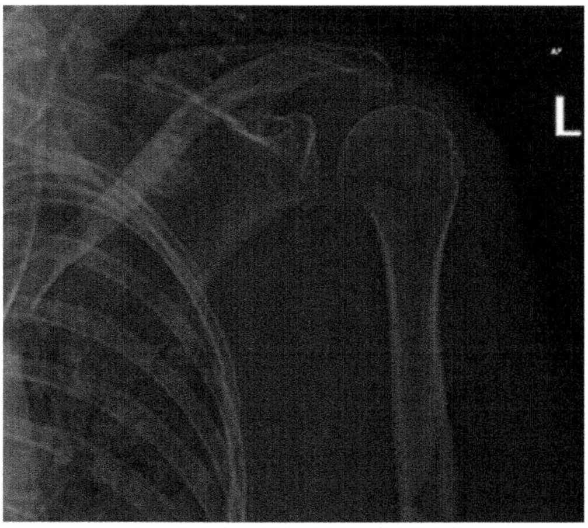

Fig. 12.4 Radiographically shoulder AP view with neuropathic arthropathy

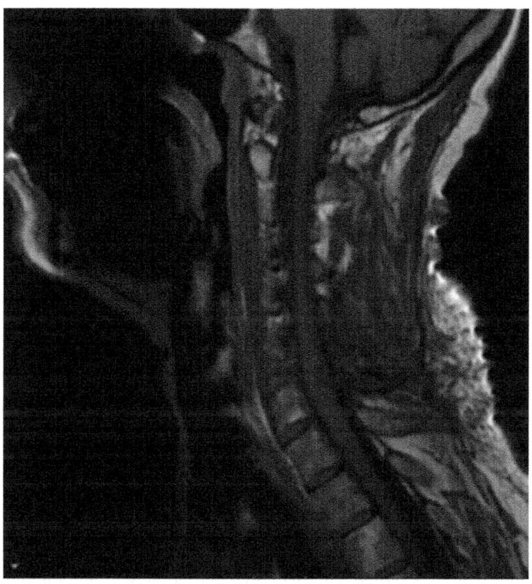

Fig. 12.5 MRI of the cervical spine with syrinx

Because the most common etiology is syrinx, the gold standard for diagnosing syrinx is to have magnetic MRI of the cervical spine in suspected cases (Fig. 12.5).

Eichenholtz classification system is applicable to the neuropathic shoulder even though it was developed for NA of the foot and ankle. The Eichenholtz classification provides a template for the natural progression of the disease on the basis of radiographic and clinical findings. The Eichenholtz system describes three stages of progression of the disease as development, coalescence, and reconstruction [30]. Stage 1 is characterized by localized warmth, erythema, and swelling about the joint on clinical exam, in addition to fragmentation, joint subluxation and dislocation, and bony debris formation may be seen on radiographs. Stage 2 shows a decrease of warmth and swelling with radiographs demonstrating absorption of the bony debris and sclerosis. Stage 3 has a complete absence of warmth, erythema, and swelling on examination, while radiographic images display osteophyte formation, decreased sclerosis, joint fusion, with uniform bone fragments. Later on, this classification has since been modified by Shibata et al. to include an inflammatory prodromal stage in which findings such as erythema, edema, and joint instability in the absence of osseous radiographic changes are present. The importance of this classification is that any operative reconstructive effort is delayed until Stage 3.

Treatment should focus on individualized management to the etiology and functional limitations of the individual patient. The treatment may be conservative or operative regarding the shoulder joint. However, syrinx decompression with drainage or shunting, posterior fossa decompression, or cervical laminectomy should be performed. Conservative treatment involves anti-inflammatory medications, joint immobilization with splinting or bracing, patient education, and if possible physical therapy to maintain ROM and strength.

If conservative management fails and symptoms persist, some surgical treatment modalities such as arthrodesis, hemiarthroplasty, or total joint arthroplasty can be performed. Underlying sensory and motor deficits limit functional expectations and increases the risk of surgical complications including infection, dislocation, and component failure. Matsuhashi et al. reported three patients who underwent uncemented hemiarthroplasty combined with open rotator cuff repair after undergoing syrinx decompression [31]. All patients had glenohumeral destruction from NAGH (neuropathic arthropathy of the glenohumeral joint) due to syringomyelia. All three patients noted an improvement in pain and active ROM after 8–10 years of follow-up. They were satisfied with their treatment and follow-up imaging did not demonstrate any evidence of prosthetic loosening. Schoch et al. presented ten Charcot shoulders they treated with hemiarthroplasty, total shoulder arthroplasty (TSA), or reverse TSA from January 2000 through December 2011. The investigators noted a statistical improvement in pain status ($p = 0.008$), although they did not find improvement in ROM [32].

12.2.2 Epilepsia

Epileptic seizures can cause shoulder dislocation and instability [33]. The incidence of shoulder dislocation during seizure is approximately 0.6% [34]. Although posterior dislocations are rarely encountered in normal populations; anterior and posterior dislocations are seen equally in epileptic patients. It has been speculated that the axial force of the adducted and internally rotated arm during the tonic phase of a seizure is the main culprit for shoulder dislocations in this setting.

Epileptic seizure can also cause fracture of the shoulder girdle. Epilepsia can lead recurrent dislocations of the shoulder because of significant bone loss of the glenoid and humeral head [35, 36].

If a patient attempts to the clinic with a first time dislocation with no history of a major trauma and the X-ray displays a deep hill sachs lesion, then the patient should be assessed for possible epileptic seizure.

The management of recurrent shoulder instability in patients with epilepsy involves a multidisciplinary team approach. In the preoperative period, epileptic attacks should be consulted with the neurology team. As a result, the focus should be on bone defects for orthopedic surgery. After surgery, a recurrence rate of 69% may be seen if postoperative seizures are not taken under control [37]. It is therefore crucial that a preoperative neurologic consultation and compliance with anticonvulsive medications is regularly assessed. Isolated arthroscopic soft tissue repair is significantly associated with high failure rate even postoperative seizures are taken under control. It is also crucial that a neurologic review is sought preoperatively and compliance with anticonvulsive medications is regularly assessed. Soft tissue repair and remplissage can be alternative treatments only if there is not a big bone defect or else arthroscopic latarjet can be an option for anterior dislocation in epileptic patients. A painful unstable shoulder with poor residual bone stock, large joint surface defects may rarely be treated with arthrodesis [38]. Arthroplasty can be option for the treatment of younger patients in selected cases; however, they have a greater risk of complications and the need for revision surgery due to their activity levels and higher life expectancy [39].

12.2.3 Hemiplegic Shoulder

Stroke, or cerebrovascular accident, is the third leading cause of death and the leading cause of adult long-term disability. Hemiplegic shoulder pain (HSP) is a common and a disabling complication following a stroke, and it may affect the quality of life [40]. It is a general term used to describe pain in the paralyzed shoulder in stroke patients within the following 3 months of stroke [41, 42]. Shoulder subluxation or rotator cuff injury can also cause HSP [43, 44]. Neurological problems such as impaired sensation, hemispatial neglect, spasticity, and flaccid paralysis can be seen in these patients. [45, 46] It occurs in approximately 65% of patients with poststroke shoulder pain (PSSP) [47].

The etiology may be multifactorial, relating to disruption of the biomechanical balance of the shoulder caused by stroke-induced weakness, spasticity, and sensory impairment. It is a multifactorial process that demands careful consideration of the contributing components, both neurologic and mechanical. Neurological factors include muscle imbalance and weakness, spasticity, brachial plexus injury, complex regional pain syndrome, and central sensitization. Weakness disrupts the stabilizers of the shoulder joint and often precedes subsequent development of spasticity. Loss of shoulder external rotation can be seen because subscapularis muscle spasticity was a significant component of a painful hemiplegic shoulder [48]. Mechanical factors include shoulder subluxation, rotator cuff injury, glenohumeral joint disorders, adhesive capsulitis, and direct trauma. Shoulder subluxation is one possible reason of hemiplegic shoulder pain (HSP). Shoulder subluxation in hemiplegic patients after stroke varies commonly in 17–64% [49].

Electrical stimulation, strapping, sling, pharmacologic therapy can be options for conservative treatment.

The development of shoulder subluxation occurs mostly during the first 3 weeks of hemiplegia [50]. Direction of the dislocation is usually displaced inferiorly and anteriorly of the shoulder girdle.

The most significant identified predictors of HSP are age (younger than 70 years), female gender, increased tone, sensory impairment, left-sided hemiparesis, hemorrhagic stroke, hemispatial neglect, and positive past medical history.

During the recovery phase of the stroke, muscle spasticity of the upper extremities is thought to cause shoulder subluxation and limited ROM, resulting in the development of shoulder pain [51]. Another important cause of HSP is frozen shoulder (adhesive capsulitis), which is indicated by a limited shoulder ROM, with a capsular type of restriction.

HSP has a significant impact on function both during and after rehabilitation. Barlak and colleagues found a significant correlation between HSP and adhesive capsulitis and complex regional pain syndrome. According to the same study there is no relationship between HSP and grade of subluxation, spasticity, impingement syndrome, or thalamic pain [52].

During the early stages following stroke, the muscles in the hemiplegic arm are usually flaccid. The most common reason is an instability of the paralyzed shoulder girdle musculature to provide dynamic stability at the joint. Downward displacement of the humerus is most common during the flaccid stage, whereas the spastic stage often leads to anterior displacement, posterior displacement, or internal rotation. Anteroposterior (AP) and oblique radiographs help diagnose and characterize shoulder subluxation. However, the association between shoulder subluxation and HSP remains controversial.

The key steps in the physical examination include observation (for asymmetry, deformity, and erythema), ROM, palpation, sensation, reflexes, strength, and special tests. The patient should demonstrate maximum active range of motion (AROM) before the examiner assesses full passive range of motion (PROM). Pain is most often the limiting factor in AROM, followed by weakness. If there is reduced PROM, contracture or anatomic block should be suspected. Strength testing in the C5-T1 myotomes (graded 0–5), sensory testing in the C5-T1 dermatomes (graded 0–21), and C5-C7 reflexes (graded 0–41) will help to localize a neurologic lesion, whether central or peripheral.

The basic components of the physical examination, such as testing of strength, sensation, and reflexes, are used regardless of the cause of HSP. Neer, Hawkins, and Jobe ("empty can") tests can assess for subacromial impingement. Apprehension and sulcus tests assess for glenohumeral joint instability.

Ultimately the diagnosis of HSP is clinical and does not necessitate diagnostic imaging. However, the use of imaging may be of benefit if the history and examination raise suspicion of underlying traumatic or structural abnormalities that may contribute to the patient's pain.

A study which evaluates of MRI findings of stroke survivors and their shoulder pain status in the chronic stage found synovial capsule thickening, synovial capsule enhancement, and enhancement in the rotator cuff interval to be more prominent in those with shoulder pain.

Efforts for prevention should be maintained throughout the course of treatment. The aims of the treatment are providing pain relief and increase in range of motion. Prevention through positioning with bracing, slings, taping, physical therapy to optimize range of motion and strength can be alternative conservative treatment methods. Transcutaneous electrical nerve stimulation (TENS) or functional electrical stimulation (FES) can be practicable. Botulinum toxin injection to the spastic muscle can be another option or else sympathetic blocks and pharmacotherapy (e.g., antispasticity, neuropathic pain) can be performed. Suprascapular nerve blocks are the other options before surgery performed.

Surgical procedures are only indicated for severe shoulder pain or stiffness, such as some cases with adhesive capsulitis which failed by all conservative treatment options. Surgery is often postponed until at least 6 months after the patient has had a stroke. Operations include release of muscle contractions, repair of rotator cuff tear. Braun and colleagues found that HSP was relieved in all 13 patients who had contracture release versus no relief in patients treated without surgery [53].

12.3 Conclusion

The exact diagnosis of the some shoulder pathologies such as rheumatological, neurological, and immunological is difficult and there are various signs that can mimic many pathologies. A multidisciplinary approach should be applied to evaluate complicated cases. In complicated cases intensive diagnosis option should be considered as biopsy. Detailed preoperative planning should be done to minimize postoperative complications.

References

1. Macchioni P, Boiardi L, Catanoso M, et al. Performance of the new 202 EULAR/ACR classification criteria for polymyalgia rheumatica: comparison with the previous criteria in a single-centre study. Ann Rheum Dis. 2014;73:1190–3.
2. Salvarani C, Cantini F, Niccoli L, et al. Acute-phase reactants and the risk of relapse/recurrence in polymyalgia rheumatica: a prospective follow-up study. Arthritis Rheum. 2005;53(1):33–8.
3. Cantini F, Salvarani C, Olivieri I, et al. Erythrocyte sedimentation rate and C-reactive protein in the evaluation of disease activity and severity in polymyalgia rheumatica: a prospective follow-up study. Semin Arthritis Rheum. 2000;30(1):17–24.
4. Kermani T, Schmidt J, Crowson C, et al. Utility of erythrocyte sedimentation rate and c-reactive protein for the diagnosis of giant cell arteritis. Semin Arthritis Rheum. 2012;41:866.
5. Dejaco C, Singh YP, Perel P, et al. Recommendations for the management of polymyalgia rheumatica: a European League Against Rheumatism/American College of Rheumatology collaborative initiative. Arthritis Rheumatol. 2015;67(10):2569–80. (Weyand CM, Fulbright JW, Evans JM, et al.: Corticosteroid requirements in polymyalgia rheumatica, Arch Intern Med 159:577, 1999.)
6. Fleming A, Crown JM, Corbett M. Early rheumatoid disease. I. Onset. Ann Rheum Dis. 1976;35(4):357–60.
7. Zumstein MA, Pinedo M, Old J, Boileau P. Problems, complications, reoperations, and revisions in reverse total shoulder arthroplasty: a systematic review. J Shoulder Elbow Surg. 2011;20(1):146–57.
8. Mease PJ, Garg A, Helliwell PS, et al. Development of criteria to distinguish inflammatory from noninflammatory arthritis, enthesitis, dactylitis, and spondylitis: a report from the GRAPPA 2013 Annual Meeting. J Rheumatol. 2014;41:1249–51.
9. Olivieri I, Barozzi L, Padula A, De Matteis M, Pavlica P. Clinical manifestations of seronegative spondylarthropathies. Eur J Radiol. 1998;27(Suppl 1):S3–6.
10. Jacobson JA, Girish G, Jiang Y, Resnick D. Radiographic evaluation of arthritis: inflammatory conditions. Radiology. 2008;248(2):378–89.
11. Randone SB, Guiducci S, Cerinic MM. Musculoskeletal involvement in systemic sclerosis. Best Pract Res Clin Rheumatol. 2008;22(2):339–50. https://doi.org/10.1016/j.berh.2008.01.008.
12. Pope JE. Musculoskeletal involvement in scleroderma. Rheum Dis Clin North Am. 2003;29:391–408.
13. Malesci D, La Montagna G, Valentini G. The arthropathy of systemic sclerosis. Rheumatism. 2005;57:232–7.
14. Zhang W, Doherty M, Pascual E, et al. EULAR evidence based recommendations for gout. Part I: diagnosis. Report of a task force of the Standing Committee for International Clinical Studies Including Therapeutics (ESCISIT). Ann Rheum Dis. 2006;65:1301–11.
15. Jordan KM, Cameron JS, Snaith M, et al. British Society for Rheumatology and British Health Professionals in Rheumatology guideline for the management of gout. Rheumatology (Oxford). 2007;46:1372–4.
16. Richette P, Bardin T, Doherty M. An update on the epidemiology of calcium pyrophosphate dihydrate crystal deposition disease. Rheumatology (Oxford). 2009;48:711–5.
17. Ivorra J, Rosas J, Pascual E. Most calcium pyrophosphate crystals appear as non-birefringent. Ann Rheum Dis. 1999;58:582–4.
18. Genta MS, Gabay C. Milwaukee shoulder. N Engl J Med. 2006;354(2):article e2.
19. McCarty DJ. Milwaukee shoulder syndrome. Trans Am Clin Climatol Assoc. 1990;102:271–83.
20. Molloy ES, McCarthy GM. Hydroxyapatite deposition disease of the joint. Curr Rheumatol Rep. 2003;5(3):215–21.
21. Paul H, Reginato AJ, Schumacher HR. Alizarin red S staining as a screening test to detect calcium compounds in synovial fluid. Arthritis Rheum. 1983;26(2):191–200.
22. Kyle RA, Therneau TM, Rajkumar SV, et al. A long-term study of prognosis in monoclonal gammopa-

thy of undetermined significance. N Engl J Med. 2002;346:564–9.
23. Alpert SK, Koval KJ, Zuckerman JD. Neuropathic arthropathy: review of current knowledge. J Am Acad Orthop Surg. 1996;4(2):100–8. (Yanik B, Tuncer S, Seckin B. Neuropathic arthropathy caused by Arnold-Chiari malformation with syringomyelia. Rheumatol Int 2004;24(4):238–41.)
24. Crowther MA, Bell SN. Neuropathic shoulder in syringomyelia treated with resurfacing arthroplasty of humeral head and soft tissue lining of glenoid: a case report. J Shoulder Elbow Surg. 2007;16(6):e38–40.
25. Richards R, Delaney J. Syringomyelia presenting as shoulder instability. J Shoulder Elbow Surg. 1992;1(3):155–61.
26. Kaynak G, Birsel O, Guven MF, Ogut T. An overview of the Charcot foot pathophysiology. Diabet Foot Ankle. 2013; https://doi.org/10.3402/dfa.v4i0.21117.
27. Jeffcoate WJ, Game F, Cavanagh PR. The role of proinflammatory cytokines in the cause of neuropathic osteoarthropathy (acute Charcot foot) in diabetes. Lancet. 2005;366(9502):2058–61.
28. Jones E, Manaster B, May D, Disler D. Neuropathic osteoarthropathy: diagnostic dilemmas and differential diagnosis. Radiographics. 2000;20:S279–93. (Hatzis N, Kaar K, Wirth M, et al. Neuropathic arthropathy of the shoulder. J Bone Joint Surg Am 1998;80(9):1314–9.)
29. Kumar S, Sharma V, Kumar S, Jain S. Imaging findings in Chiari I malformation with syringomyelia in a case of Charcot shoulder. J Clin Imaging Sci. 2011;1:46.
30. Rosenbaum AJ, DiPreta JA. Classifications in brief: Eichenholtz classification of Charcot arthropathy. Clin Orthop Relat Res. 2015;473(3):1168–71.
31. Matsuhashi T, Nagahama K, Suenaga N, et al. Midterm outcomes after humeral head replacement with rotator cuff repair in patients with syringomyelia shoulder neuroarthropathy: a report on three cases. J Shoulder Elbow Surg. 2011 Dec;20(8):e8–15.
32. Schoch B, Werthel JD, Sperling J, et al. Shoulder arthroplasty for Charcot arthropathy. Presented at the annual meeting of the American Academy of Orthopaedic Surgeons, Las Vegas, Nevada, March 25–27, 2015.
33. Buhler M, Gerber C. Shoulder instability related to epileptic seizures. J Shoulder Elbow Surg. 2002;11:339–44. https://doi.org/10.1067/mse.2002.124524.
34. Goudie EB, Murray IR, Robinson CM. Instability of the shoulder following seizures. J Bone Joint Surg Br. 2012;94:721–8. https://doi.org/10.1302/0301-620x.94b6.28259.
35. Provencher MT, Bhatia S, Ghodadra NS, Grumet RC, Bach BR Jr, Dewing CB, et al. Recurrent shoulder instability: current concepts for evaluation and management of glenoid bone loss. J Bone Joint Surg Am. 2010;92(Suppl 2):133–51. https://doi.org/10.2106/jbjs.j.00906.
36. Raiss P, Lin A, Mizuno N, Melis B, Walch G. Results of the Latarjet procedure for recurrent anterior dislocation of the shoulder in patients with epilepsy. J Bone Joint Surg Br. 2012;94:1260–4. https://doi.org/10.1302/0301-620x.94b9.29401.
37. Blackman AJ, Krych AJ, Kuzma SA, Chow RM, Camp C, Dahm DL. Results of revision anterior shoulder stabilization surgery in adolescent athletes. Arthroscopy. 2014;30:1400–5.
38. Mizuno N, Denard PJ, Raiss P, Melis B, Walch G. Long-term results of the Latarjet procedure for anterior instability of the shoulder. J Shoulder Elbow Surg. 2014;23:1691–9. https://doi.org/10.1016/j.jse.2014.02.015.
39. Thangarajah T, Lambert S. The management of recurrent shoulder instability in patients with epilepsy: a 15-year experience. J Shoulder Elbow Surg. 2015;24(11):1723–7.
40. Pan R, Zhou M, Cai H, Guo Y, Zhan L, Li M, Yang Z, Zhu L, Zhan J, Chen H. A randomized controlled trial of a modified wheelchair arm-support to reduce shoulder pain in stroke patients. Clin Rehabil. 2017;32(1):37–47.
41. Poduri KR. Shoulder pain in stroke patients and its effects on rehabilitation. J Stroke Cerebrovasc Dis. 1993;3:261–6.
42. Adey-Wakeling Z, Arima H, Crotty M, Leyden J, Kleinig T, Anderson CS, Newbury J. Incidence and associations of hemiplegic shoulder pain poststroke: prospective population-based study. Arch Phys Med Rehabil. 2015;96:241–7.
43. Yamaguchi K, Sher JS, Andersen WK, Garretson R, Uribe JW, Hechtman K, Neviaser RJ. Glenohumeral motion in patients with rotator cuff tears: a comparison of asymptomatic and symptomatic shoulders. J Shoulder Elbow Surg. 2000;9:6–11.
44. Chantraine A, Baribeault A, Uebelhart D, Gremion G. Shoulder pain and dysfunction in hemiplegia: effects of functional electrical stimulation. Arch Phys Med Rehabil. 1999;80:328–31.
45. Vuagnat H, Chantraine A. Shoulder pain in hemiplegia revisited: contribution of functional electrical stimulation and other therapies. J Rehabil Med. 2003;35:49–54.
46. Meskers CG, Koppe PA, Konijnenbelt MH, Veeger DH, Janssen TW. Kinematic alterations in the ipsilateral shoulder of patients with hemiplegia due to stroke. Am J Phys Med Rehabil. 2005;84:97–105.
47. Lindgren I, Jönsson AC, Norrving B, Lindgren A. Shoulder pain after stroke: a prospective population-based study. Stroke. 2007;38(2):343–8. https://doi.org/10.1161/01.STR.0000254598.16739.4e.
48. Bender L, Mckenna K. Hemiplegic shoulder pain: defining the problem and its management. Disabil Rehabil. 2001;23(16):698Y705. https://doi.org/10.1080/09638280110062149.
49. Park GY, Kim JM, Sohn SI, Shin IH, Lee MY. Ultrasonographic measurement of shoulder subluxation in patients with post-stroke hemiplegia. J Rehabil Med. 2007 Sep;39(7):526–30. https://doi.org/10.2340/16501977-0099.
50. Stolzenberg D, Siu G, Cruz E. Current and future interventions for glenohumeral subluxation in hemi-

plegia secondary to stroke. Top Stroke Rehabil. 2012;19:444–56.
51. Van Ouwenaller C, Laplace PM, Chantraine A. Painful shoulder in hemiplegia. Arch Phys Med Rehabil. 1986;67:23–5.
52. Barlak A, Unsal S, Kaya K, Onat SS, Ozel S. Poststroke shoulder pain in Turkish stroke patients: relationship with clinical factors and functional outcomes. Int J Rehabil Res. 2009;32(4):309–15. https://doi.org/10.1097/MRR.0b013e32831e455f.
53. Braun RM, West F, Mooney V, Nickel VL, Roper B, Caldwell C. Surgical treatment of the painful shoulder contracture in the stroke patient. J Bone Joint Surg Am. 1971;53(7):1307–12.

Nerve Entrapments Around Shoulder

13

Onur Basci, Onur Gursan, and Mustafa Özkan

13.1 Suprascapular Neuropathy

Suprascapular neuropathy has an increasing popularity considering the pain and weakness symptoms around shoulder. The distinctive anatomy of the nerve makes it susceptible to dynamic and mechanics changes on its pathway. It is first mentioned by Andre Thomas in 1936 [1] followed by Schilf in 1952 [2] and Thompson and Kopell in 1959 [3]. Numerous causes were described in literature as ganglion cysts [4], rotator cuff tears [5], trauma [6], and overhead activities [7, 8]. The challenging diagnosis of the given condition has promising treatment results when the location and the mechanism of the entrapment are identified properly.

13.1.1 Anatomy

The suprascapular nerve is a mixed sensory and motor nerve arising from the C5, C6, and variationally C4 nerve roots (76% from C5 and C6, C4, 18% from C5 and C6, 6% only C5) [9]. After arising it emerges posterolaterally to the posterior cervical triangle. The nerve obliquely and posteriorly crosses the clavicle and reaches to suprascapular notch. The shape and depth of the notch are widely variable which is clearly described by Rengachary as six different morphologies [10]. The nerve passes through this narrow notch beneath the transverse scapular ligament and reaches the posterior scapular surface. The ossification or hypertrophy of this ligament might cause entrapment in this localization [11]. After proceeding posteriorly the nerve passes through the spinoglenoid notch. This notch is 1.8–2.1 cm medial to the glenoid and the roof is made up of the spinoglenoid ligament [12]. This ligament lies 4.6 mm superior to the nerve and is tightened in adduction and internal rotation of shoulder (throwing movement) (Fig. 13.1). This entity might be another reason of an entrapment in throwing athletes [13].

Suprascapular nerve motor branches mainly innervate the supraspinatus and infraspinatus muscle after passing through the suprascapular notch together with the suprascapular artery suprascapular nerve sensory branches innervating certain localizations in the shoulder, such as coracohumeral and coracoacromial ligaments and glenohumeral and acromioclavicular joints including the skin area [9, 14].

O. Basci (✉) · O. Gursan · M. Özkan
Department of Orthopaedics and Traumatology, Dokuz Eylul University School of Medicine, Izmir, Turkey
e-mail: onur.basci@deu.edu.tr; onur.gursan@deu.edu.tr; mustafa.ozkan@deu.edu.tr

G. Huri et al. (eds.), *Fundamentals of the Shoulder*, https://doi.org/10.1007/978-3-030-94702-6_13

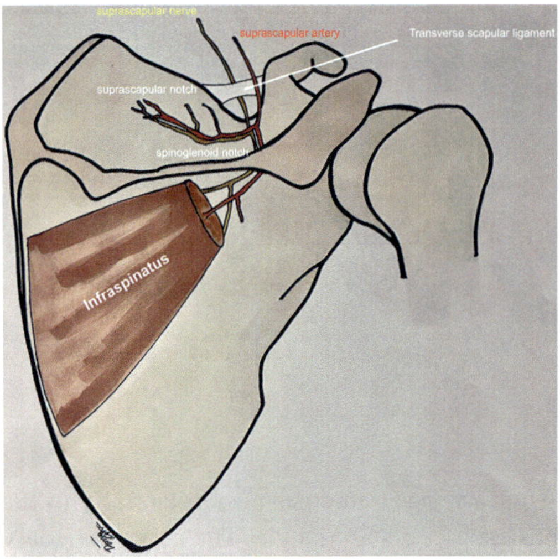

Fig. 13.1 Suprascapular nerve entrapment, main compression localizations are the suprascapular notch under the transverse scapular ligament and the spinoglenoid notch as shown on the anatomic drawing

13.1.2 Pathophysiology

The anatomic course of suprascapular nerve and the structures around is a potential reason to its injury or entrapment with various primary and secondary mechanisms. A wide range of causes that lead to suprascapular nerve injury or compression have been emphasized, including variations in the anatomy, traumatic conditions such as massive rotator cuff tears, space occupying lesions such as ganglion cysts or repetitive microtrauma.

The nerve is mostly affected by its course passing through the suprascapular notch and the spinoglenoid notch. Anatomic variations on these locations such as attenuated suprascapular or spinoglenoid notch or an abnormally tight or ossified transverse scapular or spinoglenoid ligaments are main reasons for the nerve entrapment [12, 15, 16]. Traction or compression of the nerve by certain conditions of shoulder in these anatomic locations results in pathologic compression.

The suprascapular nerve also might be injured either by compression or traction due to dynamic changes around shoulder. This mechanism is mainly common in overhead athletes or people have increased overhead activities. Nearly about 30% of elite volleyball players have suprascapular neuropathy in their dominant arms [17, 18]. Various theories have been offered for the mechanisms of this situation. When one's shoulder is in abduction and external rotation position during overhead throwing motion, the nerve might be jammed between the infraspinatus muscle fibers and the bony spinoglenoid notch. Conversely, the reason for the compression might also be the tightened spinoglenoid ligament as a result of the given motion [19, 20].

There are several static causes of entrapment. Space occupying lesions, such as paralabral cysts, caused by labral lesions, can affect the nerve in the spinoglenoid notch and in the suprascapular notch if the cyst is larger [4, 21, 22]. Massive rotator cuff tears that are retracted alter the space around the suprascapular notch causing a traction injury to the nerve [5, 23]. Variations in the anatomy on the course of the nerve, traumatic conditions as scapular fractures, posttraumatic arthritis, and heterotopic ossifications are other static conditions that lead to compression. Iatrogenic causes that might affect the nerve are surgical procedures including distal clavicular resection and posterior approaches to shoulder, injections and regional anesthetic procedures. [10, 24–29]. Furthermore, microemboli in the vasa nervorum of suprascapular nerve due to intimal injury of the suprascapular or axillary artery caused by traumatic conditions and dilatation of the suprascapular veins might either be additional causes of the compression [30–33].

13.1.3 Clinics and Diagnosis

The diagnosis of suprascapular neuropathy might be quite tough because of the similar history and complaints with the other pathologies around shoulder region. In order to be the neuropathy diagnosis consistent, younger patients with overhead activities and patients with associated conditions mentioned above should be considered. Distinctly, patients complain about dull aching pain superoposteriorly and laterally around shoulder region. Muscle weakness with overhead throwing motions with or without night pain is common among these patients [32]. Adduction and internal rotation which cause traction to the nerve worsen the pain [34]. If the compression is around the suprascapular notch, based on the supraspinatus affection, up to 75% of weakness in abduction and external rotation is expected [35]. However, if the compression is around the spinoglenoid notch due to the posterior deltoid and teres minor compensation, the infraspinatus weakness is concealed [20]. Following a careful anamnesis focusing on overhead throwing activities, trauma, previous surgeries, anesthetic shoulder injections step-by-step physical examinations should be performed. In differential diagnosis neuralgic amyotrophy and the viral neuritis or Parsonage Turner Syndrome should be considered [36].

A careful comparative inspection to shoulder is the first step of the examination. Atrophy on the supraspinatus and infraspinatus muscles is a sign for a rather proximal compression or a suprascapular notch involvement, whereas an isolated infraspinatus atrophy is a sign for a rather distal compression or spinoglenoid notch involvement [19]. In a suprascapular compression supraspinatus fossa atrophy is pathognomonic [37]. In palpation, tenderness over area between posterior clavicle and scapular spine resembles a proximal compression, whereas tenderness deep and posterior to the acromioclavicular joint resembles a distal compression.

Both passive and active range of motion should be assessed including forward flexion, external rotation, and internal rotation. All of the rotator cuff muscle strengths should be assessed paying extra attention to supraspinatus and infraspinatus by Jobe's test and external rotation lag test, respectively. Stretching the spinoglenoid ligament by cross-body adduction and internal rotation provoking the compression in this region might increase the pain at the posterior aspect of the shoulder. This test was described by Plancher as the "cross arm adduction test" [34]. Suprascaular nerve stretch test, described by Lafosse, is also a provocative test which inreases pain at the posterior aspect of the shoulder when positive. To perform the test, standing behind the patient, the head is rotated to the contralateral shoulder, while shoulder is retracted with the other hand [38]. Local lidocaine injection test is another useful test for distinguishing nerve pathologies from other shoulder problems by placing the needle 4 cm medial to the posterolateral corner of the acromion. After the injection the provocative tests are repeated to confirm that the pain is relieved.

If in the history and examination there is a suspected entrapment radiologic assessment is mandatory. The standard AP, true AP, transscapular are performed to confirm other pathologies of shoulder, a Stryker notch view is performed to assess the suprascapular and spinoglenoid notch anatomy and a Zanca view is performed to evaluate the acromioclavicular joint [39]. For further assessment of the bony anatomy for variations or traumatic lesions three-dimensional computed tomography (3D CT) scan is of great use. CT – Arthrography to confirm a ganglion cyst presence and CT – angiography to evaluate if there is a suprascapular artery compression are also available imaging modalities [33]. In the long run, a magnetic resonance image (MRI) of the shoulder is the best technique which is helpful to confirm any space occupying lesion on the course of the nerve and the concomitant shoulder pathologies. The severity of the entrapment might be assessed with MRI by edema, atrophy, and fatty degeneration of the muscles [33, 40] (Fig. 13.2).

Besides all of the useful imaging techniques mentioned above to evaluate the cause of an entrapment, the electrodiagnostic tests remain the gold standard for confirming the suprascapular neuropathies. Even though there are variable results, the accuracy of the electromyography and nerve conduction velocity studies are shown to be 91% accurate [41]. In recent approach, EMG findings may not change the surgical intervention in a suspected suprascapular nerve entrapment, on the basis of the neuropathy being a dynamic phenomenon which is not always detectable in EMG and an addition of a suprascapular release in an arthroscopic surgery being a safe and easy technique [38].

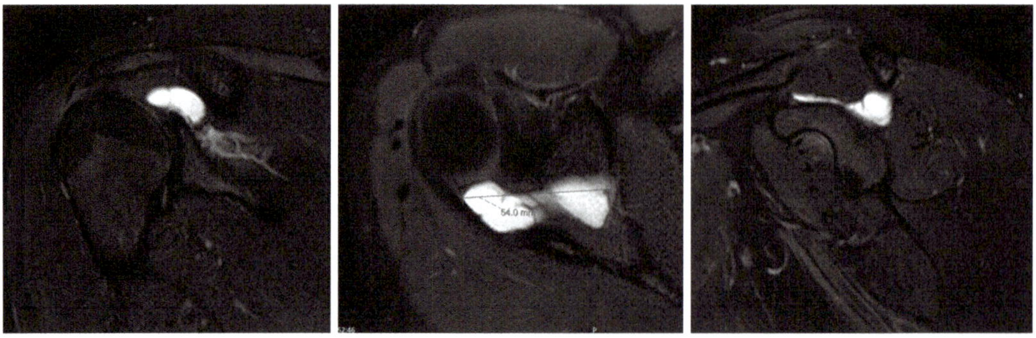

Fig. 13.2 A suprascapular nerve entrapment case caused by a paralabral cyst. Coronal, transverse, sagittal MRI sections are shown in the figure

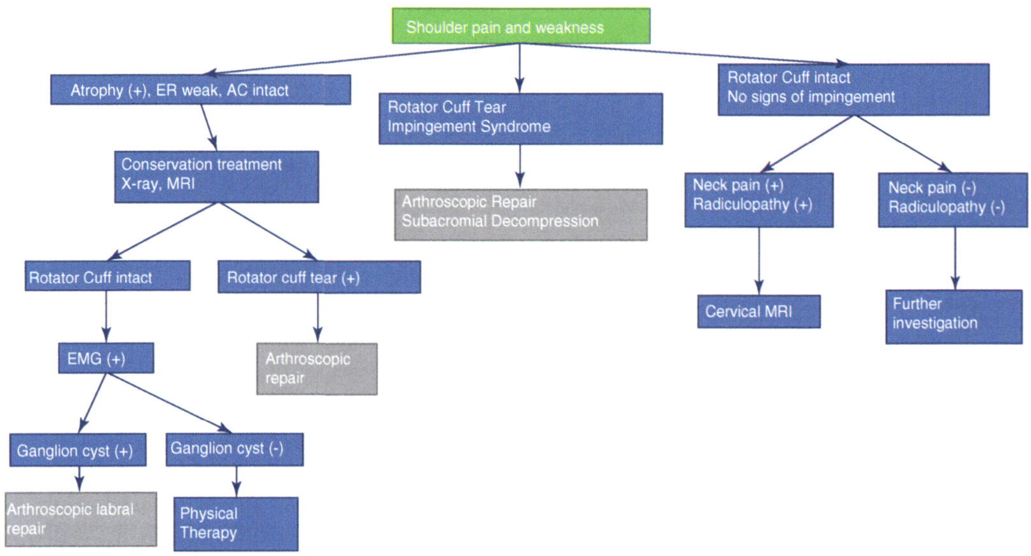

Fig. 13.3 Suprascapular nerve entrapment management algorithm

13.1.4 Treatment

The most distinctive factor that affects the selection of the optimum treatment modality is the cause of the entrapment. In a vigilant approach, nonoperative management of suprascapular neuropathies is advisable. Dynamic etiologies are often more suitable for nonoperative approaches. However, secondary causes such as space occupying lesions require operative interventions. The decision precisely depends on the duration of the symptoms and atrophy. In order to avoid irreversible changes early intervention might be mandatory (Fig. 13.3).

13.1.4.1 Conservative Treatment

In the literature, if there are no secondary lesions affecting the nerve course, high quality outcomes of conservative remedy are compromised. It consists of Nonsteroidal anti-inflammatory drugs (NSAID), shoulder motion modifications, and scapular stabilization by rotator cuff, deltoid and periscapular muscle exercises. The results of conservative management have promising results leaving a question that if the improvements are due to nerve healing or compensatory muscles that strengthen and substitute the affected ones [42]. The aspiration of the ganglion cyst under the guidance of ultrasound or CT, has a high recurrence rate as 75–100% [43]. The dispute on the mechanism of symptom relief leads others to early surgical decompression to prevent irreversible nerve damage [44].

13.1.4.2 Surgical Treatment

If the conservative management fails and/or an apparent space occupying lesion leading to the entrapment is present, surgical treatment is indicated. If an isolated case of nerve entrapment is adressed, direct nerve decompression is performed either in athroscopic or open surgical interventions. The nerve is mainly entrapped in the supracapular or spinoglenoid notch. If there is a concomminant pathology leading to entrapment such as rotator cuff tears or paralabral cysts the treatment of the concomitant lesion might be succesful. The treatment of a concomitant lesion with or without an addition of nerve decompression is still a debate [45–50].

13.2 Axillary Neuropathy and Quadrilateral Space Syndrome

Quadrilateral space (QS) is an anatomic passage in which the posterior circumflex artery and axillary nerve are contained. Any pathologic condition that leads to the entrapment of these two anatomical structures is called the quadrilateral space syndrome (QSS). Cahill and Palmer described the syndrome in 1983 [51]. It is characterized by a dull indistinct pain on the posterolateral aspect of shoulder which generally appears after overhead or throwing activities [51]. The diagnosis requires quite attention and awareness on the indistinct symptoms, with accurate diagnostic work-up.

13.2.1 Anatomy

The QS lies posteriorly on the intersection of the scapula and humerus, forming a gap just inferior to teres minor and surrounded by long head of triceps posteriorly, humerus anteriorly, and teres major and latissimus dorsi muscles inferiorly. It

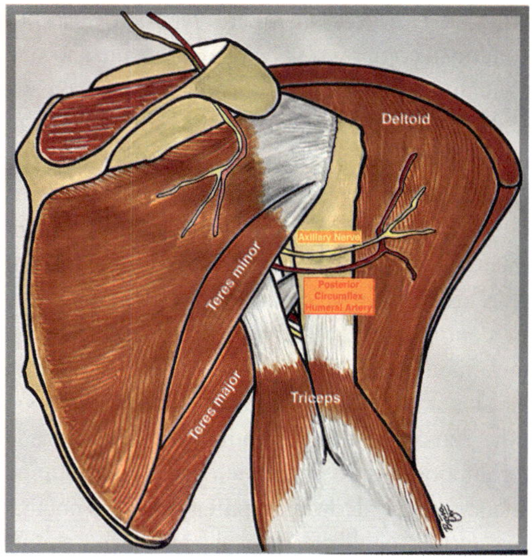

Fig. 13.4 Quadrilateral space syndrome (QSS) anatomical considerations as shown in the anatomic drawing

is an anatomic passage for the axillary nerve and the posterior circumflex humeral artery (PCHA), branch of the axillary artery, anterior to posterior aspect of the shoulder [52]. The axillary nerve innervates the teres minor and deltoid muscle in this region. The course of the nerve starts from the posterior cord of the brachial plexus on the anterior aspect of shoulder, anteriorly to the inferior one-third of subscapularis, then entering to the axillary pouch and finally to the QS. The nerve separates into anterior and posterior branches in the space. The anterior branch has its course subsided with the posterior circumflex humeral vessels. There are fibrous bands within the space described in literature that might be responsible for the entrapment [52] (Fig. 13.4).

13.2.2 Pathophysiology

Although the ethiopathogenesis of QSS remains uncertain, any lesion or anatomic structure that occupies the QS leads to an entrapment of the contents of QSS. If the affected content is axillary nerve, it is then called a neurogenic quadrilateral space syndrome (nQSS) and if posterior circumflex humeral artery is the one affected it is then called a vascular quadrilateral space syndrome (vQSS). However there are reported cases of combined compression being neurogenic and vascular together. The two distinct theories on the QSS are based on either the repetitive trauma to the PCHA due to increased frequencies of overhead activities in athletes and heavy duty workers, or space occupying lesions due to structural or traumatic entities [52]. Most accused reasons for QSS are the dynamic causes of the compression. In continuous abduction and external rotation movements, the PCHA is most likely stretched in its anatomic location causing a turbulent blood flow further leading to intimal hyperplasia and thrombotic occlusion which finalizes with an aneurysm [53]. However, axillary nerve is not that susceptible to this mechanical stress because of its excursion and strain responses [54].

The static conditions or extrinsic causes of the entrapment include most frequently the fibrous bands, resulting from repetitive microtrauma to connective tissue around the QS region, by causing a reduction in the cross-sectional area of QS [52]. Other causes of static lesions are muscular hypertrophy [55], anatomical variations [56], bony formations [57], and any space occupying lesions including labral cysts [58], osteochondromas, and fracture hematoma [59].

13.2.3 Clinics and Diagnosis

QSS is often a pathology for the second to fourth decades of life with a male predominance. It usually affects athletes with overhead activities such as volleyball, baseball, and swimming [60]. Patients with intermittent, indistinct, paresthetic pain on the posterolateral aspect of the shoulder provocated with abduction and external rotation of shoulder are candidates for an investigation of QSS [61]. The clinical presentation of the disease depends on the etiopathogenesis as mentioned above. nQSS is characterized by nonspecific neurogenic pain, paresthesia, weakness, while vQSS is characterized by acute ischemic symptoms and thrombosis or distal embolism. Both conditions might cause muscle weakness and atrophy by chronic denervation. Tenderness or tinnel sign over QS might be apparent and distinguishing in patients with QSS [53, 62]. Holding the arm on abduction and external rotation position for 1 or 2 min will exacerbate the pain caused by entrapment [60] (Fig. 13.5).

In a suspected case, MRI is the firstline imaging technique to examine the anatomy and the focal fatty degeneration and atrophy of the teres minor muscle [63]. A lidocaine block test, with or without the guidance of ultrasound just 2–3 cm inferior to the standard posterior shoulder portal, might be helpful for the diagnosis of the condition when a sudden relief of pain is obtained after injection [64].

Digital subtraction angiography, computed tomography angiography, and magnetic resonance angiography are the following imaging modalities for a suspected vQSS, visualizing the PCHA [65]. A dynamic arteriography in abduction and external rotation position of shoulder has also been described to confirm the entrapment of the PCHA in low specificity [66, 67].

While electrodiagnostic tests are an important diagnostic tool for neuropathies, EMG for axillary neuropathy has a high false negativity rate. However it is still helpful on excluding other causes of neuropathies mimicking QSS such as brachial plexus pathologies, cervical pathologies, and thoracic outlet syndrome [68].

13.2.4 Treatment

The management of QSS is still questionable with insufficient literature support; however, conservative measures in the preliminary management are favored.

13.2.4.1 Conservative Treatment

Firstline management of QSS is given as at least 6 months of conservative management, in the literature [66, 68]. The conservative management includes NSAIDs, activity modification, manual therapy, and therapeutic exercises. Other beneficial glenohumeral range of motion (ROM) exercises, rotator cuff and periscapular muscle strengthening, posterior capsule stretching, and finally transverse friction and active release soft tissue massages [66].

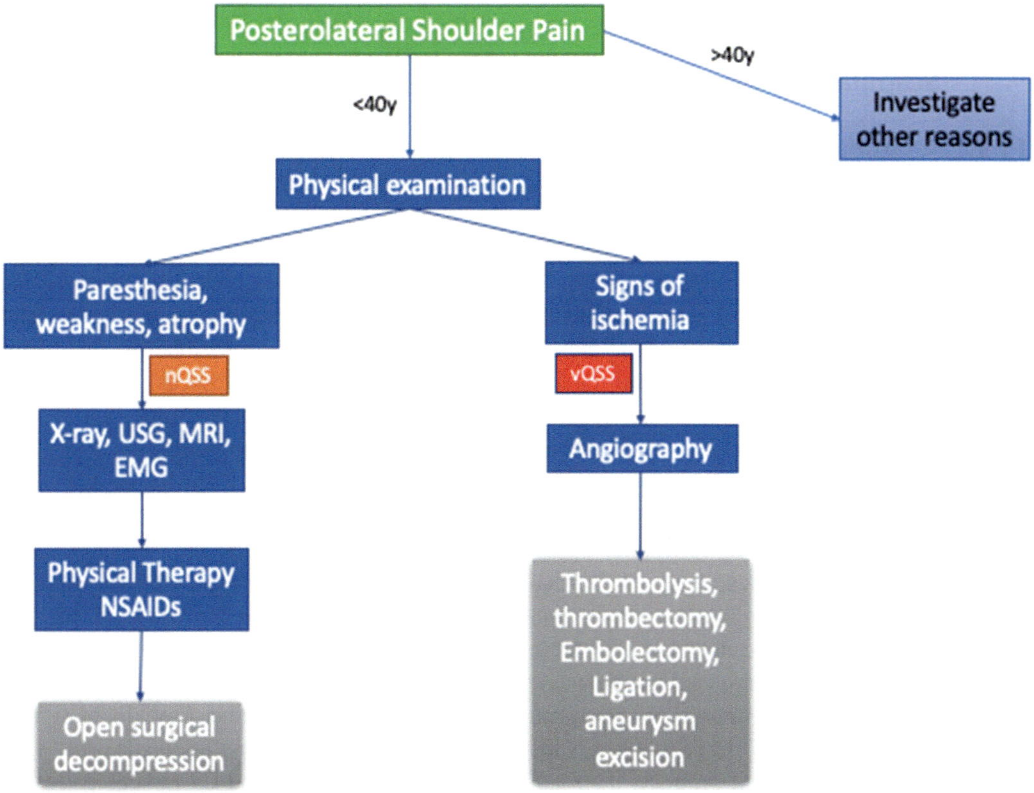

Fig. 13.5 Quadrilateral space syndrome management algorithm [52]

Even though there is no literature support, another option for a nonsurgical intervention is the steroid injection to the QS with the guidance ultrasound. Briefly it is performed directly on the QS just 2 cm below to standard posterior arthroscopic portal of the shoulder. Relief of the symptoms is either diagnostic or therapeutic [53].

13.2.4.2 Surgical Treatment

Surgical treatment is indicated when symptom relief is not accomplished in 6 months and a confirmed space occupying lesion is present. Open surgical decompression is the option for the patients that unachieve the conservative measures and have space occupying lesions in QS. Before performing the surgery, other entities that imitate QSS should be excluded with further investigation such as arteriography for the PCHA. The surgery is performed either on lateral decubitus or prone position, with a 4–5 cm incision over the Langer's lines reflecting the posterior fibers of deltoid muscle, leaving its acromial attachments intact. Axillary nerve is dissected between the teres minor and major muscles in the QS, from the fibrous bands, and both the nerve and PCHA are palpated and tested if they are compressed in external rotation and abduction of the shoulder by the excursion of the nerve and the pulsating artery, respectively [68]. An arthroscopic debridement might be added to the procedure in the presence of an intraarticular lesion such as paralabral cysts and/or labral tears that affect the QS [53]. The authors advise against the application of the arthroscopic procedure before the open surgical decompression, in order to prevent the fluid extravasation that negatively affects the surgical dissection.

In cases of PCHA aneurysm, surgical resection, and ligation or endovascular coiling, embolectomy for distal emboli, catheter-directed thrombolysis, or thrombectomy for thrombosed PCHA are other options of surgical interventions [53].

13.3 Long Thoracic Nerve Neuropathy and Medial Scapular Winging

13.3.1 Anatomy

Long thoracic nerve originating from the vertical branch of C5–6–7 cervical nerves innervates m. serratus anterior. Palsy of the serratus anterior is the most common etiology of scapular winging [69]. Musculus serratus anterior is a flat muscle that originates from the outer surface of the first nine ribs. Muscle passes posterosuperiorly along the thoracic wall and inserts onto the costal surface of the medial scapula. Muscle has three components with different roles. The proximal component which is generally innervated by C5 and C6 nerve roots is responsible for lateral rotation of inferior scapular angle. After passing brachial plexus posteriorly, long thoracic nerve receives fibers from C7 nerve root. Intermediate and distal component of muscle is innervated with the contribution of C7 nerve root to C5 and C6. Intermediate component protracts the scapula. Distal portion of muscle acts to protract and rotate inferior angle of scapula superior and laterally [69–71].

As it is mentioned above, the nerve has fibers from C5 and C6 nerve roots initially, then passes anteriorly through the scalenus medius. After gaining C7 nerve root based fibers, the nerve courses inferiorly through clavicle and brachial plexus. Long thoracic nerve courses superficially on the lateral thorax wall after passing over first costa [72]. It has an average of 21.4 cm in length [73].

13.3.2 Pathophysiology

Neurapraxia development is generally as a result of blunt trauma such as vehicle accidents or excessive traction (strech) of the nerve that can be seen in athletes and many sports injuries. Repetitive microtraumas as a result of activities with the head tilted opposite site of nerve and the arm overhead such as javelin throwers and tennis servers may cause increase in tensile force on nerve [74–76]. If the nerve has more than 10% increase in length, the neuropraxia may occur. The superficial course of nerve along thoracic wall paves the way for compression injuries and contusion. Middle scalene muscle, upper parts of first rib, inferior angle of scapula, and the interval between clavicle and second rib are most probably points where the compression of nerve takes place. In their cadaveric study Hester et al. described bow-stringing phenomenon across the fascial band with progressive abduction and external rotation [72]. Long thoracic nerve neuropraxia has been reported especially in positions where the arm is held in abduction for a long time (arm abduction, hand supporting chin, book reading position) [77, 78].

Serratus anterior paralysis can be seen due to penetrating injuries of the long thoracic nerve during surgical procedures, especially during radical mastectomy, first rib resection, axillary lymph node dissection thoracostomy tube placement, and transaxillary sympathectomy [79].

Guillain-Barré syndrome, Arnold-Chiari malformation, systemic lupus erythematosus, viral issues, Lyme disease, and C7 radiculopathy are other non-traumatic conditions that should be kept in mind in terms of serratus anterior palsy [80–82].

13.3.3 Diagnosis

Patients typically suffer from pain around the shoulder which may radiate to arm or to scapula. Absence of serratus anterior contraction, compensatory mechanism of rhomboid, and levator scapula muscles produces spasm and finally pain [69, 83]. In patients with severe pain, neuritis, such as Parsonage–Turner syndrome should be considered. Typically, the pain usually resolves spontaneously after several weeks, but the patient is left with a winged scapula [84]. Weakness especially in athletes is additional complaint to pain.

Physical examination should be done after the patient adequately exposed, with both arms in the resting position, and if there is atrophy in the shoulder, it should be evaluated. The patient's scapulothoracic rhythm should be evaluated during active elevation of the upper extremities. In patients with serratus anterior insufficiency after long thoracic nerve injury, the scapula takes the position of superior elevation, medial translation with medial rotation of the inferior corner. Medial scapular winging can be accentuated when the patient is asked to forward flex his arms to the horizontal plane and/or push on a wall in a push-up motion. Manually stabilizing the scapula to the thorax wall may help examiner to detect any deformity with motion of shoulder. Patients with serratus anterior muscle palsy have difficulty with active forward elevation beyond 120° (Fig. 13.6a). To differentiate serratus anterior palsy from muscle dysfunction due to posterior instability, forward elevation and external rotation tests are performed. The dysfunction is due to posterior instability if winging is corrected by external rotation. Muscle weakness is confirmed in 90° shoulder flexion in push-up position. Medial winging may be immediately obvious, or it may manifest with muscle fatigue after 5–10 repetition [69, 70, 85, 86].

The initial management starts with conventional X-ray of the shoulder, scapula, chest, and cervical spine in order to exclude other disorders such as osteochondromas, malunions, or accessory ribs. CT may help to better characterize osteochondromas. MRI is reserved for patients with cervical disc disease, shoulder instability, or rotator cuff tears. EMG evaluation including the shoulder girdle muscles should be done and be repeated every 3 months in order to investigate the presence of nerve recovery. Clinical recovery may or may not be related with the results of serial electromyographic examination [76, 87, 88].

13.3.4 Treatment

13.3.4.1 Conservative Treatment

If there is no evidence about the presence of direct long thoracic nerve laceration injury, conservative treatment should be approved for a period of 12–24 months. Progressive clinical and electromyographic recovery should be observed [69, 76, 89] (Fig. 13.7).

Watson et al. have described the methodology of conservative approach to serratus anterior palsy. In the first step, the objective is mainly pain relief and saving the range of motion. It is advised to make ROM exercise in supine position in order to stabilize scapula by body weight. Another important point is to avoid overhead use of arm. Second step includes painless period and nerve recovery is expected to be started in this phase. İt is important to keep up full ROM and

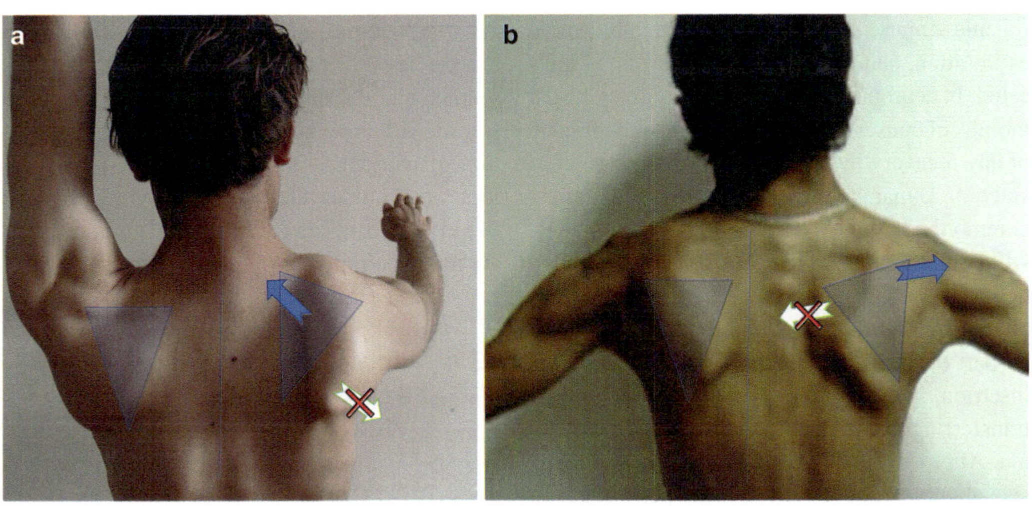

Fig. 13.6 (**a**) Medial scapular winging, the serratus anterior muscle is not working as the *white arrow* and the scapula shifts medially as the *blue arrow*. (**b**) Lateral scapular winging, the trapezius or rhomboid muscle is not working as the *white arrow* and the scapula shifts laterally as the *blue arrow*

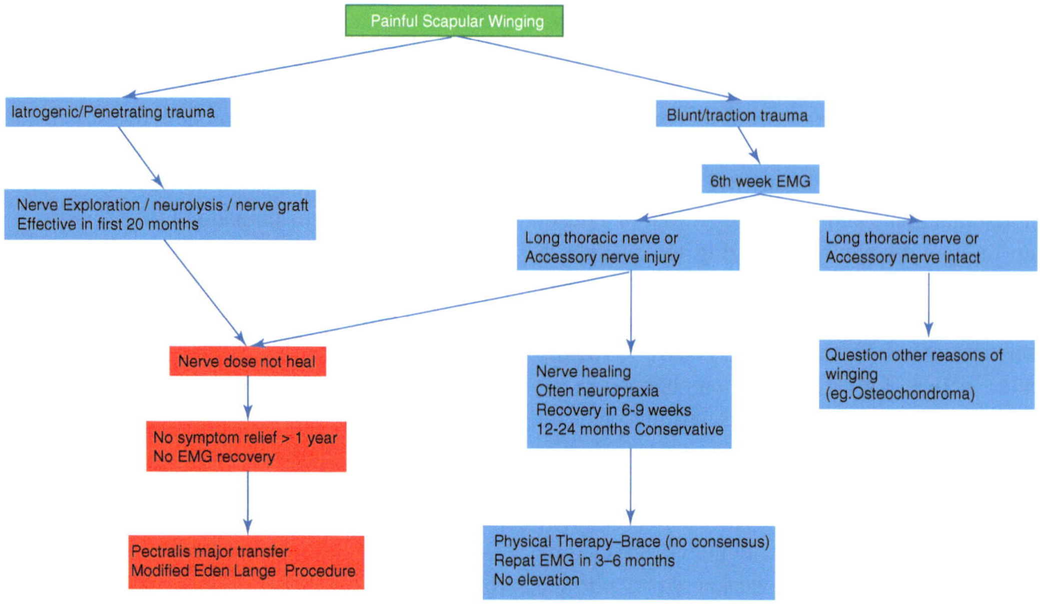

Fig. 13.7 Scapular winging management algorithm

stretch the rhomboids, levator scapulae, pectoralis minör to prevent contracture related problems. In the last step, strengthening exercises for all shoulder girdle muscles are advised [90].

Conservative treatment is more successful in serratus anterior palsies than trapezius palsies.

13.3.4.2 Surgical Treatment

If scapular winging is associated with either iatrogenic injury or penetrating trauma, nerve exploration, neurolysis, or nerve grafting is indicated. In neuropraxia cases, after failure of 12–24 months of conservative management and absence of any recovery in EMG, surgery should be considered. Dynamic muscle transfer is the most commonly used surgical procedure today. Although the transfer of many muscles has been described, it has been reported that successful results are obtained by separating the sternocostal part of the pectoralis major from the humeral insertion, strengthening with the fascia lata and transferring it to the inferior corner of the scapula. After 4 weeks of sling immobilization, passive shoulder exercises should be performed. In weightlifters and individuals participating in contact sports, rehabilitation period should be longer. It has shown that this procedure has good functional outcomes and better pain relief and sweep winging away. Facial sling operations, which are applied by tethering the fascial grafts between the medial edge of the scapula and the spinous processes of the thoracic vertebra, have lost its popularity due to postoperative recurrence. Successful results have been reported with fusions between the scapula and thorax in cases where muscle transfers and facial sling operations had failed. As a salvage procedure, scapulothoracic fusions are performed between the medial edge of the scapula and the ribs with cerclage wire or plate-screws with bone graft support. Although resolution of winging is successful in this procedure, loss of shoulder elevation and vital capacity, pseudoarthrosis, and pulmonary complications are not uncommon [69, 91–98].

13.4 Accessory Nerve Neuropathy and Lateral Scapular Winging

13.4.1 Anatomy

Accessory nerve neuropathy may produce trapezius palsy. By the way, it is important to have detailed knowledge about course of nerve and function of trapezius muscle. Trapezius muscle which originates from external occipital protuberance, one-third of the nuchal line and the spines of the 7 cervical and 12 thoracic vertebrae, elevates, retracts, rotates, and depress the scapula by three different insertions. Superior 1/3 of the muscle elevates the scapula and rotates the lateral angle upwardly, middle 1/3 adducts and retracts and inferior 1/3 depresses the scapula and rotates the inferior angle laterally. Superior and inferior portion of the muscle accommodates scapula with the mechanism of moving the glenoid upward during abduction [69, 74, 99, 100].

Accessory nerve is the only cranial nerve that enters (foramen magnum) and exits (foramen jugulare) the skull. After leaving the skull, the spinal accessory nerve penetrates to the deep surface of the sternocleidomastoid muscle, entering the posterior triangle of the neck with close association associated with a chain of five to ten lymph nodes. After crossing posterior cervical triangle superficially gives deep fibers along trapezius [99, 100].

13.4.2 Pathophysiology

Iatrogenic injuries during lymph node biopsy or cervical mass excision procedures are known as the most common etiologic factor. After median nerve injury during carpal tunnel surgery, accessory nerve injury during lymph node biopsy takes second place in terms of neural iatrogenic injuries. Another etiologic factor is direct trauma to the posterior cervical region [101].

The superficial course of the nerve makes it prone to injury. Heavy lifting and penetrating trauma to posterior cervical triangle are other mechanisms for accessory nerve injury. Spontaneous or idiopathic trapezius paralysis has also been reported [102].

13.4.3 Diagnosis

Patients typically complaint about shoulder or upper back pain, fatigue, stiffness, or muscle weakness, especially with overhead activity. Absence of trapezius contraction, compensatory mechanism of other muscles produces spasm. This spasm and strain may be a reason for severe posterior shoulder pain. In some cases, pain radiates to arm or proximally to paraspinous cervical region. In chronic cases, pain can be felt in forearm, hand, face, and head, even contralateral sidearm. Daily life activities such as writing for a long time, driving, or heavy lifting become restricted. Scapular rotation deficiency due to trapezius palsy can be connected with the pain of subacromial impingement of ipsilateral shoulder. Overhead activity such as throwing may aggravate the symptoms [69, 74, 103, 104].

Physical examination should be done after the patient is adequately exposed, with both arms in the resting position, and if there is atrophy in the shoulder, it should be evaluated. The patient's scapulothoracic rhythm should be evaluated during active elevation of the upper extremities. As a result of deficiency in muscle tone of trapezius, scapula moves inferiorly and inferior angle rotates laterally. By the way, drooping of ipsilateral neckline should call up trapezius palsy. Lateral winging is minimal in contrast with serratus anterior palsy (Fig. 13.6b). With the abduction of shoulder winging can become more visible, conversely winging may disappear with forward flexion. Commonly, limitation of abduction is present and abduction range may catch up to 80–90° [104]. An alternative test to identify accessory nerve neuropathy is proposed by Chan et al. as glenohumeral external rotation against a resistance [105]. Any lateral winging of the scapula is indicative of accessory nerve palsy including paralysis of the rhomboid and serratus anterior muscles, herniated nucleus pulposus, scoliosis, progressive neuromuscular disease, scapular osteochondroma (Fig. 13.8), fracture malunion, stroke, herpes zoster infection [106].

The management should start with conventional radiography as usual to eliminate other disorders of shoulder, cervical spine, and scapula such as accessory ribs, space occupying bone tumors (e.g. Osteochondroma), and posttraumatic changes. CT may help to better characterize osteochondromas. MRI is reserved for

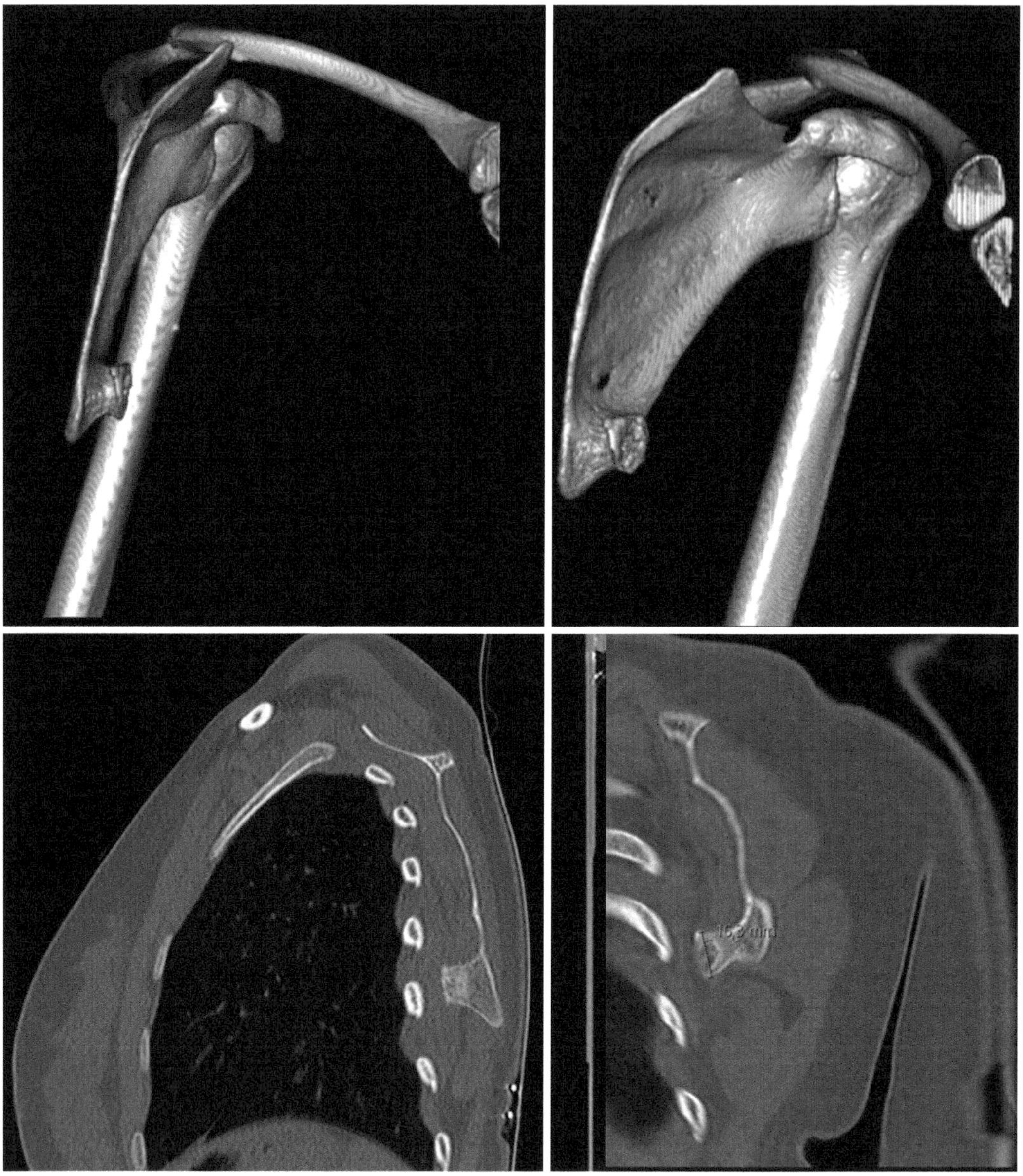

Fig. 13.8 Other causes of scapular winging might be space occupying lesions sraound scapulotoracic area such as an osteochondroma

patients with cervical disc disease, shoulder instability, or rotator cuff tears. EMG evaluation including the shoulder girdle muscles should be done and be repeated every 3 months in order to investigate the presence of nerve recovery. Clinical recovery may or may not be related with the results of serial electromyographic examination [76, 87, 88] (Fig. 13.7).

13.4.4 Treatment

13.4.4.1 Conservative Treatment

Conservative treatment options such as physical therapy, transcutaneous nerve stimulation, external support, chiropracty, NSAIDS, and narcotic analgesics are often unsuccesful in patients with trapezius palsy. Poor outcome of conservative treatment is generally as a result of inability of physical therapy that aims to improve the strength and function of the compensatory muscles [74, 106]. Shoulder brace/orthosis may be used to press the scapula against the thoracic wall especially in radical neck dissection patients. Although bracing seems to be successful in pain relief, range of motion of shoulder joint remains limited. Nonoperative management is promising with high success regarding serratus anterior palsy over trapezius palsies [107].

13.4.4.2 Surgical Treatment

In iatrogenic or penetrating injuries of spinal accessory nerve, surgical exploration, neurolysis, and if needed nerve repair, or nerve grafting should be considered. These procedures have variable outcomes. In their retrospective study, Kim et al. reported the outcomes of surgical procedures that included 58 nerve grafting, 26 end-to-end repair, and 19 neurolysis. Almost all injuries were iatrogenic (93%) and neuorolysis was performed if spinal accessory nerve was intact and intraoperative electrical evidence of regeneration was present. Neurolysis resulted in mostly excellent results, while end-to-end suture repair was reported as slightly better outcomes than graft repair [108]. Teboul et al. summarize that better results from nerve repair can be expected if the procedure is performed in first 20 months. Regarding the nerve repair site; the shorter the distance of the repair site to the end plates of nerve, the better the outcomes should be expected [102].

Especially in non-traumatic cases, after the failure of 12–24 months of conservative management and absence of any recovery in EMG, dynamic muscle transfer surgery should be considered. Eden-Lange muscle transfer procedure is the most favored for isolated chronic trapezius palsy. This procedure includes the transfer of the insertion sites of the levator scapulae and rhomboid muscles laterally along the scapula. Hereby, new vectors of these muscles allow them to stabilize the scapula and support the shoulder girdle, instead of the denervated trapezius [109]. Outcomes generally range from good to excellent and success in pain relief is satisfactory. Some authors recommend the Eden-Lange procedure only in the case of spontaneous palsy of the trapezius, failed nerve repair, or a duration of 12–20 months have elapsed post-injury [69, 102].

With the presence of serratus anterior palsy or rhomboid weakness additional to trapezius palsy, poorer outcomes should be expected. In this situation, scapulothoracic fusion may be an option [69]. Patients with muscular distrophy or those who do not obtain relief with dynamic muscle transfer are candidates for scapulothoracic stabilization. Giannini et al. reviewed the results in nine patients with muscular dystrophy treated with wire fixation to the thoracic ribs without fusion. They reported that winging had resolved in all patients, and statistically significant increase in abduction strength was found at the end of first year. Patient developed pneumothorax that resolved spontaneously within 48 h [110]. Krishnan et al. recently reported their results of scapulothoracic arthrodesis for muscular dystrophy and refractory winging using plates and wires in 22 patients (24 shoulders). Pulmonary complications developed in nearly half of shoulders, and 7 shoulders (29%) developed pseudoarthrosis. They have advocated for routine thoracostomy tube placement and iliac crest autograft to minimize complications [95]. Teboul et al. suggested neurolysis, nerve graft, or repair within 6–12 months in injuries resulting from surgery or penetrating trauma [102]. In their meta-analysis, Nath et al. reported that both long thoracic or spinal accessory nerve decompression and neurolysis are effective techniques in correcting winging scapula in comparison with muscle and tendon transfer operations [111].

References

1. La TA. Paralysie du muscle sous-épineux. Presse Med. 1936;64:1283–4.
2. Schilf E. Unilateral paralysis of the suprascapular nerve. Nervenarzt. 1952;23(8):306–7.
3. Thompson WA, Kopell HP. Peripheral entrapment neuropathies of the upper extremity. N Engl J Med. 1959;260(25):1261–5.
4. Lee BC, Yegappan M, Thiagarajan P. Suprascapular nerve neuropathy secondary to spinoglenoid notch ganglion cyst: case reports and review of literature. Ann Acad Med Singapore. 2007;36(12):1032–5.

5. Mallon WJ, Wilson RJ, Basamania CJ. The association of suprascapular neuropathy with massive rotator cuff tears: a preliminary report. J Shoulder Elbow Surg. 2006;15(4):395–8.
6. Yoon TN, Grabois M, Guillen M. Suprascapular nerve injury following trauma to the shoulder. J Trauma. 1981;21(8):652–5.
7. Ringel SP, Treihaft M, Carry M, Fisher R, Jacobs P. Suprascapular neuropathy in pitchers. Am J Sports Med. 1990;18(1):80–6.
8. Witvrouw E, Cools A, Lysens R, et al. Suprascapular neuropathy in volleyball players. Br J Sports Med. 2000;34(3):174–80.
9. Shin C, Lee SE, Yu KH, Chae HK, Lee KS. Spinal root origins and innervations of the suprascapular nerve. Surg Radiol Anat. 2010;32:235–8.
10. Rengachary SS, Neff JP, Singer PA, Brackett CE. Suprascapular entrapment neuropathy: a clinical, anatomical and comparative study. Part 1: clinical study. Neurosurgery. 1979;5:441–6.
11. Bayramoğlu A, Demiryürek D, Tüccar E, Erbil M, Aldur MM, Tetik O, et al. Variations in anatomy at the suprascapular notch possibly causing suprascapular nerve entrapment: an anatomical study. Knee Surg Sports Traumatol Arthrosc. 2003;11:393–8.
12. Warner JJP, Krushell RJ, Masquelet A, Gerber C. Anatomy and relationships of the suprascapular nerve: anatomical constraints to mobilization of the supraspinatus and infraspinatus muscles in the management of massive rotator-cuff tears. J Bone Joint Surg Am. 1992;74:36–45.
13. Plancher KD, Peterson RK, Johnston JC, Luke TA. The spinoglenoid ligament: anatomy, morphology, and histological findings. J Bone Joint Surg Am. 2005;87:361–5.
14. Vorster W, Lange CP, Briet RJ, Labuschagne BC, du Toit DF, Muller CJ, et al. The sensory branch distribution of the suprascapular nerve: an anatomic study. J Shoulder Elbow Surg. 2008;17:500–2.
15. Bigliani LU, Dalsey RM, McCann PD, April EW. An anatomical study of the suprascapular nerve. Arthroscopy. 1990;6:301–5.
16. Demaio M, Drez D Jr, Mullins RC. The inferior transverse scapular ligament as a possible cause of entrapment neuropathy of the nerve to the infraspinatus. A brief note. J Bone Joint Surg Am. 1991;73:1061–3.
17. Holzgraefe M, Kukowski B, Eggert S. Prevalence of latent and manifest supra scapular neuropathy in high-performance volleyball players. Br J Sports Med. 1994;28:177–9.
18. Lajtai G, Pfirrmann CW, Aitzetmuller G, Pirkl C, Gerber C, Jost B. The shoulders of professional beach volleyball players: high prevalence of infraspinatus muscle atrophy. Am J Sports Med. 2009;37:1375–83.
19. Ferretti A, Cerullo G, Russo G. Suprascapular neuropathy in volleyball players. J Bone Joint Surg Am. 1987;69(260-3):45.
20. Ferretti A, De Carli A, Fontana M. Injury of the suprascapular nerve at the spinoglenoid notch: the natural history of infraspinatus atrophy in volleyball players. Am J Sports Med. 1998;26:759–63.
21. Westerheide KJ, Dopirak RM, Karzel RP, Snyder SJ. Suprascapular nerve palsy secondary to spinoglenoid cysts: results of arthroscopic treatment. Arthroscopy. 2006;22:721–7.
22. Abboud JA, Silverberg D, Glaser DL, Ramsey ML, Williams GR. Arthroscopy effectively treats ganglion cysts of the shoulder. Clin Orthop Relat Res. 2006;444:129–33.
23. Albritton MJ, Graham RD, Richards Ii RS, Basamania CJ. An anatomic study of the effects on the suprascapular nerve due to retraction of the supraspinatus muscle after a rotator cuff tear. J Shoulder Elbow Surg. 2003;12:497–500.
24. Solheim LF, Roaas A. Compression of the suprascapular nerve after fracture of the scapular notch. Acta Orthop Scand. 1978;49:338–40.
25. Maquieira GJ, Gerber C, Schneeberger AG. Suprascapular nerve palsy after the Latarjet procedure. J Shoulder Elbow Surg. 2007;16:e13–5.
26. Shaffer BS, Conway J, Jobe FW, Kvitne RS, Tibone JE. Infraspinatus muscle-splitting incision in posterior shoulder surgery. Am J Sports Med. 1994;22:113–20.
27. Wijdicks CA, Armitage BM, Anavian J, Schroder LK, Cole PA. Vulnerable neurovasculature with a posterior approach to the scapula. Clin Orthop Relat Res. 2009;467:2011–7.
28. Hussain N, Goldar G, Ragina N, Banfield L, Laffey JG, Abdallah FW. Suprascapular and interscalene nerve block for shoulder surgery: a systematic review and meta-analysis. Anesthesiology. 2017;127:998–1013.
29. Draeger RW, Messer TM. Suprascapular nerve palsy following supraclavicular block for upper extremity surgery: report of 3 cases. J Hand Surg Am. 2012;37:2576–9.
30. Travlos J, Goldberg I, Boome RS. Brachial plexus lesions associated with dislocated shoulders. J Bone Joint Surg Br. 1990;72(1):68–71.
31. Visser CP, Coene LN, Brand R, Tavy DL. The incidence of nerve injury in anterior dislocation of the shoulder and its influence on functional recovery. A prospective clinical and EMG study. J Bone Joint Surg Br. 1999;81(4):679–85.
32. Vachon T, Rosenthal M, Dewing CB, Solomon DJ, Shin AY, Provencher MT. Acute painless shoulder weakness during high-intensity athletic training. Am J Sports Med. 2009;37:175–80.
33. Blum A, Lecocq S, Louis M, Wassel J, Moisei A, Teixeira P. The nerves around the shoulder. Eur J Radiol. 2013;82:2–16.
34. Plancher KD, Luke TA, Peterson RK, Yacoubian SV. Posterior shoulder pain: a dynamic study of the spinoglenoid ligament and treatment with arthroscopic release of the scapular tunnel. Arthroscopy. 2007;23(9):991–8.
35. Gerber C, Blumenthal S, Curt A, Werner CML. Effect of selective experimental suprascapular nerve block

on abduction and external rotation strength of the shoulder. J Shoulder Elbow Surg. 2007;16:815–20.
36. Misamore GW, Lehman DE. Parsonage-Turner syndrome (acute brachial neuritis). J Bone Joint Surg Am. 1996;78(9):1405–8.
37. Boykin RE, Friedman DJ, Zimmer ZR, Oaklander AL, Higgins LD, Warner JP. Suprascapular neuropathy in a shoulder referral practice. J Shoulder Elbow Surg. 2011;20:983–8.
38. Lafosse L, Piper K, Lanz U. Arthroscopic suprascapular nerve release: indications and technique. J Shoulder Elbow Surg. 2011;20(2):S9–S13.
39. Plancher KD, Petterson SC. Posterior shoulder pain and arthroscopic decompression of the suprascapular nerve at the transverse scapular ligament. Oper Tech Sports Med. 2014;22:58–72.
40. Inokuchi W, Ogawa K, Horiuchi Y. Magnetic resonance imaging of suprascapular nerve palsy. J Shoulder Elbow Surg. 1998;7:223–7.
41. Nardin RA, Rutkove SB, Raynor EM. Diagnostic accuracy of electrodiagnostic testing in the evaluation of weakness. Muscle Nerve. 2002;26:201–5.
42. Martin SD, Warren RF, Martin TL, Kennedy K, O'Brien SJ, Wickiewicz TL. Suprascapular neuropathy. results of non-operative treatment. J Bone Joint Surg Am. 1997;79:1159–65.
43. Moen TC, Babatunde OM, Hsu SH, Ahmad C, Levine WN. Suprascapular neuropathy: what does the literature show? J Shoulder Elbow Surg. 2012;21:835–46.
44. Callahan JD, Scully TB, Shapiro SA, Worth RM. Suprascapular nerve entrapment. A series of 27 cases. J Neurosurg. 1991;74:893–6.
45. Costouros JG, Porramatikul M, Lie DT, Warner JJP. Reversal of suprascapular neuropathy following arthroscopic repair of massive supraspinatus and infraspinatus rotator cuff tears. Arthroscopy. 2007;23:1152–61.
46. Schroder CP, Skare O, Stiris M, Gjengedal E, Uppheim G, Brox JI. Treatment of labral tears with associated spinoglenoid cysts without cyst decompression. J Bone Joint Surg Am. 2008;90:523–30.
47. Youm T, Matthews PV, El Attrache NS. Treatment of patients with spinoglenoid cysts associated with superior labral tears without cyst aspiration, debridement, or excision. Arthroscopy. 2006;22(5):548–52.
48. Hosseini H, Agneskirchner JD, Tröger M, Lobenhoffer P. Arthroscopic release of the superior transverse ligament and SLAP refixation in a case of suprascapular nerve entrapment. Arthroscopy. 2007;23:1134.e1–4.
49. Lafosse L, Tomasi A, Corbett S, Baier G, Willems K, Gobezie R. Arthroscopic release of suprascapular nerve entrapment at the suprascapular notch: technique and preliminary results. Arthroscopy. 2007;23:34–42.
50. Tashjian RZ, Burks RT. Arthroscopic aspiration and labral repair for treatment of spinoglenoid notch cysts. Am J Orthop (Belle Mead NJ). 2009;38:94–6.
51. Cahill BR, Palmer RE. Quadrilateral space syndrome. J Hand Surg Am. 1983;8(1):65–9.
52. McClelland D, Paxinos A. The anatomy of the quadrilateral space with reference to quadrilateral space syndrome. J Shoulder Elbow Surg. 2008;17:162–4.
53. Brown SA, Doolittle DA, Bohanon CJ, Jayaraj A, Naidu SG, Huettl E, et al. Quadrilateral space syndrome: the Mayo Clinic experience with a new classification system and case series. Mayo Clin Proc. 2015;90:382–94.
54. Topp KS, Boyd BS. Structure and biomechanics of peripheral nerves: nerve responses to physical stresses and implications for physical therapist practice. Phys Ther. 2006;86(1):92–109.
55. Redler MR, Ruland LJ III, McCue FC III. Quadrilateral space syndrome in a throwing athlete. Am J Sports Med. 1986;14(6):511–3.
56. Friend J, Francis S, McCulloch J, Ecker J, Breidahl W, McMenamin P. Teres minor innervation in the context of isolated muscle atrophy. Surg Radiol Anat. 2010;32(3):243–9.
57. Amin MF, Berst M, el-Khoury GY. An unusual cause of the quadrilateral space impingement syndrome by a bone spike. Skeletal Radiol. 2006;35(12):956–8.
58. Robinson P, White LM, Lax M, Salonen D, Bell RS. Quadrilateral space syndrome caused by glenoid labral cyst. AJR Am J Roentgenol. 2000;175(4):1103–5.
59. Cirpar M, Gudemez E, Cetik O, Uslu M, Eksioglu F. Quadrilateral space syndrome caused by a humeral osteochondroma: a case report and review of literature. HSS J. 2006;2(2):154–6.
60. Aval SM, Durand P Jr, Shankwiler JA. Neurovascular injuries to the athlete's shoulder: Part I. J Am Acad Orthop Surg. 2007;15:249–56.
61. Cummins CA, Schneider DS. Peripheral nerve injuries in baseball players. Neurol Clin. 2008;26:195–215.
62. Jackson MR. Upper extremity arterial injuries in athletes. Semin Vasc Surg. 2003;16(3):232–9.
63. Cothran RL Jr, Helms C. Quadrilateral space syndrome: incidence of imaging findings in a population referred for MRI of the shoulder. AJR Am J Roentgenol. 2005;184(3):989–92.
64. Chen H, Narvaez VR. Ultrasound-guided quadrilateral space block for the diagnosis of quadrilateral syndrome. Case Rep Orthop. 2015;2015:378627.
65. Chautems RC, Glauser T, Waeber-Fey MC, Rostan O, Barraud GE. Quadrilateral space syndrome: case report and review of the literature. Ann Vasc Surg. 2000;14(6):673–6.
66. Hoskins WT, Pollard HP, McDonald AJ. Quadrilateral space syndrome: a case study and review of the literature. Br J Sports Med. 2005;39(2):e9.
67. Mochizuki T, Isoda H, Masui T, Ohkawa Y, Takahashi M, Takehara Y, Ichijo K, Kodaira N, Kitanaka H. Occlusion of the posterior humeral circumflex artery: detection with MR angiography in healthy volunteers and in a patient with quadrilateral space syndrome. Am J Roentgenol. 1994;163(3):625–7.

68. McAdams TR, Dillingham MF. Surgical decompression of the quadrilateral space in overhead athletes. Am J Sports Med. 2008;36(3):528–32.
69. Wiater JM, Flatow EL. Long thoracic nerve injury. Clin Orthop Relat Res. 1999;368:17–27.
70. Gregg JR, Labosky D, Harty M, Lotke P, Ecker M, DiStefano V, Das M. Serratus anterior paralysis in the young athlete. J Bone Joint Surg. 1979;61:825–32.
71. Bertelli JA, Ghizoni MF. Long thoracic nerve: anatomy and functional assessment. J Bone Joint Surg Am. 2005;87:993–8.
72. Hester P, Caborn D, Nyland J. Cause of long thoracic nerve palsy: a possible dynamic fascial sling cause. J Shoulder Elbow Surg. 2000;9:31–5.
73. Kauppila LI. The long thoracic nerve: possible mechanisms of injury based on autopsy study. J Shoulder Elbow Surg. 1993;2(5):244–8.
74. Bigliani LU, Perez-Sanz JR, Wolfe IN. Treatment of trapezius paralysis. J Bone Joint Surg Am. 1985;67(A):871–7.
75. Foo CL, Swann M. Isolated paralysis of the serratus anterior. A report of 20 cases. J Bone Joint Surg Br. 1983;65(B):552–7.
76. Galano GJ, Bigliani LU, Ahmad CS, Levine WN. Surgical treatment of winged scapula. Clin Orthop Relat Res. 2008;466(3):652–60.
77. Leffert RD. Nerve injuries about the shoulder. In: Rowe CR, editor. The shoulder. New York: Churchil Livingstone; 1988. p. 435–6.
78. Leffert RD. Neurological problems. In: Rockwood CA, Matsen FA, editors. The shoulder, vol. 2. Philadelphia: WB Saunders; 1990. p. 750–73.
79. Kauppila LI, Vastamäki M. Iatrogenic serratus anterior paralysis: long-term outcome in 26 patients. Chest. 1996;109(1):31–4.
80. Makin G, Brown WF, Ebers GC. C7 radiculopathy: importance of scapular winging in clinical diagnosis. J Neurol Neurosurg Psychiatry. 1986;49:640–4.
81. Bischel OE, Hempfing A, Rickert M, Loew M. Operative treatment of a winged scapula due to peripheral nerve palsy in Lyme disease: a case report and review of the literature. J Shoulder Elbow Surg. 2008;17(6):e24–7.
82. Sivan M, Hassan A. Images in emergency medicine: winged scapula as the presenting symptom of Guillain-Barre syndrome. Emerg Med J. 2009;26(11):790.
83. Burkhart SS, Morgan CD, Kibler WB. The disabled throwing shoulder: spectrum of pathology. Part III: the SICK scapula, scapular dyskinesis, the kinetic chain, and rehabilitation. Arthroscopy. 2003;19(6):641–61.
84. Wood VE, Marchinski L. Congenital anomalies of the shoulder. In: Rockwood CA, Matsen FA, editors. The shoulder, vol. 1. Philadelphia: WB Saunders; 1990. p. 131–2.
85. Kibler WB, McMullen J. Scapular dyskinesis and its relation to shoulder pain. J Am Acad Orthop Surg. 2003;11(2):142–51.
86. Kuhn JE, Plancher KD, Hawkins RJ. Scapular winging. J Am Acad Orthop Surg. 1995;3(6):319–25.
87. Safran MR. Nerve injury about the shoulder in athletes, Part 2: long thoracic nerve, spinal accessory nerve, burners/stingers, thoracic outlet syndrome. Am J Sports Med. 2004;32(4):1063–76.
88. Friedenberg SM, Zimprich T, Harper CM. The natural history of the long thoracic and spinal accessory neuropathies. Muscle Nerve. 2002;25:535–9.
89. Perlmutter GS, Leffert RD. Results of transfer of the pectoralis major tendon to treat paralysis of the serratus anterior muscle. J Bone Joint Surg Am. 1999;81(3):377–84.
90. Watson CJ, Schenkman M. Physical therapy management of isolated serratus anterior muscle paralysis. Phys Ther. 1995;75:194–202.
91. Connor PM, Yamaguchi K, Manifold SG, Pollock RG, Flatow EL, Bigliani LU. Split pectoralis major transfer for serratus anterior palsy. Clin Orthop Relat Res. 1997;341:134–42.
92. Noerdlinger MA, Cole BJ, Stewart M, Post M. Results of pectoralis major transfer with fascia lata autograft augmentation for scapula winging. J Shoulder Elbow Surg. 2002;11(4):345–50.
93. Streit JJ, Lenarz CJ, Shishani Y, McCrum C, Wanner JP, Nowinski RJ, Warner JJ, Gobezie R. Pectoralis major tendon transfer for the treatment of scapular winging due to long thoracic nerve palsy. J Shoulder Elbow Surg. 2012;21(5):685–90. Epub 2011 Jul 1
94. Elhassan B. Pectoralis major transfer for the management of scapula winging secondary to serratus anterior injury or paralysis. J Hand Surg Am. 2014;39(2):353–61.
95. Krishnan SG, Hawkins RJ, Michelotti JD, Litchfield R, Willis RB, Kim YJ. Scapulothoracic arthrodesis: indications, technique, and results. Clin Orthop Relat Res. 2005;435:126–33.
96. Hawkins RJ, Willis RB, Litchfield RB. Scapulothoracic arthrodesis for scapular winging. In: Post M, Morrey BF, Hawkins RJ, editors. Surgery of the shoulder. St. Louis: Mosby Year Book; 1990. p. 346–59.
97. Rhee YG, Ha JH. Long-term results of scapulothoracic arthodesis of facioscapulohumeral muscular dystrophy. J Shoulder Elbow Surg. 2006;15:445–50.
98. Novak CB, Mackinnon SE. Surgical treatment of a long thoracic nerve palsy. Ann Thorac Surg. 2002;73:1643–5.
99. Moore KL. Clinically oriented anatomy. 6th ed. Philadelphia: Lippincott Williams & Wilkins; 2010.
100. Barron OA, Levine WN, Bigliani LU. Surgical management of chronic trapezius dysfunction. In: Wamer JP, Lanotıi JP, Gerher C, editors. Complex and revision problems in shouider surgery. Philadelphia: Lippincot Williams & Wilkins; 1997. p. 377–84.
101. Boström D, Dahlin LB. Iatrogenic injury to the accessory nerve. Scand J Plast Reconstr Surg Hand Surg. 2007;41(2):82–7.

102. Teboul F, Bizot P, Kakkar R, Sedel L. Surgical management of trapezius palsy. J Bone Joint Surg Am. 2004;86:1884–90.
103. Williams WW, Donell ST, Twymann RS, Birch R. The posterior triangle and the painful shoulder. Ann R Coll Surg Engl. 1996;78:521–5.
104. Martin RM, Fish DE. Scapular winging: anatomical review, diagnosis, and treatments. Curr Rev Musculoskelet Med. 2008;1(1):1–11.
105. Chan PKH, Hems TEJ. Clinical signs of accessory nerve palsy. J Trauma. 2005;60:1142–4.
106. Bigliani JU, Compito CA, Duralde XA, Wolfe IN. Transfer of the levator scapulae, rhomboid major, and rhomboid minor for paralysis of the trapezius. J Bone Joint Surg. 1996;78:1534–40.
107. Kizilay A, Kalcioglu MT, Saydem L, Ersoy Y. A new shoulder orthosis for paralysis of the trapezius muscle after radical neck dissection: a preliminary report. Eur Arch Otorrhinolaryngol. 2006;263:477–80.
108. Kim DH, Cho YJ, Tiel RL, Kline DG. Surgical outcome of 111 spinal accessory nerve injuries. Neurosurgery. 2003;53:1106–13.
109. Romero J, Gerber C. Levator scapulae and rhomboid transfer for paralysis of trapezius. The Eden-Lange procedure. J Bone Joint Surg Br. 2003;85(8):1141–5.
110. Giannini S, Faldini C, Pagkrati S, Grandi G, Digennaro V, Luciani D, Merlini L. Fixation of winged scapula in facioscapulohumeral muscular dystrophy. Clin Med Res. 2007;5(3):155–62.
111. Nath RK, Somasundaram C. Meta-analysis of long thoracic nerve decompression and neurolysis versus muscle and tendon transfer operative treatments of winging scapula. Plast Reconstr Surg Glob Open. 2017;5(8):e1481.

Fractures Around Shoulder 14

Suleyman Semih Dedeoglu

14.1 Clavicle Fractures

Clavicular fractures, an important cause of morbidity, constitute approximately 3% of all fractures in adults [1]. The overall incidence of these fractures has increased, reaching 60 per 100,000 people [2]. Although there is a 70% male dominance in terms of gender, this situation varies according to age. The most frequently affected group is young men, especially between the ages of 15 and 24, but in people over the age of 60, the effect of gender is eliminated and even female patient domination may exist [3, 4]. Clavicle fractures occur most commonly due to the fall on the shoulders during sport activities or traffic accidents. Fracture mechanism also varies by gender and age. Male and young patients are injured mostly due to traffic accidents while using bicycles, vehicles, etc., women and elderly are more likely to develop clavicular fracture after fall as a frequent etiological factor [3].

14.1.1 Classification

There are multiple classifications for clavicle fractures. Some of these classifications are more specific for certain type of fractures, but they have no obvious superiority over each other [5].

S. S. Dedeoglu (✉)
Department of Orthopedics and Traumatology, Prof. Dr. Cemil Tascioglu City Hospital, Istanbul, Turkey

Clavicle fractures can be categorized into three groups by anatomic site as described by Allman, one of the most recognized classifications: Midshaft fractures in the middle third of the clavicle as Group 1; distal clavicle or lateral fractures in the distal third of the clavicle as Group 2; and medial clavicle or proximal fractures in the medial third segment as Group 3 [6]. AO/OTA classification is also one of the commonly used classifications. In this classification, the proximal and distal end fractures are subdivided into extra-articular, partial articular, and full articular, whereas diaphyseal fractures are categorized as simple, wedge, and multifragmentary [7]. The Neer classification further evaluates distal clavicle fractures in five separate groups, as Type 1–Type 5 (Fig. 14.1) [8]. Midshaft fractures are most common subtype, constituting 65–80% of all clavicular fractures, followed by distal (lateral) fractures with approximately 20–30% and proximal (medial) fractures with up to 5% [2]. While midshaft fractures are the most common type in men and women, distal fractures may reach up to 40–45% in women and the elderly [3].

14.1.2 Clinical Anatomy

The clavicle, as an S-shaped bone, makes the sternoclavicular joint with the sternum in the proximal part, and the acromioclavicular joint

G. Huri et al. (eds.), *Fundamentals of the Shoulder*, https://doi.org/10.1007/978-3-030-94702-6_14

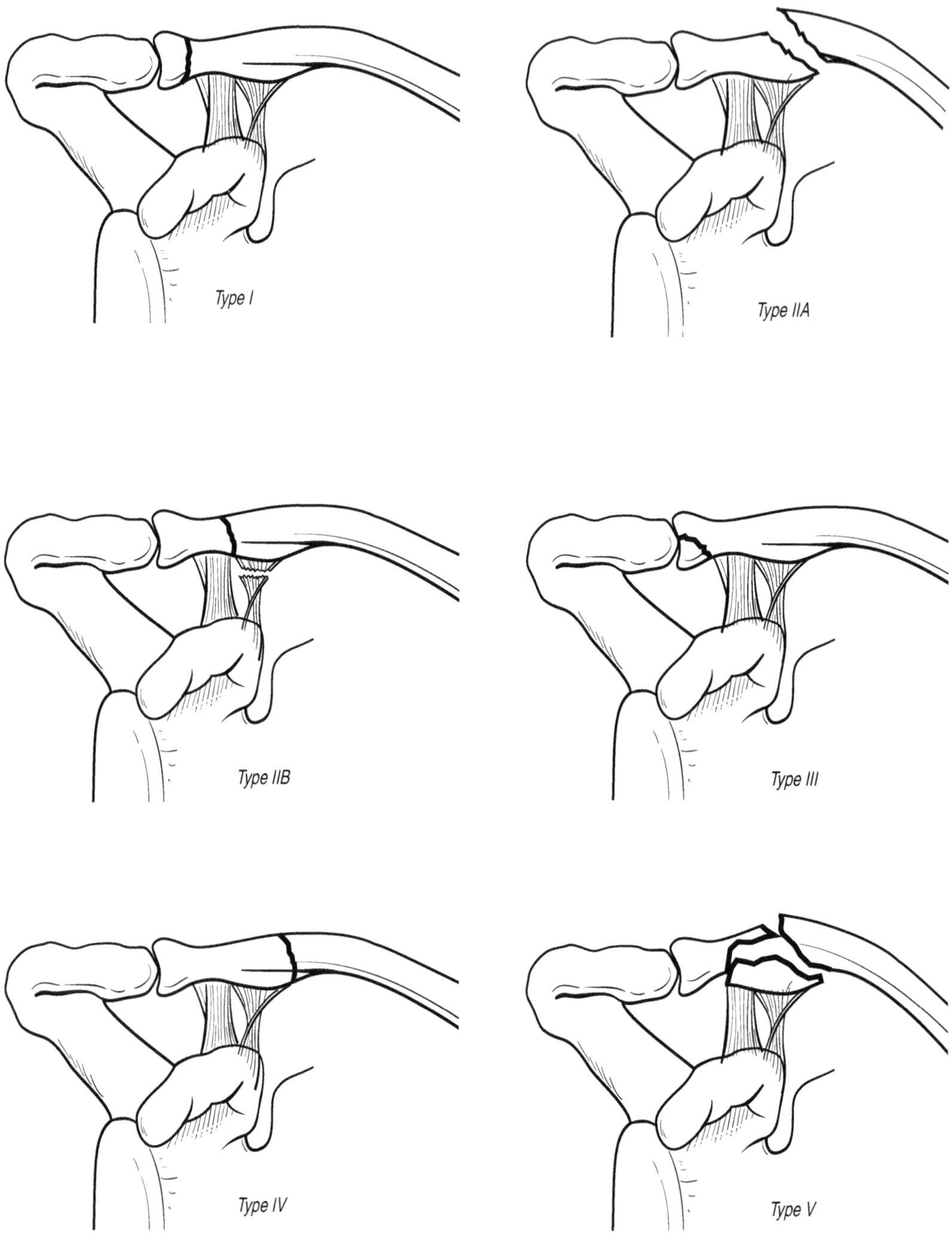

Fig. 14.1 Neer classification in distal clavicle fractures

with the acromion in the distal part. While it attaches to the sternum with the sternoclavicular ligaments, and to the scapula with the acromioclavicular and coracoclavicular ligaments. Its distal half attaching to the scapula has a concave structure, whereas the proximal half is convex, creating a space for neurovascular structures to supply the upper limb. The part where concavity and convexity fuses is the thinnest section of the clavicle and also lacks ligaments. For this reason, most of the clavicular fractures occur in this region, also called midshaft. When the bone becomes displaced, the proximal part is pulled upwards with the strength of the sternocleidomastoid muscle that is attached to the clavicle. Conversely, the distal segment is displaced inferiorly with the weight of the arm. Shortening is often seen as fracture fragments pass each other due to the forceful pulling effect of subscapularis and pectoral muscles.

14.1.3 Midshaft Fractures

14.1.3.1 Diagnostic Approach

In midshaft fractures, there is pain and edema localized at the fracture site, and the pain gets aggravated with arm movement. On physical examination, hematoma and ecchymosis, bone angulation, and displaced bone tips can be detected. In addition, crepitus can be heard and bone fragments can be felt by palpation to the affected side. Neurovascular and lung examination should not be overlooked during the evaluation not to miss other fractures and complications, especially if high-energy events are accompanied.

In the radiological evaluation of the midshaft clavicular fracture, only one-sided plain AP radiography is actually sufficient. Transverse or oblique fracture and vertical plane displacement can be seen on AP radiography. If shortness is suspected, the length between affected and unaffected clavicle should be compared on plain chest radiography. The scapula and other structures should also be extensively examined for concomitant damage and complications as part of the complete radiological evaluation.

14.1.3.2 Treatment Approach

The treatment goal is to restore shoulder functions to pre-injury state in midshaft fractures. Ensuring clavicle healing with minimal deformity also minimizes pain and movement loss as much as possible. Apart from specific clinical conservative or surgical indications for midshaft fractures, ideal treatment strategy is based on patients' individual requirements and expectations [9]. For example, open fractures or fractures accompanied by neurovascular injury or multiple trauma are usually treated with surgical intervention, whereas conservative follow-up may be sufficient for non-displaced fractures [10]. As the long-term results of the surgical and conservative approach are generally similar for well-recognized certain indications, the advantages and possible complications of surgical intervention should be evaluated on a case-by-case basis.

Conservative Treatment

Non-displaced fractures and fractures where cortical alignment is not disrupted can be managed with conservative approaches. In addition, displaced fractures up to 2 cm or fractures accompanied by a shortening of <2 cm can be treated without the need for surgery. In cases where the displacement or shortening exceeds 2 cm, surgical interventions are mostly considered, since non-union rates increase against conservative treatment [11]. However, since the long-term functional results of both approaches are similar, in cases where there is no obvious surgical indication, cosmetic expectations, return to work, operative risks, etc. should be discussed with the patient to make a shared decision.

In the conservative approach, the shoulder on the injured side is immobilized. However, this immobilization cannot be performed very strictly due to several factors, including the tension force of the muscles around the shoulder, positional changes during the day, and the effects of the respiratory muscles. Basic sling and FOE bandage to be applied for 2–6 weeks are the most frequently used methods. Though these two approaches have similar outcomes overall, sling immobilization seems to have better tolerability as it has been reported as less painful during the early period [12]. It is generally recommended to restrict internal rotation for 3–4 weeks. Care should be taken to remove the sling several times a day to prevent elbow rigidity. Free movement is usually allowed after 6 weeks. Isometric physiotherapy and resistance exercises can be done as much as pain allows.

Surgical Treatment

Surgical indications for midshaft fractures include presence of >2 cm displacement, open fracture, fractures accompanied by neurovascular complications, shoulder compression, and floating shoulder. The presence of an open fracture and neurovascular complication is considered a complicated fracture, requiring an emergent surgical intervention. Neurovascular risks should be considered before internal fixation of clavicle fractures is planned. Although rare, there is a risk of creating axillary and brachial plexus vascular lesions, especially on the inferior side of the clavicle. For this reason, plates should be placed superior part of the medial segment and anterior of the middle segment as much as possible [10].

In displaced midshaft fractures, plate osteosynthesis is the gold standard treatment (Fig. 14.2). This procedure can be performed with minimally invasive techniques with locked or unlocked anatomic plate or dynamic compression plate (Fig. 14.3) [10, 13]. Another method is intramedullary fixation. While this method has advantages such as less invasive nature, small scar area, and shorter hospital stay, it has higher risks such as implant migration due to poor rotational stability [14]. Although the risk of infection appears slightly higher in plate osteosynthesis, the functional outcomes and complication rates of both approaches are generally similar [15, 16].

14.1.4 Distal Clavicular Fractures

14.1.4.1 Diagnostic Approach

The clinical presentation of distal clavicular or lateral fractures mainly consists of pain and tenderness around the acromioclavicular joint; edema and ecchymoses are often accompanied by this condition. The clinic of distal segment fractures might be confused with various pathologies of the shoulder such as osteoarthritis and septic arthritis, and especially acromioclavicular joint dislocation. Acromioclavicular joint dislocation also has similar manifestations; in both cases the pain is exacerbated in the cross-arm test. While maximal pain is more medial to the acromioclavicular joint in fractures, tenderness is directly on the joint in dislocation. However, radiological examination is required for definitive differential diagnosis [17].

A plain AP shoulder radiography is usually sufficient for the radiological diagnosis of distal clavicle fractures. Conversely, a 15° cephalad angle AP image and stress radiography may also be required to assess the integrity of the coracoclavicular ligament. The importance of AP radiography in distal clavicle fractures is that it allows the Neer classification to be used to determine the prognosis and treatment specific to these fractures. This classification is based on the place of the fracture in relation to coracoclavicular ligament and the integrity of these structure. In Type 1, the fracture fragment is lateral to the

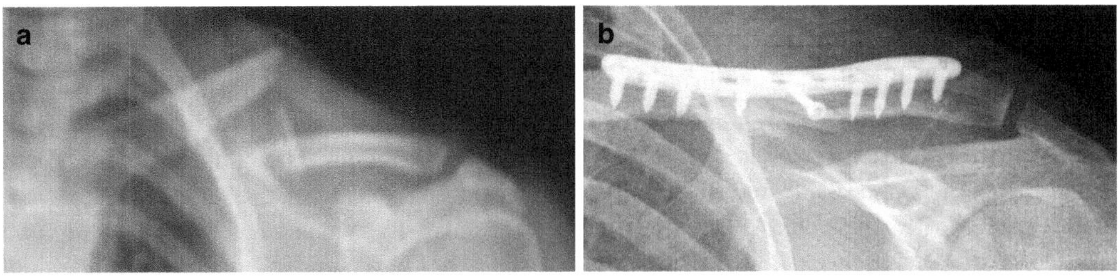

Fig. 14.2 Open reduction internal fixation of midshaft clavicle fractures: (**a**) preoperatively, (**b**) postoperatively

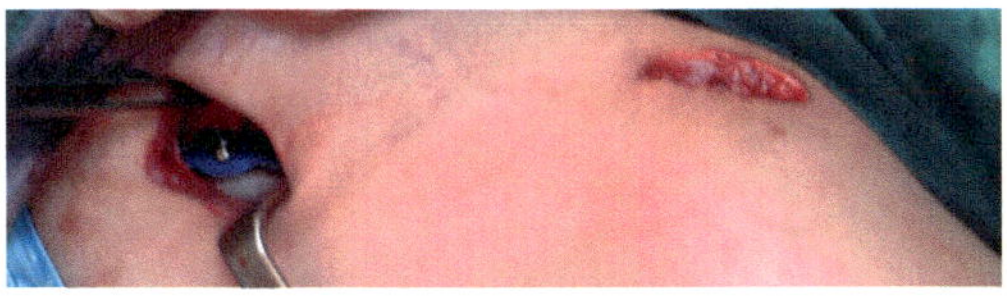

Fig. 14.3 Minimal invasive plate osteosynthesis of mid-shaft clavicle fractures

attachment site of the ligament, whereas it is medial to the attachment site in Type 2. Type 2 fractures further divide into two groups as Type 2A and Type 2B: while the former is characterized by attachment of both conoid and trapezoid ligaments to the distal fragment; Type 2B corresponds to the fractures where conoid ligament is separated from the proximal fragment. Type 3 fractures extend into the acromioclavicular joint. Type 4 is seen in children, where avulsion with coracoclavicular ligament occurs with upwards displacement of medial segment. Type 5 fracture features avulsion in addition to the characteristics observed in Type 2 fracture. These two types of fractures gain an unstable structure, which compels their treatments. The instability originates from the pulling of distal fragment to distal and medial direction and proximal fragment to posterior direction with the effect of surrounding muscle groups [18, 19].

14.1.4.2 Treatment Approach

In distal clavicle fractures, the treatment and its outcomes are determined by the presence of displacement that may cause the instability of the fracture and the injury to the coracoclavicular ligament. In general, stable Type 1 and Type 3 fractures and Type 4 fractures accompanied by periosteal disintegration in children are treated with a conservative approach. Instability and displacement in Type 2 and Type 5 fractures mostly require surgical intervention, but although many surgical techniques are used in this regard, there is no agreed gold standard approach.

Conservative Treatment

The majority of distal clavicle fractures, especially Type 1 and Type 3 fractures, are managed with a non-operative approach. In these fractures, immobilization is achieved with sling, and shoulder movements are allowed as the pain subsides. After 6 weeks of follow-up, shoulder radiography is repeated and these patients usually recover completely without developing sequelae. Non-displaced Type 2 fractures may also be sometimes followed up conservatively, on an individual basis. In these cases, the weight of the arm should be supported so that the edge of the proximal fragment is as close to the coracoid process as possible, and this can only be achieved with a sling. FOE bandage should not be used in Type 2 fractures as it gives no support to the weight of the arm [10, 17].

Surgical Treatment

Surgical treatment is preferred for Type 2 fractures, especially in the presence of displaced fractures. However, there is no gold standard technique for surgical treatment. Two main surgical principles can be followed that eliminate the forces acting on the bone ends to properly align them. Alignment of fragments by internal fixation could be provided with or without involvement of the acromion. The major problem in stabilizing these fractures is that the distal fragment may be too small to be attached with the implant. When acromion is stabilized to overcome such situation, it may create a predisposition to arthritis in the long-term as the normal rotation of the acromioclavicular joint is prevented. Therefore, the implant needs to be removed after a while. The second principle is the fixation of the displaced proximal fragment with the coracoid. In this invasive surgical procedure, which includes reconstruction of torn coracoclavicular ligaments, attention should be paid to neurovascular structures adjacent to the coracoid. Both surgical principles can be used together or alone in an intervention [20].

Surgical techniques where rigid fixation is applied include locked plate osteosynthesis, hook plate fixation, distal radius locked plate fixation, coracoclavicular screwing, and Knowles pinning (Figs. 14.4 and 14.5). Conversely, classical K-wire fixation, tension band wiring, suture anchors, coracoclavicular fixation with double button lift-up, and Dacron artery graft for coracoclavicular ligament reconstruction can be listed

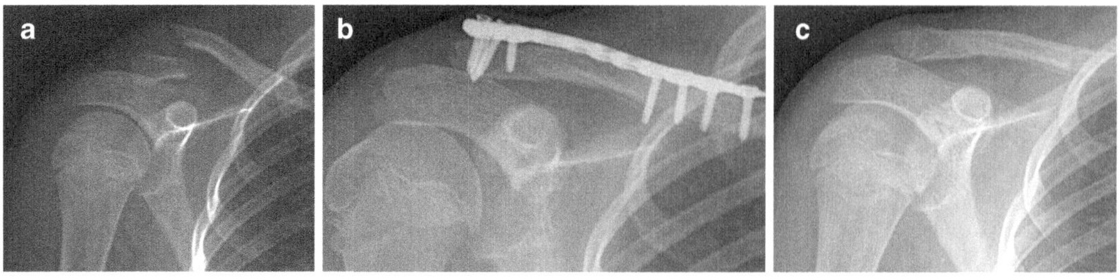

Fig. 14.4 Locking plate osteosynthesis fixation of the in Neer Type 2A distal clavicle fractures: (**a**) preoperatively, (**b**) postoperatively, (**c**) union after plate removal

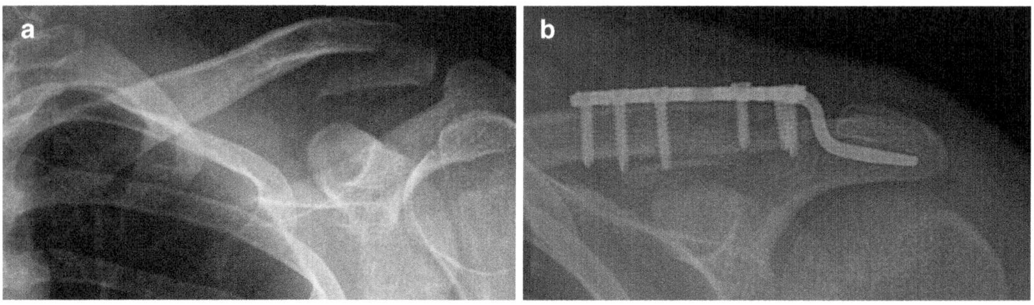

Fig. 14.5 Hook plate in Neer Type 2A distal clavicle fractures: (**a**) preoperatively, (**b**) postoperatively

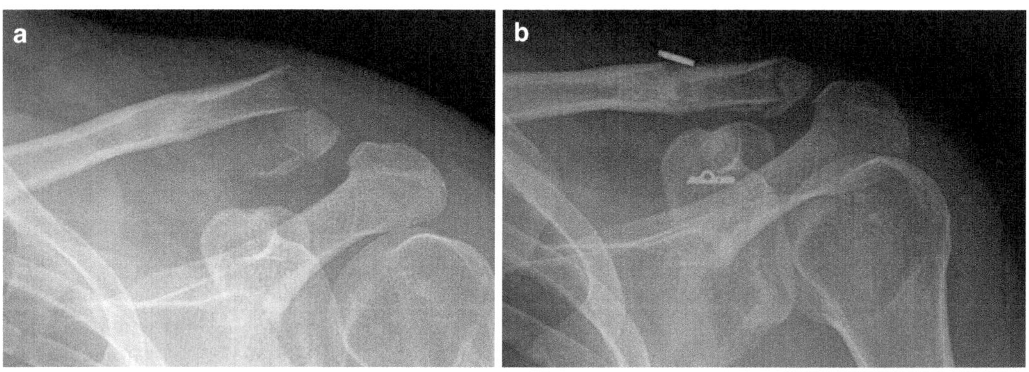

Fig. 14.6 Tightrope application in Neer Type 2B distal clavicle fractures: (**a**) preoperatively, (**b**) postoperatively

as flexible fixation techniques (Fig. 14.6). Overall, there is not much difference between techniques in terms of time required for union. Besides the slightly increased risk with K-wire tension band wiring, the risk of implant failure is similar in other techniques and all techniques achieve generally good to excellent functional outcomes. Complications are more common in rigid fixation techniques, especially those associated with the implant (Fig. 14.7). On the contrary, there are fewer complications in coracoclavicular screw fixation and flexible coracoclavicular fixation repairs. The risk of infection is generally similar and depends on the invasiveness of the intervention [20–22].

14.1.5 Medial Clavicular Fractures

14.1.5.1 Diagnostic Approach

Medial or proximal fractures, the least common type of clavicle fractures, are mostly caused by

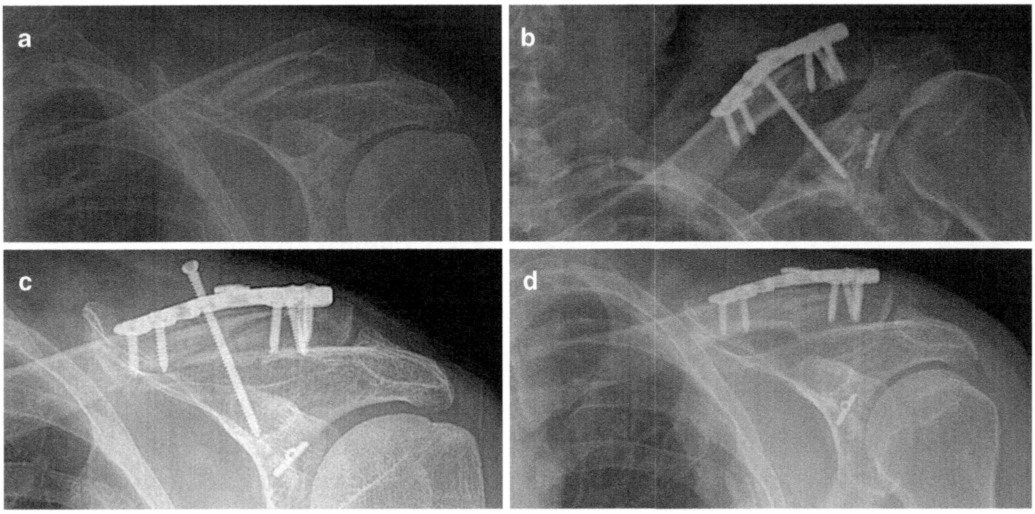

Fig. 14.7 Multifragmentary distal clavicle fracture: (**a**) preoperatively, (**b**) locking plate osteosynthesis with Tightrope-supported coracoclavicular screw, (**c**) coracoclavicular screw pull-out, (**d**) union after removal of the screw

high-energy traumas such as accidents and are included in the clinical picture as part of a multi-trauma. Another reason comprises the stress fractures secondary to repetitive activity, which requires the exclusion of other causes of bone pain, unlike to the diagnostic approach in acute trauma. In both cases, there is pain in the sternoclavicular region exacerbated by shoulder movement. As the pain usually increases in the supine position, patients feel more comfortable in a sitting position to support their arms. In acute fractures, ecchymoses can be seen, especially if there is ligament injury or displaced fragment. Anterior or posterior displacement of the medial clavicle could imply a sternoclavicular dislocation. In multi-trauma patients, additional possible injuries of scapula, rib, brachial plexus or ipsilateral upper limb and hemothorax or pneumothorax should be carefully investigated [23, 24].

Since medial fractures can be missed due to overlapping bone shadows on plain AP radiography, shoulder radiography should be taken at a 45° angled cephalic plane. The visualization of both shoulders is more suitable for comparison. If there is clinical suspicion on plain radiography, computed tomography (CT) should be performed. In acute fractures, CT and magnetic resonance imaging (MRI) should be considered as an additional test to better assess other structures.

14.1.5.2 Treatment Approach

Conservative follow-up is deemed to be adequate in stress fractures and non-displaced acute fractures, whereas proximal fractures that accompany displacement and multi-trauma require a more comprehensive and well-planned multidisciplinary interventions. Possible head, neck, thorax, and visceral injuries should be evaluated and surgical interventions should be considered, especially in the presence of posterior displacement or dislocation [25].

Conservative Treatment

In non-operative treatment, immobilization is sufficient with ice application, analgesic, and sling support. In stress fractures, cessation of the repetitive activity and physiotherapy support increases the success of the treatment. Recovery is usually expected within 6–8 weeks. Though achievement of good functional outcomes may take up to 3 years, conservative follow-up could be regarded as overall successful with 5% of non-union and 10% of delayed union rates [24, 26].

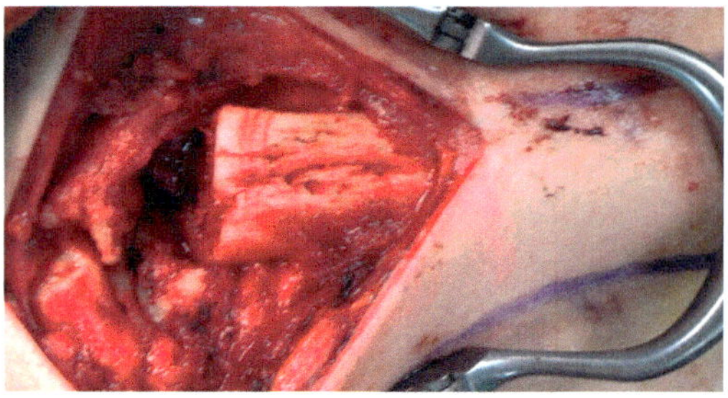

Fig. 14.8 Medial clavicle fracture

Surgical Treatment

Open reduction plus internal fixation is one of the most commonly used techniques for the surgical intervention, which is rather preferred in displaced medial fractures (Fig. 14.8) [27, 28]. This technique yields successful functional outcomes, and when performed in the early period, could also minimize the risk of painful non-union [28]. In periarticular medial fracture associated with dislocation, T-locked plates also demonstrate successful results [29–31]. In addition, it is reported that successful outcomes are achieved with inverted distal clavicle plate technique in a small number of patients [32].

14.2 Proximal Humerus Fractures

Proximal humerus fractures account for approximately 6% of all extremity fractures and the third most common fractures following hip and distal radius in elderly [33]. While it occurs in high-energy traumas such as motor vehicle accidents in young patients, it is often secondary to low-energy trauma like simple falls in the elderly [34]. Osteoporosis, diabetes, epilepsy, and female gender are important risk factors for these fractures [35]. With an increased life expectancy, the percentage of elderly population raises in these fractures. In this century, the incidence of proximal humerus fractures has also increased with osteoporotic hip and distal radius fractures. Poor bone quality causes more complex fracture types despite low-energy trauma. In last decades, the fact that more elderly has to live alone and meet their own demands like personal care increases the need for fulfillment of patients' satisfaction level from the outcomes of treatment. The patient's expectations and lifestyle should be taken into account in the choice of treatment modality.

14.2.1 Clinical Anatomy

The proximal humerus is divided into four parts. These are the greater tuberosity, lesser tuberosity, head (articular surface), and the diaphyseal regions [36]. The rotator cuff tendons attach to tubercles and contributes to glenohumeral movement and dynamic stability. The subscapularis tendon inserts onto the lesser tubercle anteriorly and the other three tendons (supraspinatus, infraspinatus, teres minor) attach to the greater tuberosity. The bicipital groove, which contains the long head of the biceps tendon is located between two tubercles. The anatomical neck is located in the part of the old epiphyseal plate between the tuberosities and humeral head, where the joint capsule attaches. The surgical neck, the weakest region, is below the tuberosities and is closely related to the axillary nerve and posterior circumflex humeral artery [34]. The average neck angle between the shaft and the humeral head is 130°. Humeral retroversion has been reported in a wide range from 18 to 30° [37, 38]. The upper border of the insertion of the pectoralis major tendon is 5.6 cm inferior from the top of the humerus head and lateral to the bicipital groove [39]. Latissimus dorsi and teres major tendons (most medial and posterior) attaches medially to the pectoralis major tendon insertion. Medial calcar is another osseous part and consists of medial humeral metaphyseal region. Intact or surgically supported medial calcar is important for stability and prognosis [40]. Although the ascending branch of the anterior humeral circumflex artery arising from the axillary artery is known to be the main vascular structure that provides perfusion of the humeral head, the incidence of osteonecrosis is low due to existing anastomoses [41]. Recent studies have reported that near two-thirds of the humeral head perfusion was derived from the posterior humeral circumflex artery. Currently, the posterior humeral circumflex artery is the main blood supply to the humeral head [42]. In terms of deforming forces, pectoralis major muscle pulls the humeral shaft anterior and medially. Supraspinatus, infraspinatus and teres minor muscles pull and externally rotate the greater tuberosity. Subscapularis muscle rotates the lesser tuberosity and/or head internally according to the configuration of the fracture, e.g. pulling the fragment medially in isolated lesser tuberosity fracture.

14.2.2 Classification

Proximal humerus fracture classifications may help to decide appropriate treatment modality and predict the prognosis. Kocher made a classification according to fracture localization [43]. In 1934, Codman classified the proximal humerus fractures into four types as articular surface, shaft, greater tuberosity, and lesser tuberosity based on the number of fragments [44]. Neer has made his classification inspired by the idea of Codman's four fragments fracture pattern and based on anatomic relations. He classified the fractures into four main segments (1 part, 2 parts, 3 parts, 4 parts) and six groups (minimal displacement, anatomic neck, surgical neck, greater tuberosity, lesser tuberosity, fracture dislocation) [45] (Fig. 14.9). This classification, which is used most frequently today, is based on the number of fracture lines and the presence of displacement. The definition of displacement entitles >45° angulation and >1 cm separation between the fragments.

The AO Classification has divided the fractures into three main types as Type A, Type B, or Type C [46]. Type A fractures are unifocal and extra-articular (involving one of the tuberosities), Type B fractures are bifocal and extra-articular (metaphyseal and tuberosity involvement), and Type C fractures are intra-articular (fracture dislocations or head split fractures). Each main group is subdivided into three subgroups (A1, A2, A3, B1, B2, B3, C1, C2, C3), which were determined according to several criteria, including displacement level, alignment, metaphyseal extension of the fracture, accompanying dislocation, or head split injury. Fracture patterns are separated into 27 subgroups. The risk of developing osteonecrosis is increased in AO Type C fractures.

Hertel classified the proximal humerus fractures according to the LEGO blocks model consisting of 12 main segments: 6 possible two-part fractures, 5 possible three-part fractures, and 1 possible four-part [47]. Less than 8 mm posteromedial metaphyseal extension of the humeral head fragment and more than 2 mm disruption in medial hinge are risk factors for ischemia. However, ischemia has shown not to cause osteonecrosis in cases where osteosynthesis is performed by preserving the humeral head with adequate reduction [48].

14.2.3 Diagnostic Approach

Severe pain during movements, swelling, and ecchymosis are common signs and symptoms. The presence of neurovascular injury, e.g. fracture dislocation, needs to be also investigated. Sensory and motor tests of the axillary nerve should be performed as it is the most commonly injured nerve, with also examination of brachial plexus and upper extremity pulses [49]. Doppler ultrasound examination of axillary or brachial artery may be required in case of suspected vascular injury. Angiography and CT angiography may also be performed upon consultation with vascular surgeon in such cases.

Radiological assessment provides information about fracture morphology and displacement level, helping with fracture classification and proper treatment plan. These include true AP (Grashey), scapular Y, and axillary lateral views. If these do not give sufficient data, apical oblique, Velpeau, West Point axillary views could be obtained. Apical oblique view is used for the diagnosis of glenohumeral dislocation and posterolateral humeral head compression fractures. Velpeau view is an alternative to axillary lateral radiography and performed with patient's arm held in

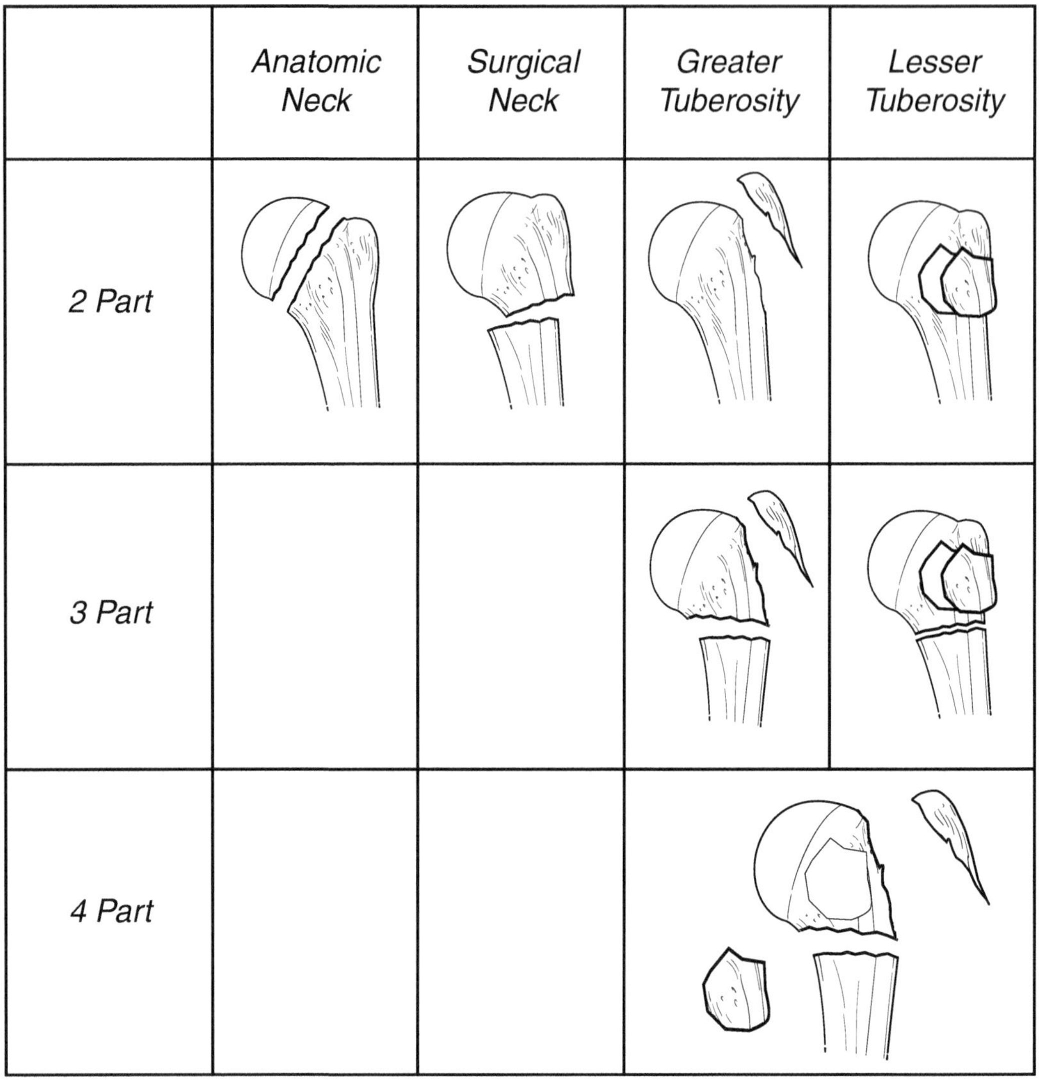

Fig. 14.9 Neer classification in proximal humerus fractures

internal rotation in a sling. West Point axillary view is specific to the evaluation of the anteroinferior rim of the glenoid. Additional CT is recommended in cases where X-ray radiographs fail to determine the fracture line, evaluation of fracture morphology in three- or four-part fractures, determination of tuberosity displacement, detection of involvement of the humeral head, and suspected concomitant glenoid fracture. It also helps to classify fractures and plan the appropriate treatment. Rarely indicated ultrasonography or MRI may be useful detection of rotator cuff injuries.

14.2.4 Treatment Approach

The goal of treatment in proximal humerus fractures is to achieve a painless and functional shoulder joint. Pain should be relieved initially and the fracture must be immobilized. After clinical and radiological evaluations, it is decided whether to apply conservative or surgical treatment. Treatment planning is made according to the appropriate indication. The determination of appropriate treatment method is based on considering age, fracture classification type, displacement level, bone and soft tissue quality, general medical status, and additional injuries [50].

14.2.4.1 Conservative Treatment

Minimally displaced one-, two-, or three-part fractures (surgical or anatomic neck fractures), <5 mm displaced greater tuberosity fractures, and patients with advanced age or additional comorbidities that do not allow for surgical treatment can be treated conservatively [49, 51, 52]. The recommended conservative method is immobilization in a sling with shoulder resting position (0° adduction and internal rotation) for 4–6 weeks [53].

Satisfactory clinical and functional results have been achieved with the progressive physical therapy program applied after 2–3 weeks of sling immobilization [54, 55]. Rehabilitation protocols include active rehabilitation and strengthening after early pendulum exercises. Early rehabilitation has been reported to improve clinical and functional outcomes, albeit with conflicting association with iatrogenic displacement [53, 56, 57]. Non-operative treatment of proximal humeral fractures is reported to yield 80–85% success rate [34]. Conversely, forced closed reduction maneuvers can cause soft tissue and neurovascular injuries, with reports of axillary nerve injury at a rate 30% [58]. Whether the injury is traumatic or iatrogenic during closed reduction is controversial. Other complications include non-union, delayed union, and humeral head osteonecrosis [49].

14.2.4.2 Surgical Treatment

Indications for surgical treatment include three- or four-part fractures, open fractures, accompanying ipsilateral shoulder girdle fractures, neurovascular injury, head split fracture, severe displacement (>5 mm) of greater tuberosity fractures, nonunions following conservative treatment [50, 51]. The primary goal of surgical treatment should be preserving the humeral head with osteosynthesis if possible. In cases where the head cannot be reconstructed, arthroplasty may be a more appropriate treatment option. In particular, multifragmentary complex fractures, fracture dislocation, elderly patients (>70 years old), poor bone quality, and head split fractures remain at high risk for development of osteonecrosis and non-union (Fig. 14.10) [59, 60]. In such cases, arthroplasty (reverse or hemiarthroplasty) sets should be available in the operating room.

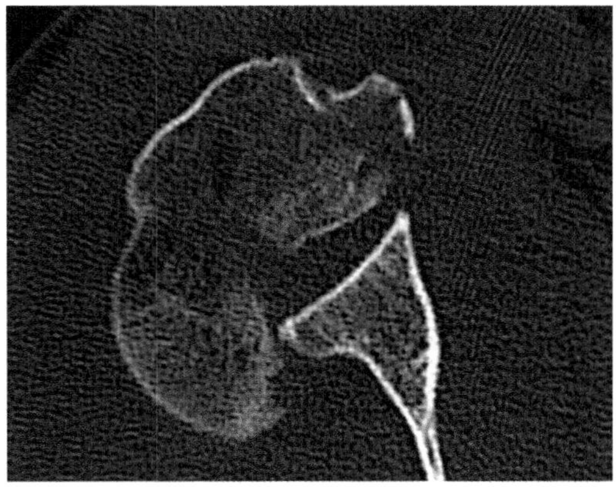

Fig. 14.10 Humeral head split fracture

Closed reduction and percutaneous pinning, open reduction and internal fixation, and intramedullary nailing are the most commonly employed fixation methods for proximal humerus fractures.

Closed reduction and percutaneous pinning was described by Bohler in 1962 [61]. Surgical indications include two-part surgical neck fractures, three-part surgical neck fractures with involvement of the greater or lesser tuberosity, and valgus-impacted four-part fractures with good bone quality [51]. There should be a very stable closed reduction, minimal metaphysical comminution, and intact medial calcar. It is recommended to use multiplanar, multiple number, and larger diameter pins to increase stability. Cortically, engaging is another factor that increases stability. In biomechanical aspect, percutaneous pinning has inferior results when compared with plate osteosynthesis. Cosmetic results are satisfactory [62]. The technique has high complication rates compared to other surgical treatment options. Pin insertion could pose a risk to several structures such as axillary nerve (lateral pin), musculocutaneous nerve (anterior pin), radial nerve (lateral pin), long head of biceps tendon (anterior pin), and cephalic vein (anterior pin) injury as well as pin track infection. Other complications include malunion, non-union, and osteonecrosis of the humeral head [51, 63].

Open reduction and plate osteosynthesis is indicated in younger patient with head splitting fractures, >5 mm displaced greater tuberosity fractures, and two-, three-, and four-part fractures [49, 64, 65]. Locking proximal humerus plates increase the rotational and angular stability in the fracture line [42]. It can be applied with open and minimally invasive techniques. The open procedures allow for more successful reduction quality and fixation stability [49]. Plate was placed into the lateral aspect of the bicipital groove and 5–8 mm inferior to the greater tuberosity ensuring that it does not cause impingement superiorly (Fig. 14.11). Locked screws decrease failure rates in osteosynthesis of osteoporotic bone and/or

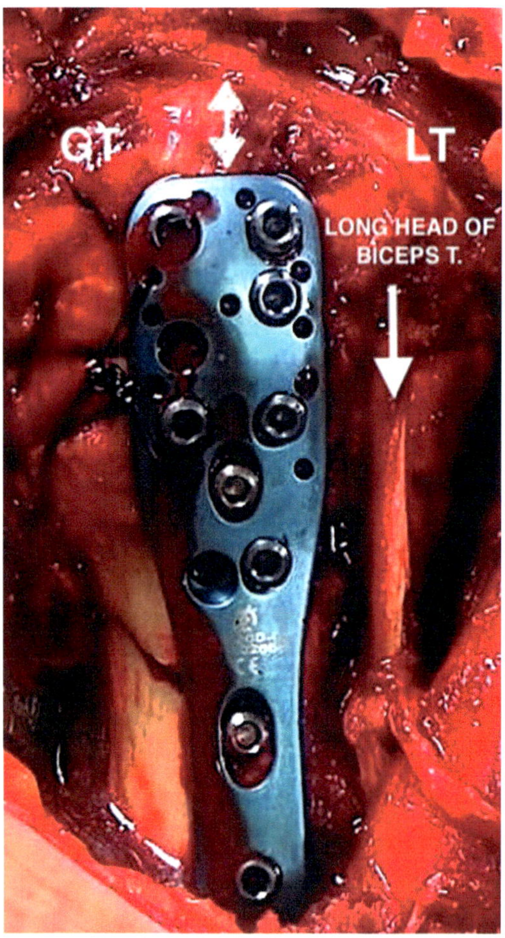

Fig. 14.11 Placement of locking plate in proximal humerus fractures. GT greater tuberosity, LT lesser tuberosity

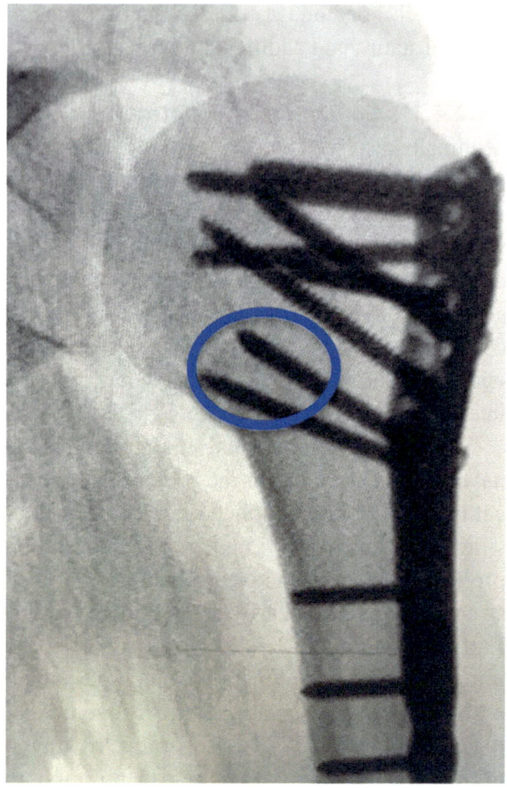

Fig. 14.12 Calcar support screws

complex fractures. Treatment outcomes are influenced by age, local bone mineral density, initial fracture displacement and complexity, initial varus deformity, quality of reduction, restoration of medial support, and vascularity of head fragment [49, 66, 67]. Calcar screw placement is recommended as it prevents varus collapse (Fig. 14.12). Medial support with fibular strut graft may be required for posteromedial comminution fractures [68]. Screw perforation is the most common complication (Fig. 14.13) [69]. Varus malunion, osteonecrosis, subacromial impingement, infection, and non-union are other complications [49–51]. Calcium phosphate cement augmentation has been shown to reduce complication rates such as displacement development and screw penetration [70].

Intramedullary nailing is indicated in two-part surgical neck fracture, three-part greater tuberosity fractures in younger patients, pathologic fractures, and those with metaphyseal com-

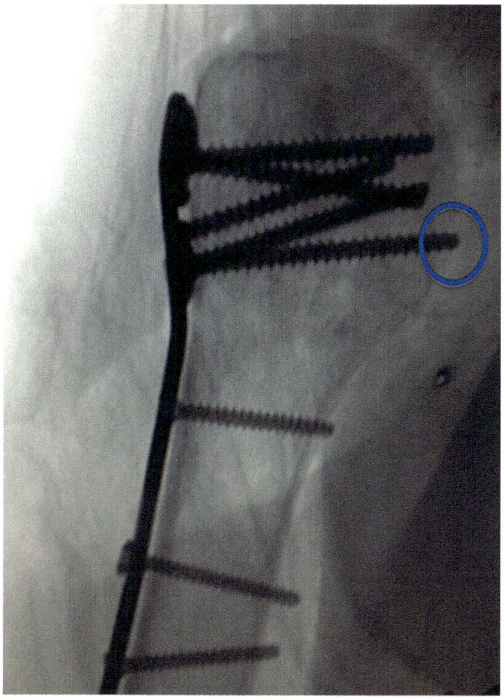

Fig. 14.13 Intra-articular penetration of the screw

minution or diaphyseal involvement [51, 69]. The technique is biomechanically inferior in torsional forces compared to plate fixation. In osteoporotic patients, early motion has been shown to increase failure rates [71]. Despite the meta-analysis by Wang et al. reporting intramedullary nailing a good treatment option in the treatment of proximal humeral fractures, several others reported high complication and reoperation rates [72–74]. Such complications include possible iatrogenic rotator cuff, biceps tendon, and cartilage injuries by nail insertion, proximal locking screw joint penetration due to shoulder pain, iatrogenic radial or axillary nerve injuries, proximal lateral cortical nail cut-out, and non-union [41, 42, 75].

Primary hemiarthroplasty is a preferred treatment method for the treatment of proximal humerus fractures with high risk of non-union or osteonecrosis: three- and four-part fracture dislocations, head splitting fractures, impaction fractures of the humeral head [76–78]. Possible indications include failure to obtain adequate reduction and fixation intraoperatively, complex proximal humerus fractures that can also be reconstructed in the younger patient (40–65 years old), head splitting fractures, tuberosities with high union potential and bone quality, and intact rotator. Intact rotator cuff and anatomically healed tuberosities are essential for satisfactory clinical results in shoulder hemiarthroplasty [51, 79, 80]. Hemiarthroplasty for treatment of complex proximal humerus fracture may have a high complication rate, including malunion or non-union of tuberosities, component malpositioning or loosening, periprosthetic fracture, heterotopic ossification, iatrogenic fractures, instability, infection, iatrogenic axillary or radial nerve injury, superior migration, and restricted range of motion [49, 81–83]. Glenoid cartilage damage and arthrosis is the most common complication affecting approximately 35% of cases [49].

In terms of treatment success, Boileau et al. reported 42% of patients were unsatisfied in their study [83]. Therapeutic outcomes could be improved with preservation of the coracoacromial ligament, anatomic reduction of the tuberosities, humeral height adjustment (5.6 cm above from upper margin of pectoralis major tendon insertion), proper retroversion of 20°, and appropriate rehabilitation program. In particular, poor bone quality usually requires cementation of the stem. Frankle et al. reported that using circumferential medial cerclage increases the potential for healing of tuberosities in the treatment of four-part fractures with hemiarthroplasty [84]. Bone grafting harvested from the head and humeral bone tunnels enhances healing and secures tuberosities-component fixation.

Reverse shoulder arthroplasty becomes popular after evidence with higher failure rate of hemiarthroplasties and secondary to expectation of increased patient satisfaction. The working principle is to medialize the center of rotation to provide greater lever arm of deltoid muscle. This eliminates the need for a rotator cuff and reduces shear forces affecting the glenoid (Grammont principles). Although the initial indication of reverse shoulder arthroplasty was rotator cuff arthropathy, its use is increasing in the treatment of complex proximal humerus fractures [76]. Its current indications in the treatment of proximal humerus fractures include >65-year-old non-

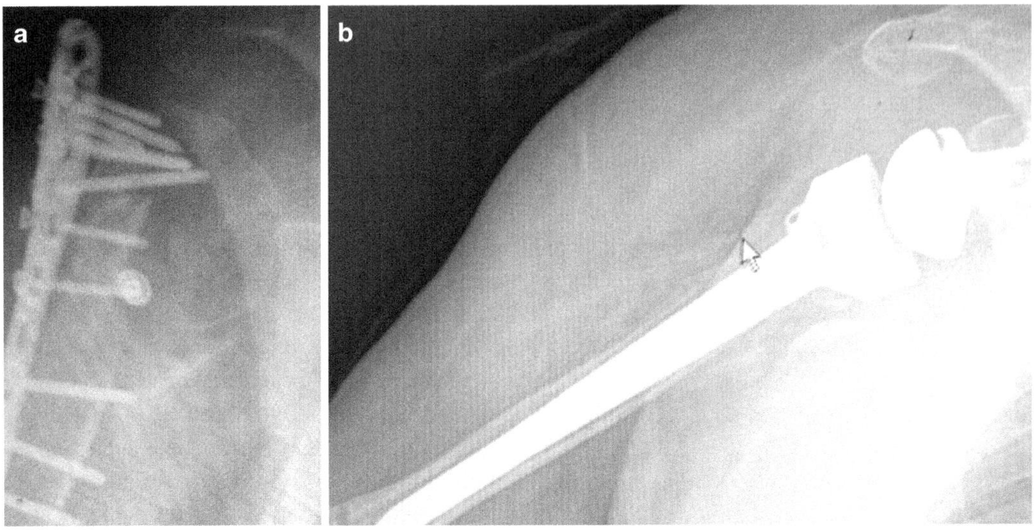

Fig. 14.14 (**a**) Failed open reduction internal fixation in proximal humerus fracture (**b**) managed with reverse shoulder arthroplasty

reconstructible tuberosities and poor bone quality in three- or four-part fractures, fracture dislocation, sequelae of fractures despite previous nonoperative treatment or osteosynthesis (malunion, non-union), and failed previous hemiarthroplasty (Figs. 14.14 and 14.15) [51, 77, 85]. In this technique, axillary nerve must be intact with a functioning deltoid muscle and adequate glenoid bone stock. The most common complication is scapular notching [86, 87], followed by others including infection, instability, periprostatic humerus or scapula neck fracture, polyethylene wear, synovitis, component loosening, acromial stress fractures or insufficiency, possible brachial plexus deficits, and other neurovascular injuries [55, 88, 89].

Reverse arthroplasty has demonstrated more satisfactory and predictable functional results compared to hemiarthroplasty [77, 80]. A systematic review involving 92 studies reported lower revision rates compared to open reduction plate fixation [90]. In multiple systematic reviews, tuberosities repair, restoration of humeral length, proper tensioning of deltoid and conjoint tendon, and autologous bone grafting are shown to be associated with improvements in clinical and functional outcomes [77, 85, 91]. CT imaging is recommended for the evaluation of the glenoid bone stock for implantation of baseplate in preoperative planning [51]. Prolonged postoperative immobilization could lead to stiffness; therefore, early passive range of motion exercises and rehabilitation process have an important role in providing the expected range of motion [92].

14.3 Scapular Fractures

The scapula fractures, which are responsible for 0.5% of all fractures, constitute only 3–5% of shoulder fractures. The category mainly consists of glenoid, acromion, coracoid process, and shaft fractures. The mean age in scapula fractures is 55 years, and male to female ratio is 3:1. While it is more common in men <50 years old, 80% of women are over 65 years old. Scapula fractures, which typically develop as a result of high-energy traumas, are almost always unilateral. Nevertheless, the presence of multiple fractures, such as other forms of shoulder fractures and rib fractures is common, particularly in males [93, 94]. In addition, head, neck, and thoracic injuries can sometimes accompany the condition. The mechanism of injury in scapula fractures varies according to the etiological cause. The fracture,

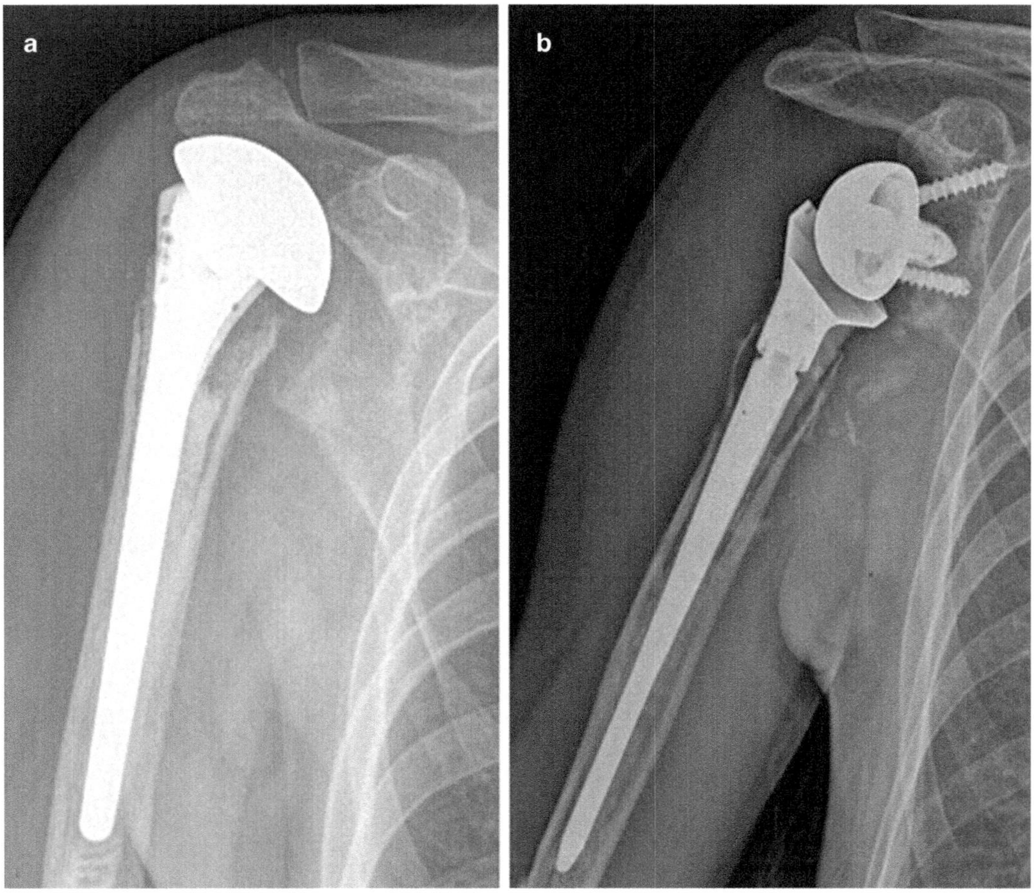

Fig. 14.15 (**a**) Failed hemiarthroplasty in proximal humerus fracture managed with (**b**) reverse shoulder arthroplasty

which may result from a traffic accident, falling from a height or a heavy object falling on the scapula, is influenced by variables such as the shape of the object, and the energy or the force vector transmitted via the impact. Other rare causes include compression fracture with severe muscle contracture, coracoid avulsion fracture caused by pulling of the coracoclavicular ligament, and fatigue fractures in athletes [95].

14.3.1 Clinical Anatomy

Connecting the clavicle and the humerus, the scapula has attachment sites for 18 muscles associated with the thorax, spine, and upper limb. The scapula is protected against impacts both with these surrounding muscle groups and with its relative mobility on the chest wall. This is also an important reason for fewer fractures compared to that of other areas of the shoulder, i.e. the proximal humerus or clavicle [96]. The factor causing damage in scapula fractures can be directly the head of the humerus. The position of the arm at the moment of injury also affects the type of fracture that the humeral head will create: When the arm is in abduction, the humeral head is forced against the inferior glenoid, which separates the lateral edge of the scapula. If the arm is in adduction, an impact on the elbow causes proximal dislocation of the humeral head, which causes damage to the acromion and the coracoid. Anterior and posterior dislocations of the humeral head may also lead to anteroinferior and posterior rupture fractures of the glenoid fossa, respectively [95].

14.3.2 Classification

There are several classifications for scapula fractures [97]. These are basically determined by taking into account the anatomical structures of the scapula: the articular segment including the glenoid fossa and the articular rim, the body of the scapula, and the processes [98]. According to the AO/OTA classification, scapular process fractures include coracoid fracture, acromion fracture, and spine fracture. Body fractures are categorized as the fractures that leave the body from ≤2 exits and fractures that leave the body from the ≥3 exits. Articular segment fractures are subdivided into three as (1) simple fractures of the glenoid fossa such as anterior rim, posterior rim, transverse or short oblique fractures; (2) glenoid neck fractures; and (3) multifragmentary fractures such as the glenoid fossa fracture and the central fracture dislocation [7]. In addition, these fractures are classified in various subgroups among themselves. Ideberg classification is used for the classification of glenoid fractures: Type 1 as the anterior rim fracture, Type 2 as the posterior rim fracture, Type 3 with fracture line extending into the lateral border of the scapula, Type 4 as superior fracture, and Type 5 with medial extension (Fig. 14.16). Coracoid fractures are divided into two as Type 1 or Type 2, where the fractures are located proximal or distal to the coracoclavicular ligament, respectively (Fig. 14.17). For acromion fractures, those non-displaced or minimally displaced fractures are classified as Type 1, and those without or with

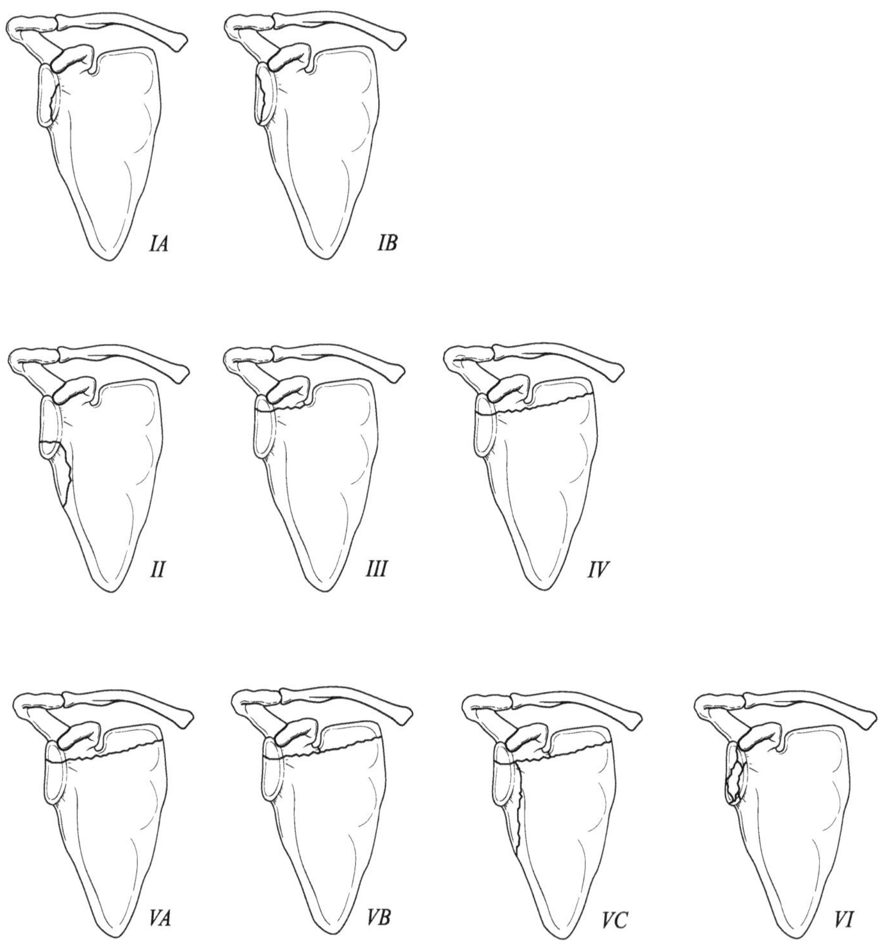

Fig. 14.16 Ideberg classification in glenoid fractures

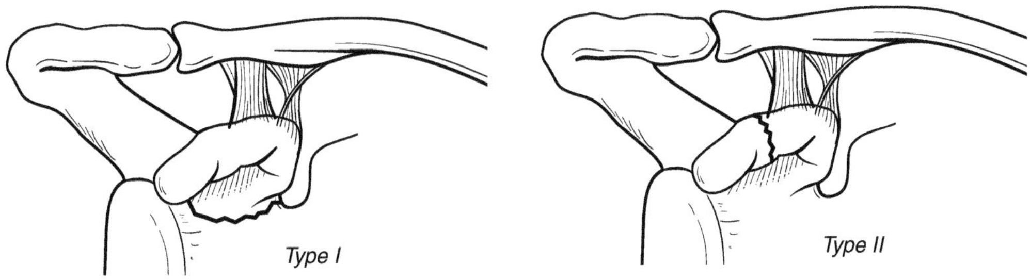

Fig. 14.17 Coracoid process fracture classification

reduction into the subacromial area are defined as Type 2 and Type 3, respectively.

14.3.3 Diagnostic Approach

The clinical evaluation of people with scapula fractures depends on the general condition of the patient. Priority should be given to the assessment of other potential life-threatening conditions in polytrauma patients. In patients with better condition without multiple trauma, physical examination of the injured side can be performed better. These patients usually apply with a position in which the affected upper limb is immobilized and supported by the contralateral hand in adduction. If possible, physical examination should be done in a sitting or standing position. On examination, all bone structures in the shoulder region are palpated and the presence of crepitus or pathological mobility is checked. The range of motion is limited during active movement, especially due to pain in abduction [99].

Although diagnostic methods are very important for diagnosing scapula fractures and determining the clinical strategy, these fractures represent one of the most difficult ones to recognize on radiography. It is reported that these fractures can be skipped in half of the chest X-rays, especially in trauma patients with scapula fractures [100]. Upright AP, Grashey view, axillary and scapula Y views should be obtained with radiography. In the Grashey view, the glenopolar angle and the medialization of the proximal fragment are calculated. Angulation is measured on the scapula Y view. These parameters are used to determine the indication for surgical intervention. Axillary view is also useful when acromion and coracoid fractures are suspected and to detect injuries such as glenohumeral dislocation [96]. 45° cephalic angled radiography can also be helpful to detect coracoid fractures. Thorax CT provides more reliable measurements than radiography, especially in patients with multiple trauma, and is also useful in revealing intra-articular glenoid fractures in more detail [101]. Three-dimensional CT scans are considered as the gold standard in the pathologic conditions of the scapula, providing the most sensitive measurements in terms of deformity and displacement [96, 100].

14.3.4 Treatment Approach

The treatment goal in scapula fractures is to regain full and painless range of motion on the shoulder and prevent late complications. There exist also other therapeutic goals specific for certain fracture types: ensuring glenohumeral joint alignment and stability in the glenoid fractures, alignment of the scapular body and the glenoid in shaft fractures, and prevention of painful non-union in acromion and coracoid process fractures. In general, non-displaced or minimally displaced scapula fractures are treated conservatively. While conservative treatment is applied in displaced fractures that do not exceed the threshold of surgical indication (<15–20 mm), regular control is performed by weekly radiography for 3 weeks and surgical intervention is considered in fractures exceeding this threshold [95, 96]. Pain-free outcomes and radiological results are better in patients undergoing surgery, whereas range of motion is better preserved in conservative treatment modalities [102].

14.3.4.1 Conservative Treatment

Conservative approach in scapula fractures includes pain relief and sling immobilization. After 2 weeks of sling application, exercises will start within 1 month to provide full passive range of motion. Then, the active range of motion is improved in the second month and the rotator cuff and other muscles are strengthened in the third month.

14.3.4.2 Surgical Treatment

The indication for surgical treatment in scapula fractures is determined by different criteria according to the type of fracture. Surgical intervention is indicated in intra-articular fractures, in glenoid fractures that result in glenohumeral instability or subluxation, or those causing ≥3–10 mm step-off. Anterior and posterior rim fractures covering more than 25–33% of the glenoid surface can be considered as a relative surgical indication. Extra-articular fractures could be surgically treated in following conditions: ≥45° angular deformity; co-occurrence of ≥35° angular deformity and ≥15 mm medialization; ≥20 mm medialization; and ≤22° glenopolar angle. Surgical intervention is required for rarely seen isolated process fractures, if the displacement is more than 10 mm, or if there is ipsilateral scapula fracture or multiple injuries that lead to instability in superior shoulder suspension complex [96].

Glenoid fractures are treated with open reduction and internal fixation, with a posterior approach in 80% of cases. In line with the increase of interest in arthroscopic fixation in recent decades, approaches with suture anchor or percutaneous screw fixation have also come into use (Fig. 14.18) [103]. In acromion fractures, stabilization can be achieved with dorsal tension band in the presence of displacement. Nevertheless, plate with osteosynthesis may also be used (Fig. 14.19). If the displacement is accompanied by complete third-degree acromioclavicular separation in coracoid fractures, these are treated with open reduction and internal fixation. Cannulated screws are used in surgical fixation (Fig. 14.20). Scapular body fractures are treated with open fixation only in the presence of associated complex fractures or neurovascular disruption; otherwise, operative fixation is not required (Fig. 14.21). If a displaced clavicle fracture accompanies glenoid neck fracture, internal fixation of the clavicle also helps the healing of the glenoid fracture [101].

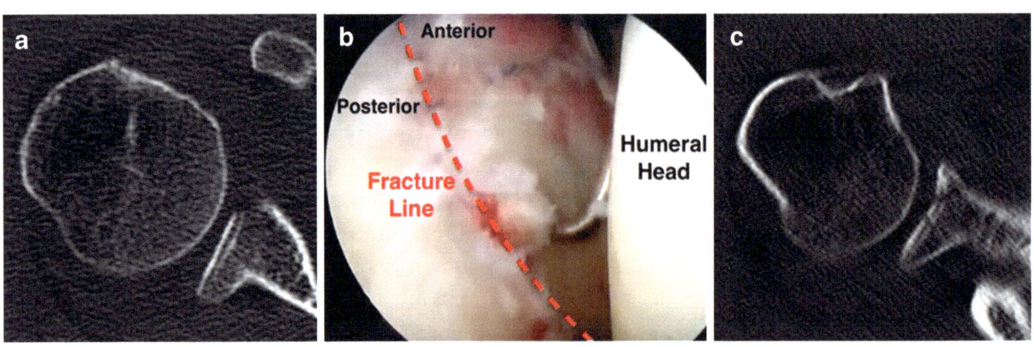

Fig. 14.18 Arthroscopic fixation of glenoid fracture: (**a**) preoperative CT image, (**b**) intraoperative view of fixated fracture, (**c**) postoperative CT image

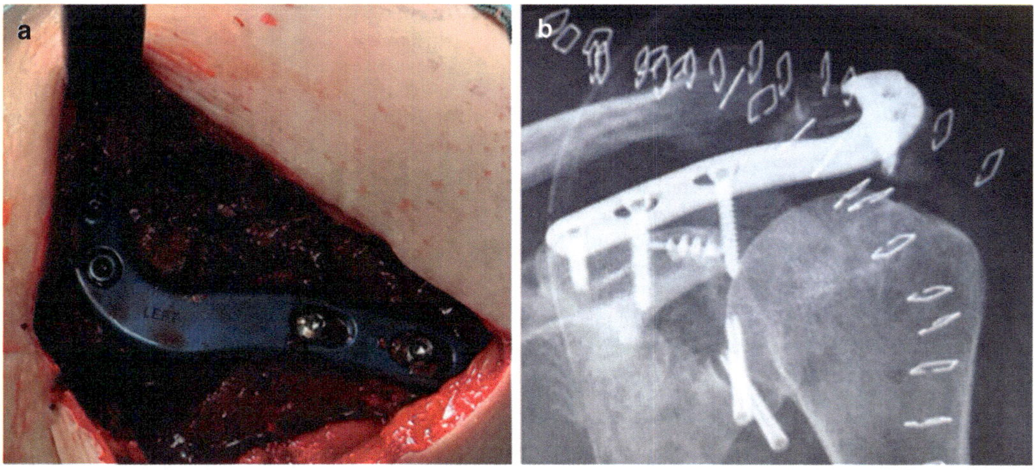

Fig. 14.19 Open reduction internal fixation of acromion fracture: (**a**) with posterior approach, (**b**) postoperative radiographic image

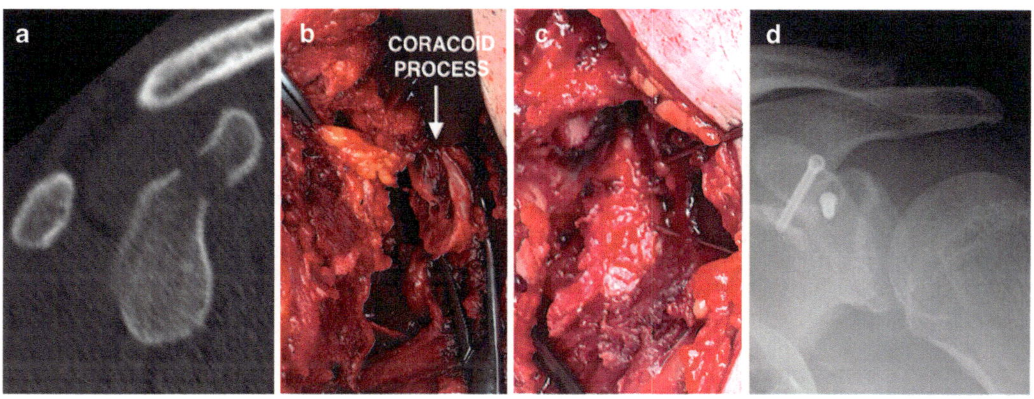

Fig. 14.20 Open reduction internal fixation of coracoid process fracture: (**a**) preoperative CT image, (**b**) intraoperative view of fracture line, (**c**) intraoperative view of fixation through K-wire-guided cannulated screw, (**d**) postoperative radiographic image

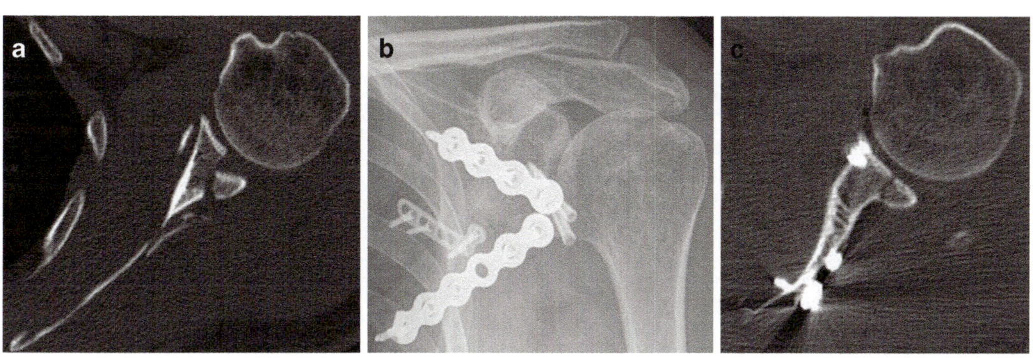

Fig. 14.21 Posteriorly- approached open reduction internal fixation of complex scapula fracture accompanying glenoid fracture: (**a**) preoperative CT image, (**b**) postoperative radiological image, (**c**) assessment of reduction of glenoid fracture on CT

References

1. Postacchini F, Gumina S, De Santis P, Albo F. Epidemiology of clavicle fractures. J Shoulder Elbow Surg. 2002;11:452–6.
2. Huttunen TT, Launonen AP, Berg HE, Lepola V, Felländer-Tsai L, Mattila VM. Trends in the incidence of clavicle fractures and surgical repair in Sweden: 2001-2012. J Bone Joint Surg Am. 2016;98:1837–42.
3. Kihlström C, Möller M, Lönn K, Wolf O. Clavicle fractures: epidemiology, classification and treatment of 2 422 fractures in the Swedish Fracture Register; an observational study. BMC Musculoskelet Disord. 2017;18:1–9.
4. Khan LAK, Bradnock TJ, Scott C, Robinson CM. Fractures of the clavicle. J Bone Joint Surg Am. 2009;91:447–60.
5. O'Neill BJ, Hirpara KM, O'Briain D, McGarr C, Kaar TK. Clavicle fractures: a comparison of five classification systems and their relationship to treatment outcomes. Int Orthop. 2011;35:909–14.
6. Allman F. Fractures and ligamenteous injuries of the clavicle and its articulation. J Bone Joint Surg Am. 1967;49:774–84.
7. Meinberg EG, Agel J, Roberts CS, Karam MD, Kellam JF. Fracture and dislocation classification compendium-2018. J Orthop Trauma. 2018;32(Suppl 1):S1–S170.
8. Holder J, Kolla S, Lehto S. Clavicle fractures: Allman and Neer classification. J Adv Radiol Med Imaging. 2017;2:102.
9. Hoogervorst P, van Schie P, van den Bekerom MPJ. Midshaft clavicle fractures: current concepts. EFORT Open Rev. 2018;3:374–80.
10. Ropars M, Thomazeau H, Huten D. Clavicle fractures. Orthop Traumatol Surg Res. 2017;103:S53–9.
11. McKee RC, Whelan DB, Schemitsch EH, McKee MD. Operative versus nonoperative care of displaced midshaft clavicular fractures: a meta-analysis of randomized clinical trials. J Bone Joint Surg Am. 2012;94:675–84.
12. Ersen A, Atalar AC, Birisik F, Saglam Y, Demirhan M. Comparison of simple arm sling and figure of eight clavicular bandage for midshaft clavicular fractures: a randomised controlled study. Bone Joint J. 2015;97B:1562–5.
13. Woltz S, Krijnen P, Schipper IB. Plate fixation versus nonoperative treatment for displaced midshaft clavicular fractures. J Bone Joint Surg Am. 2017;99:1051–7.
14. Bachoura A, Deane AS, Kamineni S. Clavicle anatomy and the applicability of intramedullary midshaft fracture fixation. J Shoulder Elbow Surg. 2012;21:1384–90.
15. Xu B, Lin Y, Wang Z, Cao J, Yang Y, Xia H, et al. Is intramedullary fixation of displaced midshaft clavicle fracture superior to plate fixation? Evidence from a systematic review of discordant meta-analyses. Int J Surg. 2017;43:155–62.
16. Houwert RM, Smeeing DPJ, Ahmed Ali U, Hietbrink F, Kruyt MC, van der Meijden OA. Plate fixation or intramedullary fixation for midshaft clavicle fractures: a systematic review and meta-analysis of randomized controlled trials and observational studies. J Shoulder Elbow Surg. 2016;25:1195–203.
17. Banerjee R, Waterman B, Padalecki J, Robertson W. Management of distal clavicle fractures. J Am Acad Orthop Surg. 2011;19:392–401.
18. Neer CS. Fracture of the distal clavicle with detachment of the coracoclavicular ligaments in adults. J Trauma. 1963;3:99–110.
19. Neer CS. Fractures of the distal third of the clavicle. Clin Orthop Relat Res. 1968;58:43–50.
20. Sambandam B, Gupta R, Kumar S, Maini L. Fracture of distal end clavicle: a review. J Clin Orthop Trauma. 2014;5:65–73.
21. Stegeman SA, Nacak H, Huvenaars KHJ, Stijnen T, Krijnen P, Schipper IB. Surgical treatment of Neer type-II fractures of the distal clavicle: a meta-analysis. Acta Orthop. 2013;84:184–90.
22. Dedeoglu SS, Imren Y, Çabuk H, Çakar M, Arslan SM, Esenyel CZ. Results of minimal invasive coracoclavicular fixation by double button lift-up system in Neer type II distal clavicle fractures. J Orthop Surg (Hong Kong). 2017;25:1–7.
23. Throckmorton T, Kuhn JE. Fractures of the medial end of the clavicle. J Shoulder Elbow Surg. 2007;16:49–54.
24. Asadollahi S, Bucknill A. Acute medial clavicle fracture in adults: a systematic review of demographics, clinical features and treatment outcomes in 220 patients. J Orthop Traumatol. 2019;20:24.
25. Bakir MS, Merschin D, Unterkofler J, Guembel D, Langenbach A, Ekkernkamp A, et al. Injuries of the medial clavicle: a cohort analysis in a level-I-trauma-center. Concomitant injuries management classification. Chirurgia (Bucur). 2017;112:586–94.
26. Salipas A, Kimmel LA, Edwards ER, Rakhra S, Moaveni AK. Natural history of medial clavicle fractures. Injury. 2016;47:2235–9.
27. Low AK, Duckworth DG, Bokor DJ. Operative outcome of displaced medial-end clavicle fractures in adults. J Shoulder Elbow Surg. 2008;17:751–4.
28. Sidhu VS, Hermans D, Duckworth DG. The operative outcomes of displaced medial-end clavicle fractures. J Shoulder Elbow Surg. 2015;24:1728–34.
29. Oe K, Gaul L, Hierholzer C, Woltmann A, Miwa M, Kurosaka M, et al. Operative management of periarticular medial clavicle fractures—report of ten cases. J Trauma Acute Care Surg. 2012;72:1–7.
30. Zheng Y, Yuan XH, Yin YH, Bin WW, Fu QS, Pang QJ. T-plate fixation for unstable proximal clavicula fractures. Acta Orthop Traumatol Turc. 2018;52:464–8.
31. Liu Z, Zhang J, Tian X, Kan S. Displaced medial-end clavicle fractures treated with locking plate osteosynthesis. Med Sci Monit. 2019;25:7591–6.
32. Titchener A, See A, Van Rensburg L, Tytherleigh-Strong G. Displaced medial end clavicular fractures treated with an inverted distal clavicle plate con-

toured through 90 degrees. J Shoulder Elbow Surg. 2019;28:e97–103.
33. Schumaier A, Grawe B. Proximal humerus fractures: evaluation and management in the elderly patient. geriatr orthop surg rehabil. Geriatr Orthop Surg Rehabil. 2018;9:2151458517750516. https://doi.org/10.1177/2151458517750516.
34. Pencle FJ, Varacallo M. Proximal humerus fracture. Treasure Island: StatPearls; 2020.
35. Chu SP, Kelsey JL, Keegan TH, Sternfeld B, Prill M, Quesenberry CP, et al. Risk factors for proximal humerus fracture. Am J Epidemiol. 2004;160(4):360–7.
36. Neer CS. Displaced proximal humeral fractures. Part I. Classification and evaluation. Clin Orthop Relat Res. 1987;52:3–10.
37. Boileau P, Walch G. The three-dimensional geometry of the proximal humerus: ımplications for surgical technique and prosthetic design. J Bone Joint Surg Br. 1997;79:857–65.
38. Randelli M, Gambrioli PL. Glenohumeral osteometry by computed tomography in normal and unstable shoulders. Clin Orthop Relat Res. 1986:151–6.
39. Murachovsky J, Ikemoto RY, Nascimento LGP, Fujiki EN, Milani C, Warner JJP. Pectoralis major tendon reference (PMT): a new method for accurate restoration of humeral length with hemiarthroplasty for fracture. J Shoulder Elbow Surg. 2006;15:675–8.
40. Ponce BA, Thompson KJ, Raghava P, Eberhardt AW, Tate JP, Volgas DA, et al. The role of medial comminution and calcar restoration in varus collapse of proximal humeral fractures treated with locking plates. J Bone Joint Surg Am. 2013;95:1–7.
41. Gerber C, Schneeberger AG, Vinh TS. The arterial vascularization of the humeral head. An anatomical study. J Bone Joint Surg Am. 1990;72(10):1486–94.
42. Hettrich CM, Boraiah S, Dyke JP, Neviaser A, Helfet DL, Lorich DG. Quantitative assessment of the vascularity of the proximal part of the humerus. J Bone Joint Surg Am. 2010;92:943–8.
43. Hinson JA. Anatomy and classification of proximal humerus fractures. In: Crosby LA, Neviaser RJ, editors. Proximal humerus fractures: evaluation and management. Cham: Springer; 2015. p. 1–22.
44. Codman E. The shoulder rupture of the supraspinatus tendon and other lesions in or about the subacromial bursa. Boston: T. Todd; 1934.
45. Neer CS. Displaced proximal humeral fractures. Part I. Classification and evaluation. J Bone Joint Surg Am. 1970;52:1077–89.
46. Müller ME, Koch P, Nazarian S, Schatzker J. The comprehensive classification of fractures of long bones. Berlin: Springer; 1990.
47. Hertel R, Hempfing A, Stiehler M, Leunig M. Predictors of humeral head ischemia after intracapsular fracture of the proximal humerus. J Shoulder Elbow Surg. 2004;13:427–33.
48. Bastian JD, Hertel R. Initial post-fracture humeral head ischemia does not predict development of necrosis. J Shoulder Elbow Surg. 2008;17:2–8.
49. Maier D, Jaeger M, Izadpanah K, Strohm PC, Suedkamp NP. Proximal humeral fracture treatment in adults. J Bone Joint Surg Am. 2014;96:251–61.
50. Laux CJ, Grubhofer F, Werner CML, Simmen HP, Osterhoff G. Current concepts in locking plate fixation of proximal humerus fractures. J Orthop Surg Res. 2017;12:1–9.
51. Kancherla VK, Singh A, Anakwenze OA. Management of acute proximal humeral fractures. J Am Acad Orthop Surg. 2017;25:42–52.
52. Lowry V, Bureau NJ, Desmeules F, Roy JS, Rouleau DM. Acute proximal humeral fractures in adults. J Hand Ther. 2017;30:158–66.
53. Lefevre-Colau MM, Babinet A, Fayad F, Fermanian J, Anract P, Roren A, et al. Immediate mobilization compared with conventional immobilization for the impacted nonoperatively treated proximal humeral fracture: a randomized controlled trial. J Bone Joint Surg Am. 2007;89(12):2582–90.
54. Calvo E, Morcillo D, Foruria AM, Redondo-Santamaría E, Osorio-Picorne F, Caeiro JR. Nondisplaced proximal humeral fractures: high incidence among outpatient-treated osteoporotic fractures and severe impact on upper extremity function and patient subjective health perception. J Shoulder Elbow Surg. 2011;20:795–801.
55. Koval KJ, Gallagher MA, Marsicano JG, Cuomo F, McShinawy A, Zuckerman JD. Functional outcome after minimally displaced fractures of the proximal part of the humerus. J Bone Joint Surg Am. 1997;79:203–7.
56. Hodgson SA, Mawson SJ, Stanley D. Rehabilitation after two-part fractures of the neck of the humerus. J Bone Joint Surg Br. 2003;85:419–22.
57. Foruria AM, De Gracia MM, Larson DR, Munuera L, Sanchez-Sotelo J. The pattern of the fracture and displacement of the fragments predict the outcome in proximal humeral fractures. J Bone Joint Surg Br. 2011;93(B):378–86.
58. Fjalestad T, Hole M, Hovden IAH, Blücher J, Strømsøe K. Surgical treatment with an angular stable plate for complex displaced proximal humeral fractures in elderly patients: a randomized controlled trial. J Orthop Trauma. 2012;26:98–106.
59. Khmelnitskaya E, Lamont LE, Taylor SA, Lorich DG, Dines DM, Dines JS. Evaluation and management of proximal humerus fractures. Adv Orthop. 2012;2012:861598.
60. Boesmueller S, Wech M, Gregori M, Domaszewski F, Bukaty A, Fialka C, et al. Risk factors for humeral head necrosis and non-union after plating in proximal humeral fractures. Injury. 2016;47:350–5.
61. Bohler J. Perkutane oisteosynthese mit dem Rontyenbildrier-Starker. Wiener Klin Wachenschr. 1962;74:485–7.
62. Magovern B, Ramsey ML. Percutaneous fixation of proximal humerus fractures. Orthop Clin North Am. 2008;39(4):405–16.
63. Sayed E-AB. Results of the percutaneous pinning of proximal humerus fractures with a modified palm tree technique. Int Orthop. 2011;35(9):1343–7.

64. Jo MJ, Gardner MJ. Proximal humerus fractures. Curr Rev Musculoskelet Med. 2012;5:192–8.
65. Murray IR, Amin AK, White TO, Robinson CM. Proximal humeral fractures: current concepts in classification, treatment and outcomes. J Bone Joint Surg Br. 2011;93(1):1–11.
66. Hardeman F, Bollars P, Donnelly M, Bellemans J, Nijs S. Predictive factors for functional outcome and failure in angular stable osteosynthesis of the proximal humerus. Injury. 2012;43:153–8.
67. Gardner MJ, Weil Y, Barker JU, Kelly BT, Helfet DL, Lorich DG. The importance of medial support in locked plating of proximal humerus fractures. J Orthop Trauma. 2007;21:185–91.
68. Gardner MJ, Boraiah S, Helfet DL, Lorich DG. Indirect medial reduction and strut support of proximal humerus fractures using an endosteal implant. J Orthop Trauma. 2008;22:195–200.
69. Zhu Y, Lu Y, Shen J, Zhang J, Jiang C. Locking intramedullary nails and locking plates in the treatment of two-part proximal humeral surgical neck fractures: a prospective randomized trial with a minimum of three years of follow-up. J Bone Joint Surg Am. 2011;93:159–68.
70. Egol KA, Sugi MT, Ong CC, Montero N, Davidovitch R, Zuckerman JD. Fracture site augmentation with calcium phosphate cement reduces screw penetration after open reduction-internal fixation of proximal humeral fractures. J Shoulder Elbow Surg. 2012;21:741–8.
71. Edwards SL, Wilson NA, Zhang LQ, Flores S, Merk BR. Two-part surgical neck fractures of the proximal part of the humerus: a biomechanical evaluation of two fixation techniques. J Bone Joint Surg Am. 2006;88:2258–64.
72. Wang G, Mao Z, Zhang L, Zhang L, Zhao Y, Yin P, et al. Meta-analysis of locking plate versus intramedullary nail for treatment of proximal humeral fractures. J Orthop Surg Res. 2015;10:122.
73. Dilisio MF, Nowinski RJ, Hatzidakis AM, Fehringer EV. Intramedullary nailing of the proximal humerus: evolution, technique, and results. J Shoulder Elbow Surg. 2016;25(5):e130–8.
74. Baltov A, Mihail R, Dian E. Complications after interlocking intramedullary nailing of humeral shaft fractures. Injury. 2014;45(Suppl 1):S9–S15.
75. Dedeoglu SS, Imren Y, Cabuk H, Tekin AC, Kir MC, Gurbuz H. Arthroscopy-assisted versus standard intramedullary nail fixation in diaphyseal fractures of the humerus. J Orthop Surg (Hong Kong). 2017;25(3):2309499017727949.
76. Spross C, Meester J, Mazzucchelli RA, Puskás GJ, Zdravkovic V, Jost B. Evidence-based algorithm to treat patients with proximal humerus fractures—a prospective study with early clinical and overall performance results. J Shoulder Elbow Surg. 2019;28:1022–32.
77. Schairer WW, Nwachukwu BU, Lyman S, Craig EV, Gulotta LV. Reverse shoulder arthroplasty versus hemiarthroplasty for treatment of proximal humerus fractures. J Shoulder Elbow Surg. 2015;24:1560–6.
78. Jaeger M, Maier D, Izadpanah K, Strohm P, Südkamp NP. Grenzen der Rekonstruktion—Prothesen. Unfallchirurg. 2011;114(12):1068–74.
79. Singh A, Padilla M, Nyberg EM, Chocas M, Anakwenze O, Mirzayan R, et al. Cement technique correlates with tuberosity healing in hemiarthroplasty for proximal humeral fracture. J Shoulder Elbow Surg. 2017;26:437–42.
80. Cuff DJ, Pupello DR. Comparison of hemiarthroplasty and reverse shoulder arthroplasty for the treatment of proximal humeral fractures in elderly patients. J Bone Joint Surg Am. 2013;95:2050–5.
81. Doursounian L, Gaillard J, Cambon-Binder A, Zbili D, Sautet A. Hemiarthroplasty for proximal humerus fractures with conservation of the whole humeral head as autograft: does it improve greater tuberosity healing? Int Orthop. 2019;43:1155–64.
82. Kontakis G, Koutras C, Tosounidis T, Giannoudis P. Early management of proximal humeral fractures with hemiarthroplasty: a systematic review. J Bone Joint Surg Br. 2008;90(11):1407–13.
83. Boileau P, Krishnan SG, Tinsi L, Walch G, Coste JS, Molé D. Tuberosity malposition and migration: reasons for poor outcomes after hemiarthroplasty for displaced fractures of the proximal humerus. J Shoulder Elbow Surg. 2002;11:401–12.
84. Frankle MA, Ondrovic LE, Markee BA, Harris ML, Lee WE. Stability of tuberosity reattachment in proximal humeral hemiarthroplasty. J Shoulder Elbow Surg. 2002;11:413–20.
85. Stahl D, De La Fuente G. Reverse total shoulder arthroplasty for a 4-part proximal humerus fracture. J Orthop Trauma. 2016;30:S9–10.
86. Friedman RJ, Barcel DA, Eichinger JK. Scapular notching in reverse total shoulder arthroplasty. J Am Acad Orthop Surg. 2019;27:200–9.
87. Tepass A, Rolauffs B, Weise K, Bahrs SD, Dietz K, Bahrs C. Complication rates and outcomes stratified by treatment modalities in proximal humeral fractures: a systematic literature review from 1970-2009. Patient Saf Surg. 2013;7:1–10.
88. Chalmers PN, Boileau P, Romeo AA, Tashjian RZ. Revision reverse shoulder arthroplasty. J Am Acad Orthop Surg. 2019;27:426–36.
89. Nicholson GP, Strauss EJ, Sherman SL. Scapular notching: recognition and strategies to minimize clinical impact. Clin Orthop Relat Res. 2011;469(9):2521–30.
90. Gupta AK, Harris JD, Erickson BJ, Abrams GD, Bruce B, McCormick F, et al. Surgical management of complex proximal humerus fractures—a systematic review of 92 studies including 4500 patients. J Orthop Trauma. 2015;29(1):54–9.

91. George MS, Khazzam M, Chin P, Dumont G, Russell R. Reverse shoulder arthroplasty for the treatment of proximal humeral fractures. JBJS Rev. 2014;2(10):01874474-201410000-00003. https://doi.org/10.2106/JBJS.RVW.M.00140.
92. Hagen MS, Allahabadi S, Zhang AL, Feeley BT, Grace T, Ma CB. A randomized single-blinded trial of early rehabilitation versus immobilization after reverse total shoulder arthroplasty. J Shoulder Elbow Surg. 2020;29:442–50.
93. Ideberg R, Grevsten S, Larsson S. Epidemiology of scapular fractures incidence and classification of 338 fractures. Acta Orthop. 1995;66:395–7.
94. Court-Brown CM. The epidemiology of fractures and dislocations. In: Court-Brown CM, Heckman JD, McQueen MM, Ricci W, Tornetta PI, editors. Rockwood, Green, Wilkins fract adults child. 8th ed. Philadelphia: Wolters Kluwer Health; 2014. p. 59–108.
95. Bartonícek J. Scapular fractures. In: Court-Brown CM, Heckman JD, McQueen MM, Ricci W, Tornetta PI, editors. Rockwood Green's Fract adults. 8th ed. Philadelphia: Wolters Kluwer Health; 2014. p. 1475–501.
96. Cole PA, Freeman G, Dubin JR. Scapula fractures. Curr Rev Musculoskelet Med. 2013;6:79–87.
97. Harvey E, Audigé L, Herscovici D, Agel J, Madsen JE, Babst R, et al. Development and validation of the new international classification for scapula fractures. J Orthop Trauma. 2012;26:364–9.
98. Jaeger M, Lambert S, Südkamp NP, Kellam JF, Madsen JE, Babst R, et al. The AO Foundation and Orthopaedic Trauma Association (AO/OTA) scapula fracture classification system: focus on glenoid fossa involvement. J Shoulder Elbow Surg. 2013;22:512–20.
99. Voleti PB, Namdari S, Mehta S. Fractures of the scapula. Adv Orthop. 2012;2012:1–7.
100. Berritto D, Pinto A, Russo A, Urraro F, Laporta A, La Porta M, et al. Scapular fractures: a common diagnostic pitfall. Acta Biomed. 2018;89:102–10.
101. Egol KA, Koval KJ, Zuckermann J. Scapula and thorax fractures. In: Egol KA, Koval KJ, Zuckermann J, editors. Handbook of fractures. 6th ed. Philadelphia: Wolters Kluwer; 2020. p. 181–93.
102. Dienstknecht T, Horst K, Pishnamaz M, Sellei RM, Kobbe P, Berner A. A meta-analysis of operative versus nonoperative treatment in 463 scapular neck fractures. Scand J Surg. 2013;102:69–76.
103. Van Oostveen DPH, Temmerman OPP, Burger BJ, Van Noort A, Robinson M. Glenoid fractures: a review of pathology, classification, treatment and results. Acta Orthop Belg. 2014;80:88–98.

Part IV

Shoulder Surgery and Complications

15 Positioning, Anesthesia, and Analgesia in Shoulder Surgery

Özer Öztürk, Selim Ergün, and Umut Akgün

Patient position selection in shoulder arthroscopy is an ongoing debate, which will continue in the coming years. Because there is no single correct, the choice is made according to the preference of the surgery and the patient characteristics. The anesthesia method to be selected may differ depending on the patient characteristics and the need for pain control after surgery. In this section, positions and anesthesia methods will be discussed and their advantages will be emphasized.

15.1 Patient Positioning

The importance of proper patient positioning and operation room setup cannot be underestimated in shoulder arthroscopy procedures. Although there are standardized positions, surgeon should do the position set-up by him/herself and make the appropriate adjustments. Otherwise, inappropriate position will make the surgery difficult and increase the operating time.

There are two main positions currently used in shoulder arthroscopy: beach-chair (BC) and lateral decubitus positions (LD). Both have advantages and disadvantages against each other (Table 15.1).

Ö. Öztürk (✉)
Orthopaedic Surgery, Marmara University Pendik Training and Research Hospital, Istanbul, Turkey

S. Ergün · U. Akgün
Orthopaedic and Traumatology, Acıbadem Kozyatağı Hospital, Istanbul, Turkey

15.2 Beach-Chair Position (BC)

The beach-chair position provides an anatomical posture for the surgeon. BC position in shoulder arthroscopy was first described by Skyhar et al. in 1988 [1]. In order to avoid the traction induced neurological complications in LD position, surgeons have developed the BC position.

Essential devices are necessary for the BC position. Operating table integrated with a back support is in the first place. It will give the desired angles of flexion to obtain the sitting position. Head-holder should be assembled to the table, and shoulder support, if present, should be removed to provide a wider space for the surgeon (Fig. 15.1). Thigh support, trunk belt, and leg belt should be applied in order to provide patient stabilization.

Following the interventions of anesthesiologist and induction of general anesthesia, it is necessary to dress the patient with compression socks and carefully pull up the patient to place the head into the head-holder (Fig. 15.2). The back of the patient is raised, slight Trendelenburg is applied, and then legs are lowered. Operating table should be positioned slowly while paying attention to the patient's airway and cervical spine stability. The back of the patient is raised then, till the thorax is 70–80° perpendicular to the floor, almost sitting position (Fig. 15.3). In addition, limp positioner can be used to keep the arm in position and apply traction when necessary (Fig. 15.4).

Cerebral blood flow is maintained by autoregulation in a person who goes from a lying position

G. Huri et al. (eds.), *Fundamentals of the Shoulder*, https://doi.org/10.1007/978-3-030-94702-6_15

Table 15.1 Advantages and disadvantages of beach-chair and lateral decubitus position

	Beach-chair	Lateral decubitus
Advantages	• More anatomic • Easy setup • Provide rotational control of shoulder • Allows for regional anesthesia • Easier for conversion to open surgery • Better in the subacromial area	• Preservation of the patient's cerebral oxygenation • Better access to the posterior and inferior glenoid • Accumulation of air bubbles in the subdeltoid space • Provides better expanding of the joint space • No specific operating table required
Disadvantages	• Higher risk of cerebral hypoperfusion • Accumulation of air bubbles in the operation area during subacromial decompression • Keeping the arm in abduction can create fatigue • Need specific operating table	• Need traction apparatus • May lead to traction-related neurological damage • Difficulty in orientation • Airway management is more difficult • Regional anesthesia cannot be easily tolerated

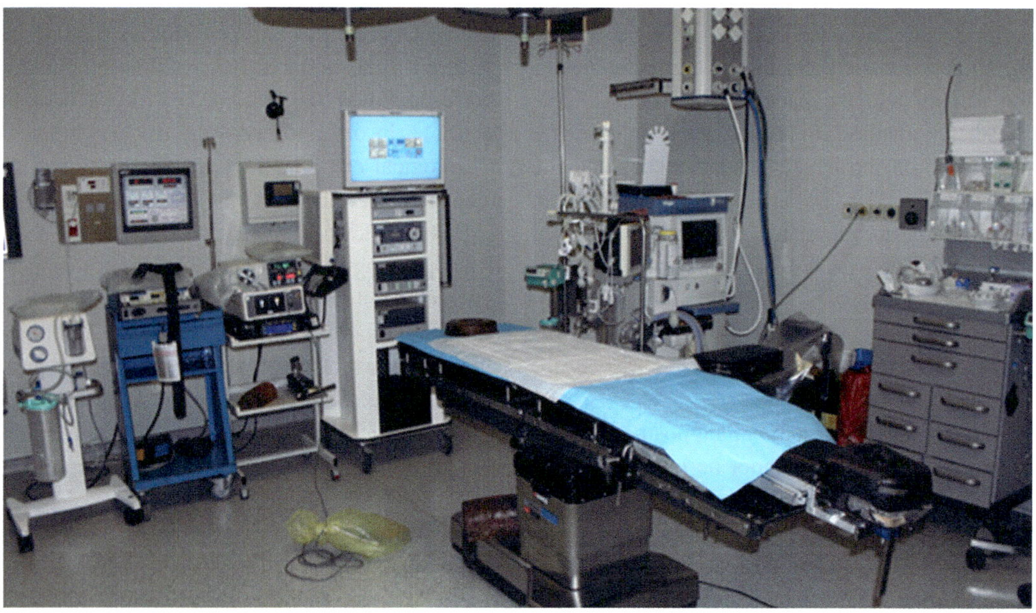

Fig. 15.1 Operating room set-up

to a sitting position. Due to the vasodilatation effect of anesthetic drugs, this autoregulation is not fully achieved in the BC position and there is a decrease in cerebral blood flow [2]. Neurological complications due to cerebral desaturation are the most important disadvantage of this position [3]. Hypotensive anesthesia with the aim of reducing bleeding during surgery may also negatively affect the patient's cerebral oxygenation.

Although BC position provides an anatomic orientation to the surgeon, which is advantageous for the procedures in subacromial area, accumulation of air bubbles with the effect of gravity during subacromial decompression is a frequent disadvantage.

Besides the disadvantages, beach-chair position provides more anatomical stance, easier installation, and working in the subacromial area, and rapid switch to open surgery when necessary. Easier access to the airway and toleration of regional anesthesia creates advantages for the anesthesia team.

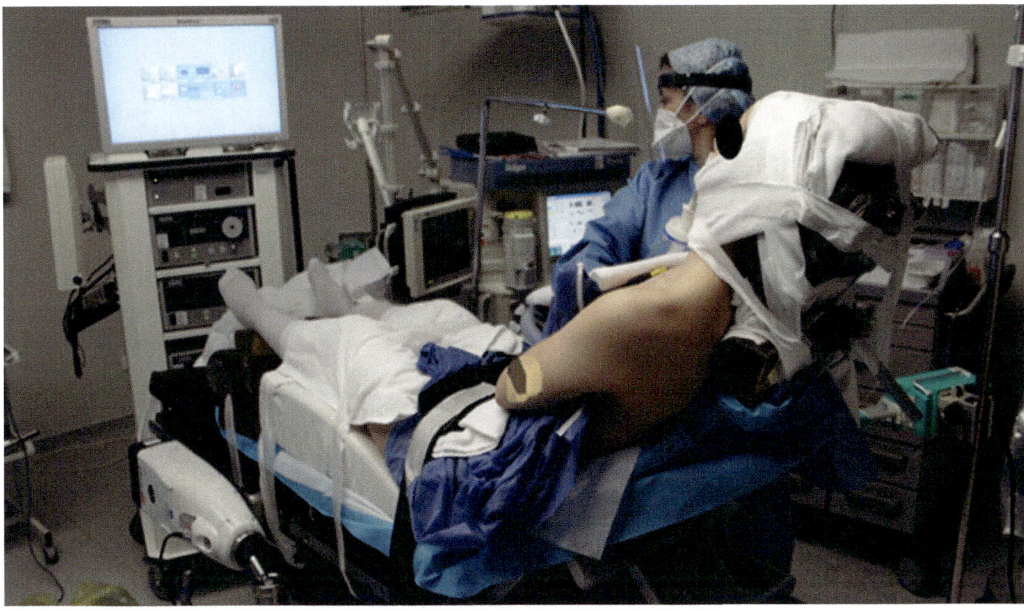

Fig. 15.2 Patient's head into the head-holder and shoulder support removed

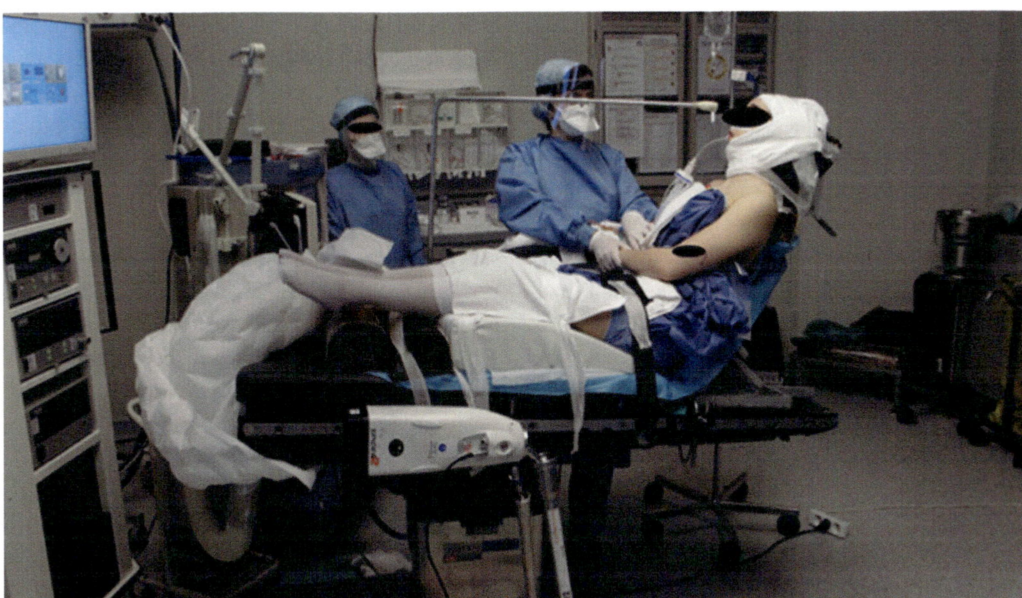

Fig. 15.3 Thorax is 70–80° perpendicular to the floor

15.3 Lateral Decubitus Position

Shoulder arthroscopy can be performed in lateral decubitus position while arm in suspension and without necessity of an adjustable shoulder table. However, a traction device that holds the arm in abduction and distraction is needed.

Gel pads should be placed on the table to prevent pressure sores before the patient is taken to the table. Following general anesthesia and endo-

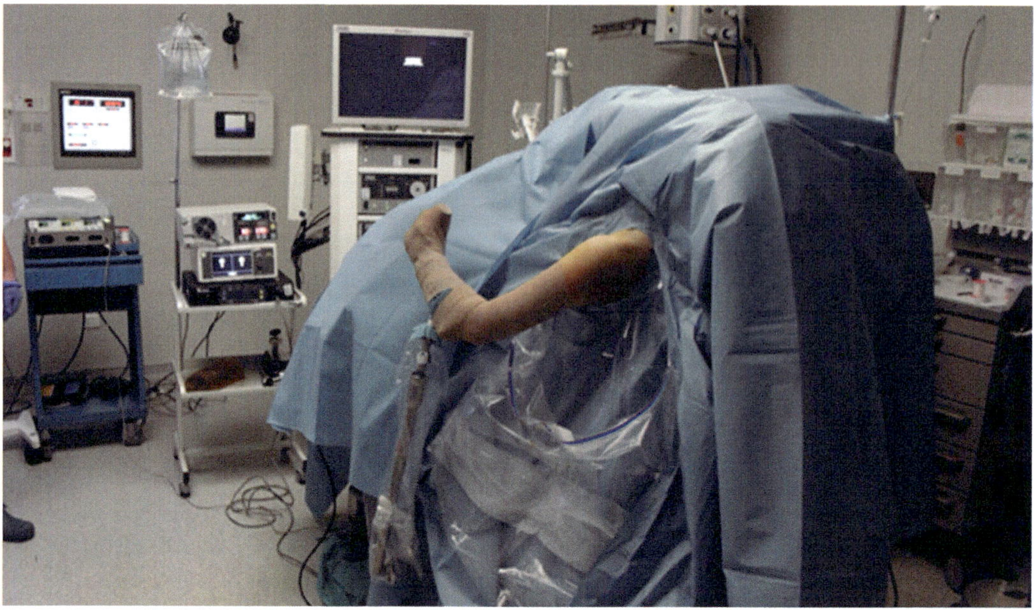

Fig. 15.4 Limp positioner keep the arm in position and apply traction when necessary

tracheal intubation, patient is turned over on the contralateral side, while pelvis and shoulders are perpendicular to the operating table, and then it is necessary to ensure that ASIS, greater trochanter, fibular head, and lateral malleolus are not under pressure. Neurovascular structures in the axillary region should be secured by placing a gel pillow about 10 cm in diameter proximal to the trunk. Front and back supports are placed and stabilization is provided with chest belt.

The table is tilted 30° posterior to make the glenoid surface parallel to the floor, then start draping (Fig. 15.5). The arm is placed on the traction device with 45° of abduction and 10° of flexion. Traction weight should be 4–6 kg; however, this may change depending on the patient's height and weight (Fig. 15.6).

The biggest advantage of this position compared to the BC position is the preservation of the patient's cerebral oxygenation during hypotensive anesthesia. The traction device provides more joint space and provides accumulation of air bubbles in the subdeltoid space which facilitates surgery. It provides better access to the posterior and inferior glenohumeral joint in instability surgeries when compared to the BC position.

The biggest disadvantage is traction-related neurological complications. Studies have shown that the most frequently affected nerve is the musculocutaneous nerve. Transient neuropraxia incidence has been reported between 10 and 30%, even if the traction weight was kept minimum [4].

Orientation is difficult due to non-anatomical position. Surgeon has to work on an abducted arm. If it is necessary to switch to open surgery, reposition and re-draping will be required. It is more difficult to access the airway than the BC position.

15.4 Comparison of Beach-Chair and Lateral Decubitus Positions

Many studies have focused on the advantages of the selected positions, their effects on the anesthesia, and postoperative outcomes. These assessments can guide surgeons for choosing the right position.

Especially the relationship between the position and cerebral perfusion was evaluated. In a sitting awake patient, the mean arterial pressure

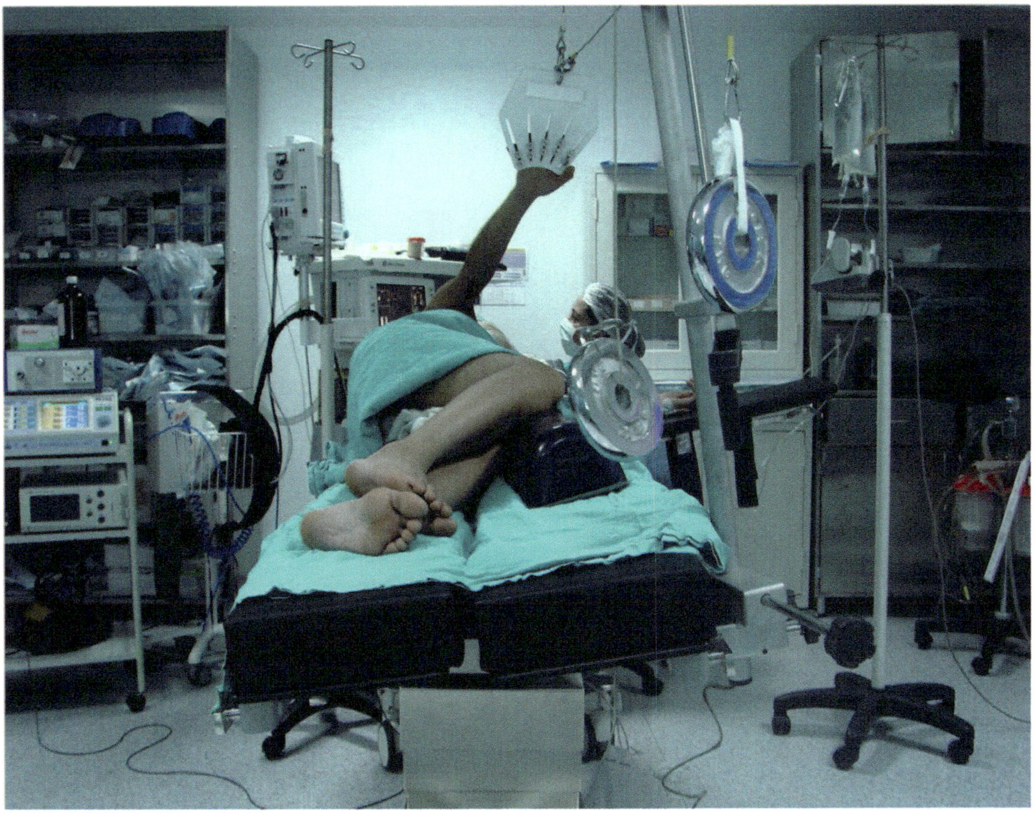

Fig. 15.5 The table is tilted 30° posterior to make the glenoid surface parallel to the floor

and associated cerebral perfusion pressures decrease. This decrease will advance with the vasodilatation effect of anesthetic agents [5].

Decrease in cerebral perfusion can lead to temporary or permanent neurological symptoms or even death. Various neurological symptoms have been reported in a review by Salazar et al., such as delirium in the early postoperative period, temporary vision loss, symptoms related to involvement of cranial nerves, cerebral ischemia, and stroke [6].

Spectroscopy can be used to evaluate cerebral perfusion (near infrared spectroscopy, NIRS). In a randomized controlled study by Cox et al., cerebral oxygenation was evaluated with NIRS in the BC position, and it was shown that cerebral desaturation was directly related to systolic arterial pressure [7]. The prospective study of Kocaoğlu et al. also supports this theory. In their study evaluating the correlation of NIRS, mean arterial pressure (MAP), heart rate, and peripheral oxygen saturation, cerebral desaturation was directly correlated with MAP, and in the absence of NIRS, monitoring the cerebral oxygenation by MAP was found to be reliable [8].

Cerebral desaturation has been shown to be associated with the sitting angle of the operating table. In the review of nine articles examining BC position and cerebral desaturation, it has been demonstrated that cerebral desaturation is more commonly observed at increased angles and as a recommendation, 30–45° angle is safe in terms of desaturation [9].

Studies that compare the effect of patient position on surgical outcomes, especially glenohumeral instability surgeries, found the recurrence rates to be lower in the LD position than the BC position [5]. In the systematic review of 25 studies on postero-inferior instability, patient satisfaction was shown to be 85–87.5% in the BC position, whereas 93–100% in the LD position, but the failure rates were higher in the LD position [10].

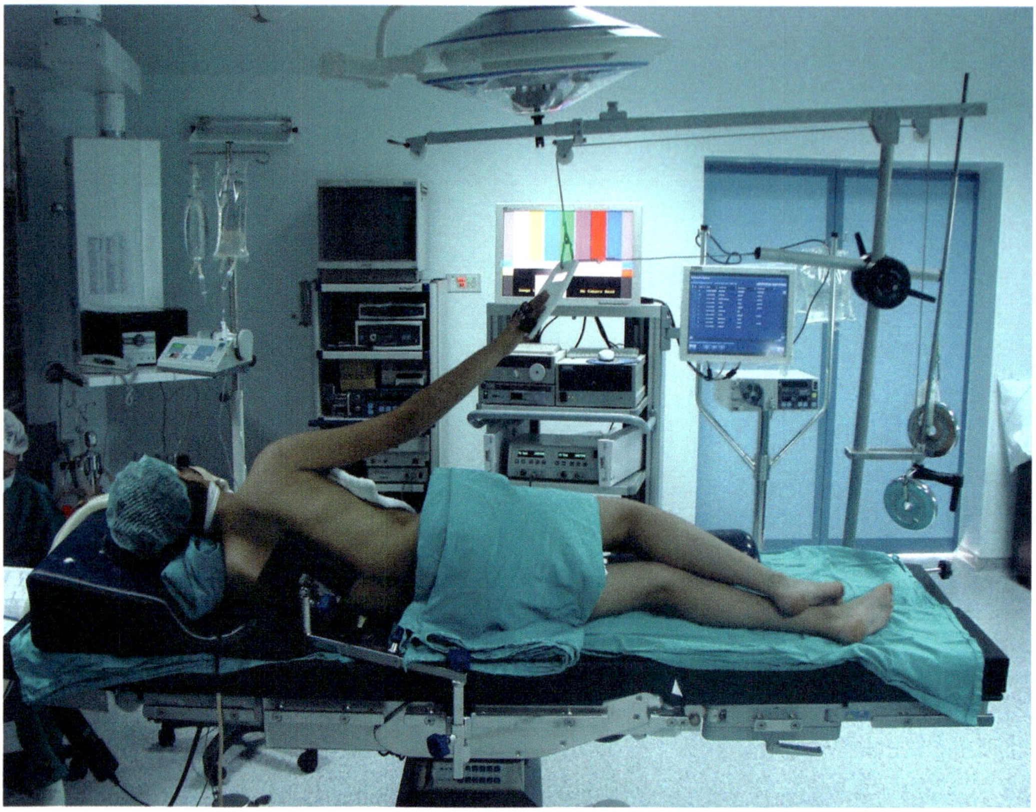

Fig. 15.6 The arm is placed on the traction device with 45° of abduction and 10° of flexion

In a review, which included 30 studies with arthroscopic capsular release due to adhesive capsulitis, it was reported that excellent clinical outcomes were obtained in both positions; however, BC position allowed advanced manipulation [11].

As a result, LD position is superior in instability procedures, particularly to visualize the posterior and inferior glenohumeral joint space. BC position provides better surgeon orientation in subacromial space pathologies. However, complications related to each position are not rare, should be kept in mind, and precautions should be taken.

15.5 Anesthesia

Many anesthetic methods can be used in shoulder arthroscopy. General and regional anesthesia and their combinations are methods that can be selected. Controlled hypotension for reducing intra-articular bleeding can be achieved more easily with general anesthesia. If general anesthesia is going to be choice, intubation should be preferred instead of laryngeal mask for airway safety.

Airway safety and cervical vertebra stability should be secured, especially in BC position. The head should be fixed to the head-holder and any motion that may occur during surgery should be prevented. Non-invasive spectroscopy (NIRS—near infrared spectroscopy) is recommended to observe cerebral oxygenation.

The nerve block procedure can be performed in two different ways: one is the shoulder block that contains the blockage of the suprascapular and axillary nerve and the other is the interscalene block. Advantages and disadvantages of these two nerve blocks are summarized in Table 15.2.

Table 15.2 Advantages and disadvantages of shoulder block and interscalene block

	Shoulder block	Interscalene block
Advantages	• Does not cause phrenic nerve blockage • Preferred in patients with respiratory system problems	• Preservation of the patient's cerebral oxygenation • Suitable for catheter use • Most proximally applied brachial plexus block • Preferred in patients who require postoperative pain control
Disadvantages	• Need blockage of two nerve: suprascapular and axillary nerve • Not suitable for catheter use • Most important complication is pneumothorax	• If phrenic nerve is effected, unilateral diaphragmatic paralysis may develop • Contraindicated in patients with low lung capacity or diaphragmatic dysfunction on the other side • Horner syndrome may develop due to stellate ganglion involvement

15.6 Shoulder Block (Fig. 15.7)

It is performed by blockage of the suprascapular nerve and axillary nerve. Suprascapular nerve innervates approximately 60% of the shoulder joint capsule [12]. The articular branch of the axillary nerve provides the innervation of the shoulder joint and the superior lateral cutaneous nerve that is originating from the posterior branch provides the innervation of the deltoid region.

The biggest advantage of the shoulder block is that it does not cause phrenic nerve blockage, so it can be preferred in patients with respiratory system problems.

It is not suitable for catheter use because block is applied to two nerves from different points. Therefore, it is not an effective method in providing long-term postoperative analgesia.

Although rarely seen (less than 1%), its most important complication is pneumothorax. Other complications include intravascular injection, local anesthetic toxicity, and nerve damage.

15.7 Interscalene Plexus Block (Fig. 15.8)

Interscalene block is performed at the level of the cricoid cartilage, between the anterior and middle scalene muscles. It is the most proximally applied brachial plexus blockage and basically blocks the C5–C7 roots.

It is suitable for catheter usage due to its location and application from a single point. Therefore, it is preferred in patients who require postoperative pain control.

Because the phrenic nerve is close to the blockage area, unilateral diaphragmatic paralysis may develop [13]. Therefore, this method is contraindicated in patients with low lung capacity or diaphragmatic dysfunction in the contralateral side. Another important complication is Horner syndrome which develops due to stellate ganglion involvement. Ptosis, decreased sweating, myosis, and hyperemia in the conjunctiva can be seen in these patients.

15.8 Postoperative Analgesia

It is important to regulate postoperative analgesia in cases requiring early physical therapy after surgery. Instead of using a single analgesic, multimodal analgesics are more effective.

Paracetamol should be the first agent to be selected in oral and parenteral treatment. Non-steroidal anti-inflammatory drugs (NSAIDs) are more effective than paracetamol in pain control and reduce opioid use. Non-selective NSAIDs should not be used in the first 24 h as they can

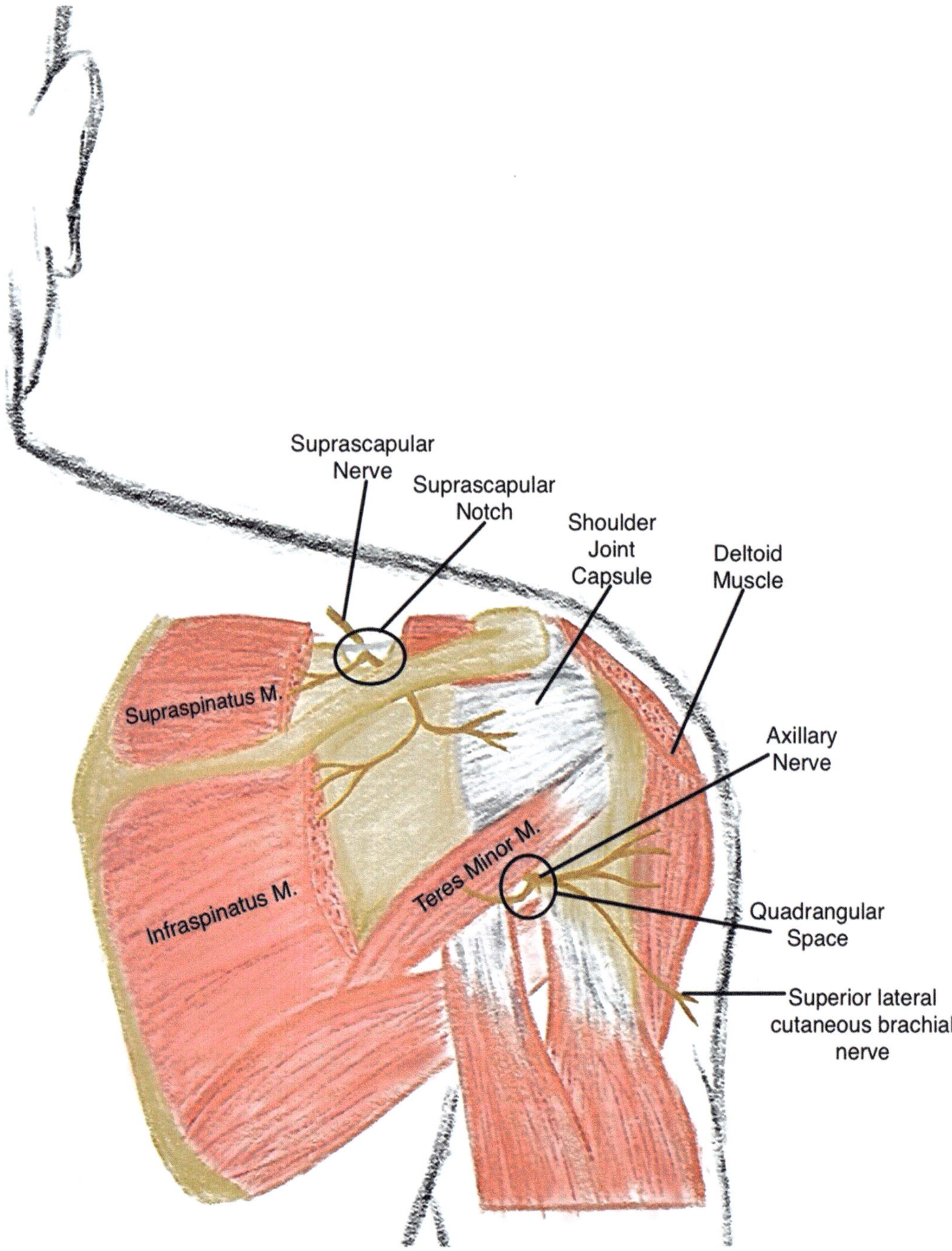

Fig. 15.7 Anatomical illustration of suprascapular nerve and axillary nerve. Nerve blockage applied on suprascapular notch area and quadrangular space area, respectively

affect coagulation and impair bleeding control. Selective Cox2 inhibitors should not be prescribed in the early postoperative period as they negatively affect bone and ligament healing.

Single dose interscalene blockage or shoulder block applied before surgery will provide an effective analgesia in the first 24 hours. Performing catheterization in addition to the

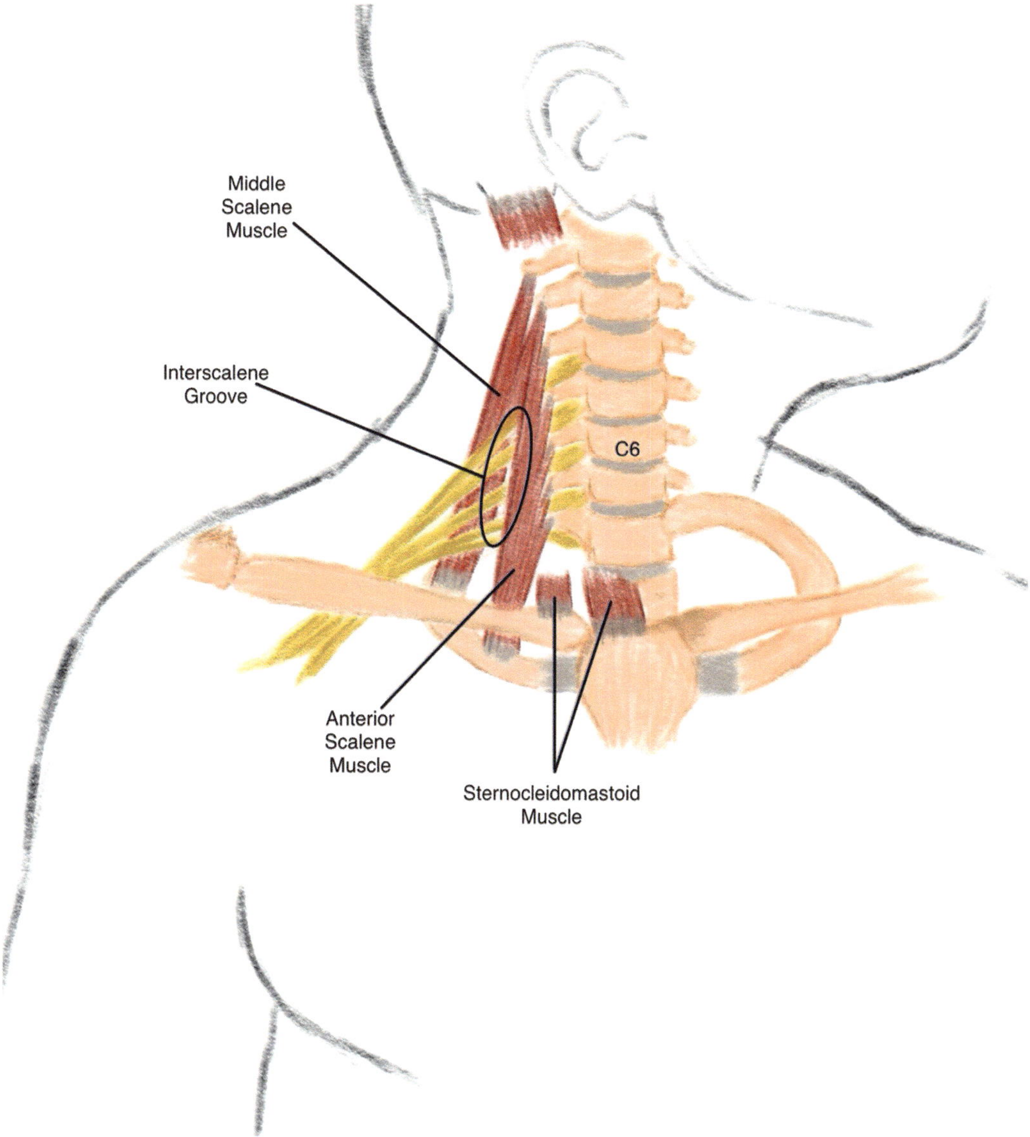

Fig. 15.8 Anatomical illustration of the brachial plexus roots that pass between anterior and middle scalene muscle. Nerve blockage applied on interscalene groove

nerve block facilitates longer-term pain control. Effective pain control is provided by a patient-controlled analgesia (PCA) device or applying local anesthetics from the catheter prior to exercise. In this way, the patient's opioid need and side effects related to opioid use are reduced.

If the patient does not have a catheter and the pain is severe, pain control should be achieved intravenously by morphine derivatives that can be placed in a PCA device.

15.9 Discussion of the Anesthesia

15.9.1 Controlled Hypotension

Hypotensive anesthesia can be applied to reduce bleeding and increase vision during shoulder arthroscopy. This may cause a risk for cerebral desaturation, especially in the BC position [14]. The general consideration is that systolic blood pressure should not be reduced below 90 mmHg, and the mean arterial pressure should not be reduced by more than 20% of the basal value in the BC position.

In a prospective study of 52 patients who were evaluated for cerebral desaturation by EEG, controlled hypotension was applied in the beach-chair position. On average, 36% decrease in systolic pressure and 42% decrease in mean arterial pressure were achieved. Ischemic changes were seen in three patients with EEG, and these changes were restored by increasing the blood pressure. No neurological sequels were observed in the postoperative period [15]. This study has shown that patients can tolerate lower blood pressure values, and that intraoperative cerebral monitoring will play an important role in preventing possible neurological damage.

15.9.2 Which Block Technique

Nerve block applications play an important role in ensuring postoperative pain control. However, debates continue in terms of complications and benefits. Comparison of shoulder block and interscalene block in a randomized controlled study has been shown that interscalene block significantly caused diaphragmatic dysfunction. In the evaluation of pain scores, it was shown that the interscalene block was more effective at the 2nd postoperative hour, and the pain scores were lower in the group with the shoulder block at the 24th hour. The more severe pain in the 24th hour of interscalane block group was thought to be related to the rebound effect [16].

The rebound pain shown in this study can be prevented by catheterization and PCA device. However, the potential of interscalene block to cause diaphragmatic dysfunction should be taken into consideration and application to patients in the risk group should be avoided.

15.10 Summary

Although there are many options, the preferred position and anesthesia method should be determined by the surgeon and the anesthesia team. Studies have shown BC position provides better surgeon orientation in subacromial space pathologies. LD position is superior in instability procedures, particularly to visualize the posterior and inferior glenohumeral joint space. Cerebral oxygenation should be closely followed by NIRS or MAP in BC position, and the development of cerebral desaturation should be prevented. It should be kept in mind that the neurological complications will be catastrophic. In the LD position, the weight of the traction should be kept minimum; otherwise, transient neuropraxia may develop. Single dose interscalene blockage or shoulder block applied before surgery will provide an effective analgesia in the first 24 h. Especially in patients requiring long-term analgesia because it allows the use of catheter, interscalene block is superior. However, the potential of interscalene block to cause diaphragmatic dysfunction should be taken into consideration and application to patients in the risk group should be avoided. Instead of using a single analgesic, multimodal analgesics are more effective to regulate postoperative analgesia. The patient's opioid need is reduced with catheterization and PCA use. If the patient does not have a catheter and the pain is severe, pain control should be achieved intravenously by morphine derivatives that can be placed in a PCA device.

References

1. Skyhar MJ, Altchek DW, Warren RF, Wickiewicz TL, O'Brien SJ. Shoulder arthroscopy with the patient in the beach-chair position. Arthroscopy. 1988;4(4):256–9.
2. Pohl A, Cullen DJ. Cerebral ischemia during shoulder surgery in the upright position: a case series. J Clin Anesth. 2005;17(6):463–9.
3. Salazar D, Hazel A, Tauchen AJ, Sears BW, Marra G. Neurocognitive deficits and cerebral desaturation during shoulder arthroscopy with patient in beach-chair position: a review of the current literature. Am J Orthop (Belle Mead NJ). 2016;45(3):E63–8.
4. Li X, Eichinger JK, Hartshorn T, Zhou H, Matzkin EG, Warner JP. A comparison of the lateral decubitus and beach-chair positions for shoulder surgery: advantages and complications. J Am Acad Orthop Surg. 2015;23(1):18–28.
5. Frank RM, Saccomanno MF, McDonald LS, Moric M, Romeo AA, Provencher MT. Outcomes of arthroscopic anterior shoulder instability in the beach chair versus lateral decubitus position: a systematic

review and meta-regression analysis. Arthroscopy. 2014;30(10):1349–65.
6. Salazar DH, Davis WJ, Ziroğlu N, Garbis NG. Cerebral desaturation events during shoulder arthroscopy in the beach chair position. J Am Acad Orthop Surg Glob Res Rev. 2019;3(8):e007.
7. Cox RM, Jamgochian GC, Nicholson K, Wong JC, Namdari S, Abboud JA. The effectiveness of cerebral oxygenation monitoring during arthroscopic shoulder surgery in the beach chair position: a randomized blinded study. J Shoulder Elbow Surg. 2018;27(4):692–700.
8. Kocaoglu B, Ozgen SU, Toraman F, Karahan M, Guven O. Foreseeing the danger in the beach chair position: are standard measurement methods reliable? Knee Surg Sports Traumatol Arthrosc. 2015;23(9):2639–44.
9. Pant S, Bokor DJ, Low AK. Cerebral oxygenation using near-infrared spectroscopy in the beach-chair position during shoulder arthroscopy under general anesthesia. Arthroscopy. 2014;30(11):1520–7.
10. de Sa D, Sheean AJ, Morales-Restrepo A, Dombrowski M, Kay J, Vyas D. Patient positioning in arthroscopic management of posterior-inferior shoulder instability: a systematic review comparing beach chair and lateral decubitus approaches. Arthroscopy. 2019;35(1):214–24.e3.
11. Houck DA, Belk JW, Vidal AF, McCarty EC, Bravman JT, Seidl AJ, et al. Outcomes of arthroscopic capsular release in the beach-chair versus lateral decubitus position: a systematic review. Orthop J Sports Med. 2019;7(12):2325967119888173.
12. Dhir S, Sondekoppam RV, Sharma R, Ganapathy S, Athwal GS. A comparison of combined suprascapular and axillary nerve blocks to interscalene nerve block for analgesia in arthroscopic shoulder surgery: an equivalence study. Reg Anesth Pain Med. 2016;41(5):564–71.
13. Albrecht E, Bathory I, Fournier N, Jacot-Guillarmod A, Farron A, Brull R. Reduced hemidiaphragmatic paresis with extrafascial compared with conventional intrafascial tip placement for continuous interscalene brachial plexus block: a randomized, controlled, double-blind trial. Br J Anaesth. 2017;118(4):586–92.
14. Murphy GS, Greenberg SB, Szokol JW. Safety of beach chair position shoulder surgery: a review of the current literature. Anesth Analg. 2019;129(1):101–18.
15. Gillespie R, Shishani Y, Streit J, Wanner JP, McCrum C, Syed T, et al. The safety of controlled hypotension for shoulder arthroscopy in the beach-chair position. J Bone Joint Surg Am. 2012;94(14):1284–90.
16. Rhyner P, Kirkham K, Hirotsu C, Farron A, Albrecht E. A randomised controlled trial of shoulder block vs. interscalene brachial plexus block for ventilatory function after shoulder arthroscopy. Anaesthesia. 2020;75(4):493–8.

16 Surgical Exposures

Nuri Aydın and Bedri Karaismailoğlu

16.1 Introduction

The number of patients with degenerative or traumatic problems of the shoulder is increasing day by day due to the aging population. Despite the current advances in arthroscopic surgery of the shoulder, arthroplasty or fracture fixation procedures necessitate mostly open surgery. However, the surgical approach might be demanding due to the surrounding structures of the shoulder. Deltoid muscle covers the shoulder almost circumferentially except the medial side, and it is needed to be whether splitted or detached to reach the joint. The rotator cuff is the other key muscular structure surrounding the shoulder which should be splitted or divided to reach the joint. The cuff should be repaired meticulously at the end of the procedure to not to compromise the functional outcome. Another important structure around the shoulder that should be given attention during open shoulder surgery is the axillary nerve. The axillary nerve lies inferior to the glenoid and around the proximal humerus. The damage to the nerve causes deltoid denervation and significant shoulder dysfunction [1]. Neurovascular structures at risk during an open shoulder approach include the cephalic vein, brachial plexus, axillary artery, musculocutaneous nerve, suprascapular nerve, and posterior circumflex artery. All these restrictions led to the investigation of different and less invasive exposure methods for open shoulder surgery.

The approaches can be roughly grouped into deltopectoral, deltoid-splitting, and posterior approaches (Table 16.1). Numerous variations end extensions are also described to increase the exposure when needed. Every approach has its advantages and disadvantages. The deltopectoral approach provides excellent exposure to joint and anterior structures, and it is the most widely used approach for both traumatic and degenerative diseases of the shoulder. The deltoid-splitting approach is more efficient in the management of some particular proximal humerus fractures and also gives chance to perform arthroplasty of the shoulder. The posterior approach is rarely needed, which is mostly used in the treatment of recurrent posterior dislocations and scapular neck fracture. The surgeons should have a grasp of all approaches to be able to manage all kinds of shoulder pathologies.

16.2 Deltopectoral (Anterior) Approach

16.2.1 Indications

The deltopectoral interval is the most commonly used plane for the open surgery of the shoulder. It

N. Aydın (✉) · B. Karaismailoğlu
Cerrahpaşa Medical Faculty, Department of Orthopaedics and Traumatology, Istanbul University-Cerrahpaşa, Istanbul, Turkey
e-mail: nuri.aydin@iuc.edu.tr; bedrikio@iuc.edu.tr

G. Huri et al. (eds.), *Fundamentals of the Shoulder*, https://doi.org/10.1007/978-3-030-94702-6_16

Table 16.1 Indications, internervous planes, and dangers of main surgical approaches of the shoulder

Approach	Indications	Internervous plane	Structures at risk
Deltopectoral	• Proximal humerus fracture • Shoulder arthroplasty • Reconstruction of recurrent dislocations • Biceps pathologies	Axillary nerve (deltoid) and medial and lateral pectoral nerves (pectoralis major)	• Cephalic vein • Axillary artery • Branches of brachial plexus • Musculocutaneous nerve • Axillary nerve
Deltoid-Splitting	• Greater tubercle fracture • Proximal humerus fracture • Shoulder arthroplasty • Calcific tendinitis • Intramedullary nailing of humerus	No true internervous plane	• Axillary nerve
Posterior	• Glenoid neck fracture • Scapular body fracture • Osseous augmentation of posterior glenoid	Suprascapular nerve (infraspinatus) and axillary nerve (teres minor)	• Axillary nerve • Posterior circumflex humeral artery • Suprascapular nerve

is an intermuscular and atraumatic interval protecting deltoid origin and insertion. It provides wide exposure of the anterior structures and an excellent view for both fracture fixation and arthroplasty procedures. It also allows the repair of anterior, inferior, and superior structures when appropriate extensions are utilized. The general indications include the fixation of proximal humerus fracture, shoulder arthroplasty, reconstruction of recurrent dislocations, and management of biceps pathologies (Table 16.1).

16.2.2 Incision and Dissection

The patient is placed in supine or beach-chair position. The deltopectoral interval might be felt by palpation in non-obese cases. The skin incision is generally around 10–14 cm, starting from the coracoid process and extending to deltoid insertion on humerus [2]. The length of the incision is dependent on the patient's arm length and the type of surgery. After the incision of skin and subcutaneous tissue, deltopectoral fascia is exposed. A fatty tissue can be present proximally around the cephalic vein. The layers can be dissected by a finger or a sponge to reveal the deltopectoral interval and cephalic vein (Fig 16.1). The bleeding can be significant in superficial dissection, thus should be taken under control before advancing to deeper layers. The internervous plane is between the axillary nerve (deltoid) and medial and lateral pectoral nerves (pectoralis major). The cephalic vein can be retracted either side according to the surgeon's preference but usually lateral retraction is preferred since the medial branches of the cephalic vein are fewer compared to lateral [3].

The deltopectoral interval is developed by blunt dissection. The small branches of the cephalic vein interfering with the exposure can be ligated. The cephalic vein should be protected when possible, to decrease postoperative swelling of the extremity. The protection of the cephalic vein can guide the surgeon during a revision approach by creating an anatomic landmark. However, it can also be divided into selected cases to ease the approach and should also be ligated when traumatized, to prevent thromboembolism risk. The correct definition of the interval is essential to prevent denervation of a part of the deltoid muscle which can limit the strength of flexion in the postoperative period. The transverse branches of the thoracoacromial arch pass deep to the cephalic vein in the upper part of the deltopectoral interval and can be ligated to prevent bleeding which might create difficulty in the exposure.

Subacromial and subdeltoid interval is developed by blunt dissection and bursal tissue is excised. Subdeltoid adhesions should be released when the arm is abducted to loosen the deltoid

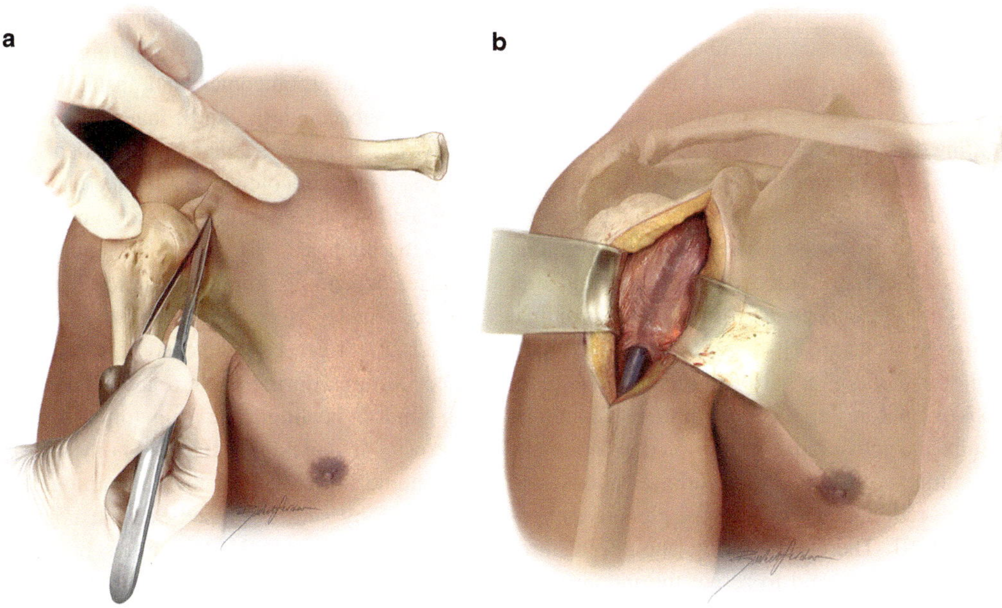

Fig 16.1 The skin incision for the deltopectoral approach (**a**) and the location of cephalic vein (**b**)

and decrease the risk of axillary nerve or rotator cuff injury [4]. Conjoined tendon is exposed and clavipectoral fascia is incised lateral to conjoined tendon. If more exposure of the medial side is needed (locked anterior dislocation, brachial plexus exploration), retracting the conjoined tendon might not be sufficient and the muscle can be detached by coracoid osteotomy (Fig 16.2). The fixation of the coracoid must be done during the closure, with the help of a screw or sutures. The screw fixation needs predrilling of coracoid before the osteotomy to prevent any malreduction.

The abduction of the arm puts tension on axillary artery and surrounding branches of brachial plexus which lie deep and medial to the conjoined tendon, under pectoralis minor muscle. Besides, these structures come closer to the surgical area with the abduction of the arm, increasing the risk of injury. Therefore, the arm should be kept adducted when possible, during the dissection around the coracoid process [5]. Excessive retraction of conjoined tendon can cause paralysis of elbow flexors by neuropraxia of the musculocutaneous nerve which enters the coracobrachialis muscle from the medial side and 5–8 cm distal to the coracoid process. The dissection around the coracobrachialis muscle should stay lateral.

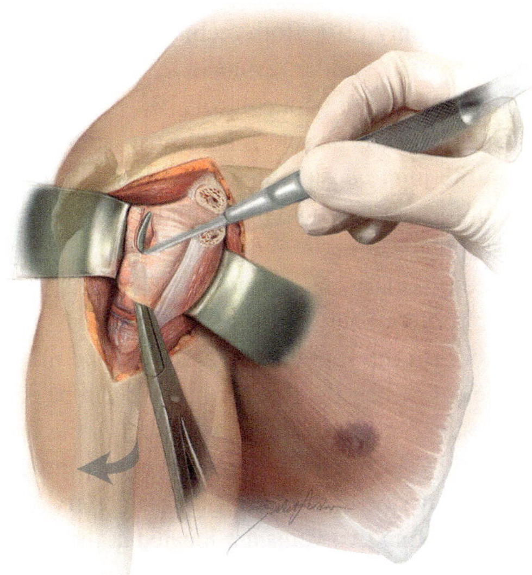

Fig 16.2 Subscapularis tenotomy after predrilling and osteotomy of coracoid process

Afterward, the long head of biceps and subscapularis tendon is identified. The long head of biceps can be followed through rotator interval and can be tenotomized or tenodesed if required (arthroplasty procedures, tendon degeneration, etc.). The functional role of the biceps tendon long head is controversial and there is no consensus on whether it is an important structure preventing the superior migration of the humeral head or a dysfunctional structure with a very limited role in the stability of the shoulder [6]. Besides, it has been also reported that the long biceps head is a source of ongoing pain after shoulder arthroplasty performed after proximal humerus fractures [7]. Regardless of the condition of the tendon, it has been reported that tenodesis or tenotomy application improves functional results and active range of motion does not make a significant difference in complications and reduces radiolucency around the glenoid component [8, 9]. Therefore, during shoulder arthroplasty, it is preferred to resect the intraarticular part of the biceps tendon rather than preserving it. Tenotomy or tenodesis should be chosen considering the age of the patient and cosmetic concerns due to the increased risk of popeye sign after tenotomy [10]. Both techniques have been shown to yield similar results in terms of postoperative function or pain control [10].

Subscapularis muscle is the last structure covering the joint capsule anteriorly. External rotation of the arm can help better find the borders of subscapularis muscle and increase the distance between axillary nerve and subscapularis. The upper border of subscapularis is continuous with the fibers of the supraspinatus muscle and might not be defined easily. Rotator interval can be palpated superior to subscapularis and it is the first arthrotomy location during arthroplasty procedures. Anterior circumflex humeral artery and accompanying two veins are located at the lower border of subscapularis muscle and usually need to be ligated or cauterized. Stay sutures are placed to the subscapularis muscle 2 cm away from its insertion. The axillary nerve passes distal and medial to subscapularis tendon and is at risk during subscapularis release.

The management of subscapularis depends on the procedure planned. The internal fixation of proximal humerus requires the protection of all muscle attachments; however, the arthroplasty requires the incision of subscapularis to expose the joint. It is an important structure for anterior stability and the fail of the repair might lead to inferior clinical outcomes and instability [11, 12]. The subscapularis can be managed in different ways when it is needed to be divided. These options include tenotomy, lesser tubercle osteotomy, tendon peeling from tuberosity, and the split of subscapularis. The healing environment changes depending on the preference. Tenotomy can be made 1 cm distal to the insertion (Fig. 16.2). It can be reattached with previously placed sutures during closure and requires tendon–tendon healing. The osteotomy of lesser tuberosity includes elevating the bony insertion and should be fixed after the completion of the procedure by sutures passed through bone tunnels and the plate [13]. It requires bone–bone healing and can be followed by direct radiography but requires a more demanding technique. Subscapularis can also be elevated by peeling and then repaired through bone tunnels but it requires tendon–bone healing, which is less favorable compared to homogenous interfaces like tendon–tendon and bone–bone [14]. Some reports claim that the risk subscapularis tear is diminished with lesser tubercle osteotomy compared to subscapularis tenotomy or peel methods, leading to increased dislocation rate [15, 16]. However, successful results were reported with each technique and the optimal method is controversial [17, 18]. Different techniques are also described to preserve the strength of subscapularis muscle, including a split in line with the subscapularis fibers or partial elevation from its insertion for later repair (Fig. 16.3) [19]. Although these methods provide similar exposure of glenoid with tenotomy, proximal humerus exposure is limited [20]. The capsule is visualized behind the subscapularis muscle and incised longitudinally (Fig. 16.4). If fracture fixation is being planned, the capsular incision should be placed in line with the fracture site.

16.2.3 Variations

The deltopectoral approach can be modified according to the surgeon's needs. The starting

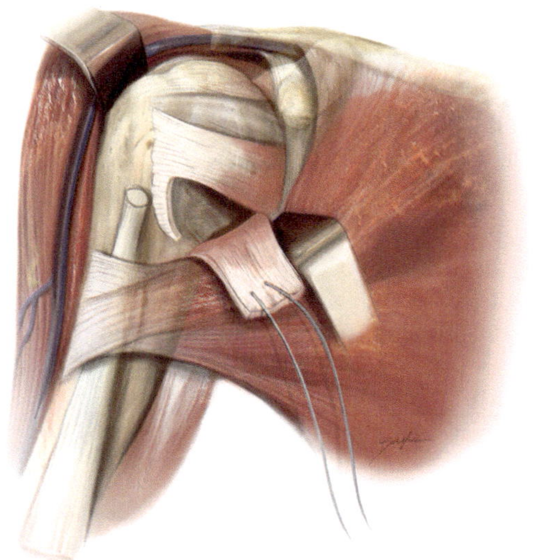

Fig 16.3 Subscapularis partial elevation from its insertion for later repair

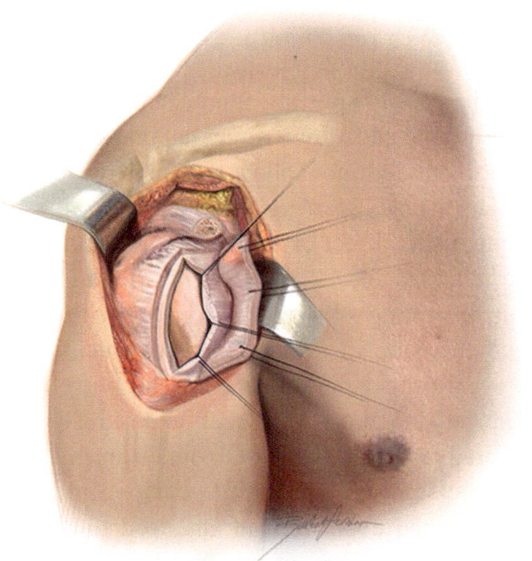

Fig 16.4 Longitudinal capsule incision exposing the glenohumeral joint. Deltoid origin can be detached from the clavicle to provide better proximal exposure

point of skin incision at coracoid process can be lateralized 2 cm and the cephalic vein can be retracted medially. This has several advantages. It can ease the exposure of glenoid and aid the centralization of the glenoid component during an arthroplasty. It also provides better exposure of anterolateral humerus and eases the plate placement for proximal humerus fractures. The lateralized incision is more distant to the axillar region compared to classical incision and this can help to decrease the bacterial contamination risk.

Inferior deltopectoral approach, which is also known as low-axillar approach, provides the exposure of anteroinferior glenoid rim. It uses the inferior part of the classical deltopectoral approach and can be used in the management of anterior instability or anterior glenoid fractures. The skin incision is placed more medial and inferior (Fig. 16.5). Inferior deltopectoral approach is cosmetically more favorable compared to the conventional approach since it is shorter and leaves most of the skin scar in the axillary fold. However, the exposure is more limited and sufficient exposure might not be obtained in muscled patients. It can also increase the risk of bacterial contamination since it is closer to axillary region. The incision starts from the apex of axillary fold and extends to axillary region vertically around 6–10 cm [5]. Deltopectoral interval is identified with the help of cephalic vein superiorly and the skin is retracted superiorly to bring incision in line with the deltopectoral interval.

Revision surgeries and internal fixation of complex fractures might require an *extended deltopectoral approach* compared to simple fractures and primary arthroplasty procedures. Deltopectoral incision can be extended to clavicle proximally or humerus shaft distally. The proximal extension can be made with a skin incision following the inferior border of the clavicle with a lateral curve. The distal extension follows the lateral border of biceps muscle which can be felt by palpation and provides better exposure of anterolateral humerus. After the retraction of biceps medially, brachialis muscle fibers can be splitted to reach humerus. These proximal and distal extensions might require the partial release of deltoid origin or insertion for better exposure, which should be repaired at the end of the procedure. The partial release of superior part of pectoralis tendon can provide better exposure in muscled patients, which does not require a repair usually.

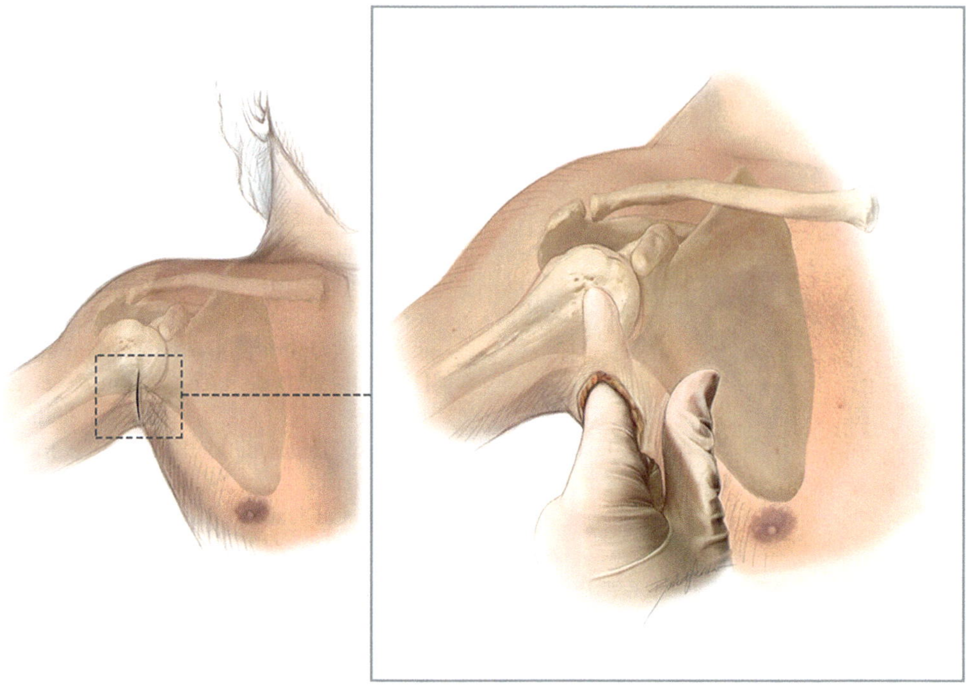

Fig 16.5 The skin incision for inferior deltopectoral (low-axillar) approach

Deltopectoral approach can also be extended more by *anteromedial approach*. It is used for shoulder arthrodesis and complex reconstructive shoulder surgeries in the presence of advanced deformity or stiffness which might lead to an uncontrolled tear of deltoid. It includes the progressive and controlled detachment of deltoid from clavicle, acromion, and scapular spine, which will require a strong repair at the end of the procedure and 4–6 weeks of immobilization [21].

16.3 Deltoid-Splitting Approach

16.3.1 Indications

Historically, the lateral deltoid-splitting approach was used for open rotator cuff repair but currently, it is largely replaced with arthroscopy. The current indications for deltoid-splitting exposure are greater tubercle or proximal humerus fractures and shoulder arthroplasty procedures (Table 16.1). The calcific tendinitis of subacromial bursae can also be managed through deltoid-splitting approach. The entry point of humerus nail is also exposed by this approach.

The deltoid-splitting approach protects anterior deltoid and allows direct lateral plate application but does not follow anatomic planes and might cause scarring between deltoid and humerus. A direct exposure facing glenoid can be obtained for glenosphere positioning without disrupting subscapularis but provides limited exposure to inferior glenoid which might prevent proper placement of glenosphere [22]. The humeral head and neck exposure is also limited making it difficult to obtain an anatomical neck cut and appropriate component sizes. The limited exposure can also lead to intraarticular penetration of screws during internal fixation [23]. The distal extension of the exposure is limited with axillary nerve passing over the deep surface of deltoid and a separate deltoid-split incision is required when more distal exposure is needed.

16.3.2 Incision and Dissection

The patient can be placed in beach-chair or lateral decubitus positions. An incision starting from the anterolateral corner of the acromion, extending

5 cm distally parallel to deltoid fibers is planned. The direction of the incision can be modified according to the type of pathology going to be addressed. Both trapezius and deltoid muscles are passed through by splitting in line with the fibers. The raphe between anterior and middle deltoid is defined and developed bluntly, which can be more prominent with distal traction of the arm. The subacromial bursae should be excised completely and supraspinatus tendon insertion on the greater tubercle is revealed (Fig. 16.6). The axillary nerve can be felt distally through the interval by finger. A stay suture can be placed to not to pass across this point and damage the axillary nerve. The axillary nerve lies 5–7 cm distal to the lateral edge of acromion but the distance can be less depending on the length and position of the arm [24, 25]. The abduction of the arm can decrease the distance between the acromion and axillary nerve and this distance is also shorter on the posterior plane compared to anterior [25].

In patients undergoing internal fixation of a proximal humerus fracture, a separate distal incision can be made distal to the axillary nerve and in line with proximal split (Fig. 16.7). An epiperiostal plane from proximal incision to distal window is developed on lateral surface of the humerus and the windows are united (Fig. 16.8). The placement of shaft screws can be applied from the distal incision. The location of the distal split can be chosen according to the extension of fracture and implant length. Aggressive retraction of the deltoid should be avoided to prevent neuropraxia of the axillary nerve. Rotator interval can be used to enter the joint in case of an arthroplasty procedure.

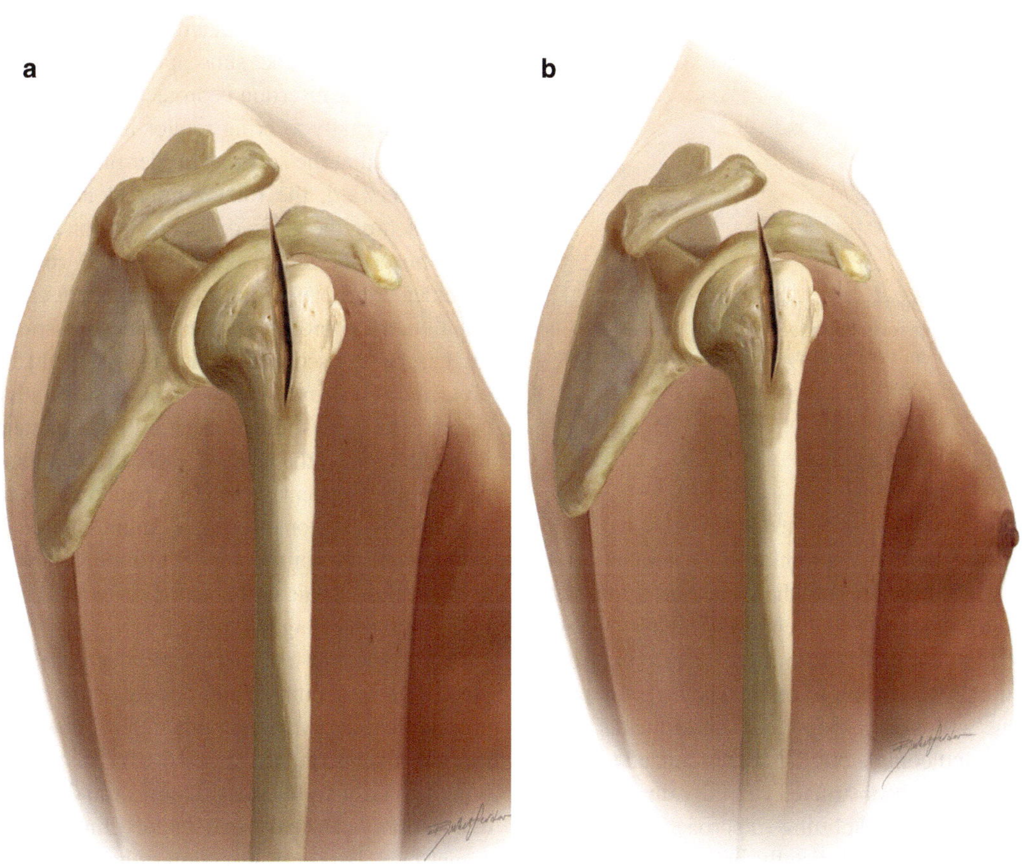

Fig 16.6 The split of deltoid (**a**) and exposure of supraspinatus insertion on greater tuberosity after subacromial bursae excision (**b**)

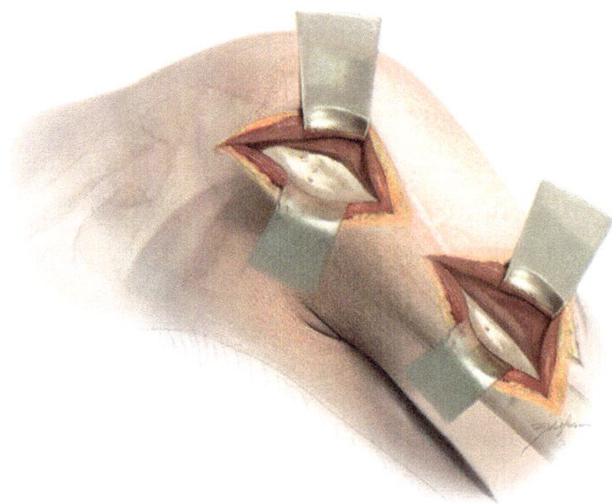

Fig 16.7 Splitted deltoid fibers exposing the periosteum of lateral aspect of humerus. Two separate incisions should be used and the level of axillary nerve should be spared

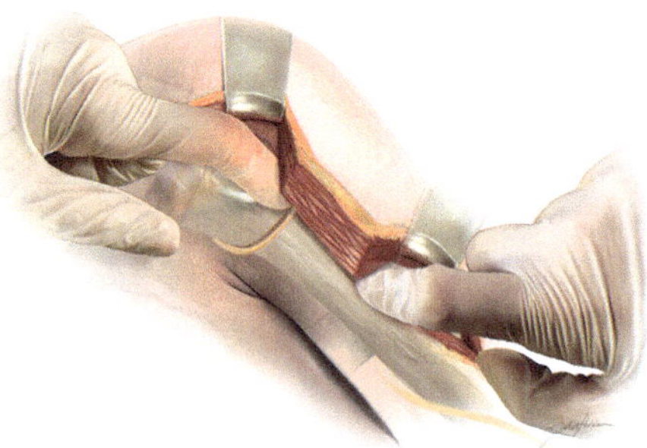

Fig 16.8 The inner section of the two-window deltoid-splitting approach. Pay attention to the location of axillary nerve. An epi-periosteal window is developed to connect the two incisions

16.3.3 Variations

The incision can be extended to superior and medial towards acromion and parallel to the superior border of the scapular spine [26]. Acromioplasty or coracoacromial ligament release can be used to increase proximal exposure [27]. Acromion can also be osteotomized in line with skin incision and supraspinatus can be exposed completely. The osteotomized acromion should be reconstructed carefully during the closure. Deltoid-splitting approach cannot be directly extended distally but can be combined with deltopectoral approach if needed. But it should be kept in mind that the raphe between anterior and middle deltoid is a watershed area since anterior portion of deltoid is supplied by thoracoacromial arch or anterior circumflex arteries, while the middle and posterior portions are supplied by posterior circumflex artery. The dissections anterior or posterior to the raphe can disrupt the blood supply of a part of deltoid [28]. Distal window of the incision can also be extended to expose middle third of humerus by peeling a part of deltoid insertion on lateral surface of the humerus.

Subscapularis sparing approach is a trending approach in total or reverse shoulder arthroplasty with reported excellent outcomes. It can be utilized via *anterosuperior approach,* which is a variation of the deltoid-splitting approach [29]. After 6–8 cm length of longitudinal incision originating from posterior part of anterolateral acromion, anterior deltoid and the coracoacromial ligament are detached together and splitting of deltoid through the anterolateral raphe is utilized (Fig. 16.9) [30]. The subscapularis tendon is exposed and its insertion is protected. The arm is rotated externally to bring rotator interval to the surgical field at the superior border of subscapularis. Afterward, the rotator interval is incised until reaching the glenoid. Biceps tendon can be tenotomized or tenodised. In case of reverse shoulder arthroplasty due to supraspinatus dysfunction, the approach is easier since supraspinatus can be excised but infraspinatus should be protected if it is intact, to maintain postoperative external rotation strength. Deltoid reattachment is crucial during the closure. Deltoid with coracoacromial ligament shoud be repaired as a single layer with nonabsorbable sutures through bone tunnels on acromion [29]. Although rare, if deltoid origin repair is not healed, functional loss might be devastating [1].

Anterosuperior approach provides excellent glenoid view facilitating appropriate reaming and proper glenoid component position without disruption of subscapularis (Fig. 16.10). The protection of anterior soft tissue structures can decrease dislocation rates and yield better clinical outcomes [16]. However, resection of humeral neck osteophytes and anatomic humeral neck cut are might be challenging due to limited humerus

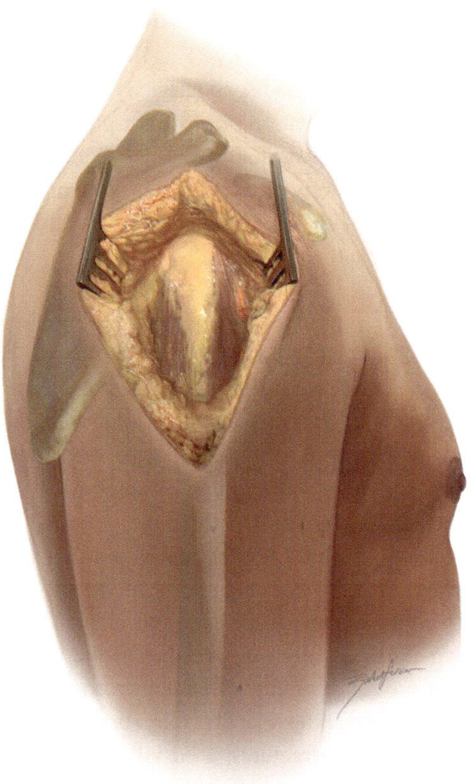

Fig 16.9 The skin incision and superficial dissection of anterosuperior approach showing the fat stripe identifying the raphe between anterior and middle portions of deltoid muscle

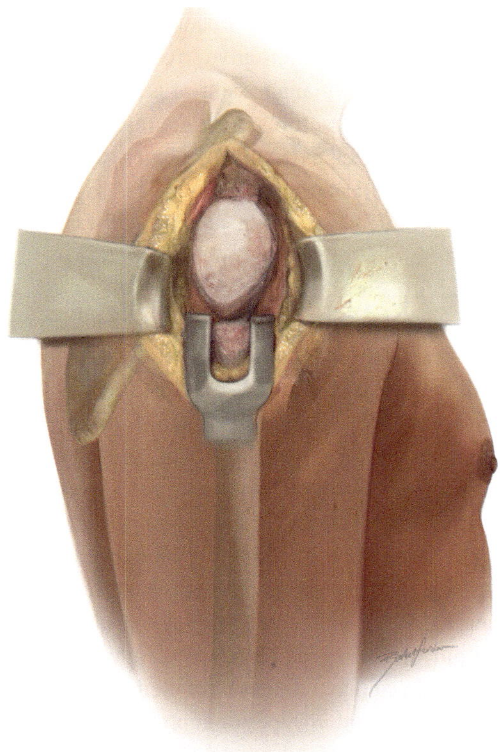

Fig 16.10 The glenoid exposure in anterosuperior approach

exposure. And despite excellent glenoid exposure, the neutral placement of glenosphere guide pin is complicated by the position of the humerus and the baseplate of the glenoid might be placed with a superior tilt which might cause scapular notching in the future [30].

16.3.3.1 Literature Review on Deltopectoral vs. Deltoid-Splitting Approaches

Both deltopectoral or deltoid-splitting approaches can be used in some particular indications such as proximal humerus fracture and shoulder arthroplasty. The research on possible advantages and superiorities of these techniques is going on.

Deltopectoral incision obliquely passes the skin cleavage lines, while deltoid-splitting is orthogonal, thus both are capable of leaving a cosmetically unpleasant scar [26]. In a prospective non-randomized study, it is reported that deltopectoral approach provides better functional outcome at first-year follow-up in the management of proximal humerus fractures compared to deltoid-splitting [31]. The functional outcome was especially increased by better active elevation and abduction values after deltopectoral approach. The reason behind this difference might be the use of anatomic planes in deltopectoral approach causing less scarring of the shoulder structures. Deltoid-splitting approach protects anterior deltoid and theoretically, it can diminish the risk of osteonecrosis, but comparative studies could not show any significant difference [23]. Although deltopectoral approach provides excellent exposure, it might be challenging to place the plate laterally enough requiring the release of deltoid and it is also hard to control greater tubercle adequately through deltopectoral incision. These problems can be overcome by deltoid-splitting approach

[26]. Deltoid-splitting approach is also reported as a valuable alternative to deltopectoral approach in the management of complex proximal humerus fractures with hemiarthroplasty, providing better exposure of posterior fracture fragments and rotator cuff tissue [32]. However, the anatomical reduction of complex fractures might be challenging through the deltoid-splitting approach [33].

Successful clinical outcomes of anatomical and reverse shoulder arthroplasties were reported with both deltopectoral and deltoid-splitting anterosuperior approaches [22]. Although extended deltopectoral approach is more frequently used for total shoulder replacement, glenoid exposure might be challenging in muscled patients and may cause anterior or anteverted glenoid baseplate insertion. Excessive retraction can cause deltoid injury and partial detachment of deltoid origin might be needed. Anterosuperior approach can avoid these problems. It is also reported that anterosuperior approach causes less postoperative instability and less acromial fracture compared to deltopectoral approach since it does not compromise subscapularis and provides a more appropriate implant position [29]. Nerve injury is also reported at higher rates in deltopectoral approach compared to anterosuperior approach in anatomic shoulder arthroplasty [34]. However, in reverse shoulder arthroplasty, some studies reported higher incidence of scapular notching with anterosuperior approach due to the difficulty in inferior placement and tilting of glenoid guide and implant [22, 35]. Additionally, anterosuperior approach requires the detachment and repair of anterior deltoid, thus might lead to the risk of deltoid dehiscence or weakening leading to postoperative pain and dysfunction [4, 30]. The revision of deltoid-splitting anterosuperior approach with a deltopectoral approach can also lead to anterior deltoid dysfunction.

16.4 Posterior Approach

16.4.1 Indications

The posterior shoulder approach provides sufficient exposure for the posterior and inferior of glenohumeral joint including posterior glenoid, scapular neck, and body. The main indications are glenoid neck and scapular body fractures and osseous augmentation of posterior glenoid in the treatment of posterior instability (Table 16.1). Since it is rarely used, the surgeons are more unfamiliar with the anatomy compared to other approaches.

16.4.2 Incision and Dissection

The patient can be placed either in a prone or lateral decubitus position. The posterior approach can include both horizontal and vertical incisions. The vertical incision alone, which is centered over the glenohumeral joint, can be preferred in posterior glenoid augmentation, in which the limited exposure is sufficient. The incision starts slightly inferomedial to the posterior corner of acromion and extends vertically [36]. When more extensile exposure is needed (e.g. scapular body fracture), a curvilinear horizontal incision can be used, providing the widest exposure. The incision extends from medial to lateral in line with the scapular spine and curves inferiorly towards axilla at the posterior corner of acromion [4]. After subcutaneous dissection and incision of the fascia, the posteroinferior edge of deltoid is identified and the internervous plane between deltoid (axillary nerve) and underlying infraspinatus (suprascapular nerve) is developed bluntly. This plane can be more easily identified in lateral region of the incision. Afterwards, deltoid is retracted superiorly. If superior retraction is not sufficient (e.g. bulky deltoid), a longitudinal split of deltoid can also be added to increase the exposure. Although this can cause a posterior deltoid injury, it is functionally less prominent than anterior deltoid injury since its function can be covered by the long head of triceps, teres minor, and infraspinatus muscles [2]. If extended exposure is being used, the origin of deltoid on scapular spine is elevated osteoperiostally, allowing better superior and lateral retraction of deltoid. The abduction of the arm also loosens the deltoid fibers.

After the retraction of deltoid, the internervous interval between infraspinatus (suprascapular nerve) and teres minor (axillary nerve) is identified (Fig. 16.11). The interval can be identified more easily at medial region since there might be a fat stripe between the muscle bellies at medial, while the tendon fibers are mixed at lateral region. After developing the interval bluntly, infraspinatus is retracted to superior and teres minor is retracted to inferior, exposing the joint capsule over the inferior part of glenoid and scapular body (Fig. 16.12). The dissection inferior to teres minor can damage axillary nerve and posterior circumflex humeral artery, which pass through the quadrangular space at inferior border of teres minor. The injury of posterior circumflex humeral artery can cause uncontrolled bleeding and can be prevented by staying in the intermuscular plane.

16.4.3 Variations

The axillary nerve and posterior circumflex artery prevent the extension of this approach distally. Dissection can be lengthened to lateral towards the glenohumeral joint. The partial release of infraspinatus from greater tubercle or infraspinatus-split can be added to provide more superior access to the glenoid. Medial retraction of infraspinatus can be made but attention must be given to not harming the suprascapular nerve by forceful retraction which can cause neuropraxia of the nerve which innervates both supraspinatus and infraspinatus muscles. The medial retraction of infraspinatus exposes the glenoid neck. The dissection should not go beyond 1 cm medial to glenoid rim to prevent suprascapular

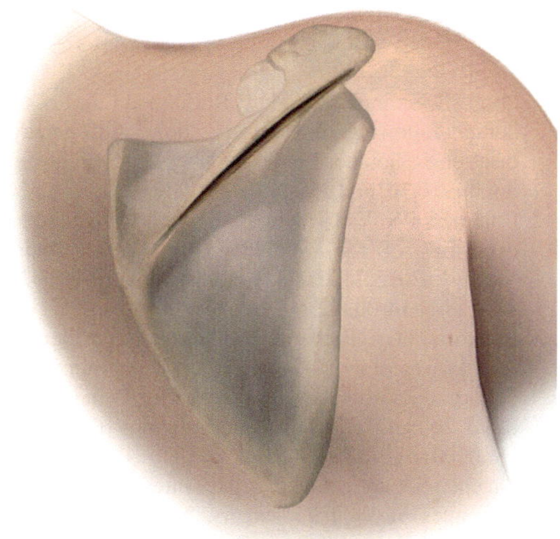

Fig 16.12 The capsular incision and the exposure of glenohumeral joint from posterior approach

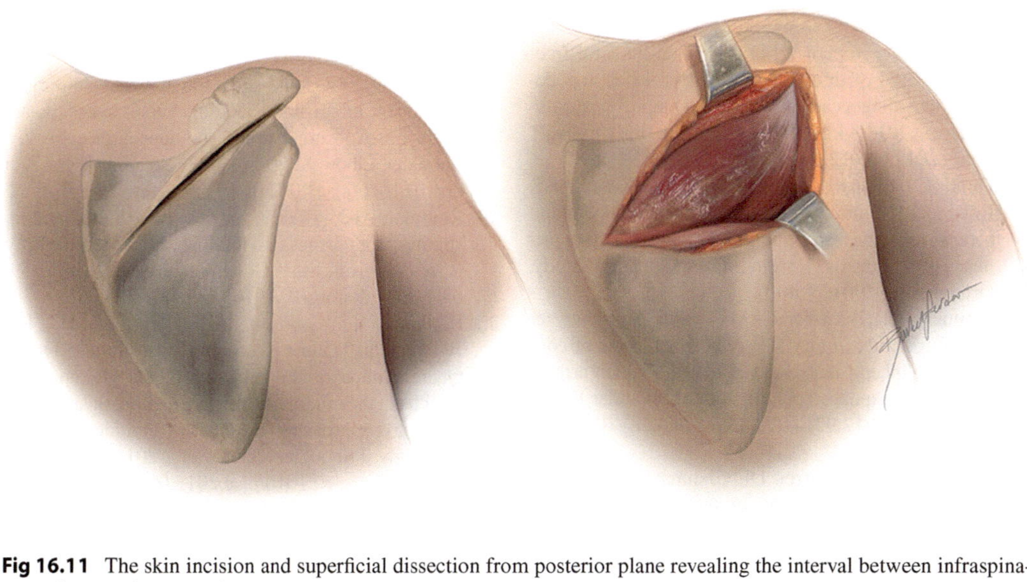

Fig 16.11 The skin incision and superficial dissection from posterior plane revealing the interval between infraspinatus and teres minor muscles

nerve injury. If more proximal exposure is needed, infraspinatus tenotomy and later repair can be utilized but it is rarely needed. The deltoid detachments made to increase exposure should be reattached by bone tunnels.

Acknowledgements We want to thank *Buket Serdar Yörükçüler* (Medical Illustrator, Istanbul University-Cerrahpasa) for drawing the illustrations of this chapter.

References

1. Groh GI, Simoni M, Rolla P, Rockwood CA. Loss of deltoid following shoulder operations: An operative disaster. J Shoulder Elbow Surg. 1995;4:S15. https://doi.org/10.1016/s1058-2746(95)80063-8.
2. Chalmers PN, Van Thiel GS, Trenhaile SW. Surgical exposures of the shoulder. J Am Acad Orthop Surg. 2016;24:250–8. https://doi.org/10.5435/JAAOS-D-14-00342.
3. Radkowski CA, Richards RS, Pietrobon R, Moorman CT. An anatomic study of the cephalic vein in the deltopectoral shoulder approach. Clin Orthop Relat Res. 2006;442:139–42. https://doi.org/10.1097/01.blo.0000181146.78434.da.
4. Sanchez-Sotelo J. Surgical exposures. Mayo Clinic principles of shoulder surgery. New York: Oxford University Press; 2018. p. 57–70.
5. Hoppenfeld S, deBoer P, Buckley R. Anterior approach to the shoulder joint. In: Surgical exposures in orthopaedics: The anatomic approach. Philadelphia: Lippincott Williams & Wilkins; 2009. p. 4–18.
6. Gartsman GM, Edwards TB. Long head of the biceps tendon. Shoulder arthroplast. 1st ed. Philadelphia: Saunders; 2008. p. 37–41.
7. Soliman OA, Koptan WMT. Proximal humeral fractures treated with hemiarthroplasty: Does tenodesis of the long head of the biceps improve results? Injury. 2013;44:461–4. https://doi.org/10.1016/j.injury.2012.09.012.
8. Fama G, Edwards TB, Boulahia A, Kempf J-F, Boileau P, Némoz C, et al. The role of concomitant biceps tenodesis in shoulder arthroplasty for primary osteoarthritis: Results of a multicentric study. Orthopedics. 2004;27:401–5.
9. Simmen BR, Bachmann LM, Drerup S, Schwyzer HK, Burkhart A, Flury MP, et al. Usefulness of concomitant biceps tenodesis in total shoulder arthroplasty: A prospective cohort study. J Shoulder Elbow Surg. 2008;17:921–4. https://doi.org/10.1016/j.jse.2008.06.006.
10. Gurnani N, van Deurzen DFP, Janmaat VT, van den Bekerom MPJ. Tenotomy or tenodesis for pathology of the long head of the biceps brachii: A systematic review and meta-analysis. Knee Surg Sports Traumatol Arthrosc. 2016;24:3765–71. https://doi.org/10.1007/s00167-015-3640-6.
11. DeFranco MJ, Higgins LD, Warner JJP. Subscapulars management in open shoulder surgery. J Am Acad Orthop Surg. 2010;18:707–17. https://doi.org/10.5435/00124635-201012000-00001.
12. Edwards TB, Williams MD, Labriola JE, Elkousy HA, Gartsman GM, O'Connor DP. Subscapularis insufficiency and the risk of shoulder dislocation after reverse shoulder arthroplasty. J Shoulder Elbow Surg. 2009;18:892–6. https://doi.org/10.1016/j.jse.2008.12.013.
13. Gerber C, Yian EH, Pfirrmann CAW, Zumstein MA, Werner CML. Subscapularis muscle function and structure after total shoulder replacement with lesser tuberosity osteotomy and repair. J Bone Joint Surg Am. 2005;87:1739–45. https://doi.org/10.2106/JBJS.D.02788.
14. Lui P, Zhang P, Chan K, Qin L. Biology and augmentation of tendon-bone insertion repair. J Orthop Surg Res. 2010;5:59. https://doi.org/10.1186/1749-799X-5-59.
15. Scalise JJ, Ciccone J, Iannotti JP. Clinical, radiographic, and ultrasonographic comparison of subscapularis tenotomy and lesser tuberosity osteotomy for total shoulder arthroplasty. J Bone Joint Surg Am. 2010;92:1627–34. https://doi.org/10.2106/JBJS.G.01461.
16. Buckley T, Miller R, Nicandri G, Lewis R, Voloshin I. Analysis of subscapularis integrity and function after lesser tuberosity osteotomy versus subscapularis tenotomy in total shoulder arthroplasty using ultrasound and validated clinical outcome measures. J Shoulder Elbow Surg. 2014;23:1309–17. https://doi.org/10.1016/j.jse.2013.12.009.
17. Lapner PLC, Sabri E, Rakhra K, Bell K, Athwal GS. Comparison of lesser tuberosity osteotomy to subscapularis peel in shoulder arthroplasty: A randomized controlled trial. J Bone Joint Surg Am. 2012;94:2239–46. https://doi.org/10.2106/JBJS.K.01365.
18. Lapner PLC, Sabri E, Rakhra K, Bell K, Athwal GS. Healing rates and subscapularis fatty infiltration after lesser tuberosity osteotomy versus subscapularis peel for exposure during shoulder arthroplasty. J Shoulder Elbow Surg. 2013;22:396–402. https://doi.org/10.1016/j.jse.2012.05.031.
19. Routman HD, Buddy Savoie FH. Subscapularis-sparing approaches to total shoulder arthroplasty: Ready for prime time? Clin Sports Med. 2018;37:559–68. https://doi.org/10.1016/j.csm.2018.05.006.
20. Bellamy JL, Johnson AE, Beltran MJ, Hsu JR. Skeletal Trauma Research Consortium (STReC). Quantification of the exposure of the glenohumeral joint from the minimally invasive to more invasive subscapularis approach to the anterior shoulder: A cadaveric study. J Shoulder Elbow Surg. 2014;23:895–901. https://doi.org/10.1016/j.jse.2013.09.013.
21. Foruria AM, Oh LS, Sperling JW, Cofield RH. Anteromedial approach for shoulder arthroplasty: Current indications, complications, and results. J Shoulder Elbow Surg. 2010;19:734–8. https://doi.org/10.1016/j.jse.2009.10.016.

22. Lädermann A, Lubbeke A, Collin P, Edwards TB, Sirveaux F, Walch G. Influence of surgical approach on functional outcome in reverse shoulder arthroplasty. Orthop Traumatol Surg Res. 2011;97:579–82. https://doi.org/10.1016/j.otsr.2011.04.008.
23. Buecking B, Mohr J, Bockmann B, Zettl R, Ruchholtz S. Deltoid-split or deltopectoral approaches for the treatment of displaced proximal humeral fractures? Clin Orthop Relat Res. 2014;472:1576–85. https://doi.org/10.1007/s11999-013-3415-7.
24. Burkhead WZ, Scheinberg RR, Box G. Surgical anatomy of the axillary nerve. J Shoulder Elbow Surg. 1992;1:31–6. https://doi.org/10.1016/S1058-2746(09)80014-1.
25. Cetik O, Uslu M, Acar HI, Comert A, Tekdemir I, Cift H. Is there a safe area for the axillary nerve in the deltoid muscle? A cadaveric study. J Bone Joint Surg Am. 2006;88:2395–9. https://doi.org/10.2106/JBJS.E.01375.
26. Robinson CM, Murray IR. The extended deltoid-splitting approach to the proximal humerus: Variations and extensions. J Bone Joint Surg Br. 2011;93(B):387–92. https://doi.org/10.1302/0301-620X.93B3.25818.
27. Lafosse L, Schnaser E, Haag M, Gobezie R. Primary total shoulder arthroplasty performed entirely thru the rotator interval: Technique and minimum two-year outcomes. J Shoulder Elbow Surg. 2009;18:864–73. https://doi.org/10.1016/j.jse.2009.03.017.
28. Hue E, Gagey O, Mestdagh H, Fontaine C, Drizenko A, Maynou C. The blood supply of the deltoid muscle. Application to the deltoid flap technique. Surg Radiol Anat. 1998;20:161–5. https://doi.org/10.1007/bf01628888.
29. Molé D, Wein F, Dézaly C, Valenti P, Sirveaux F. Surgical technique: The anterosuperior approach for reverse shoulder arthroplasty. Clin Orthop Relat Res. 2011;469:2461–8. https://doi.org/10.1007/s11999-011-1861-7.
30. Gillespie RJ, Garrigues GE, Chang ES, Namdari S, Williams GR. surgical exposure for reverse total shoulder arthroplasty. Differences in approaches and outcomes. Orthop Clin North Am. 2015;46:49–56. https://doi.org/10.1016/j.ocl.2014.09.015.
31. Hepp P, Theopold J, Voigt C, Engel T, Josten C, Lill H. The surgical approach for locking plate osteosynthesis of displaced proximal humeral fractures influences the functional outcome. J Shoulder Elbow Surg. 2008;17:21–8. https://doi.org/10.1016/j.jse.2007.03.029.
32. Chou YC, Tseng IC, Chiang CW, Wu CC. Shoulder hemiarthroplasty for proximal humeral fractures: Comparisons between the deltopectoral andanterolateral deltoid-splitting approaches. J Shoulder Elbow Surg. 2013;22(8):e1–7. https://doi.org/10.1016/j.jse.2012.10.039.
33. Röderer G, Erhardt J, Kuster M, Vegt P, Bahrs C, Kinzl L, et al. Second generation locked plating of proximal humerus fractures—A prospective multicentre observational study. Int Orthop. 2011;35:425–32. https://doi.org/10.1007/s00264-010-1015-7.
34. Lynch NM, Cofield RH, Silbert PL, Hermann RC. Neurologic complications after total shoulder arthroplasty. J Shoulder Elbow Surg. 1996;5:53–61. https://doi.org/10.1016/S1058-2746(96)80031-0.
35. Lévigne C, Garret J, Boileau P, Alami G, Favard L, Walch G. Scapular notching in reverse shoulder arthroplasty: Is it important to avoid it and how? Clin Orthop Relat Res. 2011;469:2512–20. https://doi.org/10.1007/s11999-010-1695-8.
36. Brodsky JW, Tullos HS, Gartsman GM. Simplified posterior approach to the shoulder joint. A technical note. J Bone Joint Surg Am. 1987;69:773–4. https://doi.org/10.2106/00004623-198769050-00021.

Basic Arthroscopy Portals of Shoulder

17

Muzaffer Agir, Okan Tezgel, Tunay Erden, and Mehmet Kapicioglu

Efficient use of portals in shoulder arthroscopy based on opening portals with the right direction from the right point. However, the location of each of these portals must be known to avoid possible neurovascular or musculotendinous damage and to minimize risk.

Shoulder arthroscopy begins with the opening of the posterior portal. Since it is starting portal, we utilize anatomic landmarks on skin without any visualization aid. After intra-articular visualization is done with this portal, it is determined the other portals we needed according to surgical procedure to be performed. Some rules should be considered while opening the portals. Joint space should be used effectively and unnecessary devices should be withdrawn from the joint. The devices should not interfere with each other. The surgeon should be able to reach the area that he wanted. In addition, it should be given importance to protect soft tissue [1]. Matthew et al. [2] have set some criteria for the placement of the portals. According to these criteria: (a) portal should pass through the most avascular area as possible as (b) portal should not endanger the surrounding neurovascular structures, (c) portal should enable comfortable use of arthroscopic instruments, (d) landmarks of portal sites should be remarkably definable. Thus, the same point should be achieved in patients of different weights and heights. If these issues are taken into consideration, complications from portal can be minimized.

Portals can be created with two different techniques: Outside-in and Inside-out. While sometimes both techniques can be used, some portals are suitable for opening with only a technique. For example, while the antero-inferior portal can be opened with both of these techniques, the Neviaser portal can only be opened with outside-in technique. In the inside-out technique, the scope, which is in the joint space, is pushed forward to the required area at anterior and is fixed here. Then the scope is withdrawn and anterior capsule is passed through the same point by using Wissinger rod. Rod is palpated through the skin and small incision is made with the help of a scalpel. Rod is passed out of the skin, thus portal is opened. Sometimes anatomical structures in the joint can restrain to reach the correct point on the skin in this technique. Today, outside-in technique is standard and up-to-date technique [3]. In this technique, portal's location is determined on the skin. The spinal needle is inserted into the joint from determined location, and needle is displayed inside the joint with scope. After displaying the correct position and angle with the needle, skin and subcutaneous of the entrance point is

M. Agir · O. Tezgel · M. Kapicioglu (✉)
Department of Orthopaedics and Traumatology, School of Medicine, Bezmialem Vakif University, Istanbul, Turkey

T. Erden
Department of Orthopaedics and Traumatology, Acibadem Fulya Hospital, Istanbul, Turkey

G. Huri et al. (eds.), *Fundamentals of the Shoulder*, https://doi.org/10.1007/978-3-030-94702-6_17

incised with a scalpel. Wissinger rod is entered into the joint in the same direction as the spinal needle. Rod is withdrawn and the instrument to be used is passed into the joint through the same point. After the position of the device is checked inside the joint, the spinal needle is withdrawn. Regardless of the technique, adrenaline saline or long-acting local anesthetic can be injected under the skin before the portal is opened [4]. This will decrease both bleeding and pain from portals.

Shoulder arthroscopy is not a procedure that remains stable at the same portal. During the operation, a continuous active working is required within the same portal or between the portals. Cannulas help us at this point. Cannulas, which can be of different sizes according to their diameters, are used to facilitate repetitive entry–exit of the instruments, to manipulate the ropes more easily, and to push and tie the knot easily. If the cannula is to be used, an incision appropriate to the width of the cannula is made. Then, the cannula is inserted into the joint according to its application technique. In narrow areas such as the subacromial region, no cannula is required [5]. The surgeon can maneuver more comfortably without a cannula. Additionally the cannulae restrict the movement of the instruments in such regions. While cannulae are generally used in the portals passing through the muscle, they should not be used in portals passing through the tendon unless if it is not required [1].

17.1 Glenohumeral Arthroscopy Portals

17.1.1 Posterior Portal

It is the beginning portal for shoulder arthroscopy and used for visualization. By its location, posterior portal has less risk of damage to neurovascular structures. Since it is a starting portal, it is very important to open it accurately. It is recommended that the posterior portal be shown as the last step in the shoulder arthroscopy training process [5]. When determining the entry point of this portal, the posterolateral corner of the acromion is used as guide. The classic posterior portal is 2 cm distal and 1 cm medial to the posterolateral corner of the acromion [1, 4–6] (Fig. 17.1). This point is also called "soft spot" point. Theoretically, it corresponds to interval between teres minor and infraspinatus [6]. Wolf et al. [7] described a portal, which they called "central posterior portal," which is 2–3 cm distal and 1–2 cm medial to posterolateral corner of acromion. In this way, they imported that IGLLC (inferior glenohumeral ligament labral complex) and inferior glenoid can be displayed much better.

The surgical procedure also changes the region that is important to reach within the joint. Therefore, the location of the posterior portal should not be the same for every patient, for each surgery, and for both positions (beach-chair, lateral decubitus). If posterior instability surgery will be performed, the surgeon must be able to place anchor at an appropriate angle to posterior glenoid. In this case, the posterior portal should

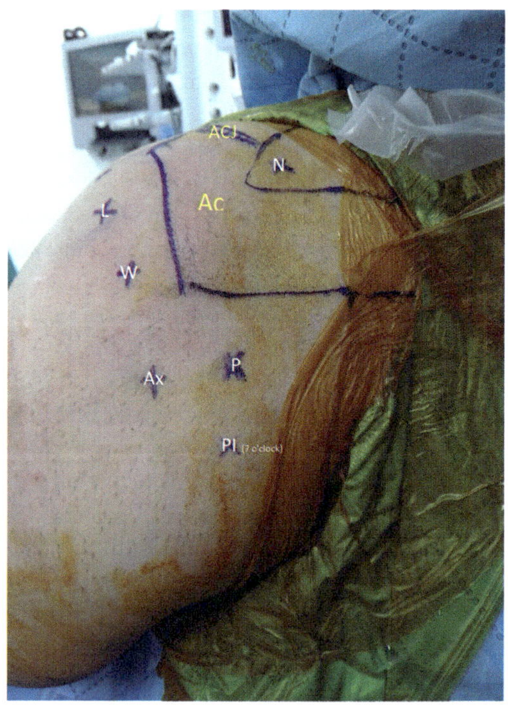

Fig. 17.1 Posterior view of shoulder portals. *P* posterior portal, *PI* posteroinferior portal (7 o'clock), *Ax* axillary pouch portal, *W* Wilmington portal, *L* lateral portal, *N* Neviaser portal, *Ac* acromion, *ACJ* acromioclavicular joint

be more laterally in line with the posterolateral corner of the acromion and more inferior [1, 4]. If rotator cuff surgery is to be performed, the surgeon must be able to reach the anterolateral of the cuff at the subacromial region. In this case, posterior portal should be more medial and more inferior than the classical posterior portal. The entry point in the lateral decubitus position should be more inferior and lateral than the point in the beach-chair position [4].

While opening the posterior portal in the beach-chair position, an assistant puts one hand on the medial proximal humerus and applies lateralization force with one hand, at the same time applies medialization force from the distal lateral of the humerus with the other hand. The joint space widens and passing into the joint of the arthroscopy trocar that is blunt becomes easier due to the stretching of the posterior capsule. 1 cm long skin and subcutaneous incision is made at the point marked on the skin for posterior portal. The blunt trocar is inserted through the skin. Sharp trocars do not need to be used, they can damage the humeral head and glenoid [7]. At this stage, direction is very important. Andrew et al. [6] adjusted direction by directing the tip of the trocar towards the coracoid in their technique. Conversely, "romeo 3-finger shuck" technique can be used to determine this [1]. For this technique, the surgeon uses the same hand as the shoulder side operated. He/She places the third finger on the coracoid and the second finger on the supraclavicular notch. Then he/she feels soft spot between teres minor and infraspinatus and puts thumb there. In this way, the direction of the glenohumeral joint in the sagittal plane is determined. In the axial plane, it should be at the equatorial level of the glenoid. After the trocar has passed the skin and subcutaneous tissue, it kept going between infraspinatus and teres minor muscles and the joint capsule is reached. The glenoid and the humerus head are palpated by tip of blunt trocar at this stage, The soft tissue in between is a capsule and it is entered into the joint. Care should be taken to be attentive. Sudden, rapid, and uncontrolled access into the joint can cause damage to the cartilage in the head of the glenoid and humerus. If entrance is difficult to capsule, the glenoid and humerus head should be palpated and examined again and the capsule should be re-determined too. If necessary, this process should be repeated and harm should be avoided. After entrance into the joint, blunt tip in the trocar is withdrawn and the scope is placed and it is ensured that it is in the joint.

The posterior portal is the safest portal that adequately displays the joint. However, it is important to open the portal correctly in order to prevent axillary nerve and suprascapular nerve (SSN) damage [8]. The posterior portal is on average 49 mm from the axillary nerve and 29 mm from the SSN, but the axillary nerve can be located close to 30 mm [9]. To avoid damage SSN, portal should not be placed medially.

17.1.2 Anterior Portal

This is the main anterior working portal in shoulder arthroscopy. Its location is 1 cm lateral to the coracoid process on the skin [2, 7]. In the technique described by Matthew et al. [2], joint is displayed through the posterior portal, firstly. The triangle formed by the humerus head, glenoid, and biceps long head is revealed. The middle glenohumeral ligament can be seen crossed at the bottom of this triangle (Fig. 17.2). Then, the scope passed through this triangle and moved forward to end as possible as. The light from tip of optic is detected on the skin from anterior side. This point is determined by the finger. Then the optic is withdrawn and the triangle is re-revealed again. The most appropriate point and angle are determined by pressing on the skin with the finger by getting help from intra-articular visualization. The spinal needle is moved forward into the joint. The spinal needle must enter this triangle above the middle glenohumeral ligament and the inferior of the biceps tendon. Arthroscopic direct imaging is used to confirm this. To achieve this direction, the spinal needle is moved forward to the caudal at an angle of about 45°. Once the appropriate angle has been found with the needle, this portal is opened with the outside-in technique.

Particular attention should be paid to staying lateral to the coracoid to protect the brachial plexus and axillary vascular bundle located in the inferomedial [8]. In a cadaver study [2], six branches of the brachial plexus and axillary artery and vein were shown to be at risk in the medial and inferior of coracoid process. The structures at risk in the inferior and lateral of the coracoid process are the musculocutaneous and axillary nerve, the subscapular artery. Therefore, the lateral of the coracoid process and this level or superior have been shown to be the safest place. The musculocutaneous nerve courses an average of 33 ± 6.2 mm inferior of the coracoid process, and the cephalic vein is also close to this portal [10]. Therefore, if the portal is placed too inferior, the risk of damage to these structures will increase.

If there is SLAP lesion, anterior portal should be close to superior part of triangle. So the glenoid superior will be reached more easier. On the other hand, if anterior instability surgery will be performed, anterior portal should close to inferior part of triangle to just above the subscapularis [1]. The distance between coracoid and portal changes depending on the procedure to be performed. Portal can slide more laterally to place anchor at an appropriate angle in instability surgery, while it can be opened a little more medially in capsular release surgeries. This portal is suitable for bone surface preparation in SLAP repair and anchor placement in instability surgery.

17.1.3 Antero-Inferior Portal (5 o'clock Portal)

It was defined by Davidson and Tibone [11]. It is also called "5 o'clock portal." Its localization on the skin is 2 cm inferior and slightly lateral to the anterior portal (Fig. 17.3). It passes through the subscapularis tendon [1, 5]. Both inside-out and outside-in techniques can be used for this portal. In the inside-out technique, the scope placed through the posterior portal is brought to the glenoid antero-inferior (at 5 o'clock) by moving from the edge of the inferior glenohumeral ligament [8]. The optic is withdrawn and Wissinger rod is placed into the trocar. The arm is positioned to the maximum adduction, the rod is moved forwarded and passed out from capsule. With the help of the scalpel, the portal is opened

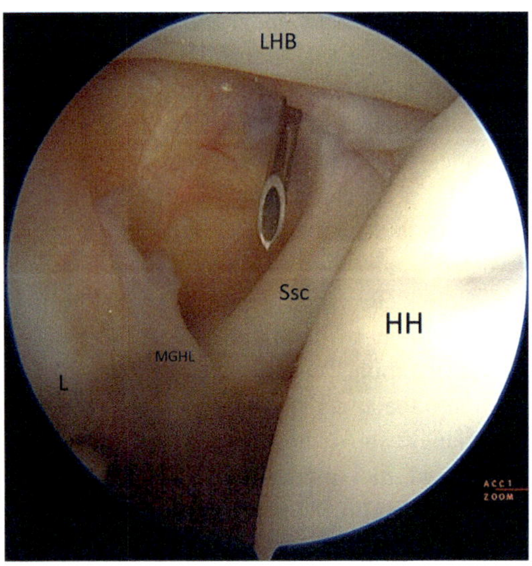

Fig. 17.2 Visualization of the joint inside the posterior portal on the right shoulder. *LHB* long head of biceps, *HH* humeral head, *Ssc* subscapularis, *MGHL* middle glenohumeral ligament, *L* labrum

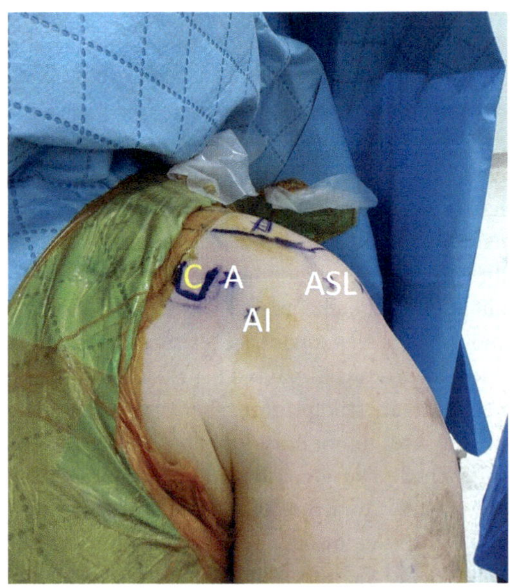

Fig. 17.3 Anterior view of shoulder portals. *C* coracoid, *A* anterior portal, *AI* antero-inferior portal (5 o'clock), *ASL* anterosuperolateral portal

laterally to the conjoint tendon and below the lower fibers of the subscapularis.

In their cadaver study [11], Davidson and Tibone showed that when opening this portal with inside-out technique, only humerus position that led to reach to lateral of the conjoined tendon is maximum adduction. The mean distance of the axillary nerve from the portal was 24.4 and 22.9 mm for the musculocutaneous nerve [11]. They studied these distances in nine different positions, including abduction, flexion, and rotational movements of the shoulder, and showed that any position of the arm did not make any significant difference to protect these nerves. Even, 90° abduction of the arm has been shown to be the position in which the nerves are closest to this portal. Some authors [9, 12] question the security of the location of the portal due to the risk of damage to the axillary nerve, musculocutaneous nerve, cephalic vein, and humerus head cartilage. Pearsall et al. [12] showed that when this portal is opened using the inside-out technique, the tough force applied to the humerus causes damage to the articular cartilage. When using the outside-in technique, the portal can be located 2 mm (medial or lateral) of the cephalic vein. For this reason, they did not recommend the use of this portal in the beach-chair position. In another cadaver study [9], similar results were found. They did not recommend this portal since the placement of the antero-inferior portal was approximately 13 mm from the axillary artery, 15 mm from the axillary nerve, and 17 mm from the cephalic vein. Cephalic vein and anterior circumflex artery also show immediate proximity (<10 mm) with this portal [11]. The placement of the portal can be in the medial or lateral of the cephalic vein in different cases; it can also be proximal or distal to the anterior circumflex artery. Additionally no damage was shown to these two structures in their series. They mentioned that the blunt obturator pushed these structures aside. In another series evaluating this portal using the outside-in technique [3], it was imported that cephalic vein is at risk. However, they said that there was no clinically significant morbidity even if it was injured. For this reason, they recommended that this portal should be used when necessary. In studies showing that this portal is not safe, the inside-out technique was used, and in this technique, the humerus head prevented the Wissinger rod from coming out far enough laterally of the joint, although it was positioned in the maximum adduction. Proximity to neurovascular structures can also be explained by this situation. Additionally, this maneuver caused cartilage damage at the head of the humerus. Furthermore, Lo et al. who advocated safety of this portal operated their patients in a lateral decubitus position, and the outside-in technique allowed lateralization of the portal placement. The surgeon should be careful to the proximity of the neurovascular structures above mentioned, and all these risks should consider when this portal is used.

With this portal, anchor can be placed at 5 o'clock of the glenoid. It is especially useful for placing an anchor in the glenoid inferior level in Bankart repair [11]. This portal is used for anchor placement and suturing tools. Because of it passes through the subscapularis tendon, it is not used the knot that requires cannula [1].

17.1.4 Antero-superior Portal

This portal has been described by Wolf et al. [7]. According to the outside-in technique, its location on the skin is the midpoint of the distance between the coracoid and acromion. This portal is at the level of the joint line and enters the joint just anterior to the biceps tendon [7]. This allows a better control of the anterior of the joint.

In procedures involving the anterior capsule, antero-inferior and antero-superior portals are the best for visualization. The combined use of these portals allows surgical triangulation of the anterior glenohumeral joint [7, 9].

Although there is a risk of injury to the axillary nerve and cephalic vein; this risk is not as high as the antero-inferior portal. It provides a good angle for anchor placement on the glenoid antero-superior during SLAP repair [8].

17.1.5 Anterosuperolateral Portal

Some authors have advocated the necessity of an anterior portal that allows direct visualization of the anterior glenoid neck. This portal has been defined by Laurencin et al. [13] to be used for this purpose. Because of the sufficient distance between the classic anterior portal, instruments can be used comfortably and easily in the joint when combined used with anterior portal. It is 1 cm lateral to the anterolateral corner of the acromion. Its location in the joint is just anterior to the supraspinatus tendon and behind the biceps tendon. Cannula can be used in this portal. If visualization is performed or an instrument is used, the cannula is taken under the biceps tendon. If the cannula remains passive, it is taken above the tendon.

This portal is very useful in the treatment of pathologies involving superior labrum, such as SLAP lesion [13]. It is useful for operations in the subacromial space, such as suprascapular nerve release, as well as allow to work at anterior of the glenohumeral joint [13, 14]. A panoramic view of the glenoid can be obtained if it is used as a visualization portal in the lateral decubitus position. During the subscapularis repair, it can be used as a working portal, it can be used for suture manipulation and anchor placement to the anterior of the tuberculum majus during supraspinatus repair.

17.1.6 Portal of Wilmington

It was first identified during SLAP repair due to the need for additional portal [15]. It is 1 cm lateral and anterior to the posterolateral corner of the acromion. This portal is opened to the glenoid superior at an angle of 45° while imaging from the posterior portal. This portal passes through the infraspinatus muscle. There are discussions about to use cannula. Stephenson et al. [16] showed rotator cuff rupture after SLAP repair in a series of six patients published. Since this problem is caused by portal, they recommended to open this portal more medially. In another study [1], they did not recommend to use cannulae as it passes through the tendon. Lo IK et al. [3] minimized the risk of damage by making a controlled incision along the rotator cuff and capsule longitudinal fibers with the number 11 scalpel during the opening of this portal.

It is very useful for the evaluation of glenoid posterior-superior. It is used in SLAP repair which has a large posterior component, especially for placing anchors on the posterior [3].

17.1.7 Posterior-Inferior Portal (7 o'clock Portal)

As an alternative to the relatively less risky antero-inferior portal, it has been defined in order to reach inferior glenohumeral joint. This portal can be opened using either the outside-in or inside-out technique [17]. Also known as the 7 o'clock portal, the portal's entry point on the skin is 2–3 cm inferior of the posterior portal [1, 5, 17]. After entering the joint from this point, moved forward to glenoid's 7 o'clock position by visualization. In the inside-out technique, the blunt-tipped obturator is forwarded to the glenoid's 7 o'clock from an anterior portal. Then the surgeon passes through the capsule, and incision is made on the skin [8]. Since the portal passes through teres minor, it is generally not necessary to use cannulae [1].

Structures at risk are the suprascapular nerve and artery, axillary nerve, and the posterior circumflex humeral artery. As theorotically, the risk of damage to the humeral head cartilage is high in the inside-out technique; in the outside-in technique, suprascapular nerve-artery damage is more likely. In the cadaver study [17], the portal was found to be 39 ± 4 mm from the axillary nerve and circumflex artery and 29 ± 3 mm from the SSN. There was no significant difference between different positions of the shoulder and its distance to neurovascular structures. Additionally, when the inside-out and outside-in techniques are compared, there is no significant difference between the average distance to the neurovascular structures. If sharp trocars are not used and the portal is opened in proper location, the surgeon will easily keep these structures safe.

This portal is used for repair of posteroinferior labral tears, removal of loose bodies, capsular shrinkage, and capsulotomy [17].

17.1.8 Axillary Pouch Portal

Inferior glenohumeral recess, where pathologies such as loose body, broken anchors, pathological synovial tissues are common, is a difficult region to reach and work with the classical described posterior portals. For this reason, axillary pouch portal has been defined by Bhatia et al. [18]. Its location on the skin is 2–3 cm inferior to the posterolateral corner of the acromion and 2 cm lateral to the standard posterior portal. While forwarding from this point, it is directed towards the medial in the axial plane. In the sagittal plan, it can be directed towards 20° superior or 20° inferior by intra-articular visualization. This portal has been shown to pass either between the muscular fibers of infraspinatus or the interval between teres minor and infraspinatus.

This portal can be used for removal of loose bodies, synovectomy, capsulotomy, anchor placement to posteroinferior glenoid rim, and evaluation and repair of HAGL lesions [18]. The most important advantage of this portal is that thanks to its superior and lateral location, the risk of neurovascular damage is lower compared to portals like the 7 o'clock portal [17].

17.2 Subacromial Portal

17.2.1 Posterior Portal

The entry point on the skin is the same as the posterior portal of the glenohumeral joint. The trocar with blunt obturator is withdrawn from the joint and got off from the capsule. The posterior corner of the acromion is palpated with the tip of the obturator, and the lower edge of the acromion is felt and forwarded towards the anterior acromion. At this stage, the bursa should be punctured. The main free space in the subacromial region is in the anterior acromion. Even, pushing the trocar laterally will increase the probability of reaching this gap. Sometimes the bursa cannot be punctured or the gap cannot be reached due to bursal hypertrophy and synovitis.

17.2.2 Lateral Portal

It is the main working portal in subacromial region. Its location on the skin is at the level of the posterior border of the acromioclavicular joint, 3–4 cm lateral to the lateral edge of the acromion [5] (Fig. 17.4).

This portal provides to attain to the coracoacromial ligament and the origin fibers of this ligament at the anteromedial acromion [19]. When rotator cuff is torn, this portal allows it to be seen directly across the tear.

17.2.3 Anterior Portal

The same point as anterior glenohumeral portal is used. It is entered through the same incision and directed towards the subacromial region.

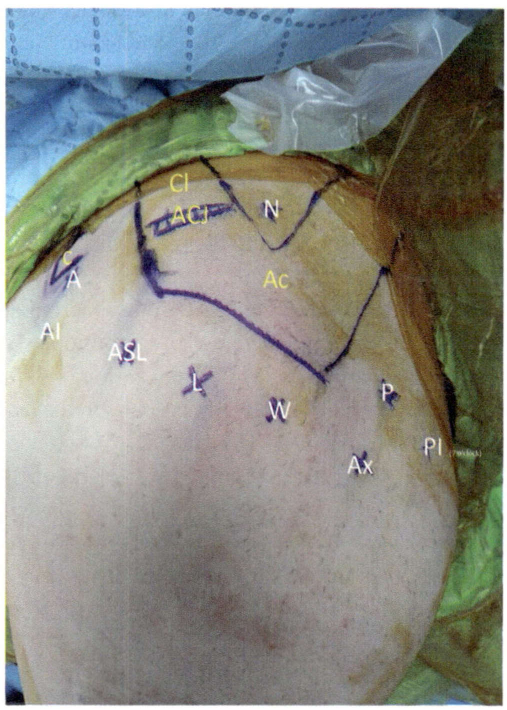

Fig. 17.4 Lateral view of shoulder portals. *P* posterior portal, *PI* posteroinferior portal, *Ax* axillary pouch portal, *W* Wilmington portal, *L* lateral portal, *ASL* anterosuperolateral portal, *AI* antero-inferior portal, *A* anterior portal, *N* Neviaser portal, *C* coracoid, *Ac* acromion, *ACJ* acromioclavicular joint, *Cl* clavicle

It is generally used for suturing tendon during cuff repair. This portal also allows working at a very suitable angle during distal clavicle resection.

17.2.4 Anterolateral Portal

Its location on the skin is 2–3 cm distal to the anterolateral border of the acromion [9, 20]. The blunt trocar is directed to lower edge of the acromion.

The axillary nerve may course up to 3.1 cm distal to the anterolateral edge of the acromion [21], so care should be taken not to open this portal inferiorly to prevent nerve damage. In addition, if the portal is superior to where it should be, the acromion blocks access to the acromioclavicular joint and medial of acromion [8].

This portal is typically used to identify acromioclavicular joint pathologies and subacromial impingement.

17.2.5 Posterolateral Portal

It was defined by Ellman [20]. The entry point on the skin is 2–3 cm distal to the posterolateral corner of the acromion [9, 20].

It is basically a portal used for subacromial decompression and imaging in acromioplasty [20]. It can also be used for visualization both in rotator cuff repairs and in labral repairs in the lateral decubitus position. The presence of two portals lateral to the acromion allows the surgeon to both view and work as the suture is passed through the cuff [8]. Opening the portal more inferiorly puts the axillary nerve at risk.

17.2.6 Neviaser Portal

It was defined by Neviaser in 1987 [22]. It is also called as supraclavicular fossa portal or superior medial portal. It is opened on soft point that is bounded scapular spine at posterior, acromioclavicular joint at anterior, and medial edge of acromion at lateral in supraclavicular notch. This point is 1 cm medial from the medial edge of the acromion. While viewing from the lateral subacromial portal, the spinal needle is forwarded through the soft tissue to 15–20° lateral and 15° anterior, and the portal is opened after the position of the needle is checked at the subacromial region.

While opening the portal in the lateral decubitus position, Nord and Mauck [23] modified this technique by making a 45° abduction to the shoulder and directing to a 20° lateral and 45° anterior. The skin entry point of the modified Neviaser portal is again similar. It is designed to allow suture passage through the rotator cuff. Therefore, while opening this portal, the shoulder is abducted more than 45° for facilitate to reaching the subacromial space.

Meyer et al. [9] did not encounter supraspinatus tendon injury during the opening of this portal in their cadaver study. Likewise, in the study of Souryal and Baker [24], this portal was found reliable. They mentioned that the arm should not be more than 45° of abduction and no forward flexion during the entry of the trocar. In another study [19], the suprascapular nerve and artery were noted to be 3 cm from the supraglenoid tubercle and at risk when using this portal [8].

This portal is particularly suitable for suture fixation during posterior SLAP repair and provides excellent visualization of the anterior glenoid [22]. It can also be used to suture through the middle of the tear during rotator cuff repair.

17.2.7 Subclavian Portal

It was described by Nord and Macuk [23]. Its location on the skin is 1–2 cm medial of the acromioclavicular joint and just under the clavicle. After the skin is passed through the inferior of the clavicle, it is directed towards the antero-inferior and the acromioclavicular joint is reached without encountering the subacromial bursa.

The purpose of this portal is to suture the anterior tendon during rotator cuff repair. After the portal is opened, the scope moves to the subacromial area, while the spurs in the lateral clavicle and acromion prevent comfortable use of the instruments. Therefore, it is recommended to open it after subacromial decompression [23].

References

1. Seroyer ST, Nho SJ, Provencher MT, Romeo AA. Four-quadrant approach to capsulolabral repair: an arthroscopic road map to the glenoid. Arthroscopy. 2010;26(4):555–62. https://doi.org/10.1016/j.arthro.2009.09.019.
2. Matthews LS, Zarins B, Michael RH, Helfet DL. Anterior portal selection for shoulder arthroscopy. Arthroscopy. 1985;1(1):33–9. https://doi.org/10.1016/s0749-8063(85)80076-1.
3. Lo IK, Lind CC, Burkhart SS. Glenohumeral arthroscopy portals established using an outside-in technique: neurovascular anatomy at risk. Arthroscopy. 2004;20(6):596–602. https://doi.org/10.1016/j.arthro.2004.04.057.
4. Mazzocca AD, Cole BJ, Romeo AA. Shoulder: Patient positioning, portal placement and arthroscopic anatomy. In: Miller MD, Cole BJ, editors. Textbook of arthroscopy. Philadelphia: Saunders; 2004. p. 65–77.
5. Burkhart SS, Lo IKY, Bradt P. Visualization. In: Burkhart SS, Lo I, Brady P, editors. Burkhart's view of the shoulder. A cowboy's guide to advanced shoulder arthroscopy. Philadelphia: Lippincott Williams & Wilkins; 2006. p. 3–6.
6. Andrews JR, Carson WG Jr, Ortega K. Arthroscopy of the shoulder: technique and normal anatomy. Am J Sports Med. 1984;12(1):1–7. https://doi.org/10.1177/036354658401200101.
7. Wolf EM. Anterior portals in shoulder arthroscopy. Arthroscopy. 1989;5(3):201–8. https://doi.org/10.1016/0749-8063(89)90172-2.
8. Paxton ES, Backus J, Keener J, Brophy RH. Shoulder arthroscopy: basic principles of positioning, anesthesia, and portal anatomy. J Am Acad Orthop Surg. 2013;21(6):332–42. https://doi.org/10.5435/jaaos-21-06-332.
9. Meyer M, Graveleau N, Hardy P, Landreau P. Anatomic risks of shoulder arthroscopy portals: anatomic cadaveric study of 12 portals. Arthroscopy. 2007;23(5):529–36. https://doi.org/10.1016/j.arthro.2006.12.022.
10. Lo IK, Burkhart SS, Parten PM. Surgery about the coracoid: neurovascular structures at risk. Arthroscopy. 2004;20(6):591–5. https://doi.org/10.1016/j.arthro.2004.04.060.
11. Davidson PA, Tibone JE. Anterior-inferior (5 o'clock) portal for shoulder arthroscopy. Arthroscopy. 1995;11(5):519–25. https://doi.org/10.1016/0749-8063(95)90126-4.
12. Pearsall AWt, Holovacs TF, Speer KP. The low anterior five-o'clock portal during arthroscopic shoulder surgery performed in the beach-chair position. Am J Sports Med. 1999;27(5):571–4. https://doi.org/10.1177/03635465990270050401.
13. Laurencin CT, Deutsch A, O'Brien SJ, Altchek DW. The superolateral portal for arthroscopy of the shoulder. Arthroscopy. 1994;10(3):255–8. https://doi.org/10.1016/s0749-8063(05)80107-0.
14. Lafosse L, Tomasi A, Corbett S, Baier G, Willems K, Gobezie R. Arthroscopic release of suprascapular nerve entrapment at the suprascapular notch: technique and preliminary results. Arthroscopy. 2007;23(1):34–42. https://doi.org/10.1016/j.arthro.2006.10.003.
15. Morgan CD, Burkhart SS, Palmeri M, Gillespie M. Type II SLAP lesions: three subtypes and their relationships to superior instability and rotator cuff tears. Arthroscopy. 1998;14(6):553–65. https://doi.org/10.1016/s0749-8063(98)70049-0.
16. Stephenson DR, Hurt JH, Mair SD. Rotator cuff injury as a complication of portal placement for superior labrum anterior-posterior repair. J Shoulder Elbow Surg. 2012;21(10):1316–21. https://doi.org/10.1016/j.jse.2011.08.054.
17. Davidson PA, Rivenburgh DW. The 7-o'clock posteroinferior portal for shoulder arthroscopy. Am J Sports Med. 2002;30(5):693–6. https://doi.org/10.1177/03635465020300051101.
18. Bhatia DN, de Beer JF. The axillary pouch portal: a new posterior portal for visualization and instrumentation in the inferior glenohumeral recess. Arthroscopy. 2007;23(11):1241.e1–5. https://doi.org/10.1016/j.arthro.2006.12.016.
19. Paulos LE, Franklin JL. Arthroscopic shoulder decompression development and application. A five year experience. Am J Sports Med. 1990;18(3):235–44. https://doi.org/10.1177/036354659001800303.
20. Ellman H. Arthroscopic subacromial decompression: analysis of one- to three-year results. Arthroscopy. 1987;3(3):173–81. https://doi.org/10.1016/s0749-8063(87)80061-0.
21. Burkhead WZ Jr, Scheinberg RR, Box G. Surgical anatomy of the axillary nerve. J Shoulder Elbow Surg. 1992;1(1):31–6. https://doi.org/10.1016/s1058-2746(09)80014-1.
22. Neviaser TJ. Arthroscopy of the shoulder. Orthop Clin North Am. 1987;18(3):361–72.
23. Nord KD, Mauck BM. The new subclavian portal and modified Neviaser portal for arthroscopic rotator cuff repair. Arthroscopy. 2003;19(9):1030–4. https://doi.org/10.1016/j.arthro.2003.09.035.
24. Souryal TO, Baker CL. Anatomy of the supraclavicular fossa portal in shoulder arthroscopy. Arthroscopy. 1990;6(4):297–300. https://doi.org/10.1016/0749-8063(90)90059-m.

Thromboembolism and Bleeding Control in Shoulder Surgery

18

Koray Şahin and Ali Erşen

18.1 Prevention of Bleeding in Shoulder Surgery

It would be more appropriate to examine this section in two different titles as preventing excessive bleeding that may occur during anatomical/reverse shoulder arthroplasty and preventing bleeding that impairs image quality during arthroscopic shoulder procedures.

18.2 Preventing Excessive Bleeding During Anatomic/Reverse Shoulder Arthroplasty

Although shoulder arthroplasty procedures are not as bleeding as lower extremity arthroplasty procedures, it has been shown in the literature that there is an average of 354–361 mL of blood loss during shoulder arthroplasty and this loss causes a need for transfusion in 2.4–9.5% of the procedures [5, 23].

It should be noted that, as in all open surgical procedures, the main method of reducing bleeding in shoulder arthroplasty is careful surgical dissection with minimal damage to soft tissues.

K. Şahin · A. Erşen (✉)
Department of Orthopaedics and Traumatology, Istanbul Medical Faculty, Istanbul University, Istanbul, Turkey
e-mail: ali.ersen@istanbul.edu.tr

Recently, the most researched product to reduce bleeding during shoulder arthroplasty is tranexamic acid. Tranexamic acid is an antifibrinolytic agent which provides effective clot formation by inhibiting the activation of plasminogen [9].

There are several studies in the literature showing the effectiveness of tranexamic acid during shoulder arthroplasty. In their Neer Award winning study, Gillespie et al. showed that the topical administration of 2 g. tranexamic acid in 100 mL saline solution provides less postoperative blood loss and less decrease in hemoglobin level [11]. In another randomized controlled level 1 study, intravenous 1 g tranexamic acid showed superior results during shoulder arthroplasty in terms of postoperative blood loss and hemoglobin level decrease compared to placebo [7].

There are also meta-analyses to evaluate the efficacy of tranexamic acid during primary shoulder arthroplasty and they found it effective to reduce the blood loss [14, 16].

18.3 Preventing Bleeding During Shoulder Arthroscopy

Avoiding excessive bleeding during arthroscopy is more important than arthroplasty because even small amount of blood can impair the clarity of the visual field and compromise the process. Therefore more researchers have worked to find

G. Huri et al. (eds.), *Fundamentals of the Shoulder*, https://doi.org/10.1007/978-3-030-94702-6_18

solutions for bleeding during shoulder arthroscopy.

- Since there is no use of tourniquet during shoulder arthroscopy, the amount of bleeding is highly affected by the patient's blood pressure. Controlled hypotension, a mean arterial pressure of slightly under 100 mmHG, is recommended to avoid excessive bleeding without compromising the cerebral perfusion in beach-chair position [6, 17]. The surgical team must be aware that a mean arterial pressure less than 83 mmHG may compromise cerebral perfusion [17].
- Selection of regional anesthesia combined with sedation has an advantage over general anesthesia in terms of induced hypotension and cerebral perfusion, which allows to reduce bleeding. Therefore regional anesthesia is recommended for suitable patients for shoulder arthroscopy in beach-chair position [15].
- Technological advancements enabled the use of thermal electrocautery device and pressurized irrigation system which played a major role in controlling bleeding during arthroscopic shoulder surgery [21].
- High velocity fluid stream during arthroscopy creates a negative pressure in the joint which is explained with Bernoulli's principle [4]. This negative pressure sucks the blood from vessels into the stream which decreases the visual clarity. To avoid this phenomenon, the fluid exit from portals should be minimized in order to decrease the fluid velocity. This could be achieved by using cannulas for all portals or plugging the portal with a finger (Dutch technique) [4].
- Adding vasoconstrictor agents like epinephrine to arthroscopy fluid is another possible solution for decreasing bleeding. Although it has been used for bleeding control in arthroscopic procedures for a long time, due to reported cardiac side effects there are contradictions for routine use [13]. Recently, in their randomized controlled trial consisting of 101 patients, van Montfoort et al. reported a significant improvement in visual clarity with decreased operation time especially in arthroscopic Bankart and SLAP repair groups [20]. In another randomized controlled study, Avery et al. showed that the addition of epinephrine to arthroscopic solution significantly improves surgeon rated visual clarity in shoulder arthroscopy without a significant difference in operation time [3].
- Finally, after its use in prosthetic surgery, utilization of tranexamic acid during shoulder arthroscopy was also investigated. There is two randomized controlled study available in the literature, where Liu et al. reported that the intravenous administration of 1000 mg tranexamic acid 10 min before surgery significantly improves clarity of the visual field without significant side effects [18]. Recently, we conducted a randomized controlled study to investigate the effect of tranexamic acid to visual clarity during double row rotator cuff repair and concluded that administration of IV tranexamic acid may be effective to improve visual clarity during rotator cuff repair. It can also be useful to reduce the need for high fluid pressure and therefore reduce the amount of irrigation fluid [10].

18.4 Thromboembolic Events After Shoulder Surgery

Every surgical procedure poses a risk for thromboembolic event. Although this risk is significantly lower for shoulder surgery than lower extremity interventions, intimal injury as a result of surgical incision combined with inherent hypercoagulabilty and decreased mobility of the patient creates a risk for thromboembolism after surgical procedures (Virchow Triad).

This topic will be discussed in three separate titles: thromboembolic events after shoulder arthroplasty, shoulder arthroscopy, and shoulder trauma (treatment of the fractures of proximal humerus).

18.5 General Risk Factors for Thromboembolic Events (TEE) After Shoulder Surgery

Systematic disorders like diabetes mellitus, rheumatoid arthritis, and heart disease are considered risk factors for TEE not only after shoulder sur-

Table 18.1 General risk factors for TEE

Older age
Active cancer
Estrogen therapy
Pregnancy
Obesity
Personal or family history of TEE
Autoimmune diseases

gery but after all surgical procedures. Other general risk factors are listed in Table 18.1 [1].

Position of the patient during shoulder surgery is controversial. Application of traction during lateral decubitus position is considered as a risk factor for TEE in upper extremity veins, whereas the immobilization in beach-chair position predisposes a risk for lower extremity TEE.

18.6 Thromboembolic Events After Shoulder Arthroplasty

Thromboembolic events after shoulder surgery are very rare according to the National Registry Database of United Kingdom [12]. Between 2005 and 2008, there were 4061 total shoulder replacements and 6168 hemiarthroplasties recorded to this database. The deep venous thrombosis, pulmonary embolus, and mortality rates were 0%, 0.2%, and 0.22% for total shoulder arthroplasty, respectively. Same rates for hemiarthroplasty were 0.10%, 0.06%, and 0.47%, respectively.

According to the study of Lung et al. with a larger patient series of over 13,000 patients, venous thromboembolism developed in 83 patients (0.62%) [19]. In the same study, the authors also investigated risk factors for VTE after shoulder arthroplasty and concluded that hypoalbuminemia (<3.5 g/dL) increased length of hospital stay and African American ethnicity are independent risk factors [19]. With the largest cohort for this topic available in the literature, Young et al. found 0.25% (1058 patients) pulmonary embolism in 422,372 patients [25]. Arthroplasty for fracture, anemia, congestive heart failure, and chronic lung disease are the top four independent risk factors according to their study [25]. As well as national registries prospective studies also investigated the VTE after shoulder arthroplasty and noticed higher incidences. Willis et al. reported an incidence of 13% [24]. Although the incidence is high, all of their patients were asymptomatic and diagnosed with prospective serial 4 limb-surveillance Doppler ultrasound. The limb of VTE also differed significantly with more than 50% in lower extremities [24].

18.7 Thromboembolic Events After Shoulder Arthroscopy

Thromboembolic events (TEE) after shoulder arthroscopy are even more rare than arthroplasty. In a systematic review, Ojike et al. reported 0.08% TEE after 23,791 arthroscopic procedures [22]. Of those TEE, 58% occurred in upper extremity. Another systematic review with a larger cohort reported 0.038% TEE and 0.017% pulmonary embolism after 92,440 shoulder arthroscopic procedures [8]. Seventy percent of these cases occurred in upper extremity veins.

Regardless of symptom status when the presence of TEE is investigated specifically with 4 limb Doppler ultrasound Takashi et al. found 5.7% TEE after 175 shoulder arthroscopies. In the vast majority of those cases (90%), TEE was found in lower extremity.

18.8 Thromboembolic Events After Treatment of Proximal Humerus Fractures

Although less researched than the other two titles, the highest rate of thromboembolic events (TEE) is found after the surgical treatment of proximal humerus fractures. In the systematic review including only three studies, the rate of TEE is 0.82% (81 TEE) in 9877 patients treated surgically for a proximal humerus fracture. When open reduction plate fixation (ORIF) and hemiarthroplasty are evaluated separately, TEE rate is significantly lower after ORIF than hemiarthroplasty (0.13% vs. 0.69%).

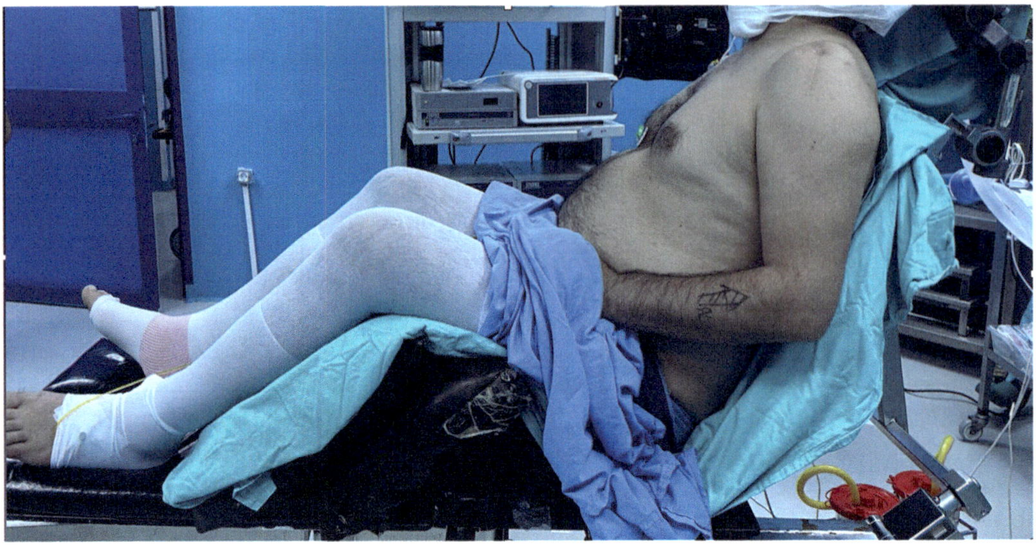

Fig. 18.1 Beach-chair position for shoulder arthroscopy patient with anti-embolic stockings

18.9 Prophylaxis

For patients without an underlying risk factor, mechanical prophylaxis with anti-embolic stockings or pneumatic compression devices is recommended for all patients planned for shoulder surgery especially in beach-chair position (Fig. 18.1). Early patient mobilization after surgery should be encouraged. Chemical prophylaxis with aspirin does not have evidence in the literature for decreasing the TEE after shoulder surgery for patients without risk factors [1].

For patients with a risk factor in Table 18.1, chemical prophylaxis with one of the available forms is recommended. Although there is not enough evidence, LMWH (low molecule weighted heparin) is the most preferred form for chemo-prophylaxis. The duration of chemical prophylaxis is limited with hospital stay and does not need to be extended [2].

18.10 Summary

The risk for TEE is low after shoulder surgery with the highest risk after hemiarthroplasty for fracture treatment. Although not fully accurate in the literature most of the TEE occur in the lower extremity. Mechanical prophylaxis is recommended for all cases, whereas chemical prophylaxis is reserved for patients with underlying risk factors.

References

1. Aibinder WR, Sanchez-Sotelo J. Venous thromboembolism prophylaxis in shoulder surgery. Orthop Clin North Am. 2018;49(2):257–63.
2. Anakwe RE, et al. Preventing venous thromboembolism in elective upper limb surgery. J Shoulder Elbow Surg. 2013;22(3):432–8.
3. Avery DM 3rd, Gibson BW, Carolan GF. Surgeon-rated visualization in shoulder arthroscopy: a randomized blinded controlled trial comparing irrigation fluid with and without epinephrine. Arthroscopy. 2015;31(1):12–8.
4. Burkhart SS, Danaceau SM, Athanasiou KA Turbulence control as a factor in improving visualization during subacromial shoulder arthroscopy. Arthroscopy. 2001;17(2):209–12.
5. Chalmers PN, et al. Predictors of early complications of total shoulder arthroplasty. J Arthroplasty. 2014;29(4):856–60.
6. Cox RM, et al. The effectiveness of cerebral oxygenation monitoring during arthroscopic shoulder surgery in the beach chair position: a randomized blinded study. J Shoulder Elbow Surg. 2018;27(4): 692–700.
7. Cvetanovich GL, et al. Tranexamic acid reduces blood loss after primary shoulder arthroplasty: a double-

blind, placebo-controlled, prospective, randomized controlled trial. JSES Open Access. 2018;2(1):23–7.
8. Dattani R, Smith CD, Patel VR. The venous thromboembolic complications of shoulder and elbow surgery: a systematic review. Bone Joint J. 2013;95-B(1):70–4.
9. Dunn CJ, Goa KL. Tranexamic acid: a review of its use in surgery and other indications. Drugs. 1999;57(6):1005–32.
10. Ersin M, et al. The effect of intravenous tranexamic acid on visual clarity during arthroscopic rotator cuff repair: a randomized, double-blinded, placebo-controlled pilot study. Acta Orthop Traumatol Turc. 2020;54(6):572–6.
11. Gillespie R, et al. Neer Award 2015: a randomized, prospective evaluation on the effectiveness of tranexamic acid in reducing blood loss after total shoulder arthroplasty. J Shoulder Elbow Surg. 2015;24(11):1679–84.
12. Jameson SS, et al. Venous thromboembolic events are rare after shoulder surgery: analysis of a national database. J Shoulder Elbow Surg. 2011;20(5):764–70.
13. Jensen KH, et al. Arthroscopic shoulder surgery with epinephrine saline irrigation. Arthroscopy. 2001;17(6):578–81.
14. Kirsch JM, et al. Tranexamic acid in shoulder arthroplasty: a systematic review and meta-analysis. JBJS Rev. 2017;5(9):e3.
15. Koh JL, et al. Neer Award 2012: cerebral oxygenation in the beach chair position: a prospective study on the effect of general anesthesia compared with regional anesthesia and sedation. J Shoulder Elbow Surg. 2013;22(10):1325–31.
16. Kuo LT, et al. Tranexamic acid in total shoulder arthroplasty and reverse shoulder arthroplasty: a systematic review and meta-analysis. BMC Musculoskelet Disord. 2018;19(1):60.
17. Lee JH, et al. Effects of beach-chair position and induced hypotension on cerebral oxygen saturation in patients undergoing arthroscopic shoulder surgery. Arthroscopy. 2011;27(7):889–94.
18. Liu YF, et al. Intravenous administration of tranexamic acid significantly improved clarity of the visual field in arthroscopic shoulder surgery. A prospective, double-blind, and randomized controlled trial. Arthroscopy. 2020;36(3):640–7.
19. Lung BE, et al. Risk factors for venous thromboembolism in total shoulder arthroplasty. JSES Open Access. 2019;3(3):183–8.
20. van Montfoort DO, van Kampen PM, Huijsmans PE. Epinephrine diluted saline-irrigation fluid in arthroscopic shoulder surgery: a significant improvement of clarity of visual field and shortening of total operation time. A randomized controlled trial. Arthroscopy. 2016;32(3):436–44.
21. Ogilvie-Harris DJ, Weisleder L. Fluid pump systems for arthroscopy: a comparison of pressure control versus pressure and flow control. Arthroscopy. 1995;11(5):591–5.
22. Ojike NI, et al. Venous thromboembolism in shoulder surgery: a systematic review. Acta Orthop Belg. 2011;77(3):281–9.
23. Saltzman BM, et al. Complication rates comparing primary with revision reverse total shoulder arthroplasty. J Shoulder Elbow Surg. 2014;23(11): 1647–54.
24. Willis AA, et al. Deep vein thrombosis after reconstructive shoulder arthroplasty: a prospective observational study. J Shoulder Elbow Surg. 2009;18(1): 100–6.
25. Young BL, et al. Factors associated with in-hospital pulmonary embolism after shoulder arthroplasty. J Shoulder Elbow Surg. 2015;24(10):e271–8.

19 Periprosthetic Infection in Shoulder Surgery

Murat Asci, Recep Kurnaz, and Taner Gunes

19.1 Introduction

Periprosthetic shoulder infection (PSI) is a rare yet crucial complication of shoulder arthroplasty. Incidence varies between 1.1 and 10% according to arthroplasty type performed [1, 2]. PSI is the major cause of revision surgery which is needed after complaints of pain, restricted movement, and loosening [3]. PSI is a devastating complication resulting in clinical and socioeconomic consequences. Therefore, until proven otherwise, any postoperative problems of shoulder prosthesis such as pain, joint stiffness, or loosening should warn us of possible periprosthetic infection.

There are some risk factors for periprosthetic shoulder infections: recurrent steroid injections, systemic steroid use, prior surgeries, posttraumatic arthrosis, coagulopathies, renal failure, rheumatoid arthritis, diabetes, immunosuppressive treatments [4]. Richards et al. reported that males were 2.5-times higher at risk for infection than females and that reverse shoulder arthroplasty was associated with a 6-times higher risk [3]. Arthroplasties following traumas were associated with a 3-times greater risk of infection. It should be noted that reverse shoulder arthroplasty is usually the preferred technique in revision surgeries which is why it is expected that the infection rates are higher.

Bacteria associated with PSI are usually the skin pathogens: Staphylococcus sp. and *Cutibacterium acnes* (*Propionibacterium acnes*). It is reported in many studies that *C. acnes* is responsible for most of the periprosthetic shoulder infections [3, 4] [5]. This is most probably because of the proximity of the surgical site to the axillary region.

These infections are classified as acute and chronic. Some authors agreed on 4 weeks after surgery as the threshold between the two yet some agreed on 3 months [6, 7]. However, infections occurring after long and problem-free years should also be regarded as acute hematogenous infections. Treatments for acute hematogenous infections are like those of acute early postoperative infections [6].

PSI is less common compared to infections seen after knee and hip arthroplasty surgeries. Eradicated microorganisms are also different. There have not been any precise and standardized methods to diagnose and to treat. In this section, diagnostic and treatment methods are defined, benefits and shortfalls are discussed.

M. Asci · R. Kurnaz · T. Gunes (✉)
Department of Orthopedics and Traumatology, Acibadem Eskisehir Hospital, Eskisehir, Turkey

G. Huri et al. (eds.), *Fundamentals of the Shoulder*, https://doi.org/10.1007/978-3-030-94702-6_19

19.2 Pathogens

Staphylococcus sp. and *C. acnes* are the most common microorganisms isolated from periprosthetic shoulder infections [8]. *C. acnes* is an anaerobic gram-positive bacterium and it is part of the skin flora in the axillary region. It is responsible for acne and it is colonized in the dermis especially in males. It is even argued to have a role in pathogenesis of glenohumeral arthrosis [9]. Hudek et al. reported two times as many positive cultures for the anterolateral approach as for the deltopectoral approach [10]. It usually causes low-grade infections with mild clinical symptoms. That's why we do not see common infection signs and symptoms. The time for incubation of culture is around 7–21 days since it is a slow-growing bacteria.

19.3 Precautions

Antibiotics treatment for shoulder arthropathies is no different than that of other joint arthroplasties. It is enough to administer first generation cephalosporin and gentamicin intravenously 30 min before skin incision. It is also important to swab surgical sites as well as antibiotics use. Chlorhexidine and alcohol mixtures are found to be more effective at eliminating bacteria than povidone-iodine [11]. Shaving the hair on surgical sites is also routinely done before orthopedic procedures. However, electric shavers are preferred over razors as they are less likely to damage the skin.

19.4 Diagnosis

For every possible case of periprosthetic infections, diagnostic tests should be considered. Preoperative diagnostics is effective at determining treatment options as it can prevent bad surprises which might happen during revision surgery.

Tools used to diagnose periprosthetic infections of the shoulder are indifferent than those of knee and hip. However, they are less common than the infections of knee and hip arthroplasties. Local and systemic inflammation signs are seen in early stage and acute hematogenous infections. Local signs are pain, redness, warmer skin, and purulent drainage. Local changes may not always show due to thick soft tissue around the shoulder joint or because of low virulence of the pathogen. Systemic signs are fever, high erythrocyte sedimentation rate (ESR), high C-reactive protein (CRP), high white blood count (WBC > 3000/mm^3 and neutrophil granulocytes (PMN) >80%), and high interleukin 6 (IL-6) levels. There is usually no purulent drainage, so synovial fluid aspiration is performed in order to isolate pathogens. Aerobic and anaerobic cultures of the synovial fluid should be ordered and those cultures need time to grow. Labs should be informed to wait for up to 3 weeks for the cultures to isolate *C. acnes* which is a slow-growing bacteria.

Sometimes it might be difficult to make a diagnosis when local or systemic signs are absent. In 2011, Musculoskeletal Infection Society (MSIS) proposed some criteria to be used in diagnosis of periprosthetic infections. In 2014, it was suggested that the presence of one major and at least three minor criteria is enough to make a diagnosis of infection.

Major criteria: fistula associated with prostheses, at least two positive cultures of synovial fluid.

Minor criteria: elevated ESR and CRP, elevated leukocyte count in synovial fluid or positive leukocyte esterase test, high PMN count in synovial fluid, one positive culture of periprosthetic tissue or fluid, positive histological findings in periprosthetic tissues.

Every suspected case of periprosthetic infection should be exactly proved or excluded prior to revision surgery. This decision should be followed by a plan of treatment and systemic and local antibiotics treatment should be started during revision surgery. Tests and cultures which are to be done during surgery should be studied in order to support the preoperative diagnosis. Most common causes of periprosthetic shoulder infections are Staphylococcus sp. and *C. acnes*. Broad-spectrum antibiotics are usually enough to eradicate these bacteria. For that reason, some surgeons start isolating pathogens during revision surgery [2]. However, these antibiotics treatments are insufficient for resistant strains of Staphylococcus and other gram-negative microorganisms. In those cases, initiating effective antibiotic treatment is delayed until the pathogen is isolated in cultures. In turn, this allows a new biofilm layer to form around the implants placed in patients. This also causes empirical antibiotics which is placed into the spacer in the joint to be ineffective in two-stage revision surgery. Isolating pathogens before revision surgery ensures effective local and systemic antibiotic treatment to be started without delay. It also prevents unnecessary use of broad-spectrum antibiotics and may also prevent drug resistance [12]. When all is considered, specific tests like aspiration and biopsy are essential in diagnosis and especially in treatment and should definitely be done before revision surgery.

There are some imaging techniques as well to be used when diagnosing periprosthetic infections. Osteolysis and loosening seen in direct imaging may be significant. There has to be at least 1 year after the first surgery for scintigraphy to be significant due to physiological adaptations. Leukocyte-tagged scintigraphy has also similar sensitivity and specificity. CT or MRI has no benefit when diagnosing infection yet they are helpful at identifying possible abscess [13].

In recent years, alpha-defensin test is recommended as a non-specific diagnostic method which is studied in synovial fluid. Its sensitivity and specificity are reported to be 97% and 100%, respectively [14]. Alpha-defensin is an antimicrobial agent released by leukocytes after contact with bacteria. Contrary to CRP or ESR, it is not affected by systemic inflammatory diseases or antibiotic use [15].

Another test studied in synovial fluid in order to isolate pathogens is PCR. It has the advantages of giving results in a couple of hours and identifying some drug resistances. However, it has high false-positive rates [16].

19.5 Treatment

Superficial wound infections are usually seen in early postoperative period and they may be handled by wound care and appropriate antibiotic treatment. However, deep tissue infections must always be suspected. Success of treatment depends on early diagnosis, early isolation of pathogen(s), appropriate surgical procedures, and effective antibiotic treatment. There are many surgical methods for PSI: debridement, resection arthroplasty, cement spacer method, single-stage revision, two-stage revision, arthrodesis, chronic antibiotic administration, and amputation.

19.5.1 Debridement

It is the first line method aiming at preserving existing prosthesis during surgical treatment of acute postoperative and acute hematogenous infections. During this intervention, radical debridement and synovectomy of periprosthetic tissues should be performed. Following that, there needs to be irrigation and cleaning with plenty of saline. Open surgery should include changing parts like the head, liner, and glenosphere. That's why open surgery is rather recommended even though arthroscopic debridement is also an option. Systemic antibiotic treatment should be ordered until inflammation parameters get to normal levels after surgery, at least 4 weeks [17].

19.5.2 Cement Spacer

Antibiotic-loaded cement spacers can be used either permanently after the removal of infected prosthesis or as the first step of a two-stage revision surgery. Either way, this method works as a repository for antibiotics. Therefore, it can deliver much bigger doses than what systemic antibiotics treatment can deliver. Also, specific antibiotics mixtures can be added according to the culture results. After removing the infected prosthesis, this method can result in a more functional joint by preserving soft tissue tension and arm length [18]. There have been, however, some complications such as broken spacer, glenoid erosion, and dislocation.

19.5.3 One-Stage Revision Arthroplasty

Based on what we know about knee and hip prosthesis infections, one-stage revision surgery is the right choice if the pathogen is isolated before surgery. This method is cost-effective and has advantages such as shorter hospital stays and shorter period of antibiotic therapy. Spacer-associated complications can be avoided as well. Antibiotics specific to the pathogens can be added to the cement during the removal of permanent prosthesis. Thus, the efficacy of local antibiotics can be improved. Many studies reported better clinical outcomes [2, 19–21]. Based on these studies, one-stage revision procedure is the better option if the pathogen is isolated before the surgery.

19.5.4 Two-Stage Revision Arthroplasty

This is the most preferred method among infected prosthesis surgery. Two-stage revision procedure is still the preferred method if the patient is stable (Fig. 19.1). Biggest advantage is that debridement can be done twice. Another reason for this method to be preferred is difficulties isolating pathogens before surgery. Extensive soft tissue debridement is done after removing the infected prosthesis during surgery. At the same time, enough tissue samples can be obtained for histologic analysis or to run cultures. Soft tissue balance can also be preserved by placing antibiotic-loaded cement spacer following irrigation with saline. Two-stage revision procedure is eased by reducing possible contractures. Since rotator cuff failure usually occurs after extensive debridement, reverse shoulder prosthesis is recommended for reimplantation [22].

Specific antibiotic treatment should be continued until postoperative CRP and IL-6 levels go back to normal. IL-6 levels normalize quicker than CRP levels and it is a sign of eradication of infection [23]. Therefore, two-stage revision surgery is not waited for any longer.

19.5.5 Resection Arthroplasty

This technique is a salvage procedure for frail patients with resistant infections. It can provide eradication with a single session. Functional results are bad yet it is enough for pain relief [24]. Antero-superior subluxation of the humerus can be avoided by preservation of tuberosities and better functional results can be expected [25].

19.6 Conclusion

PSIs are rare yet devastating complications with bad functional results. Acute infections can be managed with irrigation, changing prosthesis parts, and soft tissue debridement as soon as possible. However, low-grade infections are seen in many cases and diagnosis may be delayed. Therefore, infections should be suspected in cases with unexpected clinical signs such as pain and stiffness. One-stage revision is recommended if the pathogen is isolated before surgery; otherwise, two-stage revision is recommended. Resection arthroplasty should be kept in mind as a salvage procedure for cases with low expectancies.

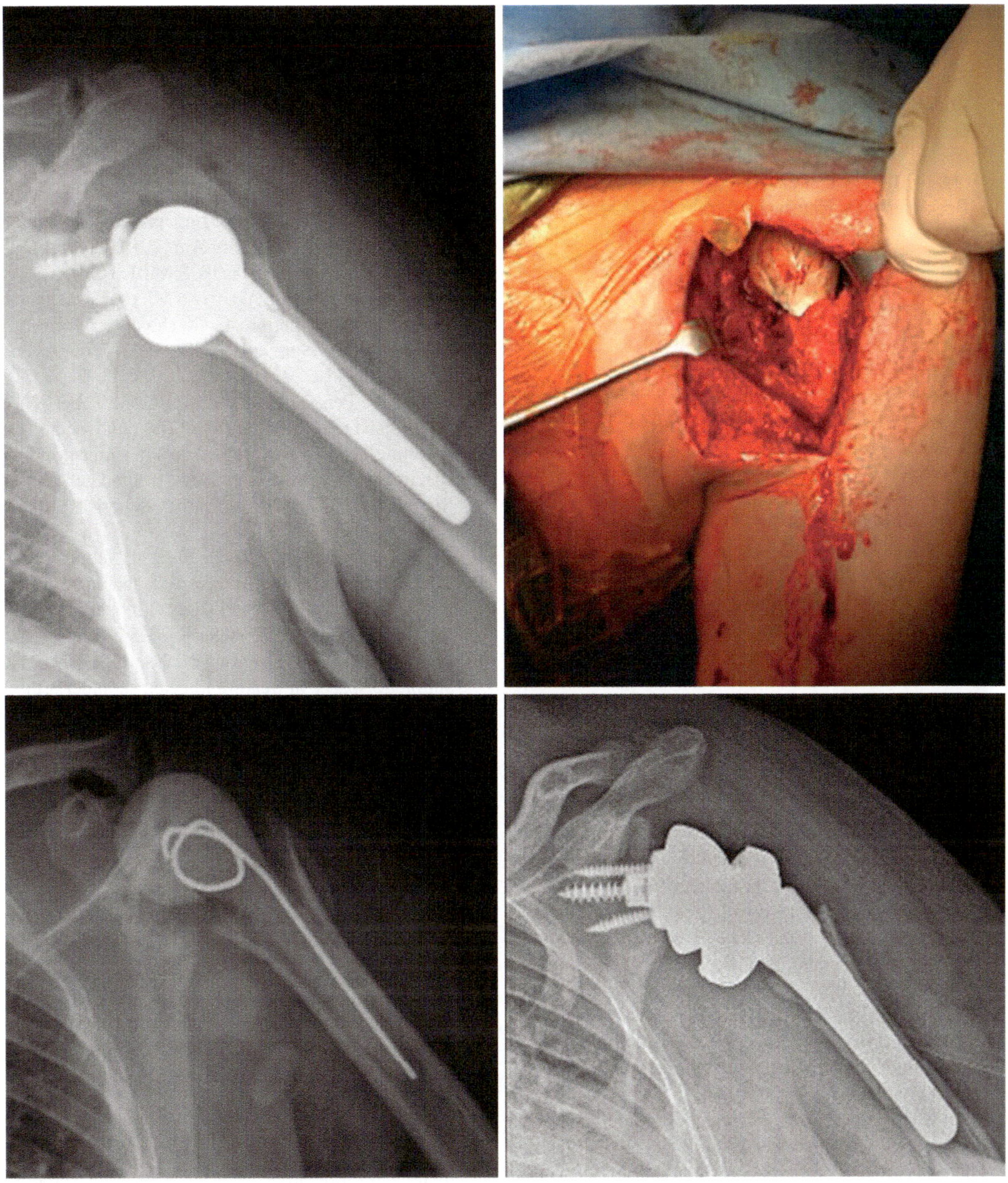

Fig. 19.1 Two-stage revision arthroplasty of an infected reverse shoulder prosthesis. Preop and postoperative graphics and intraoperative photo (Dr. KeremBilsel's archive)

References

1. Singh JA, Sperling JW, Schleck C, Harmsen WS, Cofield RH. Periprosthetic infections after total shoulder arthroplasty: A 33-year perspective. J Shoulder Elbow Surg. 2012;21(11):1534–41. https://doi.org/10.1016/j.jse.2012.01.006.
2. Beekman PDA, Katusic D, Berghs BM, Karelse A, De Wilde L. One-stage revision for patients with a chronically infected reverse total shoulder replacement. J Bone Joint Surg Br. 2010;92(6):817–22. https://doi.org/10.1302/0301-620X.92B6.23045.
3. Richards J, Inacio MCS, Beckett M, et al. Patient and procedure-specific risk factors for deep infection after primary shoulder arthroplasty. Clin Orthop Relat Res. 2014;472(9):2809–15. https://doi.org/10.1007/s11999-014-3696-5.
4. Pottinger P, Butler-Wu S, Neradilek MB, et al. Prognostic factors for bacterial cultures positive for Propionibacterium acnes and other organisms in a large series of revision shoulder arthroplasties performed for stiffness, pain, or loosening. J Bone Joint Surg Am. 2012;94(22):2075–208.
5. Piper KE, Jacobson MJ, Cofield RH, et al. Microbiologic diagnosis of prosthetic shoulder infection by use of implant sonication. J Clin Microbiol. 2009;47(6):1878–84. https://doi.org/10.1128/jcm.01686-08.
6. Tsukayama DT, Estrada R, Gustilo RB. Infection after total hip arthroplasty: a study of the treatment of one hundred and six infections. J Bone Joint Surg Am. 1996;78(4):512–23. https://doi.org/10.2106/00004623-199604000-00005.
7. Martínez-Pastor JC, Muñoz-Mahamud E, Vilchez F, et al. Outcome of acute prosthetic joint infections due to gram-negative bacilli treated with open debridement and retention of the prosthesis. Antimicrob Agents Chemother. 2009;53(11):4772–7. https://doi.org/10.1128/AAC.00188-09.
8. Saltzman MD, Marecek GS, Edwards SL, Kalainov DM. Infection after shoulder surgery. J Am Acad Orthop Surg. 2011;19:208–18.
9. Levy O, Iyer S, Atoun E, et al. Propionibacterium acnes: an underestimated etiology in the pathogenesis of osteoarthritis? J Shoulder Elbow Surg. 2013;22:505–11.
10. Hudek R, Sommer F, Kerwat M, et al. Propionibacterium acnes in shoulder surgery: true infection, contamination, or commensal of the deep tissue? J Shoulder Elbow Surg. 2014;23:1763–71.
11. Saltzman MD, Nuber GW, Gryzlo SM, Marecek GS, Koh JL. Efficacy of surgical preparation solutions in shoulder surgery. J Bone Joint Surg Am. 2009;91-A:1949–53.
12. Fink B, Makowiak C, Fuerst M, Berger I, Schäfer P, Frommelt L. The value of synovial biopsy, joint aspiration and C-reactive protein in the diagnosis of late peri-prosthetic infection of total knee replacements. J Bone Joint Surg Br. 2008;90(7):874–8. https://doi.org/10.1302/0301-620X.90B7.20417.
13. Moroder P, Gerhardt C, Renz N, Trampuz A, Scheibel M. Diagnostik und Management des Endoprotheseninfekts am Schultergelenk. Obere Extremität. 2016;11(2):78–87. https://doi.org/10.1007/s11678-016-0361-5.
14. Deirmengian C, Kardos K, Kilmartin P, Cameron A, Schiller K, Parvizi J. Combined measurement of synovial fluid a-defensin and C-reactive protein levels: Highly accurate for diagnosing periprosthetic joint infection. J Bone Joint Surg Am. 2014;96(17):1439–45. https://doi.org/10.2106/JBJS.M.01316.
15. Frangiamore SJ, Saleh A, Grosso MJ, et al. α-Defensin as a predictor of periprosthetic shoulder infection. J Shoulder Elbow Surg. 2015;24(7):1021–7. https://doi.org/10.1016/j.jse.2014.12.021.
16. Holmes S, Pena DA, M., Athwal G. S., Faber K. J., O'Gorman D. B. Neer Award 2017: A rapid method for detecting Propionibacterium acnes in surgical biopsy specimens from the shoulder. J Shoulder Elbow Surg. 2017;26(2):179–85. https://doi.org/10.1016/j.jse.2016.10.001.
17. Jacquot A, Sirveaux F, Roche O, et al. Surgical management of the infected reversed shoulder arthroplasty: a French multicenter study of reoperation in 32 patients. J Shoulder Elbow Surg. 2015;24:1713–22.
18. Levy JC, Triplet J, Everding N. Use of a functional antibiotic spacer in treating infected shoulder arthroplasty. Orthopedics. 2015;38:e512–9.
19. Klatte TO, Junghans K, Al-Khateeb H, et al. Single-stage revision for periprosthetic shoulder infection: outcomes and results. Bone Joint J. 2013;95-B:391–5.
20. George DA, Volpin A, Scarponi S, Haddad FS, Romanò CL. Does exchange arthroplasty of an infected shoulder prosthesis provide better eradication rate and better functional outcome, compared to a permanent spacer or resection arthroplasty? a systematic review Orthopedics and biomechanics. BMC Musculoskelet Disord. 2016;17:52. https://doi.org/10.1186/s12891-016-0901-6.
21. Ince A, Seemann K, Frommelt L, Katzer A, Loehr JF. One-stage exchange shoulder arthroplasty for peri-prosthetic infection. J Bone Joint Surg Br. 2005;87-B:814–8.
22. Li FL, Jiang CY, Lu Y, Zhu YM, Li X. Efficacy analysis of two-stage reverse total shoulder arthroplasty for treating postoperative deep infection after surgeries for proximal humeral fractures. Beijing Da Xue Xue Bao. 2016;48(2):263–7.
23. Coffey MJ, Ely EE, Crosby LA. Treatment of glenohumeral sepsis with a commercially produced antibiotic-impregnated cement spacer. J Shoulder Elbow Surg. 2010;19:868–73.
24. Rispoli DM, Sperling JW, Athwal GS, Schleck CD, Cofield RH. Pain relief and functional results after resection arthroplasty of the shoulder. J Bone Joint Surg Br. 2007;89(9):1184–7.
25. Verhelst L, Stuyck J, Bellemans J, Debeer P. Resection arthroplasty of the shoulder as a salvage procedure for deep shoulder infection: does the use of a cement spacer improve outcome? J Shoulder Elbow Surg. 2011;20:1224–33.

Part V

(Bio)technological Applications in Shoulder Treatment

Anchors

20

Mehmet Chodza, Mehmet Demirhan,
and Olgar Birsel

20.1 History

The first suture anchor was patented by Goble and Somers in 1985 by the name as *Statak* (Zimmer, Warsaw, USA). *Statak* was composed of titanium and designed for use in shoulder Bankart repair [5]. Another anchor design was developed by *Mitek* (Mitek Inc. USA). Over time, its design evolved and Mitek G2 was subsequently introduced to the market as a successor. And the last of this group of "first generation" (Fig. 20.1) anchor design was the *Acufex TAG rod* and *wedge* (Acufex Inc., USA) which were polyactal (plastic) devices that generate traction in bone using an interference fit [6].

In the early 1990s, several biomechanical studies with first generation suture anchors were published [5]. Failure of a suture anchor construct was described based on the concept of suture anchor chain suggested by Barber et al. in 1993 [6]. The components of this chain were the tendon to suture, the suture itself, the suture to eyelet, and the anchor to bone. Several studies have been reported that these anchors do not have a significant advantage over the traditional technique. Goble et al. demonstrated that their original anchor design has an equivalent reattaching effect with transosseous tunnels or staples in a sheep model [7]. Hecker et al. showed that in models of Bankart lesions and rotator cuff repairs suture anchors were not significantly different to sutures alone [8]. Subsequent testing with first generation anchors and new reinforced anchor designs demonstrated that suture anchors could be stronger than transosseous tunnels in rotator cuff repair [9, 10]. These suture anchors were metallic and nonabsorbable. Although the initial results of metallic anchors were good, later studies reported many complications including loosening, chondral damage, displacement, and interference with MRI [11, 12]. Because of the reported complications of metallic anchors, bioabsorbable anchors have been introduced as an alternative and have recently been used more often than metallics [13].

20.2 Anchor Design

A suture anchor consists of three parts, namely the anchor which is inserted into the bone with interference fitting or a screw mechanism; the eyelet which is a hole that provides the coupling of suture on the anchor, and the suture that is composed of absorbable or nonabsorbable material.

M. Chodza
Department of Orthopaedics and Traumatology, VKV American Hospital, Istanbul, Turkey
e-mail: mehmetch@amerikanhastanesi.org

M. Demirhan (✉) · O. Birsel
Department of Orthopaedics and Traumatology, Koç University School of Medicine, Istanbul, Turkey
e-mail: mdemirhan@ku.edu.tr; olgarbirsel@ku.edu.tr

G. Huri et al. (eds.), *Fundamentals of the Shoulder*, https://doi.org/10.1007/978-3-030-94702-6_20

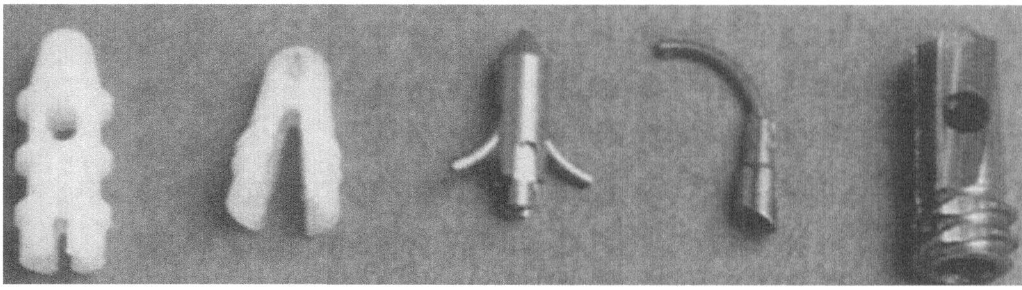

Fig. 20.1 First generation suture anchors (*left* to *right*) Acufex ROD TAG, Acufex TAG, Mitek GII, Mitek GI and Statak. (With permission from [5])

Anchors can be divided into different groups based on their composition and morphology that contributes to each anchor's particular features affecting its usability. Screw type anchors have both large (threads) and small (core) diameters with threads that advance the anchor into the bone. Impaction type anchors are threadless and the ultimate strength depends on its geometry, resistance of the anchor material, and bone quality. Impaction anchors can be categorized by the mechanism of fit such as wedging, press-fit, or tilting types.

Owing to its threads, screw type anchors have improved contact area with the bone compared to the impaction type. Contrarily, impaction type anchors are threadless but can be composed of various plastics that expand after impaction. Some anchors, such as tacs, have a passage for the suture material through tissue for fixation. The top of the tack holds the tendon and advances it into the bone. The advantages of such anchors are their ease of application and that they do not require an arthroscopic knot.

20.3 Anchor Materials

The material, shape, and the basic design of an anchor are the principal features to be considered during implant selection. Anchor materials can be classified as biodegradable and non-biodegradable. Non-biodegradable anchors are metallic in terms of material and often composed of stainless steel or titanium alloys. Each material type has its inherent advantages and disadvantages.

The first suture anchors used in shoulder surgery were made of metal since it is the predominant constituent in most of the other orthopedic implants. Thus, stainless steel and titanium anchors were widely used in glenohumeral surgery. However, it is reported that these metals have minimal osseous integration and that stainless steel becomes encapsulated by a fibrous layer in time [14]. Metals have been used in both screw and impaction type anchors providing them a self-drilling and self-tapping ability. Longer metallic screws with wider threads provide improved resistance to pull-out forces in osteoporotic bones [15, 16]. Despite the forementioned advantages of metallic anchors, these implants become an obstacle for subsequent anchor placement in revision scenarios. Despite the fact that metallic anchors allow the easy tracking of positional changes on X-rays during follow-up, they can also interfere with MRI imaging in revision cases [17]. In the first generation metallic suture anchors, the rigid structure and the sharp edges of the connection point lead to abrasion and rupture of the suture in the eyelet; hence, the failure occurred usually at this point. However, this weak junction was addressed later with the introduction of the anchors with a design including an internal crossbar providing a low friction suture attachment. This technical amelioration brought the looping of the sutures into the internal area of the metallic anchors. Anchor loosening and fatigue failure leading to chondral damage have also been associated with metallic implants [18]. These drawbacks have led many authors to investigate bioabsorbable materials as the main constituent of the anchors allowing for the complete absorption of the foreign material following soft tissue integration.

Bioabsorbable anchors developed rapidly and became more commonly used in glenohumeral surgery than metal anchors. These anchors are radiolucent and hence minimize distortion on MRI imaging. But not all bioabsorbable materials could be used. These implants must be able to meet a number of criteria: (1) They must have adequate initial fixation strength of soft tissues to bone, (2) the bioabsorption profile must enable implants to retain adequate strength while tissues regain mechanical integrity, (3) implants must be absorbed at an adequate rate to avoid the breakage and migration associated with metal implants, (4) they must be made of materials that are completely safe, with no toxicity, antigenicity, pyrogenicity, or carcinogenicity [3, 19].

Most biodegradable anchors are polymer based implants, degradation of these implant is dependent on their composition, molecular weight, and crystallinity [3]. Early bioabsorbable implants were made of polylactic acid (PLA) and polyglycolic acid (PGA) polymers. Afterwards, it was observed that PGA implants were rapidly degraded and result in a clinically significant inflammation and drainage. Degradation speed is important. PGA breaks down much more quickly than PLLA. The rapid absorption of PGA causes early loss of fixation power. Demirhan et al. in comprehensive study shows wedge shaped PGA anchor loses 75% of its pull-out strength within 3 weeks, 84% in 6 weeks, and 85% in 12 weeks [20]. However, PLA and its poly-L-lactic-acid (PLLA) form had long degradation rates [21–24]. These implants show comparable pullout strength to metallic implants. Kilicoglu et al. have been reported that effects of in vivo degradation of PLLA screws for 12 weeks did not cause decrease in fixation strength and excessive inflammation was not observed in first 12 weeks [25]. To decrease this degradation rate PLA and PGA combinations were developed but these stereoisomers and copolymers were rapidly degraded too [26].

Considering the complications (e.g. early fixation loss) associated with the rapid absorption of some bioabsorbable anchoring materials, alternative materials have been sought and biostable or inert materials have been described. The structure that resist chemical, thermal, or radiation induced degradation [27]. Polyetheretherketone (PEEK) is the most commonly used of them. It has been shown that PEEK implants are associated with minimal inflammatory response and do not produce artifacts like metal materials on imaging. PEEK is also plastic, nonabsorbable, can be drilled during revisions.

The most recent advancement in biodegradable implant technology is the introduction of biocomposite materials. These compounds are a combination of a standard biodegradable polymer such as lactic acid with a bioceramic material. Biocomposite anchors were suggested to provide initial fixation allowing final bone formation at the anchor insertion site without inducing major osteolysis or synovitis [28, 29]. Resorption of the anchor followed by bone formation would be an ideal scenario for an adequate bioabsorbable anchor. Bioabsorbable material and bone formation promote ceramic material such as β-tricalcium phosphate (β-TCP). Other bioceramics include hydroxyapatite, calcium sulfate, and calcium carbonate. Poly lactide-co-glycolic acid/β-tricalcium phosphate (PLGA/ β-TCP) is a biocomposite implant material developed to promote absorption at a controlled rate [25]. β-TCP is known to have the highest reported osteoconductivity rates in the literature, exhibiting grade 3–4 osteoconductivity in 50% implants [28–30], while other biocomposites could not provide a rate greater than 33% [29, 30]. Barber et al. [26] investigated the degradation and performance outcomes of PLGA/β-TCP as a systematic review. It demonstrated that PLGA/β-TCP biocomposite implants lost 88% of their original volume at a median 30-month follow-up and promoted osteoconductivity at 63% to 27 months of follow-up with low complication rates (tunnel widening 3%, effusion 5%, cyst formation 4%, synovitis was not report).

While the arthroscopic surgical technique evolves, knot tying has been considered as a rate limiting step and research is focused on this drawback. In 2001, Thal et al. described the first knotless suture anchors used in arthroscopic bankart repair (Mitek Knotless anchor) [31]. This was called the "suture first" design anchor where the suture is first passed through the tissue and engaged in the distal groove and then inserted into a predrilled hole [4]. These anchors were suggested to decrease the surgical time by resolving technical difficulties in tying knots and also reduce complications related to knots, such as impingement. Recent studies reported similar clinical and biomechanical results for the knotless systems in both instability and rotator cuff repair surgery [32–38].

In order to minimize the anchor induced chondral damage and related arthritic changes, the "suture only" or "all-suture" anchors were developed. These anchors are composed of one or more UHMWPE containing sutures. Anchor portion of the implant consists of a sleeve which is made of suture material. Following the insertion of the anchor into the bone, the main suture pulled and the sleeve becomes a "ball" larger than the drill hole, which serves as the anchor. All-suture anchors are attributed to have a bone preserving ability due to smaller cortical defects. However, the pull-out force of these anchors depends on the thickness of the bone cortex. Therefore, intraoperative decortication may increase the risk of anchor pull-out [39]. Recent clinical and biomechanical studies report promising outcomes of all-suture anchors as well as similar results with conventional suture anchors [39–44].

There are many kinds of anchor companies in the market, such as Arthrex®, Depuy-Mitek® Sport Medicine, Smith and Nephew®, Tornier®-Wright Medical Group, Zimmer-Biomet Sport Medicine. Their designs, materials, and placement mechanisms are different from each other. A firm is briefly mentioned below as an example.

20.3.1 Arthrex®

Has a several different anchor types loaded with FiberWire. Anchors can be made of PEEK, β-TCP-PLLA (biocomposite), PLLA (bioanchor), or titanium. *Corkscrew®* (Fig. 20.2a–c), *Swivelock®* (Fig. 20.3), and *Suturetak® are* three basic anchor group of this company. *Corkscrew®* anchors are mostly used for medial row repairs, *Swivelock®* anchors for lateral row repairs, and Suturetak® used in instability surgery for the glenoid. Different sizes of diameter anchors are available according to the repair configuration and size. (© Arthrex GmbH)

20.4 Suture Materials

Suture–tendon interface is an important point for a reliable repair. Therefore, choosing a high quality suture material is as important as the anchor in shoulder surgery. Demirhan et al. reported in comparative biomechanical study on nonfrozen cadaveric sheep shoulders that pull-out of simple sutures from tendon was the main failure mode in anchor repairs [45]. An arthroscopic suture should have good strength, adequate knot security, and biocompatibility. These sutures can be monofilament braided or

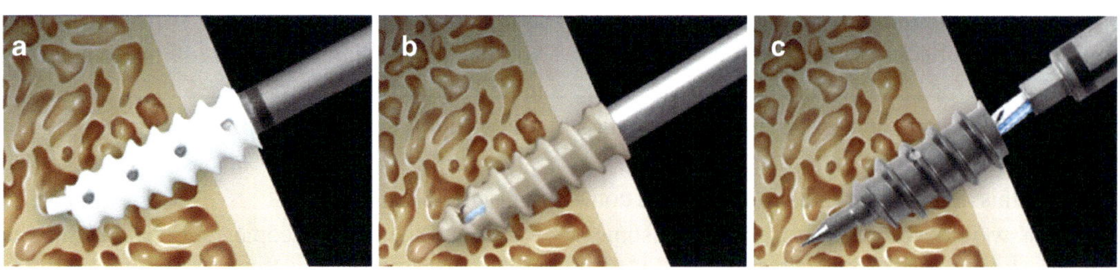

Fig. 20.2 (**a**) BioComposite Corkscrew® FT. (© Arthrex GmbH). (**b**) PEEK Corkscrew® FT. (© Arthrex GmbH). (**c**) Titanium Corkscrew® FT. (© Arthrex GmbH)

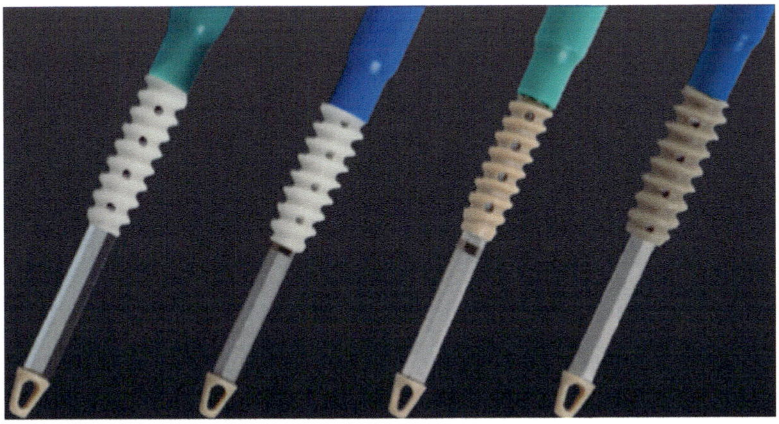

Fig. 20.3 Swivelock® C biocomposite, PEEK. (© Arthrex GmbH)

blended, as well as absorbable or nonabsorbable. While polydioxanone (PDS) was the most preferred bioabsorbable suture, many years braided polyester suture such as Ethibond has been used in anchors as nonabsorbable fashion. Recently, with the improvements in the suture material technology, stronger options composed of ultrahigh molecular weight polyethylene (UHMWPE) are now frequently utilized (first of this group was Arthrex®-FiberWire). FiberWire® has a braided polyester coat around a central core of multistrand long chain of UHMWPE. In addition of having clear advantages, it has a number of problems like mechanical irritation and impingement.

20.5 Anchor Placement

Angle of insertion: There have been many studies evaluating the anchor's angle of insertion and its relation to the pull-out force. The most popular theory is based on the "deadman concept" that is introduced by Burkhart in 1995 [46]. Burkhart has noticed that farmers in Texas were using a rock (the deadman) to hold the fence line tight against the pulling force. If the deadman is too close to the fence post, the post may lean away from the deadman until the horizontal component of the force through the safety wire and the force exerted on the fence wire reach an equilibrium (Fig. 20.4). At this point the deadman wire is inclined approximately 45° to the ground based on this observation. Burkhart applied the Deadman angle to suture anchor fixation technique and termed as the "deadman theory" (Figs. 20.5 and 20.6).

According to the deadman theory, an anchor inserted at a less than 45° of pullout angle shows the greatest pull-out strength. However, more recent studies have revisited the ideal insertion angle in rotator cuff repair as reported by Burkhart. The biomechanical studies have shown that the anchors inserted at 90° [47], 135° [48], or between 105° and 135° [49] show the greatest pullout strength. Liporace et al. have reported in their cadaveric study that the pull-out strength was similar for all anchor orientations between 30° and 90° [50]. The authors inserted Mitek SuperAnchors (Mitek) at angles of 90°, 75°, 45°, and 30° relative to the cortical border at the junction of the greater tuberosity and the articular surface there was no statistically significant difference detected in the comparison of failure strength of the anchors at varying degrees of insertion angle. Of the 4 groups, anchors inserted at 75° showed the highest load to failure (219 N) and anchors inserted at 45° showed the lowest load to failure (169 N). Strauss et al. have found that previously reported value of deadman angle may not provide the ideal

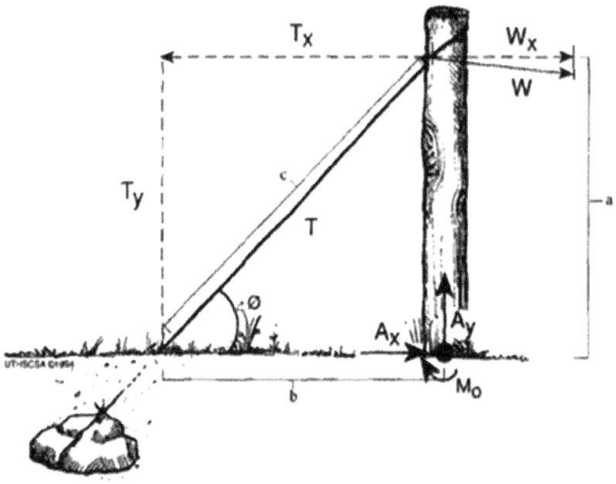

Fig. 20.4 Representation of the anchoring effect of a corner fence post by a deadman. The deadman is a large rock buried under the ground (T tension in deadman wire, W pull of fence wire A_x and A_y ground reaction forces, M_o ground reaction moment, T_x component of tension in x direction T_y component of tension in y direction, W_x pull of fence wire x direction). (With permission from [46])

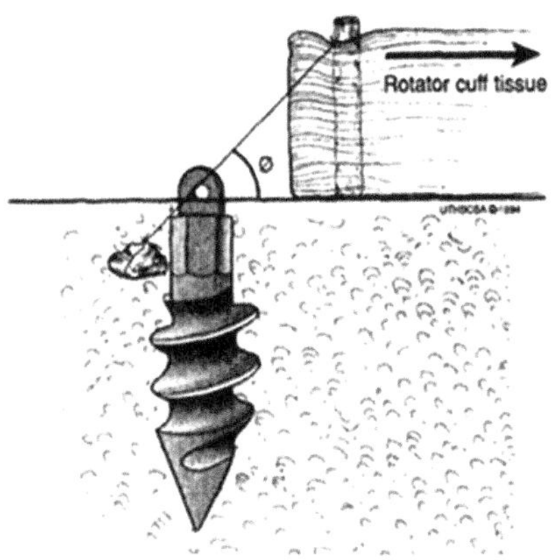

Fig. 20.5 Analogy of the deadman system to the suture anchor rotator cuff. The deadman wire is analogous to the suture; the pull of the fence wire on the corner post is analogous to the pull of the rotator cuff and the fence post is analogous to the compressed rotator cuff tissue between the suture and the bone. (With permission from [46])

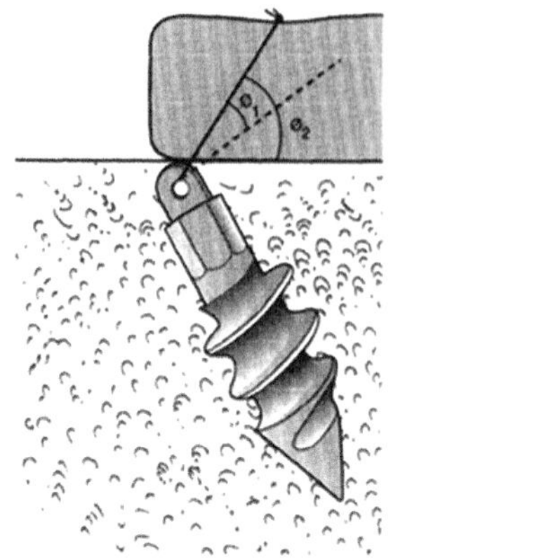

Fig. 20.6 Representation of θ_1 and θ_2, the pullout angle fort the angle θ_2, the tension reduction angle. Ideally, θ_1 and θ_2 should both be less than or equal to 45°. (With permission from [46])

soft tissue fixation in their cadaver study [47]. They found that (using screw type anchors, Depuy-Mitek Spiralock) the mean number of cycles to failure, defined as the occurrence of a 3mm gap formation in tendon repair site, was 380 cycles for anchors inserted at 90° which was significantly higher than the 297 cycles at 45°. Also cycles to complete failure were higher in the 90° group than 45°. Itoi et al. have tested three different insertion angles (45°, 90°, and 135°) and two different anchors (threaded and threadless) with one fixed pulling direction. They have found insertion angle of 45° is the strongest for a threadless anchor, but 90° is the strongest for a threaded anchor [51]. The authors concluded that the pullout strength depends on the inclination of the anchor, friction of the anchor–bone interface, and quality of the bone.

Consequently, placement pattern will differ with the evolution of anchor designs and materials.

20.6 Summary

Over the past several years, there has been a major shift in the types of anchors. In the course of time metallic implants are replaced by bioabsorbable, PEEK, biocomposite, and new all-suture anchors. Anchor materials, properties, types of repair (knotless concept), and suture materials (high strength, UHMWPE) affect ultimate repair success, besides innovations in anchor designs and suture materials may make arthroscopic shoulder surgery less technically demanding. Therefore, clinical and biomechanical studies should continue with the innovations.

References

1. Pulavarti RS, Symes TH, Rangan A. Surgical interventions for anterior shoulder instability in adults. Cochrane Database Syst Rev. 2009;4:CD005077.
2. Morse K, et al. Arthroscopic versus mini-open rotator cuff repair: a comprehensive review and meta-analysis. Am J Sports Med. 2008;36(9):1824–8.
3. Diduch DR, et al. Tissue anchor use in arthroscopic glenohumeral surgery. J Am Acad Orthop Surg. 2012;20(7):459–71.
4. Visscher LE, et al. The history of suture anchors in orthopaedic surgery. Clin Biomech (Bristol, Avon). 2019;61:70–8.
5. Carpenter JE, et al. Pull-out strength of five suture anchors. Arthroscopy. 1993;9(1):109–13.
6. Barber FA, Cawley P, Prudich JF. Suture anchor failure strength—an in vivo study. Arthroscopy. 1993;9(6):647–52.
7. Goble EM, et al. The development of suture anchors for use in soft tissue fixation to bone. Am J Sports Med. 1994;22(2):236–9.
8. Hecker AT, et al. Pull-out strength of suture anchors for rotator cuff and Bankart lesion repairs. Am J Sports Med. 1993;21(6):874–9.
9. Reed SC, Glossop N, Ogilvie-Harris DJ. Full-thickness rotator cuff tears: a biomechanical comparison of suture versus bone anchor techniques. Am J Sports Med. 1996;24(1):46–8.
10. Craft DV, et al. Fixation strength of rotator cuff repair swith suture anchors and the transosseous suture technique. J Shoulder Elbow Surg. 1996;5(1):32–40.
11. Silver MD, Daigneault JP. Symptomatic interarticular migration of glenoid suture anchors. Arthroscopy. 2000;16(1):102–5.
12. Gaenslen ES, Satterlee CC, Hinson GW. Magnetic resonance imaging for evaluation of failed repairs of the rotator cuff. Relationship to operative findings. J Bone Joint Surg Am. 1996;78(9):1391–6.
13. Marquardt B, et al. Arthroscopic Bankart repair in traumatic anterior shoulder instability using a suture anchor technique. Arthroscopy. 2006;22(9):931–6.
14. Piattelli A, et al. Histologic studies on osseointegration: soft tissues response to implant surfaces and components. A review. Dent Mater. 2011;27(1):53–60.
15. McFarland EG, et al. Suture anchors and tacks for shoulder surgery, part 1: biology and biomechanics. Am J Sports Med. 2005;33(12):1918–23.
16. Barber FA, et al. Sutures and suture anchors—update 2006. Arthroscopy. 2006;22(10):1063–1069.e2.
17. Ma R, et al. Arthroscopic rotator cuff repair: suture anchor properties, modes of failure and technical considerations. Expert Rev Med Devices. 2011;8(3):377–87.
18. Ekelund A. Cartilage injuries in the shoulder joint caused by migration of suture anchors or mini screw. J Shoulder Elbow Surg. 1998;7(5):537–9.
19. Speer KP, Warren RF. Arthroscopic shoulder stabilization. A role for biodegradable materials. Clin Orthop Relat Res. 1993;291:67–74.
20. Demirhan M, et al. Time-dependent reduction in load to failure of wedge-type polyglyconate suture anchors. Arthroscopy. 2000;16(4):383–90.
21. Pietrzak WS, Kumar M. An enhanced strength retention poly (glycolic acid)-poly (L-lactic acid) copolymer for internal fixation: in vitro characterization of hydrolysis. J Craniofac Surg. 2009;20(5):1533–7.
22. Kang RW, et al. Complications associated with anterior shoulder instability repair. Arthroscopy. 2009;25(8):909–20.
23. Böstman OM, Pihlajamäki HK. Adverse tissue reactions to bioabsorbable fixation devices. Clin Orthop Relat Res. 2000;371:216–27.
24. Böstman O, Pihlajamäki H. Clinical biocompatibility of biodegradable orthopaedic implants for internal fixation: a review. Biomaterials. 2000;21(24): 2615–21.
25. Kilicoglu O, et al. Failure strength of bioabsorbable interference screws: effects of in vivo degradation for 12 weeks. Knee Surg Sports Traumatol Arthrosc. 2003;11(4):228–34.
26. Barber FA, et al. Biocomposite implants composed of poly (Lactide-co-Glycolide)/β-tricalcium phosphate: systematic review of imaging, complication, and performance outcomes. Arthroscopy. 2017;33(3): 683–9.
27. Suchenski M, et al. Material properties and composition of soft-tissue fixation. Arthroscopy. 2010;26(6):821–31.
28. LeGeros RZ. Properties of osteoconductive biomaterials: calcium phosphates. Clin Orthop Relat Res. 2002;395:81–98.
29. Barber FA, Dockery W, Hrnack SA. Long-term degradation of a poly-lactide co-glycolide/β-tricalcium phosphate biocomposite interference screw. Arthroscopy. 2011;27(5):637–43.
30. Barber FA, Dockery W. Long-term degradation of self-reinforced Poly–Levo (96%)/Dextro (4%)–Lactide/β-Tricalcium phosphate biocomposite interference screws. Arthroscopy. 2016;32(4):608–14.
31. Thal R. Knotless suture anchor: arthroscopic bankart repair without tying knots. Clin Orthop Relat Res. 2001;390:42–51.
32. Hayashida K, et al. Arthroscopic Bankart repair with knotless suture anchor for traumatic anterior shoulder instability: results of short-term follow-up. Arthroscopy. 2006;22(6):620–6.
33. Zheng N, Harris HW, Andrews JR. Failure analysis of rotator cuff repair: a comparison of three double-row techniques. J Bone Joint Surg Am. 2008;90(5):1034–42.
34. Kocaoglu B, et al. No difference between knotless sutures and suture anchors in arthroscopic repair of Bankart lesions in collision athletes. Knee Surg Sports Traumatol Arthrosc. 2009;17(7):844–9.
35. Sileo MJ, et al. Biomechanical comparison of a knotless suture anchor with standard suture anchor in the repair of type II SLAP tears. Arthroscopy. 2009;25(4):348–54.

36. Matache BA, et al. Knotted versus knotless anchors for labral repair in the shoulder–a systematic review. Arthroscopy. 2021;37(4):1314–21. https://doi.org/10.1016/j.arthro.2020.11.056.
37. Wu IT, et al. Comparable clinical outcomes using knotless and knot-tying anchors for arthroscopic capsulolabral repair in recurrent anterior glenohumeral instability at mean 5-year follow-up. Knee Surg Sports Traumatol Arthrosc. 2021;29(7):2077–84. https://doi.org/10.1007/s00167-020-06057-7.
38. Zwolak P, et al. The functional outcome of arthroscopic rotator cuff repair with double-row knotless vs knot-tying anchors. Arch Orthop Trauma Surg. 2022;142(1):25–31. https://doi.org/10.1007/s00402-020-03584-3.
39. Ntalos D, et al. All-suture anchor pullout results in decreased bone damage and depends on cortical thickness. Knee Surg Sports Traumatol Arthrosc. 2021;29(7):2212–9. https://doi.org/10.1007/s00167-020-06004-6.
40. Lacheta L, et al. Biomechanical evaluation of knotless and knotted all-suture anchor repair constructs in 4 bankart repair configurations. Arthroscopy. 2020;36(6):1523–32.
41. Ergün S, et al. The clinical and biomechanical performance of all-suture anchors: a systematic review. Arthrosc Sports Med Rehabil. 2020;2(3):e263–75.
42. Frank RM, et al. Biomechanical analysis of all-suture suture anchor fixation compared with conventional suture anchors and interference screws for biceps tenodesis. Arthroscopy. 2019;35(6):1760–8.
43. Bernardoni E, et al. Biomechanical analysis of all-suture anchor fixation for rotator cuff repair. Orthop J Sports Med. 2018;6(7 Suppl 4):2325967118S00175.
44. Barber FA, Herbert MA. All-suture anchors: biomechanical analysis of pullout strength, displacement, and failure mode. Arthroscopy. 2017;33(6):1113–21.
45. Demirhan M, Atalar AC, Kilicoglu O. Primary fixation strength of rotator cuff repair techniques: a comparative study. Arthroscopy. 2003;19(6):572–6.
46. Burkhart SS. The deadman theory of suture anchors: observations along a south Texas fence line. Arthroscopy. 1995;11(1):119–23.
47. Strauss E, et al. The effect of the angle of suture anchor insertion on fixation failure at the tendon–suture interface after rotator cuff repair: deadman's angle revisited. Arthroscopy. 2009;25(6):597–602.
48. Green RN, et al. Biomechanical study: determining the optimum insertion angle for screw-in suture anchors—is deadman's angle correct? Arthroscopy. 2014;30(12):1535–9.
49. Clevenger TA, et al. The effect of insertion angle on the pullout strength of threaded suture anchors: a validation of the deadman theory. Arthroscopy. 2014;30(8):900–5.
50. Liporace FA, et al. The mechanical effects of suture anchor insertion angle for rotator cuff repair. Orthopedics. 2002;25(4):399–402.
51. Itoi E, et al. Deadman theory revisited12. Biomed Mater Eng. 2016;27(2-3):171–81.

21 Tissue Engineering and Graft Options

Haluk Celik, Ismail Turkmen, and Mustafa Karahan

21.1 Introduction

Tissue engineering and regenerative medicine are the most common terms that are used to describe the approach to generate tissues and organs. Both approaches are multidisciplinary and emerging in biotechnology and medicine, which are expected to treat patients with generating and regenerating tissues instead of repair [1]. Tissue engineering, is first defined in 1993 by Langer and Vacanti, is regarded as a more technical concept of tissue reconstruction by the use of scaffolds which is defined as an artificial structure used to support three-dimensional tissue formation and biomolecules [2]. The term regenerative medicine is commonly more focused on the support of self-healing capabilities and the use of stem cells. The goal of musculoskeletal tissue engineering is to produce tissue that initially provides the mechanical strength to restore function while promoting further tissue growth and remodeling over time [3].

The typical approach to tissue engineering utilizes a scaffold to promote tissue formation [4]. Tissue engineering investigations in shoulder were commonly performed for rotator cuff (RC) surgery [5–9]. The products are utilized in augmentation or interposition of the RC repair. Besides, a limited number of studies also evaluated the use of tissue engineering products in capsular reconstruction and glenohumeral arthritis treatment [10–13]. The objective of the present chapter is to review the use of tissue engineering and grafts in shoulder surgery.

21.2 Extracellular Matrix and Scaffolds

Extracellular matrix (ECM) has important functions such as providing structural support, creating a physiological environment for human cells [14, 15]. Besides, ECM consists of growth factors and offers bioactive signals to the cells for cellular activities. Extracellular stimuli is interpreted by cells using the particular signals such as cytokines, growth factors, or hormones. Interleukins and interferons are among the cytokines and maintain the cell homeostasis. Growth factors such as transforming growth factor-beta (TGF-β) and insulin-like growth factor (IGF) interact with cells for regulation of cell growth

H. Celik (✉)
Department of Orthopaedics and Traumatology, Hisar Intercontinental Hospital, Istanbul, Turkey

I. Turkmen
Department of Orthopaedics and Traumatology, Goztepe City Hospital, Istanbul Medeniyet University, Istanbul, Turkey

M. Karahan
Department of Orthopaedics and Traumatology, Acıbadem Mehmet Ali Aydinlar University, Istanbul, Turkey

G. Huri et al. (eds.), *Fundamentals of the Shoulder*, https://doi.org/10.1007/978-3-030-94702-6_21

and proliferation. Specific receptors called integrins present important role for both outside-in and inside-out signaling [15]. Tissue engineering therefore aims at optimizing the interaction among cells, constructing an extracellular matrix scaffold for tissue regeneration.

Scaffolds in shoulder surgery may be derived from ECM or synthetic materials [6]. ECM-derived scaffolds can be xenogeneic or allogeneic. Dermal tissues of human, porcine or bovine, human fascia lata, equine pericardium, porcine small intestinal submucosa (SIS) are defined sources for ECM-derived scaffolds [1, 6, 8]. Besides more biological, the presence of ECM has several disadvantages such as generating an immune response, possible contamination, and ease of degradation [12].

Synthetic scaffolds are produced using synthetic materials such as polypropylene, polyarylamid, dacron, silicone, and nylon [16]. Synthetic scaffolds present superior biomechanical features than ECM-derived scaffolds [6]. Nevertheless, complications such as infection, synovitis, and foreign body reactions are raising issues although synthetic scaffolds provide strong biomechanical features [16]. Recently, combinations of ECM and synthetic scaffolds have been initiated to utilize. In those, ECM molecules are incorporated into artificial biomaterials to create additional functionality. In addition to strong biomechanical features, biocompatibility and flexible designs are the main characteristics of these implants. Poly-L-lactic acid (PLLA), poly-lactic-co-polydioxanone, polycaprolactone, and polydioxanone are among new generation scaffolds [17].

21.3 Host Response and Scaffold Remodeling

The products of tissue engineering are foreign structures for human body and an immediate host response occurs following the implantation [18]. The severity of the response depends on the contents of the scaffolds and the anatomical location. Once the product is implanted, proteins related to coagulation and plasma derived proteins cover the surface of the material and particular factors are released for the migration of the inflammatory cells [19, 20]. Inflammatory cells, particularly neutrophils start to release chemoattractant factors. Phagocytosis by macrophages and neutrophils may lead to a destruction of implanted material in a short term [20, 21]. In addition to acute response, a chronic inflammation occurs by activated macrophages. In further, foreign body giant cells and angiogenesis remain at the forefront of the inflammation [20]. Conversely, particular foreign molecules in xenogeneic scaffolds may lead to hypersensitivity immune response in humans. Porcine small intestine submucosa grafts were reported to encounter hypersensitivity immune response in 20–30% of patients due to presence of galactose-a (1,3)-galactose (a-Gal) terminal disaccharide [7, 22]. Cellular components are not the single reason for the host response. Chemical structure of the scaffold, processing methods, methods of terminal sterilization, surgical exposure, and mechanical loading also influence the severity of the host response [6].

Therefore, the ECM scaffolds are usually prepared by decellularization of the tissue and removal of the immunogenic antigens [23]. Decellularization may be performed using gamma irradiation, physical, chemical, and enzymatic methods. Freezing procedures and mechanical disruption of the harvested tissues are among the physical methods. Chemical wash-out using detergents or hypotonic solutions aim to remove the cellular remnants. Some particular enzymes such as trypsin are also used to lyse the cellular components [24, 25]. It has been shown that extensive decellularization of the native tissues leads to disruption of the ECM and poor mechanical properties of the graft [12, 26]. In addition, preservation of the native ECM architecture such as tissue morphology, alignment of the collagen fibers, and thermal stability is critical in facilitating recellularization of the scaffolds [26].

During the remodeling process, ECM-derived scaffolds have been shown to induce a large macrophage response via an M1 (proinflammatory) or M2 (proremodeling) response [27, 28]. It seems that M2 macrophage component of the response is the major determinant in the remodeling and integration of an ECM scaffold device. The scaffolds are expected to resolve in time, while a new synthesized ECM, cells, and tissues are replaced on the degraded scaffold without any biological reaction by immune system. Scaffold degradation and remodeling are influenced by several factors including material of scaffold, mechanical loading, and physical rehabilitation. In a study that evaluated the host response and remodeling of several ECM-derived scaffolds, non-cross-linked SIS were rapidly degraded and replaced with host tissues, whereas scaffolds derived from dermis presented slower remodeling. ECM scaffolds that were cross-linked were minimally degraded and were associated with the presence of foreign body giant cells, chronic inflammation, or poorly organized fibrous tissue [28]. Valentin et al. reported that the Restore-patch was completely degraded in between 3 and 4 months and the other ECM scaffolds either partially degrade (GraftJacket, Cuff Patch, and Tissue Mend) or do not degrade at all (Zimmer Patch) [28].

Synthetic scaffolds have different immune responses than ECM scaffolds. Commonly a chronic inflammation with a granulation tissue and fibrous capsule is encountered [29]. The duration and intensity of the host response to a synthetic scaffold are influenced by its composition and morphology (size, shape, porosity, and roughness) and likely biologic and mechanical factors at the implantation site. Polymer products such as lactic and glycolic acid have been shown to inhibit matrix mineralization and were found to decrease cellular proliferation [30]. Polymers are cleared away from the body via either chemical reactions on polymer chain or enzymatic reactions by cells [31].

Derwin et al. investigated poly-L-lactide used for RC repair in a canine model and reported the presence of macrophages and foreign body giant cells without neutrophils and lymphocytes observation [32]. Synthetic scaffolds have been reported that the presence of those still might continue in knee joints after 15 years with poor host tissue integration [33]. However synthetic scaffolds made from aliphatic polyesters were reported to be degraded [8].

21.4 Mechanical Features

The aim of the use of scaffolds is to provide sufficient mechanical support during the healing process. Besides the longevity of the degradation, time zero characteristics are also important for the mechanical features. In general, ECM scaffolds are approved to be more biologic rather than strength [8].

Smith et al. compared the macro and nano-micro mechanical properties of seven different scaffolds to those of the human supraspinatus tendons [34]. The products were subjected to scanning electron microscopy, tensile testing, rheometer testing, and scanning probe microscopy. All scaffolds failed to approximate the mechanical properties of human supraspinatus tendons. The authors concluded that synthetic scaffolds better approximated the macro mechanical properties of supraspinatus tendon and ECM scaffolds approximated the micro mechanical properties; however, none of the scaffolds fully approximated the properties of native tissue [34]. Beitzel et al. [35] evaluated the strength of rotator cuff augmentation using collagen or dermis based scaffolds and reported an increased ultimate load to failure for both scaffolds compared to non-augmented group.

It has to be considered that several possible factors may influence the biomechanical performance of the scaffolds. The method of the fixation, location of the graft, suture retension properties, pre-tensioning at the time of the fixation, number and type of the sutures are important for mechanical stability of the scaffold. Tissue engineering products may have poor suture retension properties despite having a strong structure [32]. The studies that evaluated the mechanics of various ECM and synthetic scaffolds demonstrated that fascia lata ECM and poly-L-lactide have similar material properties in uniaxial mechanical tests [6, 32, 36]. However, fascia lata ECM had poor suture retention properties [37]. Sahoo et al. [38] reported that acellular human dermal grafts presented unrecoverable elongation at low physiologic loads. In order to reduce compliance of construct the authors utilized cyclical stretching, reverse-cutting needles and applied 20 N of pretension.

In an in vitro study, human dermis-derived scaffolds (GraftJacket, Permacol, TissueMend) were reported to have greater load to failure than SIS-derived scaffolds (CuffPatch, Restore). The reason for the failure was mostly reported to be suture pull-out [39]. In another cadaveric study, GraftJacket Extreme was reported to be more strong to control group. Failures were seen at the tendon–suture interface in six of ten models, while suture breakage was observed in four cases [40]. In similar, Omae et al. [41] reported higher load to failure compared to control group and the most common reason for the failure was tendon cut-out (70%). McCarron et al. [42] evaluated a woven poly-L-lactic acid device and showed that the scaffold increased the yield load and ultimate load but did not affect initial repair stiffness.

21.5 Clinical Applications of Tissue Engineering in Rotator Cuff Surgery

21.5.1 Extracellular Matrix- Derived Scaffolds

ECM-derived scaffolds contain organic substitutes such as collagen, lipid, or carbohydrate depending on the type of product used [43, 44]. Proteins may be brought together by a different binding structure in order to provide diversity.

ECM-derived scaffolds will be discussed in two main headings as xenografts and allografts.

21.5.1.1 Xenografts

These animal-derived products are usually produced from porcine intestine [22, 28, 44–47], porcine dermis [48–52], bovine pericardium [53], bovine dermis [54], and equine pericardium [55] (Table 21.1).

Porcine Intestine Submucosa Derived Scaffolds

Porcine small intestine submucosa contains type-I collagen, fibronectin, chondroitin sulfate, heparin, heparin sulfate, hyaluronan, and growth factors [45]. These scaffolds have also been used in vascular grafting, bladder, and dural reconstruction surgeries [46, 47]. CuffPatch Bioengineered Tissue Reinforcement (Arthrotek) and Restore Orthobiologic Implant (DePuy Orthopaedics) are the most frequently used products of this group and available as hydrated or dry packaged. Although good functional results have been reported in the studies, graft failures and retears are frequent [22, 45–47]. While Iannotti, Walton, and Sclamberg reported a high rate of failure in their studies with a limited number of patients, Metcalf reported that early clinical results were successful after 2 years of follow-up [45]. The major problem in these scaffolds is that strength, which is high in time zero, loses in the following period.

Porcine Dermis-Derived Scaffolds

Zimmer Collagen Repair Patch (Zimmer), Conexa Reconstructive Tissue Matrix (Tornier), Arthrex DX Reinforcement Matrix (Arthrex) are the most frequently used products of this group in the market. There are studies reporting the failure of all grafts used on the patients despite the progress of clinical results during the early period after rotator cuff surgery in the literature [48]. Cho and Badhe reported low failure rates and advanced functional scores at least 8 months of follow-up [49, 50]. None of these studies reported intraoperative complications during surgery. In a study by Gupta et al. [51], patients monitored using USG in the follow-up and good clinical results were obtained. Similarly, in this study, over 70% intact construct was obtained. However, in a study comparing transosseous-equivalent repair with additional xenograft patch, porcine dermal xenograft augmentation was shown to have no additional benefit to the patients in terms of reducing the risk of a recurrent tendon defect or improving shoulder function up to 24 months after surgical repair [52].

Bovine Pericardium Derived Scaffolds

TUTOPATCH (Tutogen Medical GmbH) is a fenestrated natural collagen matrix which is first used for hernia repair [53]. In a comparative clinical study Ciampi et al. [53] retrospectively compared clinical results of the patients with chronic

Table 21.1 Xenografts

Source/Tissue	Company	Product	Comments
Porcine intestine	DePuy Orthopaedics	Restore Orthobiologic Implant	Low healing rates, severe inflammatory reaction
Porcine intestine	Arthrotek	CuffPatch	
Porcine dermis	Zimmer	Zimmer Collagen Repair	Relatively improved functional outcomes, ineffective in massive tears
Porcine dermis	Tornier	Connexa	
Porcine dermis	Arthrex	Arthrex DX Reinforcement Matrix	
Bovine pericardium	Tutogen Medical GmbH	TUTOPATCH	Improved results Higher retear when comparing open repair only or synthetic augmentation
Bovine dermis	Stryker Orthopaedics	TissueMend	No clinical study
Equine Pericardium	Pegasus Biologics	OrthADAPT Bioimplants	No clinical study

Table 21.2 Allografts

Source/Tissue	Company	Product	Comment
Human dermis	Wright Medical Technology	GrafJacket	Safe for revision cuff surgery Low inflammatory reaction Higher satisfaction rates Lower retear rates
Human dermis	Arthrex	ArthroFlex	
Human dermis	Musculoskeletal Tissue Foundation	AlloPatch	

rotator cuff tear treated with TUTOPATCH or synthetic scaffolds augmentation. In this study, in which the functional scores of 152 patients were compared, the result of the 36-month follow-up was found to be superior to the results of the synthetic graft. Retears have been observed in 25 of 49 cases with TUTOPATCH.

Bovine Dermis-Derived Scaffolds

TissueMend (TEI Biosciences, Boston, MA) is a single layer graft derived from fetal bovine skin [54]. To date, no clinic study has been available in reference to this product.

Equine Pericardium Derived Scaffolds

The OrthADAPT bioimplant is a versatile collagen scaffold that is used to be modified in a variety of ways to provide augmentation and strength. Although there are no clinical studies on rotator cuff surgery related to this product, its use in other ligament repairs has been reported [55].

21.5.1.2 Allografts

After the production of the human derived extracellular matrix scaffolds, new treatment strategies have evolved to address the biological healing difficulty associated with primary repair of massive rotator cuff tears (Table 21.2) [56–61]. Although allografts are recommended for an augmentation of rotator cuff repair, many shoulder surgeons use it as a gap filling tool (interpositional bridging) in rotator cuff surgery.

Human Dermis-Derived Scaffolds

GraftJacket Regenerative Tissue Matrix (Wright Medical Technology), ArthroFlex (Arthrex), and AlloPatch HD (Musculoskeletal Tissue Foundation) are products available in the market. GraftJacket is the most studied scaffold in this group [56–59]. Although the results cannot be evaluated homogeneously since there are different methods such as single row, double row, and partial repair methods, lower implant failure and graft rejection rates are reported with improved functional outcomes. In a study of Gupta et al. [51], one case of infection was reported among 45 patients. In the study of Wong et al. [59] including 24 patients, retear was reported in one case, while no infection and graft rejection were reported.

21.5.2 Synthetic Scaffolds

Synthetic grafts are produced as an alternative to biological scaffolds. A synthetic scaffold applied onto the repaired tendon has the potential to both support the suture and protect the repair, providing healing process [62–70]. Current materials used for rotator cuff augmentation after repair include poly-L-lactic acid, polyethylene terephthalate, polycarbonate polyurethane-urea, poly-4-hydroxybutyrate, expanded polytetrafluoroethylene, and polyurethane-urea (Table 21.3).

LARS Ligament is the best known and leading product of this group. Petrie et al. [66] demonstrated significantly increased acromiohumeral interval distance in the patients with Goutallier grade 3–4 supraspinatus muscle atrophy after augmentation and reconstruction procedure in 31 shoulders. Only 2 patients required revision among this cohort. Nada et al. augmented the polyester ligament (5 × 1 cm) to rotator cuff in 21 cases and 2 complications (1 failure and 1 infection) were reported. At the final follow-up of the remaining cases the MR scans confirmed intact and thickened repair in 15 of 17 patients [67].

21.5.3 New Generation Scaffolds

Today, with the development of nanotechnology, many unique properties of nano-sized materials have been revealed. With their superior properties, nanofibers find a wide range of uses and/or applications in many industrial, medical, and military products. The electrospinning technique for the production of this material is the most widely known and easily applicable method. With this method, ultra-thin (10–100 nm) nanofiber membranes are produced from various polymers in liquid form (PVA, PU, PA, etc.) [71]. If bioactive growth factors are added to these nanofibers or embedding the stem cells into the scaffold devices, very powerful hybrid products can be obtained. In the literature, studies on these products are at an in vivo or in vitro experimental level.

Zhao et al. [72] in his study compared local application of bFGF–PLGA fibrous membranes with repair alone. The bone tendon junction after rotator cuff repair in the experiment model was found to strengthen the healing enthesis and improve collagen organization and healing volume compared with control group. Another hybrid scaffold is doxycycline-incorporated nanofibrous membranes. Administration of doxycycline orally has been demonstrated to improve collagen fibril organization through inhibition of local matrix metalloproteinase activity [73]. In the experimental study of Weng et al. [73] nanofibers loaded with doxycycline demonstrated great potential for the repair of rat Achilles tendon rupture and gives promising results. In similar, experimental RC tear models have been studied with xenograft patch impregnated with poly-glycolic acid (PGA) sheets and autologous cultured mesenchymal stem cell and human PRP [74, 75]. Although these studies are encouraging, they are still insufficient in daily clinical practice for surgeons to perform these on patients.

Table 21.3 Synthetic scaffolds

Material	Company	Product	Comment
Poly-L-lactic acid	Synthasome	X-Repair	Improved clinical outcomes Decreased acromiohumeral distance Lower retear rates Biomechanically strong material Needed further comparative clinical studies
Polyethylene terephthalate	Xiros Ltd., Neoligaments	Poly-Tape	
Polyethylene terephthalate	LARS	LARS Ligament	
Polycarbonate polyurethane-urea	Biomerix	Biomerix RCR Patch	
Poly-4-hydroxybutyrate	Tornier	BioFiber	
Expanded polytetraflouroethylene	Gore Medical	Gore-Tex Patch	
Polyurethane-urea	Biomet Sports Medicine	SportMesh	

21.6 Clinical Applications of Tissue Engineering in Different Shoulder Issues

21.6.1 Superior Capsular Reconstruction

Superior capsular reconstruction was developed for treatment of irreparable RC tears and initial studies reported good early results with the use of autografts (fascia lata) [76]. However, because of concerns about donor site morbidity, additional time, and effort associated with graft harvest, an SCR procedure using dermal allografts has been developed recently [10, 77]. Although dermal allografts do not have any donor site morbidity and their use can shorten the operation time, there are concerns about low viability, graft rejection, infections, and high cost [78].

Denard et al. [77] reported the results of 59 patients underwent an SRC procedure using an acellular dermal graft. The VAS score decreased from 5.8 to 1.7, the ASES score improved from 43.6 to 77.5, and the Subjective Shoulder Value (SSV) improved from 35.0 to 76.3 at a minimum 1-year follow-up. Postoperative MRI scans showed healing of the graft in 45% of patients, thinner grafts (1-mm) had approximately 60% failure rate. Eleven patients (18.6%) required a revision procedure including seven reverse shoulder replacement. Pennington et al. [10] used 3-mm-thick dermal allograft in 86 patients. The patients showed improvements in the VAS score from 4.0 to 1.5 and the ASES score from 52 to 82 at 1-year follow-up. Complication rate was reported as 4.5%. Three graft tears were revealed by MRI scans, and a revision surgery was required in 1 patient. Alternatively, Polacek [12] reported the results of the patients underwent an arthroscopic SCR with an acellular porcine dermal xenograft. A successful outcome in 60% of cases reported at 1 year follow-up. The procedure showed a quite high complication rate and the most severe cases were related to acute immunologic rejection of the xenograft.

A recent systematic review compared the use of auto and allografts in SCR surgery. For autografts and allografts, respectively, the mean gain in forward elevation (FE) was 48.7 and 33.3, the ASES score increased by 47.3 and 31.9, and the acromiohumeral distance increased by 1.2 and 1.8 mm. The rate of graft tears was 10.0% and 12.9%, the rate of other complications was 7.5% and 3.9%, and the rate of reoperations was 3.1% and 8.2% for autografts and allografts, respectively [78]. Despite short term outcome reports, SCR using a tissue engineering product has promising results and it has been expected that different type of materials will be utilized for this procedure.

21.6.2 Glenoid Resurfacing

Glenoid biologic resurfacing has been reported for the young patients who had severe glenohumeral arthritis. In addition to fascia lata autograft, meniscal allografts, Achilles tendon allografts, or shoulder joint capsule tissues, acellular human dermal grafts are also used for resurfacing. Lo et al. [13] evaluated 55 patients underwent hemiarthroplasty and human acellular dermal allograft (GraftJacket MaxForce Extreme; Wright Medical, Arlington, TN, USA) with an average of 60 months follow-up. Eighty-one percent of the patients were satisfied (10/47) or highly satisfied (28/47) with their result and a total of 5 cases (9.1%) were revised to anatomic total shoulder arthroplasty with implantation of a glenoid component. However, despite the evaluation of a small size of patients, other studies used acellular human dermal grafts with humeral head arthroplasty, did not favor this technique because of the limited improvement in patient outcomes and the relatively high revision rate [79–81]. The procedure has been recommended for well indicated situations with caution and appropriate counseling to the patient [82].

21.6.3 Capsular Reconstruction for Shoulder Instability

The use of acellular dermal allografts has been described in technical reports for anterior or posterior capsular reconstruction. Especially for the patients with soft tissue pathologies for which primary repair of the capsulolabral complex is not possible because of absent capsular tissue, severe subscapularis tendon deficiency or the patients with collagen disorders can be a candidate for the capsular reconstruction using an allograft. Open or arthroscopic dermal graft applications were reported for anterior instability [83, 84]. Posterior capsule reconstruction was also performed using a GraftJacket allograft in case of persistent posterior instability despite a previous plication surgery [11, 85]. Despite promising results, cohort studies are needed to evaluate the efficiency of the usage of tissue engineering products for instability surgeries.

21.7 Summary

Different ECM-derived and synthetic scaffolds have been used in shoulder surgery for augmentation or interposition of the structures. Besides RC surgery, use of the scaffolds for capsular reconstructions has become popular. Human dermal allograft is the most common scaffold and associated with good functional outcomes. Xenografts were reported with increased retear rates and less improvement in patient-reported outcomes, rather than synthetic grafts and allografts. The synthetic grafts have the lowest retear rates and did not exhibit any tissue reactions or osteolysis. In order to establish clear recommendations, prospective, randomized controlled trials comparing the various scaffolds are required. In future, gene therapy and nanotechnology are expected to improve the mechanical properties and biocompatibility of the scaffolds.

References

1. Tsiridis E, Velonis S, Limb D, Giannoudis PV. Tissue engineering approaches to rotator cuff tendon deficiency. Connect Tissue Res. 2008;49(6):455–63. https://doi.org/10.1080/03008200802325383.
2. Langer R, Vacanti JP. Tissue engineering. Science. 1993;260:920–6.
3. Ikada Y. Challenges in tissue engineering. J R Soc Interface. 2006;3:589–601.
4. Laurencin CT, Freeman JW. Ligament tissue engineering: an evolutionary materials science approach. Biomaterials. 2005;26(36):7530–6. https://doi.org/10.1016/j.biomaterials.2005.05.073.
5. Thangarajah T, Pendegrass CJ, Shahbazi S, Lambert S, Alexander S, Blunn GW. Augmentation of rotator cuff repair with soft tissue scaffolds. Orthop J Sports Med. 2015;3(6):2325967115587495. https://doi.org/10.1177/2325967115587495.
6. Ricchetti ET, Aurora A, Iannotti JP, Derwin KA. Scaffold devices for rotator cuff repair. J Shoulder Elb Surg. 2012;21(2):251–65. https://doi.org/10.1016/j.jse.2011.10.003.
7. Xu H, Wan H, Zuo W, Sun W, Owens RT, Harper JR, et al. A porcine-derived acellular dermal scaffold that supports soft tissue regeneration: removal of terminal galactose-alpha-(1,3)-galactose and retention of matrix structure. Tissue Eng Part A. 2009;15(7):1807–19. https://doi.org/10.1089/ten.tea.2008.0384.
8. Karuppaiah K, Sinha J. Scaffolds in the management of massive rotator cuff tears: current concepts and literature review. EFORT Open Rev. 2019;4(9):557–66. https://doi.org/10.1302/2058-5241.4.180040.
9. James KS, Cornwell KC, Greenburg AG. Extracellular matrix bioscaffolds for orthopaedic applications. J Bone Joint Surg Am. 2010;92(5):1316–7; author reply 1317-1318
10. Pennington WT, Bartz BA, Pauli JM, Walker CE, Schmidt W. Arthroscopic superior capsular reconstruction with acellular dermal allograft for the treatment of massive irreparable rotator cuff tears: short-term clinical outcomes and the radiographic parameter of superior capsular distance. Arthroscopy. 2018;34(6):1764–73. https://doi.org/10.1016/j.arthro.2018.01.009.
11. Karpyshyn J, Gordey EE, Coady CM, Wong IH. Posterior glenohumeral capsular reconstruction using an acellular dermal allograft. Arthrosc Tech. 2018;7(7):e739–45. https://doi.org/10.1016/j.eats.2018.03.011.
12. Polacek M. Arthroscopic superior capsular reconstruction with acellular porcine dermal xenograft for the treatment of massive irreparable rotator cuff tears. Arthrosc Sports Med Rehabil. 2019;1(1):e75–84. https://doi.org/10.1016/j.asmr.2019.08.001.
13. Lo EY, Flanagin BA, Burkhead WZ. Biologic resurfacing arthroplasty with acellular human dermal allograft and platelet-rich plasma (PRP) in young patients with glenohumeral arthritis-average of 60 months of at mid-term follow-up. J Shoulder Elb Surg. 2016;25(7):e199–207. https://doi.org/10.1016/j.jse.2015.11.063.
14. Ide J, Kikukawa K, Hirose J, Iyama K, Sakamoto H, Mizuta H. Reconstruction of large rotator-cuff tears with acellular dermal matrix grafts in rats. J Shoulder Elb Surg. 2009;18(2):288–95. https://doi.org/10.1016/j.jse.2008.09.004.
15. Petreaca M, Martins-Green M. The dynamics of cell–ECM interactions with implications for tissue engineering (Chap. 9). In: Lanza R, Langer R, Vacanti

J, editors. Principles of tissue engineering. 4th ed. Amsterdam: Elsevier; 2014. p. 175–80.
16. Longo UG, Lamberti A, Khan WS, Maffulli N, Denaro V. Synthetic augmentation for massive rotator cuff tears. Sports Med Arthrosc Rev. 2011;19(4):360–5. https://doi.org/10.1097/JSA.0b013e318224e359.
17. Hakimi O, Mouthuy PA, Carr A. Synthetic and degradable patches: an emerging solution for rotator cuff repair. Int J Exp Pathol. 2013;94(4):287–92. https://doi.org/10.1111/iep.12030.
18. Karahan M, Nuran R. Tissue engineering and new biomaterials (Chap. 5). In: Gobbi A, Espregueira-Mendes J, Lane JG, Karahan M, editors. Bio-orthopaedics. A new approach. 1st ed. Berlin: Springer; 2017. p. 65–71.
19. Koo H, Yamada KM. Dynamic cell-matrix interactions modulate microbial biofilm and tissue 3D microenvironments. Curr Opin Cell Biol. 2016;42:102–12.
20. Brown BN, Badylak SF. The role of the host immune response in tissue engineering and regenerative medicine (Chap. 25). In: Lanza R, Langer R, Vacanti J, editors. Principles of tissue engineering. 4th ed. Amsterdam: Elsevier; 2014. p. 497–509.
21. Hu WJ, Eaton JW, Ugarova TP, Tang L. Molecular basis of biomaterial-mediated foreign body reactions. Blood. 2001;98(4):1231–8. https://doi.org/10.1182/blood.v98.4.1231.
22. Walton JR, Bowman NK, Khatib Y, Linklater J, Murrell GA. Restore orthobiologic implant: not recommended for augmentation of rotator cuff repairs. J Bone Joint Surg Am. 2007;89(4):786–91. https://doi.org/10.2106/JBJS.F.00315.
23. Reing JE, Zhang L, Myers-Irvin J, Cordero KE, Freytes DO, Heber-Katz E, et al. Degradation products of extracellular matrix affect cell migration and proliferation. Tissue Eng Part A. 2009;15(3):605–14. https://doi.org/10.1089/ten.tea.2007.0425.
24. Cheung EV, Silverio L, Sperling JW. Strategies in biologic augmentation of rotator cuff repair: a review. Clin Orthop Relat Res. 2010;468(6):1476–84. https://doi.org/10.1007/s11999-010-1323-7.
25. Gilpin A, Yang Y. Decellularization strategies for regenerative medicine: from processing techniques to applications. Biomed Res Int. 2017;2017:9831534. https://doi.org/10.1155/2017/9831534.
26. Wong ML, Wong JL, Vapniarsky N, Griffiths LG. In vivo xenogeneic scaffold fate is determined by residual antigenicity and extracellular matrix preservation. Biomaterials. 2016;92:1–12. https://doi.org/10.1016/j.biomaterials.2016.03.024.
27. Brown BN, Valentin JE, Stewart-Akers AM, McCabe GP, Badylak SF. Macrophage phenotype and remodeling outcomes in response to biologic scaffolds with and without a cellular component. Biomaterials. 2009;30(8):1482–91. https://doi.org/10.1016/j.biomaterials.2008.11.040.
28. Valentin JE, Badylak JS, McCabe GP, Badylak SF. Extracellular matrix bioscaffolds for orthopaedic applications. A comparative histologic study. J Bone Joint Surg Am. 2006;88(12):2673–86. https://doi.org/10.2106/JBJS.E.01008.
29. Cole BJ, Gomoll AH, Yanke A, Pylawka T, Lewis P, Macgillivray JD, et al. Biocompatibility of a polymer patch for rotator cuff repair. Knee Surg Sports Traumatol Arthrosc. 2007;15(5):632–7. https://doi.org/10.1007/s00167-006-0187-6.
30. Meyer F, Wardale J, Best S, Cameron R, Rushton N, Brooks R. Effects of lactic acid and glycolic acid on human osteoblasts: a way to understand PLGA involvement in PLGA/calcium phosphate composite failure. J Orthop Res. 2012;30(6):864–71. https://doi.org/10.1002/jor.22019.
31. Bajaj P, Schweller RM, Khademhosseini A, West JL, Bashir R. 3D Biofabrication strategies for tissue engineering and regenerative medicine. Annu Rev Biomed Eng. 2014;16:247–76. https://doi.org/10.1146/annurev-bioeng-071813-105155.
32. Derwin KA, Codsi MJ, Milks RA, Baker AR, McCarron JA, Iannotti JP. Rotator cuff repair augmentation in a canine model with use of a woven poly-L-lactide device. J Bone Joint Surg Am. 2009;91(5):1159–71. https://doi.org/10.2106/JBJS.H.00775.
33. Debnath UK, Fairclough JA, Williams RL. Long-term local effects of carbon fibre in the knee. Knee. 2004;11(4):259–64. https://doi.org/10.1016/j.knee.2003.10.003.
34. Smith RDJ, Zargar N, Brown CP, Nagra NS, Dakin SG, Snelling SJB, et al. Characterizing the macro and micro mechanical properties of scaffolds for rotator cuff repair. J Shoulder Elb Surg. 2017;26(11):2038–46. https://doi.org/10.1016/j.jse.2017.06.035.
35. Beitzel K, Chowaniec DM, McCarthy MB, Cote MP, Russell RP, Obopilwe E, et al. Stability of double-row rotator cuff repair is not adversely affected by scaffold interposition between tendon and bone. Am J Sports Med. 2012;40(5):1148–54. https://doi.org/10.1177/0363546512437835.
36. Derwin KA, Baker AR, Spragg RK, Leigh DR, Farhat W, Iannotti JP. Regional variability, processing methods, and biophysical properties of human fascia lata extracellular matrix. J Biomed Mater Res A. 2008;84(2):500–7. https://doi.org/10.1002/jbm.a.31455.
37. Aurora A, Mesiha M, Tan CD, Walker E, Sahoo S, Iannotti JP, et al. Mechanical characterization and biocompatibility of a novel reinforced fascia patch for rotator cuff repair. J Biomed Mater Res A. 2011;99(2):221–30. https://doi.org/10.1002/jbm.a.33179.
38. Sahoo S, Greeson CB, McCarron JA, Milks RA, Aurora A, Walker E, et al. Effect of pretension and suture needle type on mechanical properties of acellular human dermis patches for rotator cuff repair. J Shoulder Elb Surg. 2012;21(10):1413–21. https://doi.org/10.1016/j.jse.2011.10.028.
39. Barber FA, Herbert MA, Coons DA. Tendon augmentation grafts: biomechanical failure loads and failure patterns. Arthroscopy. 2006;22(5):534–8. https://doi.org/10.1016/j.arthro.2005.12.021.
40. Barber FA, Herbert MA, Boothby MH. Ultimate tensile failure loads of a human dermal allograft rotator

cuff augmentation. Arthroscopy. 2008;24(1):20–4. https://doi.org/10.1016/j.arthro.2007.07.013.
41. Omae H, Steinmann SP, Zhao C, Zobitz ME, Wongtriratanachai P, Sperling JW, et al. Biomechanical effect of rotator cuff augmentation with an acellular dermal matrix graft: a cadaver study. Clin Biomech (Bristol, Avon). 2012;27(8):789–92. https://doi.org/10.1016/j.clinbiomech.2012.05.001.
42. McCarron JA, Milks RA, Chen X, Iannotti JP, Derwin KA. Improved time-zero biomechanical properties using poly-L-lactic acid graft augmentation in a cadaveric rotator cuff repair model. J Shoulder Elb Surg. 2010;19(5):688–96. https://doi.org/10.1016/j.jse.2009.12.008.
43. Derwin KA, Baker AR, Spragg RK, Leigh DR, Iannotti JP. Commercial extracellular matrix scaffolds for rotator cuff tendon repair. Biomechanical, biochemical, and cellular properties. J Bone Joint Surg Am. 2006;88(12):2665–72. https://doi.org/10.2106/JBJS.E.01307.
44. Gillespie RJ, Knapik DM, Akkus O. Biologic and synthetic grafts in the reconstruction of large to massive rotator cuff tears. J Am Acad Orthop Surg. 2016;24(12):823–8. https://doi.org/10.5435/JAAOS-D-15-00229.
45. Metcalf MH, Savoie FH III, Kellum B. Surgical technique for xenograft (SIS) augmentation of rotator-cuff repairs. Oper Tech Orthop. 2002;12(3):204–8.
46. Sclamberg SG, Tibone JE, Itamura JM, Kasraeian S. Six-month magnetic resonance imaging follow-up of large and massive rotator cuff repairs reinforced with porcine small intestinal submucosa. J Shoulder Elb Surg. 2004;13(5):538–41. https://doi.org/10.1016/j.jse.2004.03.005.
47. Iannotti JP, Codsi MJ, Kwon YW, Derwin K, Ciccone J, Brems JJ. Porcine small intestine submucosa augmentation of surgical repair of chronic two-tendon rotator cuff tears. A randomized, controlled trial. J Bone Joint Surg Am. 2006;88(6):1238–44. https://doi.org/10.2106/JBJS.E.00524.
48. Soler JA, Gidwani S, Curtis MJ. Early complications from the use of porcine dermal collagen implants (Permacol) as bridging constructs in the repair of massive rotator cuff tears. A report of 4 cases. Acta Orthop Belg. 2007;73(4):432–6.
49. Badhe SP, Lawrence TM, Smith FD, Lunn PG. An assessment of porcine dermal xenograft as an augmentation graft in the treatment of extensive rotator cuff tears. J Shoulder Elb Surg. 2008;17(1 Suppl):35S–9S. https://doi.org/10.1016/j.jse.2007.08.005.
50. Cho CH, Lee SM, Lee YK, Shin HK. Mini-open suture bridge repair with porcine dermal patch augmentation for massive rotator cuff tear: surgical technique and preliminary results. Clin Orthop Surg. 2014;6(3):329–35. https://doi.org/10.4055/cios.2014.6.3.329.
51. Gupta AK, Hug K, Boggess B, Gavigan M, Toth AP. Massive or 2-tendon rotator cuff tears in active patients with minimal glenohumeral arthritis: clinical and radiographic outcomes of reconstruction using dermal tissue matrix xenograft. Am J Sports Med. 2013;41(4):872–9. https://doi.org/10.1177/0363546512475204.
52. Flury M, Rickenbacher D, Jung C, Schneider MM, Endell D, Audige L. Porcine dermis patch augmentation of supraspinatus tendon repairs: a pilot study assessing tendon integrity and shoulder function 2 years after arthroscopic repair in patients aged 60 years or older. Arthroscopy. 2018;34(1):24–37. https://doi.org/10.1016/j.arthro.2017.06.024.
53. Ciampi P, Scotti C, Nonis A, Vitali M, Di Serio C, Peretti GM, et al. The benefit of synthetic versus biological patch augmentation in the repair of posterosuperior massive rotator cuff tears: a 3-year follow-up study. Am J Sports Med. 2014;42(5):1169–75. https://doi.org/10.1177/0363546514525592.
54. Coons DA, Alan BF. Tendon graft substitutes-rotator cuff patches. Sports Med Arthrosc Rev. 2006;14(3):185–90. https://doi.org/10.1097/00132585-200609000-00011.
55. Johnson W, Inamasu J, Yantzer B, Papangelou C, Guiot B. Comparative in vitro biomechanical evaluation of two soft tissue defect products. J Biomed Mater Res B Appl Biomater. 2021;109(1):149–54. https://doi.org/10.1002/jbm.b.30816.
56. Bond JL, Dopirak RM, Higgins J, Burns J, Snyder SJ. Arthroscopic replacement of massive, irreparable rotator cuff tears using a GraftJacket allograft: technique and preliminary results. Arthroscopy. 2008;24(4):403–409 e401. https://doi.org/10.1016/j.arthro.2007.07.033.
57. Wong I, Burns J, Snyder S. Arthroscopic GraftJacket repair of rotator cuff tears. J Shoulder Elb Surg. 2010;19(2 Suppl):104–9. https://doi.org/10.1016/j.jse.2009.12.017.
58. Barber FA, Burns JP, Deutsch A, Labbe MR, Litchfield RB. A prospective, randomized evaluation of acellular human dermal matrix augmentation for arthroscopic rotator cuff repair. Arthroscopy. 2012;28(1):8–15. https://doi.org/10.1016/j.arthro.2011.06.038.
59. Gupta AK, Hug K, Berkoff DJ, Boggess BR, Gavigan M, Malley PC, et al. Dermal tissue allograft for the repair of massive irreparable rotator cuff tears. Am J Sports Med. 2012;40(1):141–7. https://doi.org/10.1177/0363546511422795.
60. Gilot GJ, Alvarez-Pinzon AM, Barcksdale L, Westerdahl D, Krill M, Peck E. Outcome of large to massive rotator cuff tears repaired with and without extracellular matrix augmentation: a prospective comparative study. Arthroscopy. 2015;31(8):1459–65. https://doi.org/10.1016/j.arthro.2015.02.032.
61. Agrawal V. Healing rates for challenging rotator cuff tears utilizing an acellular human dermal reinforcement graft. Int J Shoulder Surg. 2012;6(2):36–44. https://doi.org/10.4103/0973-6042.96992.
62. Encalada-Diaz I, Cole BJ, Macgillivray JD, Ruiz-Suarez M, Kercher JS, Friel NA, et al. Rotator cuff repair augmentation using a novel polycarbonate polyurethane patch: preliminary results at 12 months' follow-up. J Shoulder Elb Surg. 2011;20(5):788–94. https://doi.org/10.1016/j.jse.2010.08.013.

63. Proctor CS. Long-term successful arthroscopic repair of large and massive rotator cuff tears with a functional and degradable reinforcement device. J Shoulder Elb Surg. 2014;23(10):1508–13. https://doi.org/10.1016/j.jse.2014.01.010.
64. Lenart BA, Martens KA, Kearns KA, Gillespie RJ, Zoga AC, Williams GR. Treatment of massive and recurrent rotator cuff tears augmented with a poly-L-lactide graft, a preliminary study. J Shoulder Elb Surg. 2015;24(6):915–21. https://doi.org/10.1016/j.jse.2014.09.044.
65. Audenaert E, Van Nuffel J, Schepens A, Verhelst M, Verdonk R. Reconstruction of massive rotator cuff lesions with a synthetic interposition graft: a prospective study of 41 patients. Knee Surg Sports Traumatol Arthrosc. 2006;14(4):360–4. https://doi.org/10.1007/s00167-005-0689-7.
66. Petrie MJ, Ismaiel AH. Treatment of massive rotator-cuff tears with a polyester ligament (LARS) patch. Acta Orthop Belg. 2013;79(6):620–5.
67. Nada AN, Debnath UK, Robinson DA, Jordan C. Treatment of massive rotator-cuff tears with a polyester ligament (Dacron) augmentation: clinical outcome. J Bone Joint Surg Br. 2010;92(10):1397–402. https://doi.org/10.1302/0301-620X.92B10.24299.
68. Williams JF, Patel SS, Baker DK, Schwertz JM, McGwin G, Ponce BA. Abrasiveness of high-strength sutures used in rotator cuff surgery: are they all the same? J Shoulder Elb Surg. 2016;25(1):142–8. https://doi.org/10.1016/j.jse.2015.07.018.
69. Hirooka A, Yoneda M, Wakaitani S, Isaka Y, Hayashida K, Fukushima S, et al. Augmentation with a Gore-Tex patch for repair of large rotator cuff tears that cannot be sutured. J Orthop Sci. 2002;7(4):451–6. https://doi.org/10.1007/s007760200078.
70. Leuzinger J, Sternberg C, Smolen D, Jakob R. Patch augmentation in rotator cuff repair surgery with elder patients. Z Orthop Unfall. 2016;154(5):504–12. https://doi.org/10.1055/s-0042-106475.
71. Li D, Xia Y. Electrospinning of nanofibers: reinventing the wheel? Adv Mater. 2004;16(14):1151–70.
72. Zhao S, Zhao J, Dong S, Huangfu X, Li B, Yang H, et al. Biological augmentation of rotator cuff repair using bFGF-loaded electrospun poly(lactide-co-glycolide) fibrous membranes. Int J Nanomedicine. 2014;9:2373–85. https://doi.org/10.2147/IJN.S59536.
73. Weng CJ, Lee D, Ho J, Liu SJ. Doxycycline-embedded nanofibrous membranes help promote healing of tendon rupture. Int J Nanomedicine. 2020;15:125–36. https://doi.org/10.2147/IJN.S217697.
74. Yokoya S, Mochizuki Y, Natsu K, Omae H, Nagata Y, Ochi M. Rotator cuff regeneration using a bioabsorbable material with bone marrow-derived mesenchymal stem cells in a rabbit model. Am J Sports Med. 2012;40(6):1259–68. https://doi.org/10.1177/0363546512442343.
75. Hee CK, Dines JS, Dines DM, Roden CM, Wisner-Lynch LA, Turner AS, et al. Augmentation of a rotator cuff suture repair using rhPDGF-BB and a type I bovine collagen matrix in an ovine model. Am J Sports Med. 2011;39(8):1630–9. https://doi.org/10.1177/0363546511404942.
76. Mihata T, Lee TQ, Watanabe C, Fukunishi K, Ohue M, Tsujimura T, et al. Clinical results of arthroscopic superior capsule reconstruction for irreparable rotator cuff tears. Arthroscopy. 2013;29(3):459–70. https://doi.org/10.1016/j.arthro.2012.10.022.
77. Denard PJ, Brady PC, Adams CR, Tokish JM, Burkhart SS. Preliminary results of arthroscopic superior capsule reconstruction with dermal allograft. Arthroscopy. 2018;34(1):93–9. https://doi.org/10.1016/j.arthro.2017.08.265.
78. Kim DM, Shin MJ, Kim H, Park D, Jeon IH, Kholinne E, et al. Comparison between autografts and allografts in superior capsular reconstruction: a systematic review of outcomes. Orthop J Sports Med. 2020;8(3):2325967120904937. https://doi.org/10.1177/2325967120904937.
79. Puskas GJ, Meyer DC, Lebschi JA, Gerber C. Unacceptable failure of hemiarthroplasty combined with biological glenoid resurfacing in the treatment of glenohumeral arthritis in the young. J Shoulder Elb Surg. 2015;24(12):1900–7. https://doi.org/10.1016/j.jse.2015.05.037.
80. Muh SJ, Streit JJ, Shishani Y, Dubrow S, Nowinski RJ, Gobezie R. Biologic resurfacing of the glenoid with humeral head resurfacing for glenohumeral arthritis in the young patient. J Shoulder Elb Surg. 2014;23(8):e185–90. https://doi.org/10.1016/j.jse.2013.11.016.
81. Hammond LC, Lin EC, Harwood DP, Juhan TW, Gochanour E, Klosterman EL, et al. Clinical outcomes of hemiarthroplasty and biological resurfacing in patients aged younger than 50 years. J Shoulder Elb Surg. 2013;22(10):1345–51. https://doi.org/10.1016/j.jse.2013.04.015.
82. Meaike JJ, Patterson DC, Anthony SG, Parsons BO, Cagle PJ. Soft tissue resurfacing for glenohumeral arthritis: a systematic review. Shoulder Elbow. 2020;12(1):3–11. https://doi.org/10.1177/1758573219849606.
83. Pogorzelski J, Hussain ZB, Lebus GF, Fritz EM, Millett PJ. Anterior capsular reconstruction for irreparable subscapularis tears. Arthrosc Tech. 2017;6(4):e951–8. https://doi.org/10.1016/j.eats.2017.03.008.
84. Whelan A, Coady C, Ho-Bun WI. Anterior glenohumeral capsular reconstruction using a human acellular dermal allograft. Arthrosc Tech. 2018;7(11):e1235–41. https://doi.org/10.1016/j.eats.2018.08.002.
85. Matthewson G, Wong IH. Posterior glenohumeral capsular reconstruction with modified McLaughlin for chronic locked posterior dislocation. Arthrosc Tech. 2019;8(12):e1543–50. https://doi.org/10.1016/j.eats.2019.08.004.

3D Printing in Shoulder Surgery

22

Seyda Gokyer, Emre Ergene, Onur Demirak,
and Pinar Yilgor Huri

22.1 Introduction

3D printing technology is one of the major innovations that is affecting the world. It is an additive manufacturing technique based on the principle of accumulation of material layers in a successive manner [1, 2]. Although 3D printing was mostly used for industrial applications in prototype production in the past, it is increasingly more frequently used as a delicate production method for many other products today. 3D printing technology is also revolutionizing the medical treatment options. It has recently been used for the production of functional tissues and organs, as well as in the cancer diagnosis and treatment and in the pharmaceutical field that allowed the development of new biomedical applications [3, 4]. In addition, 3D printers are used in the production of surgical guides and pre-surgical models. The models that are produced by 3D printing are becoming increasingly more common as surgical planning tools for tumor removal operations [5, 6] and in arthroplasty. Development of patient-specific implants is also possible with 3D printing technology [3, 7, 8].

Here, we will examine four major classes of 3D printing techniques and their importance and use in orthopedics, in particular in shoulder surgery.

22.1.1 Stereolithography (SLA)

SLA 3D printers are based on the principle of photopolymerization of a photosensitive resin material that polymerize layer-by-layer to create a 3D structure. Photocuring technology in 3D printing is attractive since it enables production of high-resolution structures, smooth part surfaces, and fast builds. Another advantage is that this technology does not require finishing processes. Briefly in SLA 3D printing, a liquid resin (photopolymer) is cured by light that solidifies in certain areas on the surface of the liquid through a chain reaction initiated by reactive species produced by exposure to light. After polymerization of the first layer, the platform is lowered by a distance determined in the z direction, allowing liquid resin to set to a new layer on top of the previously solidified part. This process continues layer-by-layer until the 3D model is build up. Final 3D model is post-cured to improve polymerization between layers and to reduce surface irregularities. Although SLA 3D printing has many advantages such as enhanced resolution and improved surface smoothness and regularity, its general limitation in medical applications is

S. Gokyer · E. Ergene · O. Demirak · P. Y. Huri (✉)
Department of Biomedical Engineering, Ankara University, Ankara, Turkey
e-mail: phuri@ankara.edu.tr

G. Huri et al. (eds.), *Fundamentals of the Shoulder*, https://doi.org/10.1007/978-3-030-94702-6_22

the scarcity of biocompatible and biodegradable photo-polymerizable liquid resins for use in the development of implants and tissues/organs. A summary of some major applications of SLA technology in the medical field is given in Table 22.1 [1, 3, 4, 15].

22.1.2 Selective Laser Sintering (SLS)

Selective laser sintering (SLS) is a technology based on selective sintering of a metal, ceramic, or polymer powder bed using a high intensity laser beam in accordance with a 3D model. As in the SLA, the new powder bed is mechanically spread with a roller above the previous one to form the layer-by-layer 3D model followed by the sintering of the powder in each layer. This process continues layer-by-layer until the 3D model is build up. The mechanical and structural properties of the 3D product are influenced by the process parameters and material properties such as the size of the powder particles, laser intensity, and bed temperature. A particle size of 10–150 μm is preferred for high-resolution 3D models. The advantage of SLS technology is the ability of use with a wide range of materials that can be used in powdered form. Besides metals, ceramics, and polymer materials, this technology can be used with composite materials such as glass reinforced polymers, metal/polymer composites, and metal/metal composites. Another advantage of SLS 3D printing technology is that it does not require the use of organic solvents during production. Therefore, it provides an advantage in medical applications, some examples of which can be seen in Table 22.1 [1, 3, 4, 16].

22.1.3 Extrusion-Based Techniques (Fused-Deposition Modeling—FDM)

Extrusion-based 3D printers (or FDM 3D printers) are the most widely used and most affordable 3D printers available today. This technology relies on layer-by-layer deposition of the material in the form of fibers that are produced via a micro-nozzle and positioning of the fibers with a computer-controlled three-axis (*x*–*y*–*z*) motion platform either of the printing heads or the collecting stage. The material loaded in metallic or plastic syringes is deposited either by pneumatic or by mechanical means (piston and screw). The extrusion-based 3D printers have the potential to provide a variety of innovative applications. Therefore, it provides an advantage in the medical applications. The resolution of extrusion-based 3D printers is in the order of 200 μm which is much cruder than laser and inkjet-based 3D printing. However, the production speed is significantly higher and anatomically shaped structures can be easily produced. A list of some major applications is provided in Table 22.1 [1, 3, 4, 17].

22.1.4 Droplet-Based Techniques

Droplet-based 3D printing technology relies on deposition of the liquid material in a droplet form instead of a continuous flow. This approach uses a non-contact strategy where the material itself or cells encapsulated in droplets are printed and patterned layer-by-layer on a substrate. Solidification of the material can occur via cooling, chemical cross-linking, or solvent evaporation. Droplet-based 3D printing technology has a high resolution as compared to laser-based systems. Conversely, printing material for droplet-based 3D printing is limited and with high cost. The physical properties of printing materials such as the viscosity, surface tension, and inertia are affecting the print quality to a great extent. Droplet-based 3D printing systems can be categorized based on the mechanism to produce droplets, as thermal and piezoelectric. Some orthopedic applications of the technique are listed in Table 22.1 [1, 3, 4, 18].

Table 22.1 Some major orthopedic applications of 3D printing

3D Printing method	Material used	Application	Product	Ref.
SLA	MED610, a commercial biocompatible resin-based material	Osteosarcoma resection. Operation time was shortened by pre-assessment of variable location of multiple tumors that are in close proximity to the surrounding tissues	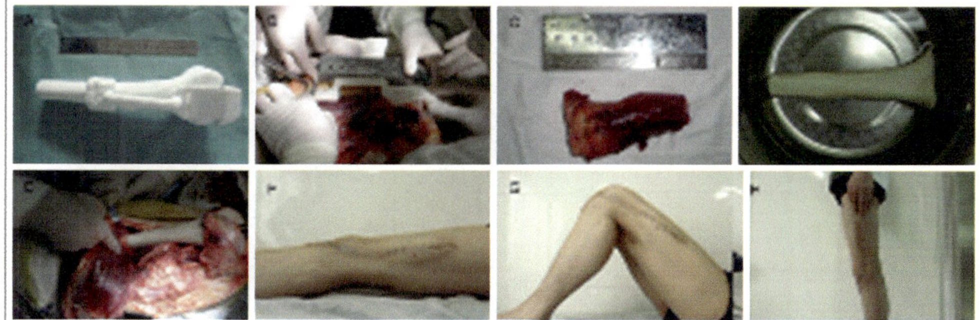	[9]
SLA and SLS	Photosensitive resin materials for proximal tibia block model, Ti6Al4V powder for patient-specific implant	3D printer system used in this case (1) to create proximal tibia block model with SLA system and (2) to manufacture the patient-specific implant with EBM (type of SLS 3D printer) system	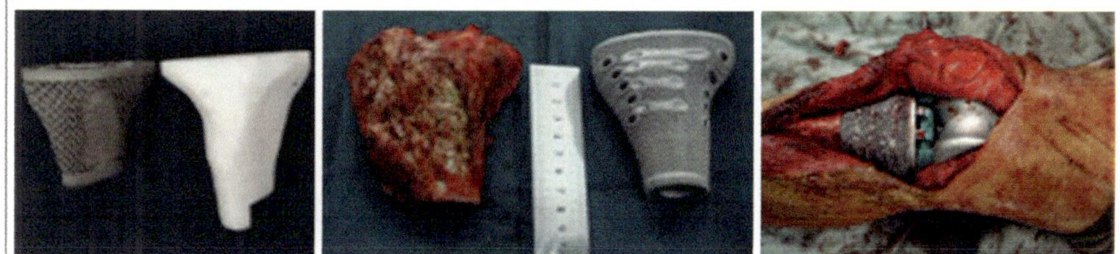	[10]
SLS	Ti–6Al–4 V powder	Sacral osteosarcoma surgically was treated with hemisacrectomy and sacral reconstruction using a 3D-printed implant	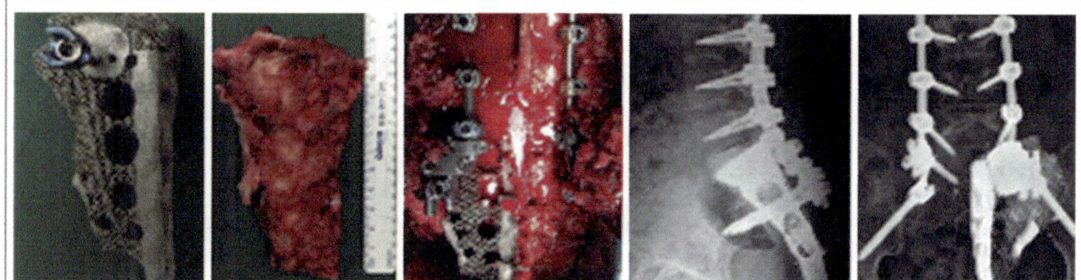	[11]

(continued)

Table 22.1 (continued)

3D Printing method	Material used	Application	Product	Ref.
İnkjet and SLS	Gypsum powder for anatomic biomodel, Ti6Al4V powder for patient-specific implant	3D printer system used in this case (1) to create anatomic biomodel with FDM system and (2) to manufacture the patient-specific implant with DMLS system	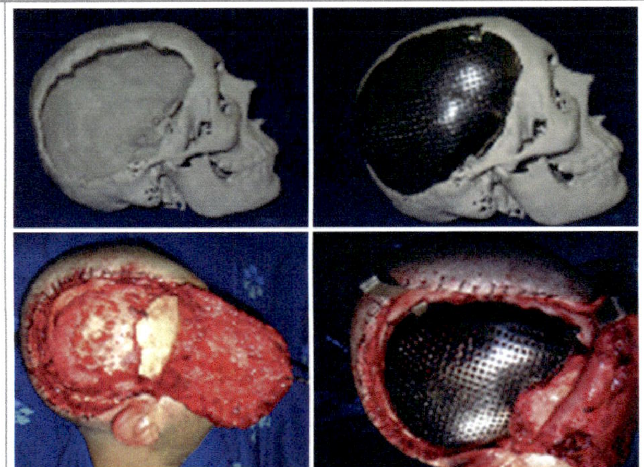	[12]
SLS	Ti–6Al–4 V powder, particle size 1–10 μm	In order to mimic the stiffness of bone to reduce stress shielding, titanium implants with different mechanical properties have been produced	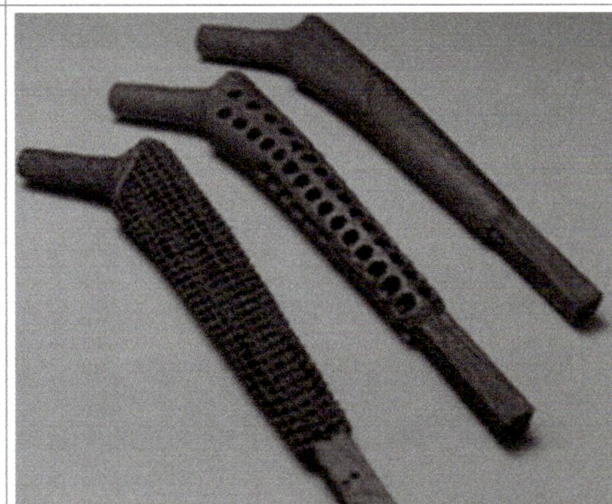	[13]

FDM	Poly (lactic acid) PLA filament	A personalized navigation template for pedicle screw fixation in the upper cervical spine	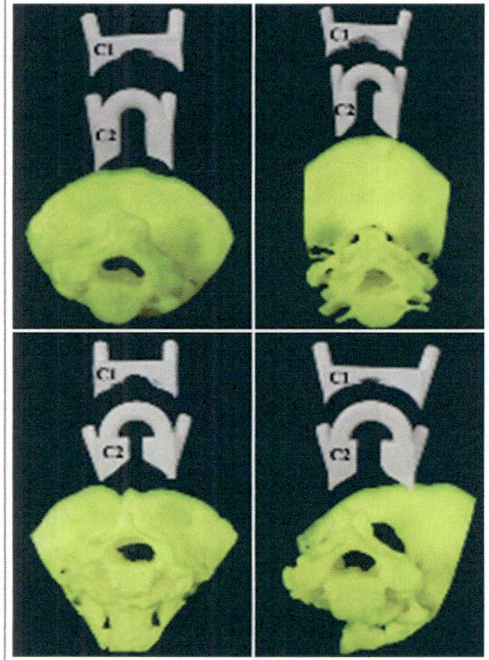 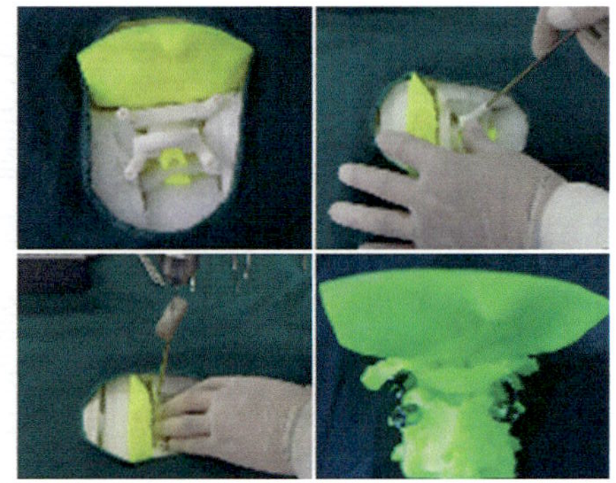	[14]

SLA stereolithography, *SLS* selective laser sintering, *FDM* fused-deposition modeling

22.2 3D Printing for Shoulder Arthroplasty

3D printing technology has greatly improved the arthroplasty procedures since implants can be produced with personalized size, structure, and features. This personalization in the implants is expected to become a golden-standard application since it reduces the risk of incompatibilities with the anatomic structures as well as reducing the operation time. Some examples of the use of 3D printed personalized implants for shoulder arthroplasty cases are discussed below.

In a study, researchers aimed to investigate the practicality of the use of 3D printed prosthesis for benign fibrous histiocytoma (BFH) of scapula [19]. A patient who has magnetic resonance imaging (MRI) results that indicate abnormal signals in the left shoulder was treated with 3D printing technology due to the fact that total scapula resection may lead to poor postoperative function and difficulties in exact reconstruction. CT images were used to obtain 3D CAD model and porous, polyetheretherketone (PEEK) scapula prosthesis generated (Fig. 22.1). Satisfactory results were obtained after implantation and observation of the implantation site with X-ray scans. CT examination indicated lack of autogenous scapula and denser shadow in scapula region with a shoulder activity of 120°. Therefore, 3D printed PEEK scapula was concluded to be an easy-to-prepare, cheap, and biocompatible alternative in shoulder arthroplasty.

A case report focused on 3D printed patient-specific glenoid implants for two reverse shoulder arthroplasty cases [20]. The first case presented rotator cuff tear arthropathy of the shoulder caused by deformed glenoid. The sec-

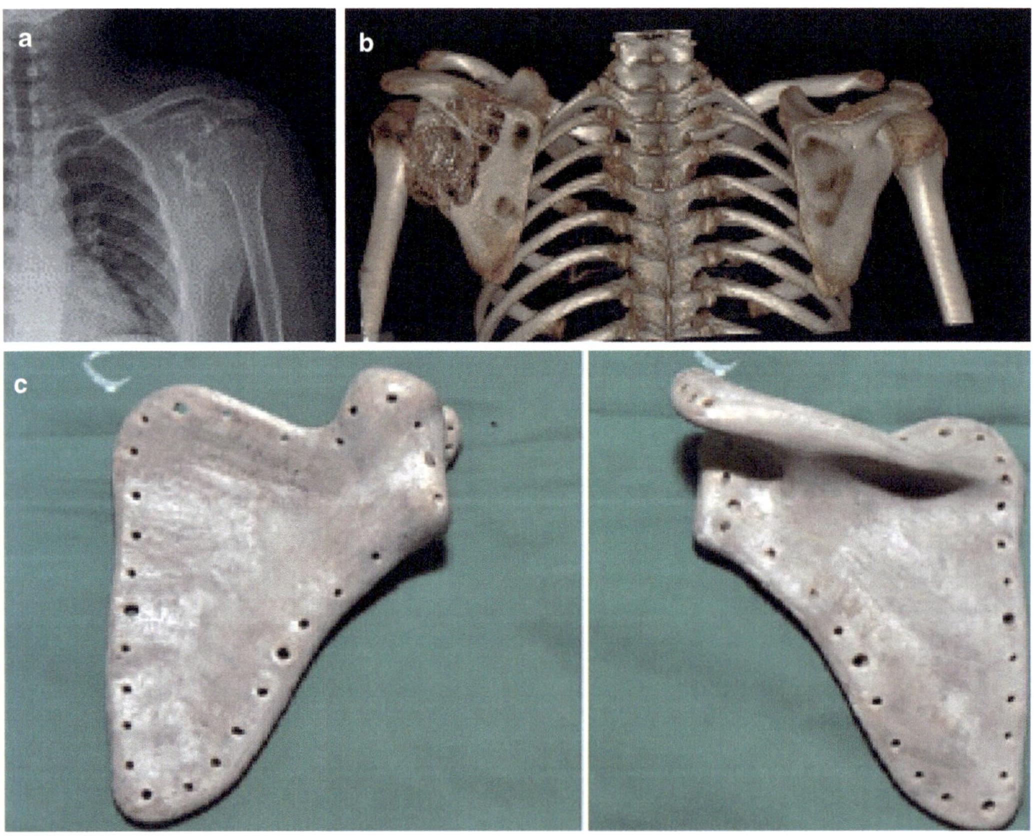

Fig. 22.1 Preoperative data and 3D printed PEEK scapula: (**a**) X-ray thin slice of left shoulder joint, (**b**) 3D reconstructed shoulder joints, (**c**) 3D printed PEEK prosthesis of scapula [19]

ond case is a patient with deformed glenoid after hemiarthroplasty of the shoulder. Instead of using standard prosthesis which promised unsatisfactory results, 3D printed custom-made titanium porous glenoid components were used to exactly fit into deformed area. Both patients were treated successfully with custom-made implants combined with non-custom glenosphere and humeral stem. However, the authors stated that there were some risks which did not occur in these cases such as changing deformity in glenoid from the time CT was performed to time of surgery and metal artifacts in implantation sites. Also they suggested that cost efficiency should be studied.

In another study emphasized on reverse shoulder arthroplasty, 3D printed glenoid prosthesis and humerus prosthesis were preferred due to high complication rate [21]. Seven patients who had proximal humerus tumors were treated with 3D printed baseplate assisted 3D printed glenoid prosthesis implantation. CT and MRI data were utilized to construct 3D models and Nylon powder was used to print the baseplate, where the prosthesis was produced from titanium alloy with 60% porosity. Surgical procedure included (1) removal of the tumor, (2) exposure of glenoid and fixation of the baseplate, (3) removal of baseplate after glenoid prosthesis location and position confirmation, (4) implantation of 3D printed glenoid prosthesis and humerus prosthesis, and (5) examination of shoulder joint. 1.5 years follow-up revealed loss of one of the patients due to pulmonary metastasis. Rest of the patients survived without disease. Three of the patients had lower range of motion, the mean Musculoskeletal Tumor Society functional score determined as 85.7%, the mean Toronto Extremity Salvage Score was 90.0%.

3D printed customized titanium alloy (Ti6Al4V) was used for total shoulder arthroplasty [22]. The CT scan data of the patient was imported to MIMICS software and a 3D reconstructed model was created (Fig. 22.2). The model was then rendered in STL format and printed with an Electron Beam 3D printer. The pore size was preferred as 70 μm, with a porosity of 60%. According to American Shoulder and Elbow Surgeons (ASES) scoring, the initial score was 36. After surgery, this score was increased to 71.4 after 1 month, to 85.8 after 24 months. This "superior" improvement has proved that 3D printed prosthesis has satisfactory short-term effects. However, long-term study must be supported to learn details about durability of customized prosthesis under longtime compression and stress and for validation of mechanical properties. It is also stated that, although porous structure has the importance for ingrowth of soft tissue, metal remnants in these pores may cause loosening of the prosthesis. Thus, pre-steps of implantation such as mechanical cleaning and disinfection techniques must be studied in detail.

22.3 3D Printing for Surgical Planning of the Shoulder

The benefit of using 3D printed models for presurgical planning of shoulder arthroplasty was demonstrated in a study where 11 3D printed glenoid models of nine patients were produced [23]. Accuracy of the models were compared with the CT images. Different 3D printing techniques and different material types were used for each patient, accordingly. 3D printers such as EOS-P, SLA 7000, Form2, ProJet 7000 HD, and ProJet 660 were preferred. The utilized materials consist of nylon powder, translucent resin, white resin, clean resin, and plaster. Clarity for CT image, humerus-subtracted volume rendering, and printed model was compared and scored ranging from 1 (not preferred) to 5 (highly preferred). Scoring was made also for the complexity and usefulness of the printed model. The results showed that CT based 3D printing can be utilized when complexity increases. It was also stated that clarity values were higher for humerus-subtracted-volume rendered evaluation compared to CT image review and better with 3D printed models for higher morphological complexity. Conversely, it was noted that the surgical planning can be made directly through the software. As a result, 3D printed glenoid models were found to be useful when higher clarity is desired and other surgical planning tools are impractical due to high complexity.

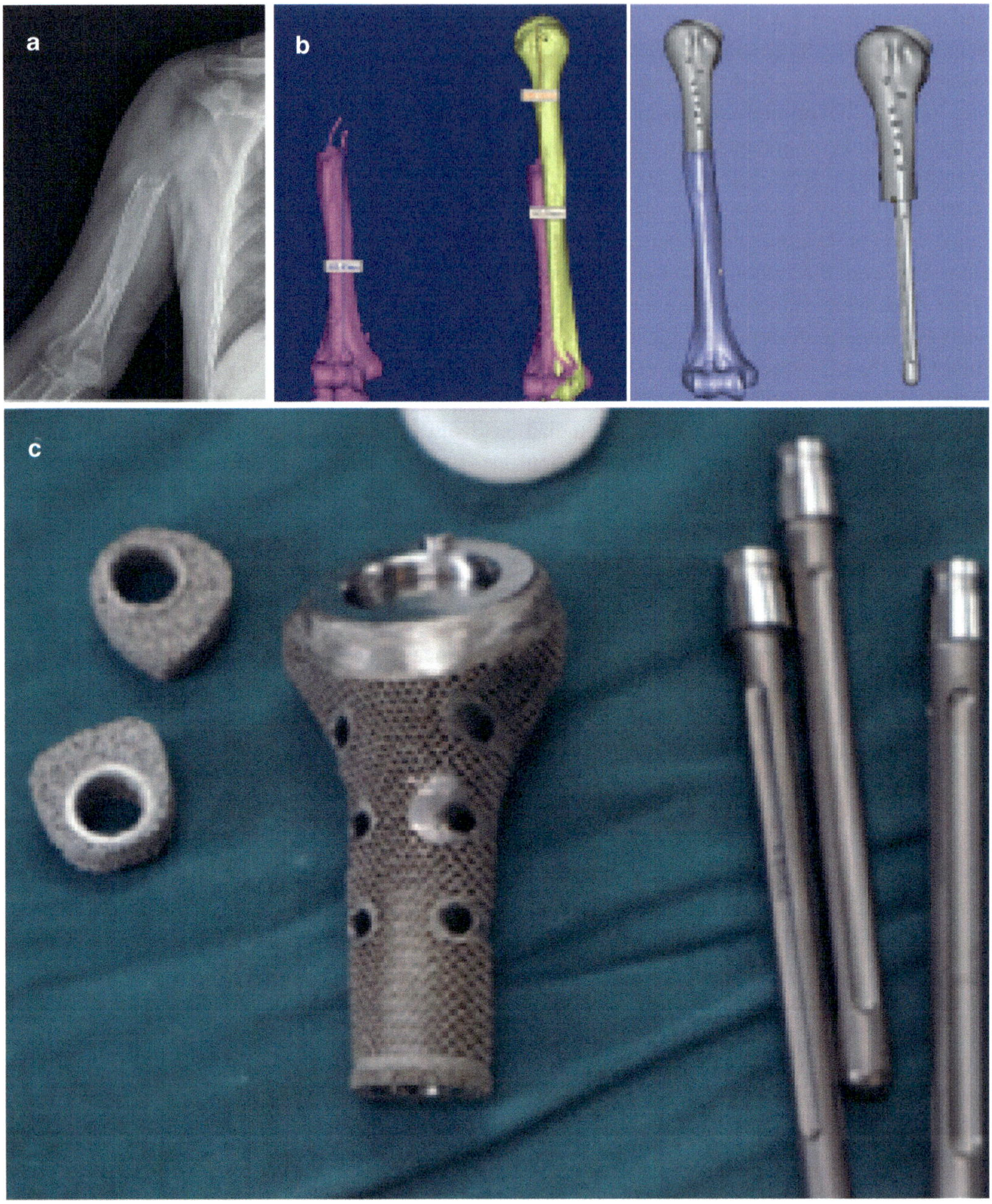

Fig. 22.2 Preoperative data and design and manufacture of prosthesis: (**a**) bone defect observation, (**b**) prosthesis design, (**c**) the shoulder parts manufactured with EBM technique [22]

Another study focused on comparison between using 3D printing technology and typical thin-layer CT scan for the treatment of humeral fractures [24]. This study was one of the studies with the highest number of cases: 34 patients in the test group and 32 patients in the control group. The test group was established to observe the benefits of 3D printed technology in diagnosis and surgical planning. To build the humerus fracture model acquired CT and 3D processing software (MIMICS) were used. Model creating process can be summarized as: (1) positioning

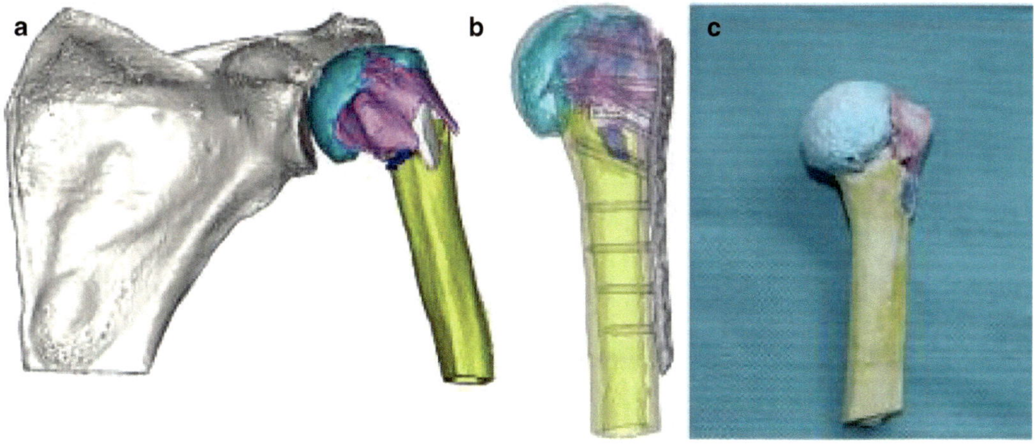

Fig. 22.3 Humerus model creating process. (**a**, **b**) Segmentation and 3D reconstruction of proximal humerus (anterior), (**c**) printed humerus model (anterior) [24]

and thresholding to define proximal humerus area, (2) region growing to separate hard and soft tissues, (3) post-processing (noise reduction, fracture smoothing), (4) segmentation of humerus fracture which will be printed. In the control group, only CT scan was utilized for the operational planning (Fig. 22.3). Results according to comparison parameters were not unexpected. First of all, 3D printed models provided omnidirectional display unlike thin slice CT scan. Lower surgery length, less blood loss, and fluoroscopy have been observed. Using 3D printing technology for the treatment of proximal humerus fracture was proved to be a better choice than the traditional methods despite the fact that there was no significant change in the healing time of the groups. The authors also stated that these models may be used as medical practice and teaching.

22.4 3D Printed Anatomical and Surgical Models, Training Material, Pre- and Post-Surgical Models

3D printing techniques, which have been increasingly used in engineering and health sciences in recent years, have made it possible to create fast and cheap prototypes with the aid of computer-aided design (CAD) [25]. In addition to the increasing use of 3D printing technology in non-medical fields, it has widely used in orthopedics, spinal surgery, craniofacial surgery, neurosurgery, and cardiac surgery in medicine [26].

Medical professionals frequently use two-dimensional X-ray images, computed tomography (CT), and magnetic resonance (MR) scans to investigate about pathologies. Although three-dimensional CT and MR technologies have developed in recent years, 3D printed objects are much more useful for studying complex cases and teaching students [27]. Furthermore, some complex surgical procedures require guidance avert from damage important parts of the body and achieve an esthetic result. In some cases, anatomical defects may also require individual prostheses to better repair the damage [28].

The need for improved imaging and surgical results has increased the interest in 3D printed anatomical and surgical models. Additionally, this technology has opened the way to produce the prototype models needed in medical education quickly and inexpensively [29].

22.4.1 3D Printed Anatomical Models

In cases that require maneuver around sensitive nerves, cerebral structures, bones, and vessels, 2D radiographic images may be insufficient. 3D printing technology makes it easier for surgeons to think about the path to follow before the operation, ana-

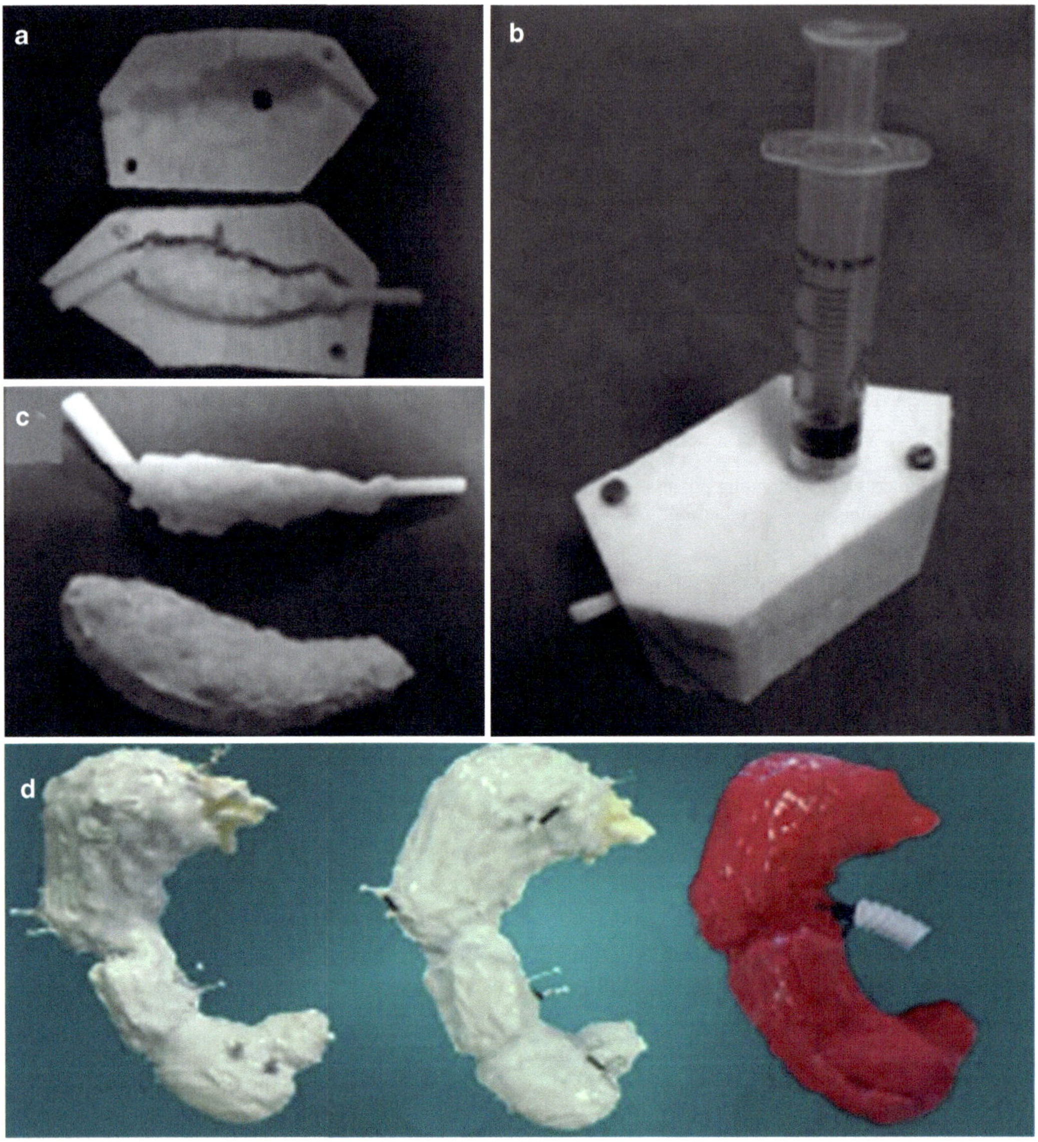

Fig. 22.4 Producing patient-specific abdominal cavity with 3D printer. (**a**) 3D printed mold parts. (**b**) Silicone was injected into the mold with a syringe. (**c**, **d**) The finished silicone model [31]

lyze the patient's anatomy, and predict the technical difficulties they may encounter. This may reduce the operation time and trauma to the patient [30]. In one of the case studies, Condino et al. printed a patient-specific abdominal cavity molds from data obtained through computerized imaging. In this way, the researchers succeeded in planning the operation with great accuracy [31] (Fig. 22.4).

3D printed anatomical models are also used in the surgical planning of complex congenital heart malformations, percutaneous valve implantation, aortic, and cranial aneurysm repairs. These studies have reported that these models are beneficial in selecting patients to undergo endovascular procedures compared to standard medical imaging [26, 32, 33] (Fig. 22.5).

Complex hip replacement studies, mentioned the benefits of making a preoperative plan on 3D printed models, showed that planning using 3D models before the operation prevents the aberra-

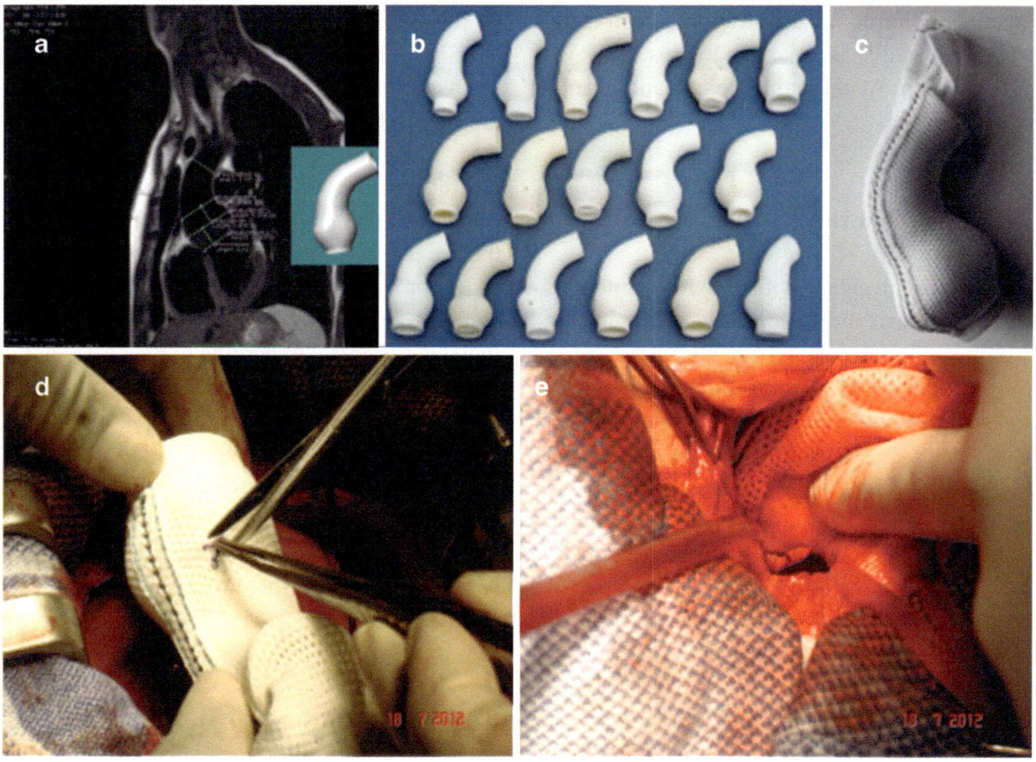

Fig. 22.5 An example of the CAD aided Marfan aortic root. (**a**, **b**) Aortic root and ascending aorta models of some patients receiving personalized external aortic root support. (**c**) Personalized external support mesh which is produced from a medical-grade polymer fabric. (**d**, **e**) The stage of placing the support around the aorta by surgeons [33]

tion of the joint placement and reduces the operation time [34, 35].

3D anatomical models printed for cranial fractures are used to shape the implant before the operation and to plan a better fit in the defect region [36]. 3D printed structures are also used in maxillofacial surgery. Digitally mirroring the healthy mandible to the defect side helps guide printing the patient-specific geometry [37]. In a study investigating the potential of 3D printing technology instead of the thin-layer CT scan, which is frequently used in the treatment of proximal humerus fractures in older people, there was no significant difference in the results compared to the control group that only CT scan was performed. Conversely, less surgery time and significantly less blood loss were observed [24].

It is also common to use 3D printed models for medical education and to express patients about their condition. In a study to examine and understand the deformation closely, clubfoot was printed in 3D, four times larger than its normal size [38].

3D printed models are also used to improve the learning experience in medical education. Usually, medical students observe a tumor in environments that are difficult to visualize, just like a textbook or a CT slice. The kidney [39], lung, liver [40], and bone tumors [41] that are printed in 3D increase the learning experience with the vessels and cavities around them. There may also be soft tissue and tumor components printed using different materials for situations that require more tactile experience. For example, soft breast with rigid mass tissue was printed by Jones et al. to teach palpation methods [40, 42].

Advances in 3D printing technology are available to produce a realistic cerebral artery aneurysm model. In a study by Ryan et al., when the skull, brain, and vasculature model were used together, the true aneurysm clipping procedure was successfully mimicked. In this way, it is aimed to improve the surgical experience through this model [43].

22.4.2 3D Printed Surgical Models

Besides the advantages of using 3D printed anatomical models for preoperative planning and training, this technology can also assist in performing surgical procedures through drill guides and templates. Unlike anatomical models, these are tools that facilitate the operation technically but are removed after the operation [32]. These tools are frequently used in the operations of hard tissues such as bones. In a study, it was successfully applied in the form of intraoperative jig in a specified area in maxillofacial surgery. Planned osteotomies and bone movements were adapted to patient anatomy [37].

Birnbaum et al. have reported that they used polycarbonate templates to place pedicle screws at a point determined by 3D images taken before the operation for use in spinal surgery [44]. In another study, the model created by researchers using 3D CT reconstruction images was used to guide where to collect cartilage plugs for knee mosaic arthroplasty. Then, the precision tools that fit the knee contour were printed in 3D, inspired by this model. This method has been reported to facilitate surgical planning and minimize inadequate defect closure [29, 45].

In a recent study conducted by Zhou et al., 3D printing patient-specific instrumentation model was used to increase the operative accuracy and safety of artificial knee arthroplasty. In this study, patients' total blood loss, latent blood loss, and hemoglobin (Hb) values were compared with the control group after the operation. They were concluded that all values decreased statistically significant. Furthermore, the researchers reported that 3D printing patient-specific instrumentation model can effectively simulate the lower limb coronal force line. Random interviews with the patients after the operation revealed that the knees recovery of patients had well [46] (Fig. 22.6).

In another study, using reverse engineering and 3D printing technology, a special surgical template was designed for the placement of thoracic pedicle screws. The researchers have created a 3D model by defining the optimum insertion direction, length, and diameter of the screws on a patient-specific CAD model during the pre-planning stage (Fig. 22.1). After that, this model produced with 3D printing and used as a template to prevent misplacement between the thoracic vertebrae. According to the research, the duration for one screw placement was 18.75 min per vertebrae in the traditional operation (without using the template); however, this time decreased to 6.25 min in the innovative operation (using the template). In parallel with these findings, it was reported that the number of X-ray shots was decreased from 20.17 to 3.67 per vertebrae when the templates were used [47].

22.5 Bioprinting of Tissues

Tissue engineering is an interdisciplinary field of study that develops biological systems designed to fulfill or support the functions of tissues and organs, using the principles of engineering, medicine, and life sciences. In tissue engineering approach, tissue pieces are obtained from the patient by biopsy or surgery [1], cells are isolated and multiplied in vitro [2], cells are encapsulated or seeded on scaffolds [3], tissue is generated and functionalized with the use of bioreactors and bioactive molecules [4, 5], the tissue produced is transferred to the defect area of the patient [6] (Fig. 22.7). In tissue engineering applications, the selection of cell source is very important and ideal approach is using primary cells that are isolated from the healthy area of the target tissue. However, isolation and proliferation of primary cells have same limitations. The use of stem cells as a source of cells provides various advantages. Stem cells have the potential to regenerate themselves and differentiate into different cell types. Scaffolds are required to keep the cells localized and to provide target tissue formation in an organized manner. The scaffold materials should be biocompatible and biodegradable in proportion

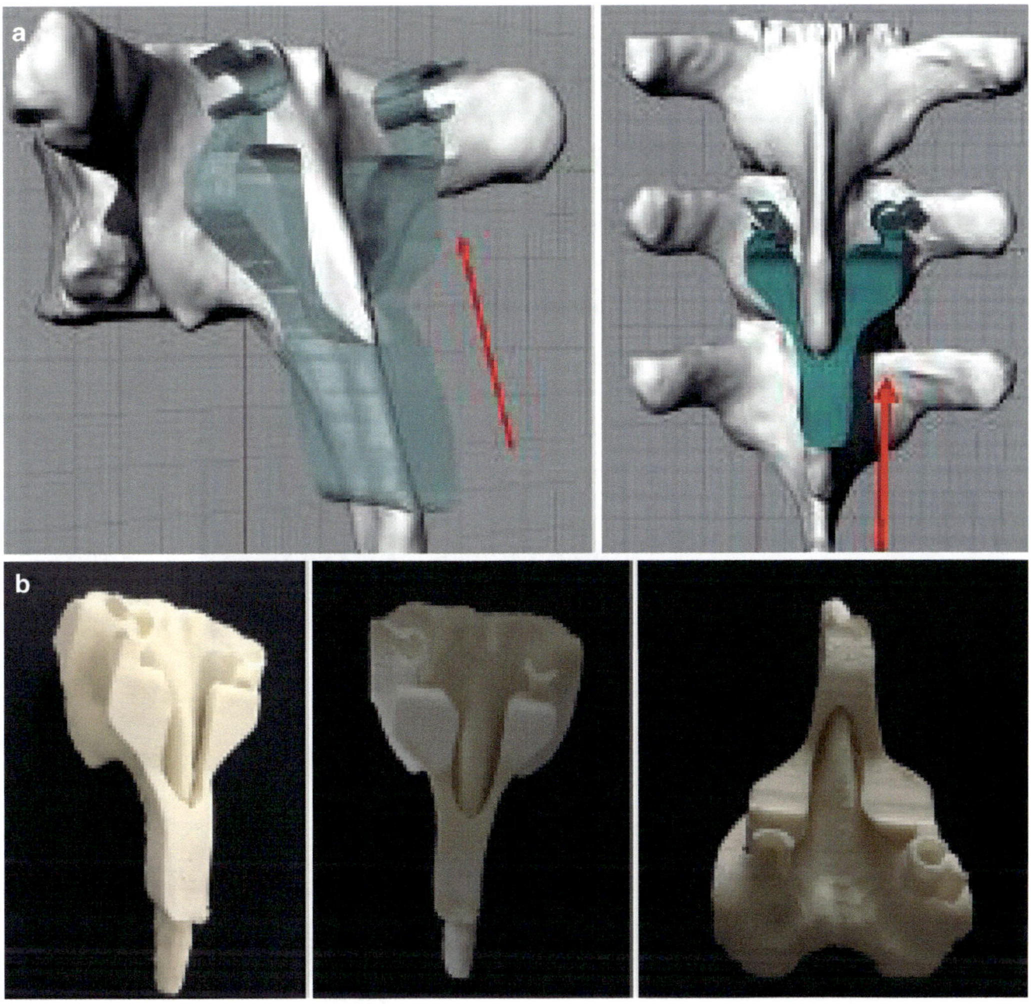

Fig. 22.6 Ad hoc surgical template design using CAD model. (**a**) Coupled with the thoracic vertebra model; (**b**) 3D printed model using the fused filament fabrication (FFF) technique [47]

to the rate at which the target tissue heals. Additionally, the mechanical properties of scaffolds should be suitable to the target region and able to mimic the extracellular matrix (ECM) of the target tissue structurally. Scaffolds that can mimic the 3D porous ECM can be produced by various methods. 3D printing technology provides significant advantages in the production of scaffold with an anatomic structure that can be designed and controlled compared to other methods. Therefore, it is possible to successfully mimic the ECM with 3D printers [49–51].

3D printing technology enables new developments in the production of surgical guides and patient-specific implants, as well as generation of complex tissues with 3D bioprinting. 3D bioprinting technology enables the production of complex biological structures with the additive manufacturing of cells, biomaterials, and biomolecules. Nowadays, although the restrictions on the functionality of 3D printed scaffold have not yet been overcome, 3D printed scaffolds in the future have the potential to prevent deaths from organ failure worldwide [3, 4, 8]. One of the examples where the target tissue is mimicked with bioprinting includes the 3D ear model developed by Kang et al. [52]. 3D printing of the ear model was fabricated by the integrated tissue-

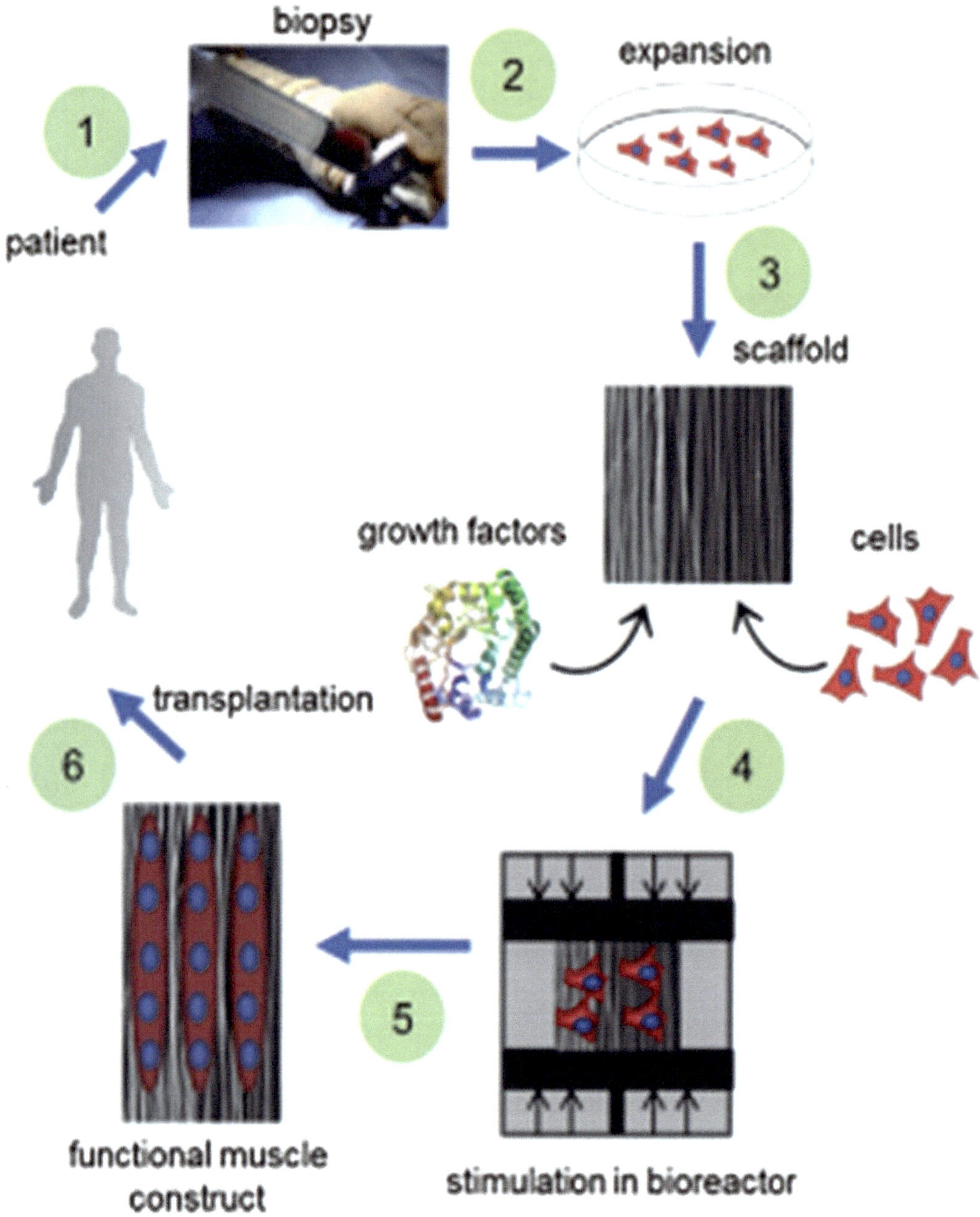

Fig. 22.7 General overview of the tissue engineering approach to produce viable and functional tissues and organs in the laboratory [48]

organ printer (ITOP) device developed at the Wake Forest University. Researchers used PCL for the structural stability and integrity of the 3D model, while the bioink made up of a natural polymer was used for the 3D bioprinting of two different cell lines [52]. Widespread studies are the production of functional scaffold with 3D printing in musculoskeletal tissue engineering applications. Osteoplug™ and Osteomesh™ are used for covering bone holes or repair various type of bone fractures in clinical applications. A 13 year-old patient was implanted with a PCL-

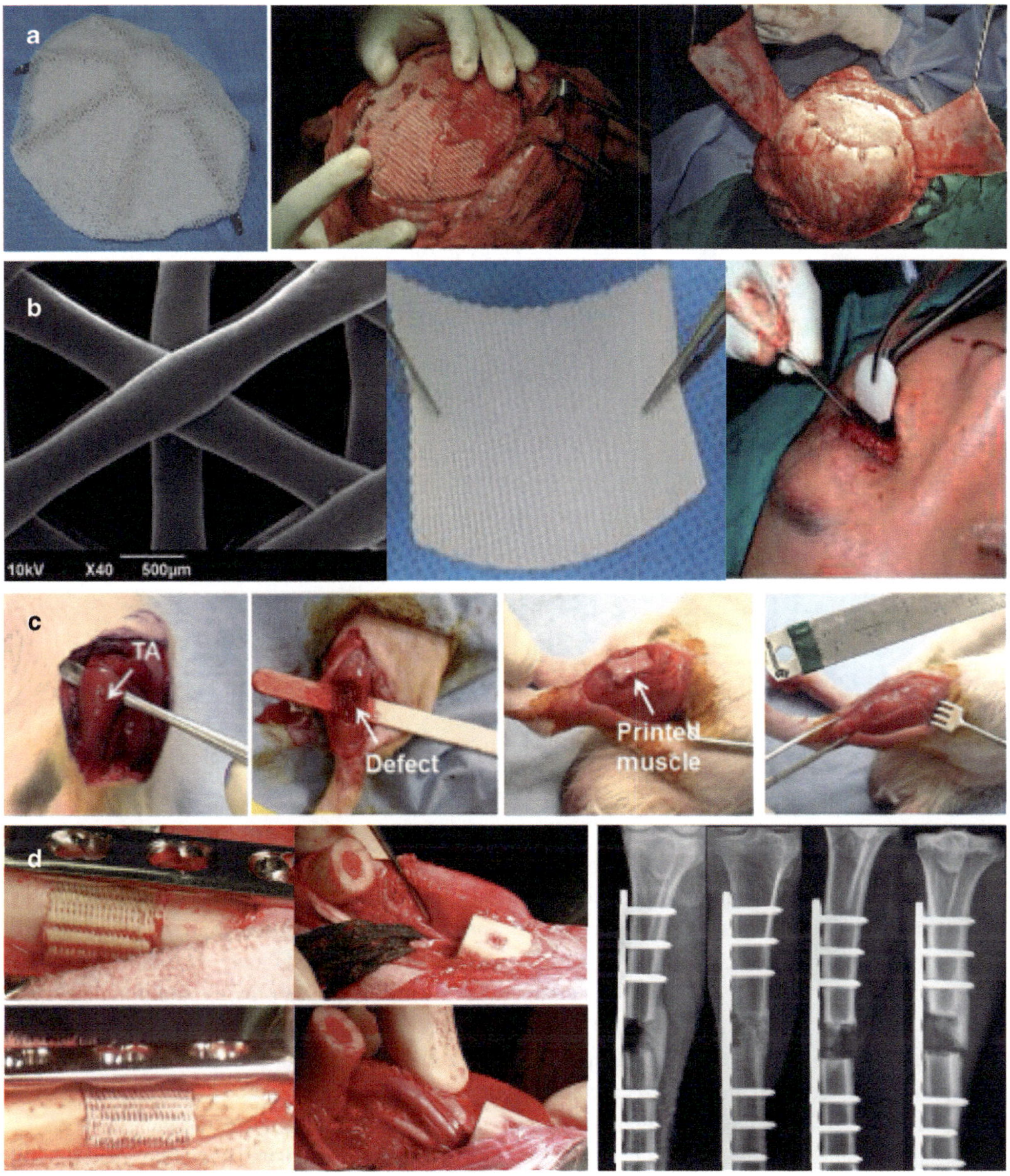

Fig. 22.8 (**a**) The Osteomesh implantation in cranioplasty surgery [53], (**b**) the Osteomesh implantation in orbital fracture repair [54]. (**c**) 3D printed muscle construct [55], (**d**) polyester composite scaffold [56]

based 3D printed implant, which was developed by Osteopore Int. (Singapore) for the cranioplasty surgery [53] (Fig. 22.8a). In another clinical application, A total of 20 patients were implanted with the same implant (Osteomesh™) prepared for orbital fracture repair. The CT scans performed 1.5 years after the surgery revealed neo-bone formation at the implant site [54] (Fig. 22.8b). There are also in vivo studies with successful results in literature. 3D printed muscle construct was fabricated by mimicking the 3D fiber structure with ITOP by Kim et al. [55]. The 3D muscle constructs were recovery in a rat model of tibialis anterior (TA) muscle defect at 8 weeks of post-implantation [55] (Fig. 22.8c). In another study, they investigated effect of polyes-

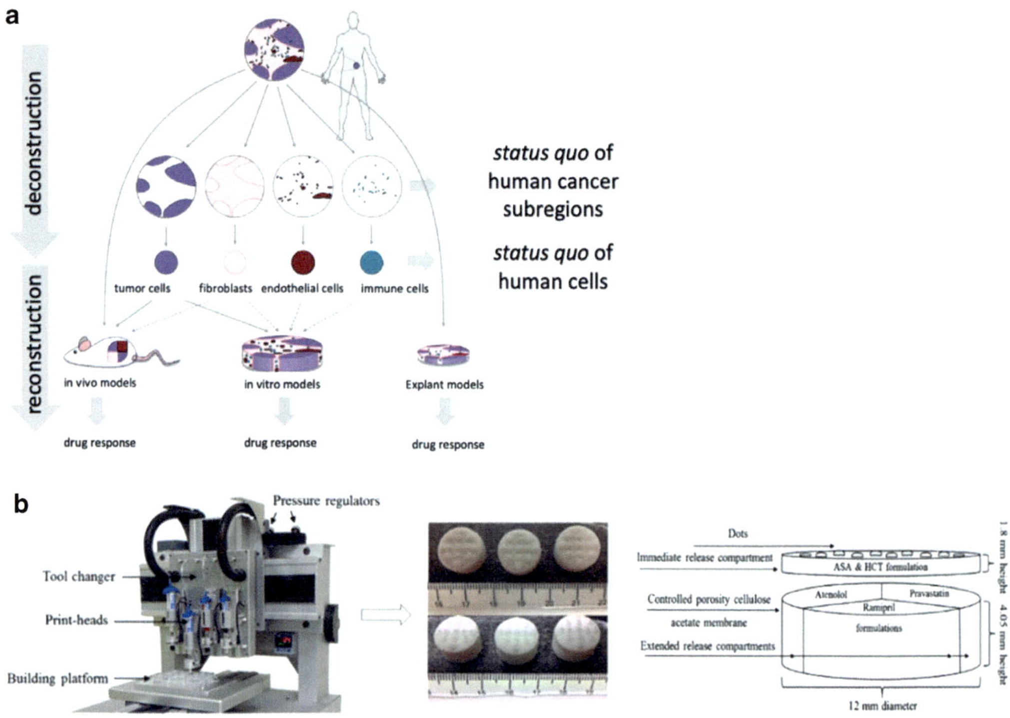

Fig. 22.9 (**a**) 3D in vitro models for the carcinoma. (**b**) The workflow for the production of 3D printed tablet-form pharmaceutics

ters and ceramics scaffold and polyester composite scaffold containing bioactive materials on new bone formation [56] (Fig. 22.8d).

Bioprinters are also used for different purposes in the medical field, including 3D cancer models and treatment methods developed on them [3, 4, 8]. Unger et al. developed carcinomas 3D in vitro model, then they investigated human tumor–stroma interactions and drug responses (Fig. 22.9a) [57]. In an application for the drug delivery systems, a 3D printed tablet was prepared to contain five seperate compartments with independently controlled, well-defined release profiles (Fig. 22.9b) [58].

22.6 Conclusion

The interest in using 3D printing technologies for the purpose of both clinical applications and training medical students and patients is gradually increasing, and this method offers potential promise for future surgical applications. Thanks to the advances in 3D imaging techniques, different 3D printing technologies, and printable material technology, it has now become possible to create 3D anatomical models and various instruments for patient-specific surgical use.

The use of imaging techniques, such as CT and MRI, used to visualize anatomy and pathology, has revolutionized surgical procedures over the past decades. 3D printing provides a 3D representation that surgical teams can contact physically before surgery and examine without time pressure. The literature we reviewed reports that this technology simplifies the planning and decision-making process of surgeons, as well as the reduced duration of operation and exposure X-rays. Many studies state that innovative operations using 3D models result in less blood loss than traditional ones. In cases where joint placement or screw placement is required, the 3D printed templates and guides used have enabled more accurate and minimally invasive surgical operations.

Although 3D printing technology is still in its early stages, it is already in use in surgical operations. Conversely, the combination of tissue engineering and 3D printing technologies has contributed to the development of bioprinting technology in recent years. In the future, based on 3D CAD design, anatomical models that will be printed with cells may enable the printing of individual-specific organs that can be transplanted.

References

1. Provaggi E, Kalaskar DM. 3D printing families: laser, powder, nozzle based techniques. In: Kalaskar DM, editor. 3D Printing in medicine. Cambridge: Woodhead; 2017. p. 21–42.
2. Skardal A. Bioprinting essentials of cell and protein viability. In: Atala A, Yoo JJ, editors. Essentials of 3D biofabrication translation. Boston: Academic Press; 2015. p. 1–17.
3. Woodfield T, Lim K, Morouço P, Levato R, Malda J, Melchels F. Biofabrication in tissue engineering. In: Ducheyne P, editor. Comprehensive biomaterials II. Amsterdam: Elsevier; 2017. p. 236–66.
4. Moroni L, Burdick JA, Highley C, Lee SJ, Morimoto Y, Takeuchi S, et al. Biofabrication strategies for 3D in vitro models and regenerative medicine. Nat Rev Mater. 2018;3:21–37. https://doi.org/10.1038/s41578-018-0006-y.
5. Rath SN, Sankar S. 3D printers for surgical practice. In: Deepak KM, editor. 3D printing in medicine. Cambridge: Woodhead; 2017. p. 139–54.
6. Yilmaz A, Badria AF, Huri PY, Huri G. 3D-printed surgical guides. Ann Joint. 2019;4:16.
7. Akilbekova D, Mektepbayeva D. Patient specific in situ 3D printing. 3D Print Med. 2017:91–113.
8. Murphy SV, Atala A. 3D bioprinting of tissues and organs. Nat Biotechnol. 2014;32(8):773–85. https://doi.org/10.1038/nbt.2958.
9. Ma L, Zhou Y, Zhu Y, Lin Z, Wang Y, Zhang Y, et al. 3D-printed guiding templates for improved osteosarcoma resection. Sci Rep. 2016;6:2–10. https://doi.org/10.1038/srep23335.
10. Luo W, Huang L, Liu H, Qu W, Zhao X, Wang C, et al. Customized knee prosthesis in treatment of giant cell tumors of the proximal tibia: application of 3-dimensional printing technology in surgical design. Med Sci Monit. 2017;23:1691–700.
11. Kim D, Lim JY, Shim KW, Han JW, Yi S, Yoon DH, et al. Sacral reconstruction with a 3D-printed implant after hemisacrectomy in a patient with sacral osteosarcoma: 1-year follow-up result. Yonsei Med J. 2017;58(2):453–7.
12. Jardini AL, Larosa MA, Filho RM, Zavaglia CADC, Bernardes LF, Lambert CS, et al. Cranial reconstruction: 3D biomodel and custom-built implant created using additive manufacturing. J Craniomaxillofac Surg. 2014;42(8):1877–84.
13. Harrysson OLA, Cansizoglu O, Marcellin-Little DJ, Cormier DR, West HA. Direct metal fabrication of titanium implants with tailored materials and mechanical properties using electron beam melting technology. Mater Sci Eng C. 2008;28(3):366–73.
14. Guo F, Dai J, Zhang J, Ma Y, Zhu G, Shen J, et al. Individualized 3D printing navigation template for pedicle screw fixation in upper cervical spine. PLoS One. 2017;12(2):1–12.
15. Catros S, Keriquel V, Fricain JC, Guillemot F. In vivo and in situ biofabrication by laser-assisted bioprinting. In: Atala A, Yoo JJ, editors. Essentials of 3D biofabrication and translation. London: Academic Press; 2015. p. 81–7.
16. Raman R, Bashir R. Stereolithographic 3D bioprinting for biomedical applications. In: Atala A, Yoo JJ, editors. Essentials of 3D biofabrication and translation. London: Academic Press; 2015. p. 89–121.
17. Pati F, Jang J, Lee JW, Cho DW. Extrusion bioprinting. In: Atala A, Yoo JJ, editors. Essentials of 3D biofabrication and translation. London: Academic Press; 2015. p. 123–52.
18. Iwanaga S, Arai K, Nakamura M. Inkjet bioprinting. In: Atala A, Yoo JJ, editors. Essentials of 3D biofabrication and translation. London: Academic Press; 2015. p. 61–79.
19. Liu D, Fu J, Fan H, Li D, Dong E, Xiao X, et al. Application of 3D-printed PEEK scapula prosthesis in the treatment of scapular benign fibrous histiocytoma: a case report. J Bone Oncol. 2018 Sep;1(12):78–82.
20. De Boer FA, Huijsmans PE. Use of a 3D-printed custom reverse shoulder arthroplasty. Tech Orthop. 2018;35(1):38–41.
21. Hu H, Liu W, Zeng Q, Wang S, Zhang Z, Liu J, et al. The personalized shoulder reconstruction assisted by 3D printing technology after resection of the proximal humerus tumours. Cancer Manag Res. 2019;11:10665–73.
22. Zou Y, Yang Y, Han Q, Yang K, Zhang K, Wang J, et al. Novel exploration of customized 3D printed shoulder prosthesis in revision of total shoulder arthroplasty a case report. Medicine (Baltimore). 2018;97(47):e13282.
23. Wang KC, Jones A, Kambhampati S, Gilotra MN, Liacouras PC, Stuelke S, et al. CT-based 3D printing of the glenoid prior to shoulder arthroplasty: bony morphology and model evaluation. J Digit Imaging. 2019;32(5):816–26.
24. You W, Liu LJ, Chen HX, Xiong JY, Wang DM, Huang JH, et al. Application of 3D printing technology on the treatment of complex proximal humeral fractures (Neer3-part and 4-part) in old people. Orthop Traumatol Surg Res. 2016;102(7):897–903.
25. Nikitichev I, Daniil I, Patel P, Patel P, Avery J, Avery J, et al. Patient-specific 3D printed printed models models for for education, education, and surgical

surgical simulation simulation. IntechOpen; 2018. p. 116–35.
26. Tack P, Victor J, Gemmel P, Annemans L. 3D-printing techniques in a medical setting: a systematic literature review. Biomed Eng Online. 2016;15(1):115. https://doi.org/10.1186/s12938-016-0236-4.
27. Wurm G, Tomancok B, Pogady P, Holl K, Trenkler J. Cerebrovascular stereolithographic biomodeling for aneurysm surgery. Technical note. J Neurosurg. 2004;100(1):139–45. https://doi.org/10.3171/jns.2004.100.1.0139.
28. D'Urso PS, Effeney DJ, Earwaker WJ, Barker TM, Redmond MJ, Thompson RG, et al. Custom cranioplasty using stereolithography and acrylic. Br J Plast Surg. 2000;53(3):200–4.
29. Malik HH, Darwood ARJ, Hons B. Three-dimensional printing in surgery: a review of current surgical applications. J Surg Res. 2015;199(2):512–22. https://doi.org/10.1016/j.jss.2015.06.051.
30. Klein GT, Lu Y, Wang MY. 3D printing and neurosurgery—ready for prime time? World Neurosurg. 2013;80:233–5.
31. Condino S, Carbone M, Ferrari V, Faggioni L, Peri A, Ferrari M, et al. How to build patient-specific synthetic abdominal anatomies. An innovative approach from physical toward hybrid surgical simulators. Int J Med Robot. 2011;7(2):202–13. https://doi.org/10.1002/rcs.390.
32. Wilasrusmee C, Suvikrom J, Suthakorn J, Lertsithichai P, Sitthiseriprapip K, Proprom N, et al. Three-dimensional aortic aneurysm model and endovascular repair: an educational tool for surgical trainees. Int J Angiol. 2008;17(03):129–33.
33. Pepper J, Petrou M, Rega F, Rosendahl U, Golesworthy T, Treasure T. Implantation of an individually computer-designed and manufactured external support for the Marfan aortic root. Multimed Man Cardiothorac Surg 2013;2013:mmt004. Available from: http://europepmc.org/abstract/MED/24413003
34. Parchi PD, Ferrari V, Piolanti N, Andreani L, Condino S, Evangelisti G, et al. Computer tomography prototyping and virtual procedure simulation in difficult cases of hip replacement surgery. Surg Technol Int. 2013;23:228–34.
35. Mao Y, Xu C, Xu J, Li H, Liu F, Yu D, et al. The use of customized cages in revision total hip arthroplasty for Paprosky type III acetabular bone defects. Int Orthop. 2015;39(10):2023–30.
36. Rosenthal G, Ng I, Moscovici S, Lee KK, Lay T, Martin C, et al. Polyetheretherketone implants for the repair of large cranial defects: a 3-center experience. Neurosurgery. 2014;75(5):523–9.
37. Ayoub AF, Rehab M, O'Neil M, Khambay B, Ju X, Barbenel J, et al. A novel approach for planning orthognathic surgery: the integration of dental casts into three-dimensional printed mandibular models. Int J Oral Maxillofac Surg. 2014;43(4):454–9.
38. Windisch G, Salaberger D, Rosmarin W, Kastner J, Exner GU, Haldi-Brändle V, et al. A model for clubfoot based on micro-CT data. J Anat. 2007;210(6):761–6.
39. Knoedler M, Feibus AH, Lange A, Maddox MM, Ledet E, Thomas R, et al. Individualized physical 3-dimensional kidney tumor models constructed from 3-dimensional printers result in improved trainee anatomic understanding. Urology. 2015;85(6):1257–62.
40. Jones DB, Sung R, Weinberg C, Korelitz T, Andrews R. Three-dimensional modeling may improve surgical education and clinical practice. Surg Innov. 2016;23(2):189–95. https://doi.org/10.1177/1553350615607641.
41. AbouHashem Y, Dayal M, Savanah S, Štrkalj G. The application of 3D printing in anatomy education. Med Educ Online. 2015 Jan;20(1):29847.
42. Li C, Kui C, Kah E, Lee M, Wu WKK. The role of 3D printing in anatomy education and surgical training: a narrative review. MedEdPublish. 2017;6(2):31.
43. Ryan JR, Almefty KK, Nakaji P, Frakes DH. Cerebral aneurysm clipping surgery simulation using patient-specific 3D printing and silicone casting. World Neurosurg. 2016;88:175–81.
44. Birnbaum K, Schkommodau E, Decker N, Prescher A, Klapper U, Radermacher K. Computer-assisted orthopedic surgery with individual templates and comparison to conventional operation method. Spine (Phila Pa 1976). 2001 Feb;26(4):365–70.
45. Kunz M, Waldman SD, Rudan JF, Bardana DD, Stewart AJ. Computer-assisted mosaic arthroplasty using patient-specific instrument guides. Knee Surg Sports Traumatol Arthrosc. 2012 May;20(5):857–61.
46. Zhou F, Xue F, Zhang S. The application of 3D printing patient specific instrumentation model in total knee arthroplasty. Saudi J Biol Sci. 2020;27(5):1217–21.
47. Naddeo F, Fontana C, Naddeo A, Cataldo E, Cappetti N, Narciso N. Novel design for a customized, 3D-printed surgical template for thoracic spinal arthrodesis. Int J Med Robot Comput Assist Surg. 2019 Aug;15(4):e2005.
48. Qazi TH, Mooney DJ, Pumberger M, Geißler S, Duda GN. Biomaterials based strategies for skeletal muscle tissue engineering: existing technologies and future trends. Biomaterials. 2015;53:502–21. https://doi.org/10.1016/j.biomaterials.2015.02.110.
49. Langer R, Vacanti JP. Tissue engineering. Science. 1993;260:920–6.
50. Khademhosseini A, Langer R. A decade of progress in tissue engineering. Nat Protoc. 2016;11(10):1775–81.
51. McClelland RE, Dennis R, Reid LM, Stegemann JP, Palsson B, Macdonald JM. Tissue engineering. In: Enderle JD, Bronzino JD, editors. Introduction to biomedical engineering. Cambridge: Academic Press; 2012. p. 273–357.
52. Kang H, Lee SJ, Ko IK, Kengla C, Yoo JJ, Atala A. A 3D bioprinting system to produce human-scale tissue constructs with structural integrity. Nat Biotechnol. 2016;34(3):312–9. https://doi.org/10.1038/nbt.3413.

53. Yogishwarappa CN, Srinivasan S, Teoh SH, Vijayakumar A, Ishwar M. Customized osteomesh cranioplasty. J Adv Plast Surg Res. 2016;2:25–32.
54. Teo L, Teoh SH, Liu Y, Lim L, Tan B, Schantz JT, et al. A novel Bioresorbable implant for repair of orbital floor fractures. Orbit. 2015;34(4):192–200.
55. Kim JH, Seol YJ, Ko IK, Kang HW, Lee YK, Yoo JJ, et al. 3D bioprinted human skeletal muscle constructs for muscle function restoration. Sci Rep. 2018;8(1):1–15. https://doi.org/10.1038/s41598-018-29968-5.
56. Reichert JC, Wullschleger ME, Cipitria A, Lienau J, Cheng TK, Schütz MA, et al. Custom-made composite scaffolds for segmental defect repair in long bones. Int Orthop. 2011;35(8):1229–36.
57. Unger C, Kramer N, Walzl A, Scherzer M, Hengstschläger M, Dolznig H. Modeling human carcinomas: physiologically relevant 3D models to improve anti-cancer drug development. Adv Drug Deliv Rev. 2014;79:50–67. https://doi.org/10.1016/j.addr.2014.10.015.
58. Khaled SA, Burley JC, Alexander MR, Yang J, Roberts CJ. 3D printing of five-in-one dose combination polypill with defined immediate and sustained release profiles. J Control Release. 2015;217:308–14.

The Digitized Shoulder: From Preoperative Planning to Patient-Specific Guides

23

Majid Mohammad Sadeghi, Mehmet Kapicioglu, Emin Faruk Kececi, and Kerem Bilsel

23.1 Introduction

Humeral head replacement first started with Neer in 1955 for degenerative joint disease and complex fractures [1]. Reverse shoulder arthroplasty (RSA), introduced by Paul Grammont in 1987, has emerged as an alternative to total shoulder arthroplasty for patients with cuff tear arthropathy, osteoarthritis, and massive irreparable cuff tear [2]. Although the results of anatomic total shoulder arthroplasty (aTSA) and reverse total shoulder arthroplasty (rTSA) are generally satisfactory, an important point that still needs to improve is preoperative planning. It provides an advantage in terms of implant positioning and potentially improving clinical outcomes [3]. The use of this technology allows surgeons to better understand variations in glenoid morphology and to also better achieve preoperatively planned targets [4].

Since the beginning of this century, the number of shoulder arthroplasties has been steadily increasing [5–7]. The correct glenoid version, joint line restoration, and appropriate soft tissue tension in total anatomic and reverse shoulder arthroplasty provide long-term and satisfactory results in implant survival. Glenoid loosening is among the main causes of implant failure [8]. Unlike the hip, the smaller size of the glenoid only allows a narrow margin of error. In aTSA, malposition of the glenoid component has been reported to be associated with poor function, early loosening, and instability and is the leading cause of long-term clinical failure [9]. Baseplate malpositioning can cause some catastrophic failures like instability, early implant loosening, scapular notching, and scapular fractures [10]. Complications of the glenoid component generally comprise 30–50% of both aTSA (loosening and wear) and rTSA (notching) [11]. In many studies, aseptic glenoid loosening is responsible for 30–100% of prosthesis revisions [12, 13]. The same precision should be demonstrated on the humeral side, but today we continue to follow the rules such as 20° retroversion instead of considering the patient-specific functional position of the prosthesis [8].

Due to difficult joint exposure and the complex geometry of the scapula, it is technically difficult to insert implants by conventional methods [14]. In traditional methods, the surgeon decides according to his own experience and experience by using conventional guides that are not specific to the patient or by investigating the patient's

M. M. Sadeghi
Department of Mechatronics Engineering, Istanbul Technical University, Istanbul, Turkey

M. Kapicioglu · K. Bilsel (✉)
Department of Orthopaedics and Traumatology, School of Medicine, Bezmialem Vakif University, Istanbul, Turkey

E. F. Kececi
Department of Mechanical Engineering, Abdullah Gul University, Kayseri, Turkey

G. Huri et al. (eds.), *Fundamentals of the Shoulder*, https://doi.org/10.1007/978-3-030-94702-6_23

preop X-ray and CT scan [15]. Glenoid version and inclination are estimated based on preoperative measurements. Computer-assisted surgical techniques have recently become popular for hip and knee arthroplasty, but are not widely applied to the shoulder [16, 17].

23.2 3D Visualization and Preoperative Planning

Plain radiographs, computed tomography (CT), 3D CT reconstruction, and magnetic resonance imaging (MRI) are valuable preoperative imaging methods to understand the patient's specific pathology [18]. Preoperative planning for shoulder arthroplasty begins with a detailed assessment of standard two-dimensional radiographs. AP X-ray view can be used to evaluate coronal glenoid bone loss and inclination, while axillary imaging can be used to evaluate glenoid bone loss, glenoid version, and humeral subluxation as well [19].

The CT scan provides an accurate imaging assessment of the glenoid version, inclination, bone loss pattern, and bone quality. Imaging of the glenoid bone loss pattern is first performed using two-dimensional (2D) CT images. 3D reconstructive CT images are more useful in understanding the patient-specific glenoid anatomy and bone loss site and also in increasing the sensitivity of the glenoid component positioning (Fig. 23.1). 3D imaging provides a better assessment of glenoid bone loss, deformity, and prediction of optimal prosthesis implantation [20]. Werner et al. in determining the glenoid version and tilt measurements stated that 3D reconstruction measurements were more sensitive than 2D CT scans, and when they re-evaluated images through 3D reconstructive imaging, surgical planning, and implant selection changed in 7 out of 50 shoulders [21].

MRI is used to show pathologies such as articular cartilage, labrum, muscle and tendon ruptures, fatty infiltration and atrophy of muscle, bursae, acromion type, acromioclavicular hypertrophy, and coracoacromial ligament thickening. MRI evaluates the glenoid version better than plain axillary radiographs, but in patients with severe glenoid deformity, CT is better than MRI to determine retroversion [22, 23].

23.3 New Concepts in Shoulder Arthroplasty

Three systems improve the accuracy of the component placement in the shoulder arthroplasty: single-use patient-specific instrumentation (PSI), reusable PSI, CT guided intraoperative navigation, and virtual reality (VR) assisted surgery [18]. Trying to achieve the placement of the glenoid component with maximum sensitivity and accuracy at a lower cost, PSI has been introduced as an attractive alternative to computerized navigation [19].

PSI uses high-resolution CT scans to preoperative templating and positioning of glenoid component. This patient-specific guide determines the version and tilt of a central guide pin placed in the glenoid. There are single-use or reusable systems. In single-use systems, a special

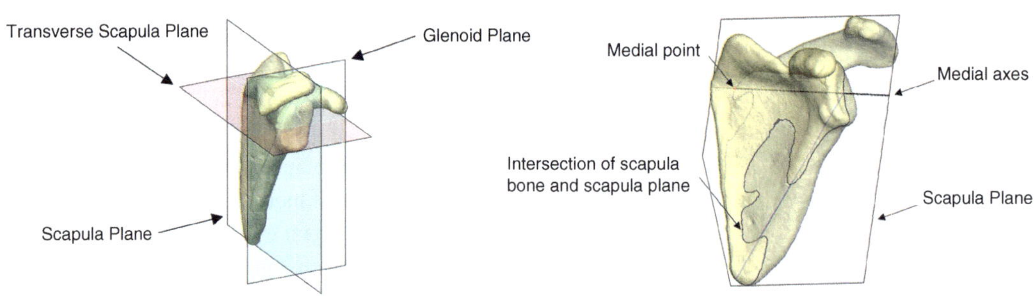

Fig. 23.1 3D visualization of the glenohumeral joint and the glenoid surface

guide should be created for each anatomy. The reusable system has an adjustable feature that can vary from case to case based on the patient's individual anatomy using preoperative measurements. Therefore, the costs of the reusable system are lower. All of these systems usually come with a 3D printed scapula/glenoid model that allows preoperative and intraoperative visualization for the surgeon [18]. Studies demonstrated that PSI correctly identified the preoperative 3D virtual plan. Levy et al. showed that 3D preoperative planning for the central pin location with PSI guides was accurate in reverse shoulder arthroplasty performed on 14 cadaver shoulders. In this study, the accuracy of the pin entry point was measured as 1.2 mm, inferior inclination 1.2°, and version 2.6° [24]. Walch et al. evaluated the accuracy of 3D planning and PSI in total shoulder arthroplasty in scapulae of 18 cadaver. In this cadaveric study, the accuracy of the guide pin placement after implantation with CT was measured and showed a 1.5 mm entry point error, 1.64° version error, and 1.42° inclination error [25]. Gauci et al. in 17 patients, it was concluded that the use of 3D preoperative planning and PSI provides more accuracy and reproducibility in aTSA. The mean standard deviation in the guidewire entry point position was <1 mm in both vertical and horizontal planes, while the mean errors in inclination and version were 1.8° and 3.4°, respectively [26]. Hendel et al. in a randomized clinical study compared 15 aTSA using PSI and 16 aTSA performing standard surgery. The authors reported improvement with a mean inclination and statistically significant decreases in medial-lateral offset deviation compared to the standard surgical group of the PSI group [27].

Today, 3D glenoid component planning and patient-specific instrumentation options in the market are: DePuy TRUMATCH Personalized Solutions System (Warsaw, IN, USA), DJO Match Point System (Lewisville, TX, USA), the Zimmer Biomet PSI Shoulder for Trabecular Metal Reverse Glenoid System (Warsaw, IN, USA), the Stryker TrueSight Personalized.

Planning System (Kalamazoo, MI, USA), the Wright Tornier BLUEPRINT planning software and PSI (Memphis, TN, USA), and the Arthrex Virtual Implant Positioning System (Naples, FL, USA). All systems mentioned above use a single-use 3D printed guide for central guide pin placement, except Arthrex. The Arthrex VIP system uses a reusable calibrator device that transfers the desired location and trajectory of the guide pin from a 3D printed glenoid model to the patient's glenoid.

Praxim (Praxim Inc., Palo Alto, CA) and Kinamed (Kinamed Incorporated, Camarillo, CA) presented the first navigation in shoulder arthroplasty in the mid-2000s. This technology has evolved and has now become a commercially available system that allows real-time, intraoperative CT navigation [26]. The system allows the surgeon to transform 2D axial, coronal, and sagittal CT images into 3D reconstruction to better understand the anatomy and location between the glenoid and the humeral head. It measures the scapular axis and the glenoid version using the Freidman's axis as a reference line for the scapular axis from the scapular trigonum to the center of the glenoid joint (fossa) [28]. This advanced preoperative planning optimizes implant placement, minimizes bone resection, and optimizes version and inclination. A tracker is placed on the coracoid intraoperatively. Then, some specific anatomical points determined by preop CT are marked with a tracker and recorded in the system. After this has been achieved, highly sensitive intraoperative CT guidance is possible [18]. Kircher et al. in a prospective, randomized clinical study comparing patients undergoing conventional TSA (10 patients) and TSA with navigation (10 patients), in terms of retroversion angle the range was from 15.4 ± 5.8 to 3.7 ± 6.3 in navigation TSA, from 14.4 ± 6.1° to 10.9 ± 6.8° in traditional TSA. Although both groups improved retroversion significantly, TSA with navigation provided a significantly better improvement than the other [29]. Sadoghi et al. in a meta-analysis compared conventional TSA and TSA with navigation patients; in conventional and navigational groups, there was an error in correcting 10.6 and 4.4 glenoid retroversion, respectively [30]. The TSA with navigation was found significantly better than the other. The advantage of intraoperative CT guided navigation over the PSI system and conventional shoulder arthroplasty is that it allows real-time adjustments and feedback with sterile instrumentation and display [18]. The disadvantages are the cost, the possibility of the coracoid fracture during the placement of the tracker pin, and the long operation time. Kircher et al. revealed that navigation-assisted TSA takes an average of 169.5 min and conventional arthroplasty takes an average of 138 min of operating time [29].

Virtual reality (VR) assisted surgery publications in orthopedics have increased steadily since its introduction in the early 1990s [31]. VR, first invented by Jaron Lanier in 1986, has expanded from the entertainment industry to clinical medicine in previous decades [32]. It is used for preoperative planning areas and intraoperative surgical simulation for educational purposes in orthopedics. The focus of these studies is on surgical training especially arthroscopy, as well as the complexity of the skill and difficult learning [31, 33, 34]. VR provides students with the ability to critically analyze technical and surgical decision-making with minimal error, without harming the patient in the cognitive assessment process. VR currently available on the market uses a combination of equipment such as a 3D rendering capable computer, head-mounted display (HMD), and controllers with position trackers. VR, augmented reality (AR), and mixed reality (MR) devices are used in a number of clinical and surgical fields, including orthopedics, as well as neurosurgery, plastic surgery, and urological surgery [35]. There are more than 60 available VR products cited in the literature [33]. Six of these products are related to shoulder arthroscopy, namely, ArthroMentor/Insight Arthro (Simbionix, Airport City, Israel), Alex Shoulder Professor (Sawbones Europe, Malmo, Sweden), Procedicus arthroscopy (Mentice Corp, Gothenburg, Sweden), ArthroS (VirtaMed, Zurich, Switzerland), and insight MIST (3D Systems, Rock Hill, SC, USA). Two products were seen to involve general arthroscopy skill training, namely Swemac/Augmented Reality Systems (Swemac, Linköping, Sweden) and Virtual Reality Tetris Game Using Arthroscopy (VirtaMed) [35]. VR is currently used for surgical training and simulation, especially in arthroscopic shoulder surgery. Although showing promising developments for the future, there is still not enough evidence for use in the real operating room environment.

23.4 3D Printing

In the digital world, the surgeon can use software to observe the shoulder joint and can plan the operation [32]. By using a 3D printer, the parts of the shoulder joint can be printed and assembled to represent the actual joint. With the help of 3D printing even the visualization can be made more physical [36]. The data set is turned into a virtual model to be printed by 3D printers. The 3D printed model is used to explain and observe the current situation of the shoulder joint [37]. The materials used in the 3D printing can be as hard as bones or can be as soft as soft tissue of the joint [38, 39]. Printing different parts of the joint allows the surgeon to show the working of the joint and the existing damages of the current patient (Fig. 23.2).

However, since the bone density and 3D printer material density are different, the weight of the actual bone and the 3D printed bone would be different (Fig. 23.3). With some engineering calculations, the weight of the 3D printed bone can be made equal to the original by using the infill concept where the inner section of the part density can be changed but the current slicer software does not have a pattern similar to the inside of a bone. Material density, infill structure, and shell thickness differences of the 3D printed part to an actual bone cause errors if the surgeon needs to operate on the bone.

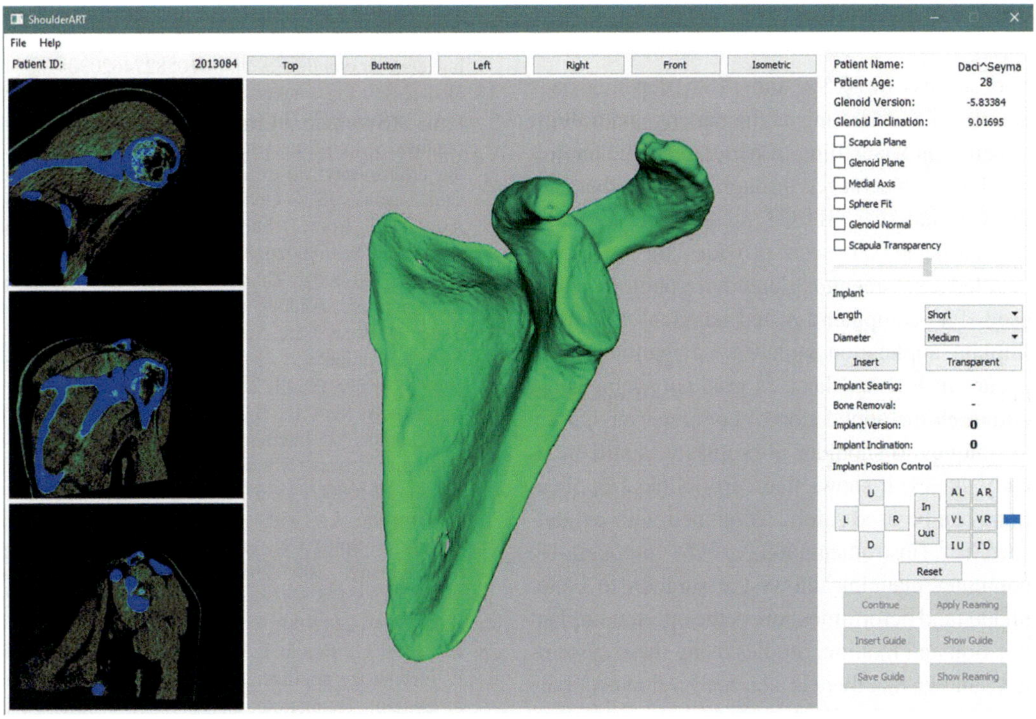

Fig. 23.2 Graphical user interface of the specific software

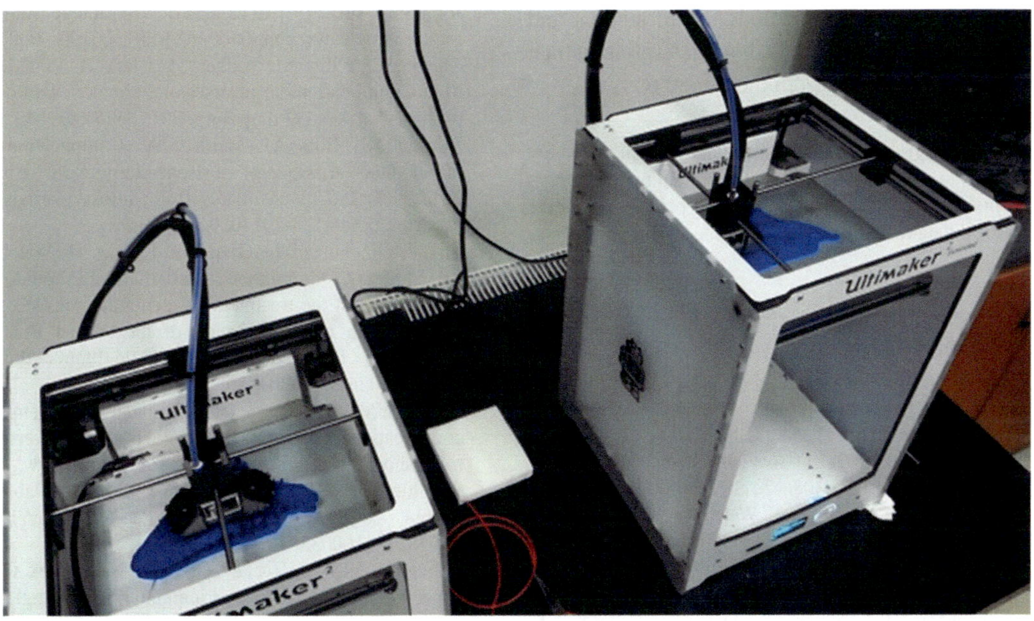

Fig. 23.3 3D printing of the bone models to be used for the experiment

23.5 Conclusion

The success of aTSA and rTSA depends on a detailed understanding of the patient's pathology. A better understanding of patient-specific pathology has been achieved through software that uses 2D CT and then 3D CT reconstruction. A 3D reconstructed CT scan provides an additional detailed assessment of glenoid morphology and is used for computer-assisted surgical planning. Implant application with CT navigation, which occurs in PSI and later, is gradually increasing. Although the application of computer navigation techniques to shoulder arthroplasty is still in its early stages, it shows that early clinical applications can be used safely. Combined with patient-specific instrumentation, 3D preoperative computer planning allows the surgeon to better understand deformities, select the optimal implant position and fixation. Studies using these systems have shown that there is a higher level of accuracy and repeatability. Further studies are needed to investigate the effect of prostheses on long-term survival to improve the anatomical location of implants with these methods. More studies and clinical results are needed to determine which patients will benefit from these methods.

Acknowledgments The CT scan data used in this study belongs to patients who received treatment by the authors of this chapter. The patients have given permission for the medical data to be used in this study.

References

1. Neer CS. The classic: articular replacement for the humeral head. Clin Orthop Relat Res. 2011;469(9):2409–21. Available from: https://journals.lww.com/00003086-201109000-00003
2. Smithers CJ, Young AA, Walch G. Reverse shoulder arthroplasty. Curr Rev Musculoskelet Med. 2011;4(4):183–90. https://doi.org/10.1007/s12178-011-9097-4.
3. Erickson BJ, Chalmers PN, Denard P, Lederman E, Horneff G, Werner BC, et al. Does commercially available shoulder arthroplasty preoperative planning software agree with surgeon measurements of version, inclination, and subluxation? J Shoulder Elbow Surg. 2021;30(2):413–20. Available from: https://linkinghub.elsevier.com/retrieve/pii/S1058274620304730
4. Denard PJ, Provencher MT, Lädermann A, Romeo AA, Parsons BO, Dines JS. Version and inclination obtained with 3-dimensional planning in total shoulder arthroplasty: do different programs produce the same results? JSES Open Access. 2018;2(4):200–4. Available from: https://linkinghub.elsevier.com/retrieve/pii/S2468602618300263
5. Kim SH, Wise BL, Zhang Y, Szabo RM. Increasing incidence of shoulder arthroplasty in the United States. J Bone Joint Surg Am. 2011;93(24):2249–54. Available from: http://journals.lww.com/00004623-201112210-00003
6. Lübbeke A, Rees JL, Barea C, Combescure C, Carr AJ, Silman AJ. International variation in shoulder arthroplasty. Acta Orthop. 2017;88(6):592–9. https://doi.org/10.1080/17453674.2017.1368884.
7. Oppermann J, Celik E, Bredow J, Beyer F, Hackl M, Spies CK, et al. Shoulder arthroplasty in Germany: 2005–2012. Arch Orthop Trauma Surg. 2016;136(5):723–9. https://doi.org/10.1007/s00402-016-2417-9.
8. Porcellini G, Tarallo L, Novi M, Spiezia F, Catani F. Technology applications in shoulder replacement. J Orthop Traumatol. 2019;20(1):27. https://doi.org/10.1186/s10195-019-0535-1.
9. Ho JC, Sabesan VJ, Iannotti JP. Glenoid component retroversion is associated with osteolysis. J Bone Joint Surg Am. 2013;95(12):e82-1–8. Available from: http://journals.lww.com/00004623-201306190-00007
10. Cheung E, Willis M, Walker M, Clark R, Frankle MA. Complications in reverse total shoulder arthroplasty. J Am Acad Orthop Surg. 2011;19(7):439–49.
11. Bohsali KI, Bois AJ, Wirth MA. Complications of shoulder arthroplasty. J Bone Joint Surg Am. 2017;99(3):256–69. Available from: https://journals.lww.com/00004623-201702010-00010
12. Sirveaux F, Favard L, Oudet D, Huquet D, Walch G, Mole D. Grammont inverted total shoulder arthroplasty in the treatment of glenohumeral osteoarthritis with massive rupture of the cuff. J Bone Joint Surg Br. 2004;86-B(3):388–95. https://doi.org/10.1302/0301-620X.86B3.14024.
13. Zumstein MA, Pinedo M, Old J, Boileau P. Problems, complications, reoperations, and revisions in reverse total shoulder arthroplasty: a systematic review. J Shoulder Elbow Surg. 2011;20(1):146–57. Available from: https://linkinghub.elsevier.com/retrieve/pii/S1058274610003381
14. Stübig T, Petri M, Zeckey C, Hawi N, Krettek C, Citak M, et al. 3D navigated implantation of the glenoid component in reversed shoulder arthroplasty. Feasibility and results in an anatomic study. Int J Med Robot. 2013;9(4):480–5. https://doi.org/10.1002/rcs.1519.

15. Nguyen D, Ferreira LM, Brownhill JR, King GJW, Drosdowech DS, Faber KJ, et al. Improved accuracy of computer assisted glenoid implantation in total shoulder arthroplasty: an in-vitro randomized controlled trial. J Shoulder Elbow Surg. 2009;18(6):907–14. Available from: https://linkinghub.elsevier.com/retrieve/pii/S1058274609001372
16. DiGioia AM, Jaramaz B, Colgan BD. Computer assisted orthopaedic surgery. Image guided and robotic assistive technologies. Clin Orthop Relat Res. 1998;354:8–16. Available from: http://www.ncbi.nlm.nih.gov/pubmed/9755759
17. Jaramaz B, DiGioia AM, Blackwell M, Nikou C. Computer assisted measurement of cup placement in Total hip replacement. Clin Orthop Relat Res. 1998;354:70–81. Available from: http://journals.lww.com/00003086-199809000-00010
18. Wanner JP, Maslow J, Byram I. CT-guided navigation in shoulder arthroplasty. Tech Shoulder Elbow Surg. 2019;20(2):68–73. Available from: https://journals.lww.com/00132589-201906000-00008
19. Mijic D, Levy J. Computer-assisted planning and patient-specific instrumentation (PSI) in shoulder arthroplasty. In: Complex and revision shoulder arthroplasty [internet]. Cham: Springer; 2019. p. 15–25. https://doi.org/10.1007/978-3-030-02756-8_2.
20. Scalise JJ, Codsi MJ, Bryan J, Brems JJ, Iannotti JP. The influence of three-dimensional computed tomography images of the shoulder in preoperative planning for total shoulder arthroplasty. J Bone Joint Surgery Am. 2008;90(11):2438–45. Available from: http://journals.lww.com/00004623-200811000-00015
21. Werner BS, Hudek R, Burkhart KJ, Gohlke F. The influence of three-dimensional planning on decision-making in total shoulder arthroplasty. J Shoulder Elbow Surg. 2017;26(8):1477–83. Available from: https://linkinghub.elsevier.com/retrieve/pii/S1058274617300198
22. Lowe JT, Testa EJ, Li X, Miller S, DeAngelis JP, Jawa A. Magnetic resonance imaging is comparable to computed tomography for determination of glenoid version but does not accurately distinguish between Walch B2 and C classifications. J Shoulder Elbow Surg. 2017;26(4):669–73. Available from: https://linkinghub.elsevier.com/retrieve/pii/S1058274616304451
23. Raymond AC, McCann PA, Sarangi PP. Magnetic resonance scanning vs axillary radiography in the assessment of glenoid version for osteoarthritis. J Shoulder Elbow Surg. 2013;22(8):1078–83. Available from: https://linkinghub.elsevier.com/retrieve/pii/S1058274612004831
24. Levy JC, Everding NG, Frankle MA, Keppler LJ. Accuracy of patient-specific guided glenoid baseplate positioning for reverse shoulder arthroplasty. J Shoulder Elbow Surg. 2014;23(10):1563–7. Available from: https://linkinghub.elsevier.com/retrieve/pii/S1058274614000962
25. Walch G, Vezeridis PS, Boileau P, Deransart P, Chaoui J. Three-dimensional planning and use of patient-specific guides improve glenoid component position: an in vitro study. J Shoulder Elbow Surg. 2015;24(2):302–9. Available from: https://linkinghub.elsevier.com/retrieve/pii/S1058274614003280
26. Gauci MO, Boileau P, Baba M, Chaoui J, Walch G. Patient-specific glenoid guides provide accuracy and reproducibility in total shoulder arthroplasty. Bone Joint J. 2016;98-B(8):1080–5. https://doi.org/10.1302/0301-620X.98B8.37257.
27. Hendel MD, Bryan JA, Barsoum WK, Rodriguez EJ, Brems JJ, Evans PJ, et al. Comparison of patient-specific instruments with standard surgical instruments in determining glenoid component position. J Bone Joint Surg. 2012;94(23):2167–75. Available from: https://journals.lww.com/00004623-201212050-00008
28. Friedman RJ, Hawthorne KB, Genez BM. The use of computerized tomography in the measurement of glenoid version. J Bone Joint Surg Am. 1992;74(7):1032–7. Available from: http://www.ncbi.nlm.nih.gov/pubmed/1522089
29. Kircher J, Wiedemann M, Magosch P, Lichtenberg S, Habermeyer P. Improved accuracy of glenoid positioning in total shoulder arthroplasty with intraoperative navigation: a prospective-randomized clinical study. J Shoulder Elbow Surg. 2009;18(4):515–20. Available from: https://linkinghub.elsevier.com/retrieve/pii/S105827460900189X
30. Sadoghi P, Vavken J, Leithner A, Vavken P. Benefit of intraoperative navigation on glenoid component positioning during total shoulder arthroplasty. Arch Orthop Trauma Surg. 2015;135(1):41–7. https://doi.org/10.1007/s00402-014-2126-1.
31. Vaughan N, Dubey VN, Wainwright TW, Middleton RG. A review of virtual reality based training simulators for orthopaedic surgery. Med Eng Phys. 2016;38(2):59–71. Available from: https://linkinghub.elsevier.com/retrieve/pii/S1350453315002799
32. Berhouet J, Gulotta LV, Dines DM, Craig E, Warren RF, Choi D, et al. Preoperative planning for accurate glenoid component positioning in reverse shoulder arthroplasty. Orthop Traumatol Surg Res. 2017;103(3):407–13. Available from: https://linkinghub.elsevier.com/retrieve/pii/S1877056817300506
33. Morgan M, Aydin A, Salih A, Robati S, Ahmed K. Current status of simulation-based training tools in orthopedic surgery: a systematic review. J Surg Educ. 2017;74(4):698–716. Available from: https://linkinghub.elsevier.com/retrieve/pii/S1931720417300089

34. Stirling ER, Lewis TL, Ferran NA. Surgical skills simulation in trauma and orthopaedic training. J Orthop Surg Res. 2014;9(1):126. https://doi.org/10.1186/s13018-014-0126-z.
35. Lohre R, Warner JJP, Athwal GS, Goel DP. The evolution of virtual reality in shoulder and elbow surgery. JSES Int. 2020;4(2):215–23. Available from: https://linkinghub.elsevier.com/retrieve/pii/S2666638320300268
36. Petzold R, Zeilhofer H-F, Kalender WA. Rapid prototyping technology in medicine—basics and applications. Comput Med Imaging Graph. 1999;23(5):277–84. Available from: https://linkinghub.elsevier.com/retrieve/pii/S0895611199000257
37. Tack P, Victor J, Gemmel P, Annemans L. 3D-printing techniques in a medical setting: a systematic literature review. Biomed Eng Online. 2016;15(1):115. https://doi.org/10.1186/s12938-016-0236-4.
38. Ratinam R, Quayle M, Crock J, Lazarus M, Fogg Q, McMenamin P. Challenges in creating dissectible anatomical 3D prints for surgical teaching. J Anat. 2019;234(4):419–37. https://doi.org/10.1111/joa.12934.
39. Ghosh U, Ning S, Wang Y, Kong YL. Addressing unmet clinical needs with 3D printing technologies. Adv Healthc Mater. 2018;7(17):1800417. https://doi.org/10.1002/adhm.201800417.

GPSR Compliance

The European Union's (EU) General Product Safety Regulation (GPSR) is a set of rules that requires consumer products to be safe and our obligations to ensure this.

If you have any concerns about our products, you can contact us on ProductSafety@springernature.com

In case Publisher is established outside the EU, the EU authorized representative is:

Springer Nature Customer Service Center GmbH
Europaplatz 3
69115 Heidelberg, Germany

Batch number: 10732801

Printed by Printforce, the Netherlands